Donkey and Mule Medicine: Clinical Medicine and Reproduction

Amy K. McLean
Francisco Javier Navas González
Erin L. Goodrich
Editors

Donkey and Mule Medicine: Clinical Medicine and Reproduction

A Practical Approach to the Treatment and Care of Long Ears
Volume 2

Springer

Editors
Amy K. McLean
Department of Animal Science
University of California
Davis, CA, USA

Francisco Javier Navas González
Department of Genetics
Faculty of Veterinary Sciences
University of Córdoba
Córdoba, Spain

Erin L. Goodrich
Department of Population Medicine
and Diagnostic Sciences
Cornell University College of Veterinary
Ithaca, NY, USA

ISBN 978-3-032-18539-6 ISBN 978-3-032-18540-2 (eBook)
https://doi.org/10.1007/978-3-032-18540-2

Cover image: Photo credit to Cassidy Cooper (pictured from left to right, a mule, hinny and donkey).

This Springer imprint is published by the registered company Springer Nature Switzerland AG
The registered company address is: Gewerbestrasse 11, 6330 Cham, Switzerland

Preface – A Long-Awaited Resource for Long-Eared Equids

Volume 2: Clinical Medicine and Reproduction

Donkeys, mules, and hinnies have served humanity for thousands of years—essential partners whose health and welfare remain vital across continents and cultures. Despite their importance, their clinical and reproductive needs have long been misunderstood or treated as identical to those of horses. *Volume 2: Clinical Medicine and Reproduction* seeks to correct that misconception by providing the most comprehensive reference ever dedicated to their medical care and breeding biology.

This volume brings together leading veterinarians, scientists, and clinicians from around the world to present current knowledge on disease mechanisms, diagnostics, pharmacology, anesthesia, surgery, and advanced reproductive techniques. Each chapter combines rigorous science with practical applications designed to improve outcomes in both field and hospital settings.

Beyond pathology and treatment, the book addresses the physiological adaptations that distinguish donkeys and mules from other equids. Understanding these differences—whether in metabolic rate, immune response, or reproductive function—is essential to delivering effective and humane veterinary care.

Our hope is that this work empowers practitioners, researchers, and owners to enhance the quality of life for long-eared equids everywhere, ensuring that their medical and reproductive welfare receives the attention and respect it has long deserved.

Davis, CA, USA Amy K. McLean
Córdoba, Spain Francisco Javier Navas González
Ithaca, NY, USA Erin L. Goodrich

Introduction – Volume 2: Clinical Medicine and Reproduction

Donkeys, mules, and hinnies present unique clinical challenges and opportunities that demand a specialized understanding of their biology and medicine. Their stoic demeanor, adaptive metabolism, and hybrid physiology can mask disease symptoms and alter therapeutic responses, requiring veterinarians to think beyond conventional equine models. *Volume 2: Clinical Medicine and Reproduction* offers a detailed, evidence-based exploration of the diagnostic, therapeutic, and reproductive dimensions of these remarkable animals, representing a culmination of decades of research and clinical insight.

This volume is organized into three major parts that together define the modern clinical and reproductive science of long-eared equids.

Part "Medical Systems and Diseases" provides a comprehensive examination of the organ systems and the disorders that affect them. Readers will find in-depth chapters on cardiovascular, respiratory, hepatic, endocrine, neurologic, gastrointestinal, urogenital, and integumentary systems, supported by the latest diagnostic and treatment protocols. The inclusion of immunology and genomic studies underscores the growing recognition that donkeys and mules possess distinct immune responses and disease susceptibilities, challenging assumptions derived from horse-based models. These insights form the basis for improved diagnostic accuracy, therapeutic efficiency, and preventive healthcare.

Part "Diagnostics, Treatment, and Surgery" focuses on the practical implementation of clinical knowledge. It outlines best practices for sample collection, laboratory testing, pharmacology, anesthesia, pain management, and surgical interventions, taking into account the species' unique physiology and pharmacokinetics. Through detailed guidance and clinical case studies, this section demonstrates how tailored diagnostic and surgical strategies can reduce complications and improve recovery outcomes. Each chapter is designed to equip practitioners with the decision-making tools necessary to provide safe, compassionate, and effective care.

Part "Reproduction and Advanced Breeding" delves into the reproductive biology and technologies that are transforming equid breeding programs. Topics include male fertility assessment, semen collection, embryo transfer, assisted reproductive technologies (ART), and neonatal management. The section also examines hybrid reproduction—addressing the complexities of mule and hinny sterility, the rare occurrences of hybrid fertility, and the molecular mechanisms that underlie these phenomena. The integration of genomics, proteomics, and advanced imaging has ushered in a new era of reproductive science, where precision tools and ethical breeding practices converge to improve both productivity and welfare.

Throughout this volume, readers are encouraged to approach donkey and mule medicine as a discipline in its own right—grounded in comparative physiology but shaped by the distinctive traits of these species. Their metabolic efficiency, unique pain responses, and differing reactions to drugs and anesthesia highlight the importance of clinical protocols developed specifically for them, rather than adapted from horses.

In uniting clinical rigor with a compassionate understanding of long-eared equids, this work aims to advance the standard of veterinary care across global contexts. Whether in rural clinics, academic hospitals, or conservation programs, the principles outlined here provide a framework for evidence-based, humane, and sustainable veterinary practice.

These animals have served humanity for millennia; it is now our responsibility to serve them with the same dedication—through science, care, and respect.

University of California, Davis, CA, USA — Amy K. McLean
University of Córdoba, Córdoba, Spain — Francisco Javier Navas González
Cornell University, Ithaca, NY, USA — Erin L. Goodrich

Contents

Part I

Medical Systems and Diseases

Cardiovascular Diseases of Donkey and Mule

Enrica Zucca, Francesca Bindi, Chiara Bozzola, and Micaela Sgorbini

Abstract

The donkey (*Equus asinus*) is an important working animal in many areas of the world. In Europe, donkeys and mules are predominantly companion or leisure animals. Donkeys have recently gained significant interest from both veterinarians and owners. The rediscovery of their abilities as working, sport, and companion animals has led to increased awareness of their well-being and care. However, the available literature on their cardiovascular characteristics and the incidence of heart and vascular diseases remains limited compared to that for horses (Burden F, Thiemann A (2015) Donkeys are different. J Equine Vet Sci 35 (5):376–382. https://doi.org/10.1016/j.jevs.2015.03.005; Mendoza FJ, Toribio RE, Perez-Ecija A. J Equine Vet Sci 65:86–97, 2018). Since donkeys and mules are not commonly used as riding animals throughout the world, cardiovascular diseases are likely underdiagnosed (Roberts SL, Dukes-McEwan J. Vet Rec 179:384, 2016a).

Several cardiac acquired and congenital anomalies have been documented in donkeys (Roberts SL, Dukes-McEwan J. Vet Rec 179:384, 2016a; Roberts SL, Dukes-McEwan J. Vet Rec 179:332, 2016b), but clinical information and echocardiographic findings are sparse and not well described (Decloedt A. Vet Rec 179:382–383, 2016). Moreover, much of the existing literature on cardiovascular diseases is based mainly on post-mortem findings (Morrow LD, Smith KC, Piercy RJ, du Toit N, Burden FA, Olmos G, Gregory NG, Verheyen KL. J Comp Pathol 144:145–156, 2011; Roberts SL, Dukes-McEwan J. Vet Rec 179:332, 2016b).

E. Zucca · C. Bozzola
Department of Veterinary Medicine and Animal Sciences, University of Milan, Lodi (MI), Italy

F. Bindi · M. Sgorbini (✉)
Department of Veterinary Sciences, University of Pisa, Pisa, Italy
e-mail: micaela.sgorbini@unipi.it

A. K. McLean et al. (eds.), *Donkey and Mule Medicine: Clinical Medicine and Reproduction*, https://doi.org/10.1007/978-3-032-18540-2_1

Keywords

Cardiovascular disease · Heart rate · ECG · Heart disease · Vascular disease · Vagal tone

1 Examination of Cardiocirculatory System

The clinical procedures for examining the cardiovascular system in donkeys and mules are the same as those applied to horses. However, it is important to interpret the clinical data while considering significant species differences, such as their stoic nature, different management practices, and heart rate, to avoid the mistake of treating donkeys, mules as small horses (Senior 2013).

1.1 History and Signalment

(Fregin 1992; Gay et al. 2000; Ware et al. 2021)

When evaluating a cardiac patient, it is crucial to gather a complete and detailed history along with a performing a comprehensive physical examination. Thorough information about clinical signs is essential to fully understand the patient's condition. Cardiac issues might be detected during routine check-ups, such as vaccinations or pre purchase assessments. Alternatively, clinical signs like a murmur or arrhythmia discovered during a routine visit, behavioral changes, poor performance, exercise intolerance, breathing difficulties, cough, weakness, collapse, swollen peripheral veins, ventral edema, epistaxis, persistent unexplained tachycardia, or a fever of unknown origin may prompt further examination and diagnostic testing for cardiac disease.

Compared to other domestic animals, cardiac disease is uncommon in horses. In this species, due to their significant cardiac reserve, clinical signs typically become evident only when dysfunction is severe or during intense exercise. Cardiovascular diseases in donkeys and mules are likely as common as in horses, but due to limited data, clinicians often underestimate their prevalence. In horses, these conditions are often detected during evaluations for poor performance, while in donkeys, who are less athletic and not typically ridden, they may frequently go undiagnosed.

Considering the signalment is crucial, as certain conditions are more prevalent or significant in specific groups.

The historical information particularly relevant for patients undergoing cardiovascular examination is listed in Table 1.

Table 1 Information relevant to the general medical history

Category	Details
Signalment	Species, breed, age, sex
Current client complaint	Current complaint or concern
Medical history	Previously established medical diagnoses
Current medical therapy	Medications currently or previously prescribed for the presenting problem Dispensing size, dosage, administration frequency Response to therapy (none, partial, or complete) Other medications or supplements
Preventive care	Vaccination status Deworming status Parasite prevention (internal, external) Dental care
Diet and appetite	Food consumed (type and volume) Nutritional supplement Appetite (normal, increased, decreased, partial, or complete anorexia) Water consumption
Environment	Current geographic location Travel history Housing and cohabitation with other animals (type and health) Potential exposure to toxins, contaminated food/water, or poisonous plants
Body	Changes in attitude or interactions with people/animals Weight gain or loss Changes in body condition including muscle mass
Activity and exercise	Recent changes in activity or exercise capacity Tiring or fatigue with exercise Exertional weakness
Collapse or syncope	Frequency Full description of episodes, and videos when possible
Respiratory system signs	Change in rate, depth, or pattern of ventilation Respiratory noise Respiratory distress Coughing
Mucous membranes	Pink or blue Changes with exercise
Organ system review	Digestive: Oral cavity or dental issues; diarrhea; recent bowel issues Urinary: Changes in urinary habits; frequency; quantity Dermatologic: Edema Musculoskeletal: Joint or bone disorders; lameness or gait abnormalities Nervous: History of seizures or possible neurologic problems Ophthalmologic: Recent changes in vision; chronic ocular conditions

1.2 Physical Examination

(Fregin 1992; Amory et al. 2004; Blissitt 2010; Roberts and Dukes-McEwan 2016a; Ware et al. 2021).

1.2.1 Visible Mucous Membranes

The color of the mucous membranes and capillary refill time (CRT) provide essential insights into peripheral perfusion and tissue oxygenation. While oral membranes are typically assessed, preputial or vaginal membranes can also be evaluated. If oral membranes are pigmented, the ocular conjunctival mucosa can be used instead. Both eyes' conjunctivae must be examined to distinguish abnormalities due to local disease from systemic clinical findings. The nasal mucosa, usually deep pink to slightly red and moist, can also be inspected.

Several cardiovascular, respiratory, and hematologic conditions can alter the normally pink color of the mucous membranes. Pallor may indicate anemia, peripheral vasoconstriction, or shock. In mild dehydration, the membranes appear dry and pale pink to slightly blanched. Hyperemia can result from polycythemia. In young animals, it is suggestive of a right-to-left shunting congenital heart defect, while in older animals, it usually indicates sepsis. Examination of the mucosa can also reveal hemostatic abnormalities. Hemostatic function is maintained by the balanced interaction of platelets, clotting factors, fibrinolytic factors, and blood vessel wall integrity. Petechiae and ecchymotic hemorrhages, due to low platelet numbers or dysfunction, are often detected on conjunctival, oral, nasal, and vaginal or preputial mucous membranes. Cyanosis, characterized by a bluish discoloration of the skin and mucous membranes, occurs due to an increased amount of desaturated hemoglobin or abnormal hemoglobin pigments in the blood perfusing these areas. Patients with congenital heart disease often exhibit cyanosis, which worsens during exertion due to the lower saturation of blood returning to the right side of the heart and the increased right-to-left shunt.

CRT is evaluated by applying digital pressure to blanch the membrane (Fig. 1); the color should return within 2 s. Dehydration and other conditions that reduce cardiac output can prolong the CRT due to high peripheral sympathetic tone and vasoconstriction. Therefore, mucous membrane pallor can indicate either anemia or peripheral vasoconstriction. Anemic animals usually show normal CRT unless hypoperfusion is also present. However, assessing CRT in severely anemic patients is challenging due to the lack of color contrast.

1.2.2 Arterial Pulses

The rate, rhythm, amplitude, and quality of peripheral arterial pressure waves are evaluated by palpating peripheral arteries. The facial or digital arteries are commonly used in horses. The facial artery is palpated where it crosses the ramus of the mandible at the front edge of the masseter muscle. To feel the pulse, pressure is applied against the artery on the inner side of the ventral border of the mandible (Fig. 2).

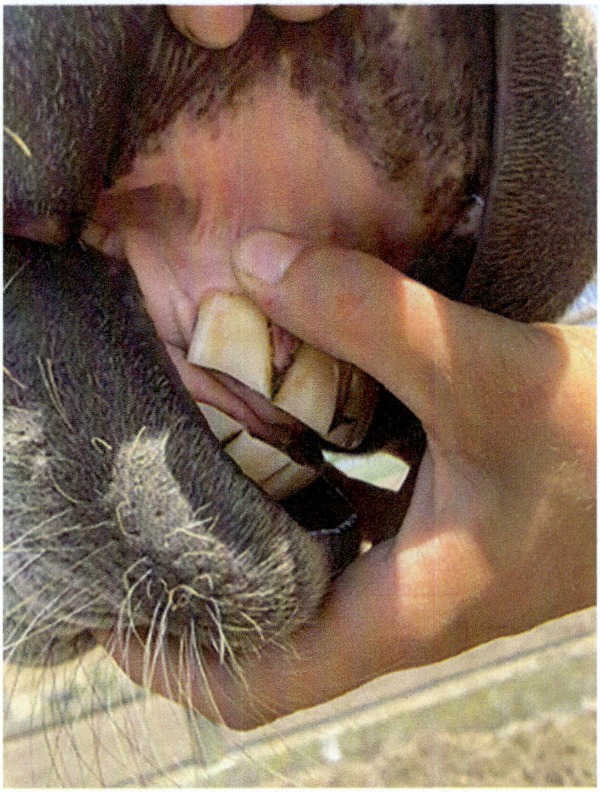

Fig. 1 Normal pink mucous membrane color in a donkey. Digital pressure allows determination of the CRT

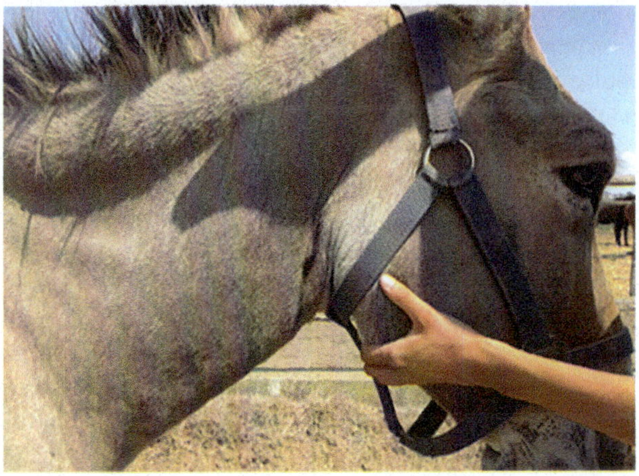

Fig. 2 Arterial pulse palpation using the facial artery in a donkey

Typically, the pulse rate is counted over 30 s, with a longer examination to assess rhythm and quality. The pulse should be monitored for at least a minute, as rhythm abnormalities may be intermittent. In healthy resting donkeys, the heart rate

(HR) ranges from 38 to 45 bpm (Fielding and Krause 1998). In mules normal heart rate ranges between 24–58 bpm (Silva et al. 2018). The hinnies are reported to have the lowest HR among equids (McLean et al. 2016). Physiologically, a regular rhythm is characterized by equal intervals between successive pulse waves. When irregularities are found, the clinician must determine if the basic rhythm is regular with occasional disruptions or if there is no consistent rhythm. A regular rhythm with occasional disruptions suggests that the heart is controlled by a single pacemaker with temporary transmission disruptions. Examples include sinoatrial block and second-degree atrioventricular block, where the excitatory impulse is sporadically blocked at the sinus node or atrioventricular node, resulting in a failure of the ventricles to contract and generate an arterial pulse. This scenario indicates a regular basic rhythm with occasional missed pulse waves (dropped pulses). If there is no predictable rhythm to the pulse, and it cannot be consistently tapped out, it indicates that the heart is not under a single pacemaker influence.

Assessing pulse amplitude and quality is subjective. Moreover, it offers the least clinical value for diagnosing heart abnormalities. The perceived strength of the pulse depends on the difference between peak systolic pressure and diastolic pressure. When diastolic pressure is abnormally low and systolic pressure remains unchanged, pulse pressure increases, resulting in a stronger perceived pulse, known as a hyperkinetic pulse. This occurs when diastolic pressure significantly drops, as seen in aortic valvular insufficiency or patent ductus arteriosus, leading to very high pulse pressure and bounding pulses. In mitral valve regurgitation, the pulse may rise rapidly because the left ventricular ejection time is shortened by mitral valve incompetence. Conversely, a hypokinetic or weak pulse occurs when left ventricular stroke volume decreases. Conditions such as hypovolemia, left ventricular failure, and cardiac tamponade commonly reduce stroke volume, often accompanied by compensatory tachycardia, which further reduces pulse pressure. In subaortic stenosis, pulses are hypokinetic and slow rising. A pulse deficit occurs when the arterial pulse rate is lower than the HR determined by auscultation. This happens when cardiac contractions generate heart sounds but lack sufficient stroke volume to produce a palpable pulse, typical in atrial fibrillation with high HR and multiple ventricular extrasystoles. Generally, periodic or beat-to-beat variations in pulse pressure detected during the examination are likely to reflect underlying heart disease, often arrhythmias.

1.2.3 Jugular Veins

The jugular veins provide an indication of systemic venous and right heart filling pressures. In a standing animal with its head in a normal position, these veins may appear distended in the distal one-quarter to one-fifth of the jugular groove. However, this proportion changes with head position; when the head is lowered below the level of the heart, such as during grazing, the jugular distension extends up to the ramus of the mandible. In healthy animals, veins located below the heart level are physiologically full.

Persistent distension of the jugular veins indicates either elevated central venous pressure or an obstruction in the cranial vena cava flow. Jugular pulse waves correspond to pressure changes associated with atrial contraction, and right heart filling. Pulsations that rise above the point of the shoulder in a standing animal are generally considered abnormal. In horses, the jugular vein may collapse when atrial pressure drops during the cardiac cycle, and the normal refilling of the vein can sometimes be mistaken for a pathological jugular pulse.

Pathological jugular pulses are primarily caused by elevated right atrial pressures, impaired right ventricular diastolic function with a compensatory strong right atrial contraction, moderate to severe tricuspid regurgitation, or arrhythmias that cause atrioventricular dissociation. The latter is most seen in complete atrioventricular block and ventricular tachycardias. Regardless of the underlying cause, an abnormal jugular pulse should prompt further assessment of heart rhythm and right-sided heart function.

1.2.4 Precordial Palpation

The heart's contraction and rotation during the early systole generate the apex beat, a low-frequency vibration that coincides with the first heart sound. It can be palpated between the third and the sixth intercostal spaces (ICS). A shift of the cardiac beat caudally suggests cardiac enlargement. Other less likely causes include thoracic masses that displace the heart or lung-lobe collapse, which allows the heart to shift. An increase in the intensity of the apex beat occurs when there is an increase in sympathetic tone, but it can also indicate cardiac hypertrophy. Diminished intensity of the precordial impulse can result from obesity, weak cardiac contractions, pericardial effusion, intrathoracic masses, pneumothorax, or pleural effusion. During precordial palpation, a "precordial thrill" may be detected. This term describes a palpable vibration on the chest wall caused by a very intense heart murmur.

In horses, the precordium can also be percussed to determine the area of cardiac dullness. Any displacement is suggestive of potential cardiomegaly. However, precordium percussion is a skill that is rarely applied to date.

1.2.5 Cardiac Auscultation

Thoracic auscultation is used to evaluate HR and rhythm, heart sounds, and respiratory sounds. Heart sounds result from sudden intracardiac changes in turbulent blood flow, pressure fluctuations, and the vibrations in contiguous tissues during the cardiac cycle. Since some heart sounds can be difficult to hear, it's crucial to have the patient's cooperation and a quiet environment during auscultation. Respiratory sounds can interfere with heart sound assessment and can mimic murmurs. This is especially true when heart and respiratory rates are synchronized, and vesicular respiratory sound is reinforced. Temporarily closing both nostrils can reduce breath sounds, making it easier to hear heart sounds clearly.

The orientation of the heart and the location of valve areas within the thorax can vary based on breed, conformation of the chest, and heart size. However, the relative position of the valve areas is fairly consistent among individuals; on the left side, from cranial to caudal, the pulmonic, aortic, and mitral valve areas are found, while the tricuspid valve area is located on the right. For optimal results, the heart should be examined systematically. Heart sounds are located with reference to the base and apex of the heart, focusing on the point of maximal intensity (PMI) for each of the four valve regions. Auscultation typically begins on the left side, starting at the PMI of the mitral valve. The stethoscope is then moved towards the base of the heart to listen to the aortic and pulmonic valves. On the right side, the examination generally starts at the PMI of the tricuspid valve, after which the stethoscope is moved dorsally and cranially to listen over the base of the heart on the right side.

The first (S1) and second (S2) heart sounds are easily heard. In some horses a third and fourth heart sound might be heard. The S1 occurs with the closing of the atrioventricular valves. The mitral valve closes slightly before the tricuspid valve, although this difference is not noticeable. Similar to S1, the S2 is made up of two components that relate to the closure of the aortic and pulmonary valves. The pulmonary valve closes after the aortic valve, and in most healthy animals, these two components of S2 merge into a single sound. However, during inspiration, the length of right ventricular ejection increases, leading to a noticeable physiological split of S2 in horses.

The third heart sound (S3) is due to the rapid filling of the ventricle during early diastole and is heard immediately after S2. Typically, it is most noticeable on the left side caudal to the tricuspid area, but it can also be heard over the base of the heart and on the right side. The S3 is common in horses and is often detected in fit racing horses.

The fourth heart sound (S4) is related to atrial contraction and occurs just before the S1. It is most clearly heard over the base of the heart on both the left and right sides. This sound is common in horses but not mules and donkeys, and its perception depends on the length of the PR interval, which varies between individuals. In the same resting patient, the gap between the sound of atrial contraction and the first heart sound often varies, leading to a clear separation in some beats and a blending of the two sounds in others. In the cardiac cycle, the sequence of the four heart sounds is S4, S1, S2, S3. Not all equids have audible S3 and S4 sounds, but when they are perceptible, their sound intensity is lower than that of S1 and S2. However, the presence of the S3 and/or S4 in horses is not an indication of cardiovascular abnormalities.

The 'gallop rhythm' presents when more than two heart sounds are present per cycle. This may be due to the splitting of the S1 or S2 or when the S3 or S4 is audible. The gallop rhythm is not considered an arrhythmia, and the clinical significance of this finding should be assessed based on the physical examination and the equid's medical history.

A split heart sound occurs when the components of a single heart sound, typically either the S1 or S2, are heard as two distinct sounds. Because the S1 is generated by the closure of the two atrioventricular valves, a delay in the closure of one valve can result in a split S1. Splitting of the S1 is not as common as splitting of S2 and the causes of a split S1 include structural lesions interfering with valve closure (i.e. infective endocarditis), conduction delay of one ventricle (i.e. bundle branch block), or premature contraction of one ventricle. Split S2 is not infrequent and is mostly caused by pulmonary hypertension. This condition slows the emptying of the right ventricle, leading to a delay in the closure of the pulmonic valve after the aortic valve. Other potential causes of a split S2 include premature ventricular contractions or delayed electrical conduction through a ventricle (i.e during bundle branch block).

1.2.6 Detection of Heart Murmurs

In donkeys, the prevalence of audible heart murmurs is much lower compared to that reported in horses (2% vs. 50–80%). This reduced prevalence could apparently be due to the lower sensitivity of cardiac auscultation in the donkey species, possibly secondary to a thicker skin and subcutaneous fat, longer and denser hair, or it could be due to the smaller population studied. Moreover, in donkeys, unlike in horses, the presence of heart murmurs generally reflects the presence of structural cardiac abnormalities. Ejection or flow murmurs are less frequently detected in donkeys, and when present, they are predominantly associated with the aortic valve.

As in horses, heart murmurs in donkeys can be classified and diagnosed based on the following criteria: location, timing, quality, radiation, and intensity.

The PMI of a heart murmur is generally found at the site of turbulence and helps determine the murmur's origin. The apex beat identifies the PMI for mitral valve sounds, with mitral regurgitation murmurs being most audible here. After locating the apex beat, other valvular areas can be assessed by changes in heart sound intensity. The S1 is best heard at the apex, while S2 is louder dorsally and cranially. This area is where the aortic valve sounds most commonly radiate. Pulmonary valve sounds are heard cranially and ventrally to the aortic valve, with the stethoscope positioned cranially under the olecranon. Sounds from the tricuspid valve are audible from the right hemithorax, cranial to the apex beat palpated on the left side. A stethoscope positioned too caudally will result in a reduction in murmur intensity, which may cause the murmur to be missed. In some cases, it is helpful to position the stethoscope well forward under the triceps muscle and/or pull the right forelimb cranially. While valvular areas are the primary sites for cardiac auscultation, the stethoscope should remain in contact with the thoracic wall while being moved between valvular areas, as this reduces the chances of missing audible sounds and murmurs.

Heart murmurs can be further classified based on their timing within the cardiac cycle, being either systolic, diastolic, or continuous. When positioning the stethoscope, the clinician should focus on the heart's frequency and rhythm, identify the heart sounds, and then listen during systole (between S1 and S2) and diastole (between S2 and S1) for any murmurs. If it is challenging to distinguish between systole and diastole due to the murmur masking the heart sounds, remember that diastole is typically longer than systole, so a very long murmur is usually diastolic. It may also help to find a spot where the heart sounds are clearer and the murmur is less prominent, or to feel the arterial pulse while listening to the heart. A systolic murmur will align with the pulse, while a diastolic murmur will alternate with it.

Systolic murmurs occur when blood regurgitates through atrioventricular valves or when it is ejected through the semilunar valves. Common causes of these murmurs include mitral and tricuspid regurgitation. Systolic ejection murmurs result from increased flow through the semilunar valves, obstruction in ventricular outflow, or damage to the valves. These ejection-type murmurs have also been detected in donkeys that are clinically normal. Another possible etiology of systolic murmur are congenital cardiac abnormalities such as ventricular septal defect (VSD) or tetralogy of Fallot. These conditions have also been reported in donkeys. Diastolic murmurs can be classified into three types: atrial systolic murmurs (between S4 and S1), ventricular filling murmurs (between S2 and S3), and regurgitant murmurs from the semilunar valves (aortic and pulmonary). While atrial systolic murmurs and ventricular filling murmurs are commonly heard in clinically normal horses, there have been no documented cases in donkeys. Atrioventricular valve stenosis is rare in horses and has never been reported in donkeys. Therefore, the most probable cause of a diastolic murmur extending beyond the S3 is semilunar valve incompetence, with the aortic valve being the most likely cause according to the literature. Continuous murmurs persist throughout both systole and diastole. They can originate from aortopulmonary, aortocardiac, or arteriovenous connections, or from a combination of two or more separate murmurs. The most common aortopulmonary connection is patent ductus arteriosus (PDA), although truncus arteriosus and pseudotruncus arteriosus have also been reported in donkeys.

Concerning the quality (or frequency), most cardiac murmurs are noisy and consist of a wide range of frequencies. Some, called musical, have a pure tone with harmonic frequencies. Terms like musical, buzzing, and honking describe harmonic murmurs at different frequencies: buzzing for low frequencies, musical for high frequencies, and honking for short-duration murmurs. The frequency or pitch of a murmur can aid in diagnosis: for example, the diastolic murmur of aortic regurgitation is often musical, while the diastolic murmur associated with ventricular filling is often a high-pitched squeak.

Radiation refers to the direction a murmur spreads across the body surface. A VSD murmur is heard over the right hemithorax near the tricuspid valve and can be distinguished from tricuspid regurgitation by its cranial and ventral radiation. Aortic regurgitation murmurs typically radiate to the left hemithorax near the aortic valve but can also be heard on the right as the dysfunction worsens. The extent of murmur radiation often correlates with clinical significance; the wider the spread, the more serious the lesion is likely to be.

Murmurs can also be distinguished by their intensity, which is graded on a 1–6 or 1–5 scale. Murmurs graded 4–5/5 or 5–6/6 are often linked to a palpable precordial thrill and significant cardiovascular dysfunction. However, the link between murmur intensity and severity is not always clear, as not all severe valvular issues cause loud murmurs, and not all loud murmurs indicate serious problems. Intensity is influenced by the amount and speed of blood flow, the distance from the stethoscope, and how well the sound transmits through surrounding tissues.

2 Hematology and Biochemistry in Donkeys and Mules

Several studies have highlighted the lack of transferability of laboratory parameter results between donkeys, mules and horses, emphasizing the importance of using species-appropriate reference intervals for clinical decision making (Caldin et al. 2005; Burden et al. 2016; Gloria et al. 2018; Goodrich and Webb 2024). However, information regarding specific hematologic and biochemical parameters for donkeys, mules, and hinnies, established with large sample sizes following the American Society for Veterinary Clinical Pathology (ASVCP) guidelines (Friedrichs et al. 2012), is very scarce or absent. For this reason, the laboratory results obtained in donkeys and their hybrids are often compared with reference ranges that have been established for horses, which is challenging given the evidence for the hematologic and biochemical differences between these species. In the literature, little information is available about donkey-specific hematological and biochemical parameters and even fewer studies have been conducted on mules (Greene et al. 2006; McLean et al. 2016; Lagos and Tadich 2019; Vernaccini et al. 2024).

Studies that involved donkeys of different breed as Pêga (Girardi et al. 2015), Ragusana (Caldin et al. 2005), Kirgyz (Kısadere et al. 2019), or Northwestern donkeys (Bature et al. 2018) highlighted breed-related differences in some blood parameters such as lymphocyte and neutrophil counts that are higher in Ragusana (Caldin et al. 2005) than in crossbred donkeys (Sgorbini et al. 2013).

Many factors may influence blood parameters. The most important of them are age, gender, physiological status, or circadian rhythms. In particular, age represents an important variable that should be considered for the evaluation of the hematological and biochemical reference range. Sgorbini et al. (2013) demonstrated age-related changes in hematological and biochemical parameters of Amiata donkey foals from birth up to the second month of life. Similar findings have been reported in mule foals (Vernaccini et al. 2024). Furthermore, the investigation of hematological and biochemical parameters in the endangered Balkan donkey, revealed significant differences in some parameters related to age (Ljubodrag et al. 2015). Recently, de Palo et al. (2018) showed that different breeding techniques in the early life affects biochemical profiles and lipid peroxidation patterns in donkey foals.

Among other factors influencing biochemical parameters, gender and physiological status are particularly important. Trimboli et al. (2020) reported the highest total protein value in periparturient or lactating female donkeys. Similarly, a study conducted on lactating donkeys found that both age and stage of lactation may induce differences in hematobiochemical profiles (Dezzutto et al. 2018).

According to these studies, the construction of reference ranges must necessarily include the effect of influencing factors (age, breed, physiological status) for the appropriate interpretation of hematological and serum biochemical results. Finally, the instrument–instrument variation of analysis methods represents a significant influencing factor for the generation of a reference range; therefore, it is also important that each laboratory develops its own normal range, and caution should be used when comparing reference interval data from different laboratories.

3 Electrocardiography

During a physical examination, the assessment of the heart is a crucial aspect of the cardiovascular system. Donkeys and mules, like horses, can have both normal or abnormal cardiac rhythms, and the electrocardiography (ECG) is the gold standard for detecting the arrhythmias (Escudero et al. 2009; De Lange et al. 2021a, b).

Electrocardiography in donkeys and mules can be performed, as described in horses, using standard electrocardiograph and a base-apex lead. The positive electrode can be placed on the left thorax in the fifth intercostal space at the level of the point of the elbow, the negative electrode can be attached to the skin on the right jugular furrow approximately in the first third of the neck, while the ground electrode attached to a location remote from the heart (Escudero et al. 2009; Kojouri et al. 2007; Bozzola et al. 2024a, b). To improve contact between electrodes and the skin alcohol may be used. The ECG recording should be set at paper speed of 25 mm/s and calibration of 10 mm/mV.

When interpreting an ECG recording one should first assess the quality of the recording and the absence of artifacts. The heart rate, heart rhythm, the correlation between P waves and QRS complexes, the morphology of P waves and QRS complexes and the duration of the different complexes and intervals should be evaluated.

Normal heart rate ranging between 36–52 bpm in adult donkeys (Svendsen 2008), 40–80 bpm in young donkeys, and 80–120 bpm in newborn donkeys have been reported by The Donkey Sanctuary factsheet (https://www.thedonkeysanctuary.org.uk/sites/default/files/2018-12/the-clinical-examination-20181214.pdf). In mules normal heart rate ranges between 24–58 bpm (Silva et al. 2018).

Electrocardiographic parameters have been published for healthy donkeys with some variations in certain parameters in different donkey breeds (Rezakhani and Yazdanmehr 1977; Kojouri et al. 2007; Escudero et al. 2009; Cruz-Aleixo et al. 2023). Electrocardiographic parameters measured in healthy donkeys of different breeds and in mules are reported in Table 2.

Recently, a pilot Study on HR and HRV Indices in a cohort of 7 athlete mules Evaluated by 24-h ECG. The authors reported an average HR of 32–42 bpm, with a mean of 36 $\pm$ 3 bpm for 23.8 $\pm$ 0.2 h. Minimum HR was 16–24 bpm, and maximum was 70–156 bpm. Individual mules showed supraventricular complexes, sinoatrial heart block, and second-degree atrioventricular block during the observation period (Maas et al. 2025).

3.1 Rhythm Disturbance Described in Donkeys and Mules

Arrhythmias, or abnormal heart rhythms, are a common concern in equine medicine, affecting both horses and their close relatives, donkeys and mules. While the epidemiology, diagnosis, and treatment of arrhythmias in horses have been extensively studied, there is a relative paucity of research on these conditions in donkeys

Table 2 Electrocardiographic parameters in healthy donkeys of different breeds and in mules

Heart rate (bpm)	P wave	RR interval	PR interval	ST segment	QRS	T wave	QT interval	Ref.
Donkeys								
46 (28–108)	0.1 (0.04–0.14)	–	0.24 (0.15–0.3)	–	0.1 (0.07–0.12)	0.12 (0.08–0.18)	0.48 (0.28–0.56)	[1]
65 ± 16.5	0.111 ± 0.018	0.974 ± 0.277	0.230 ± 0.041	–	0.112 ± 0.015	0.137 ± 0.032	0.432 ± 0.061	[2]
52 ± 1.29	0.095 ± 0.004	1.218 ± 0.042	0.273 ± 0.009	0.260 ± 0.007	0.068 ± 0.001	0.139 ± 0.006	0.328 ± 0.069	[3]
55.1 ± 10.68	0.113 ± 0.01	1.10 ± 0.21	0.21 ± 0.03	0.35 ± 0.08	0.10 ± 0.01	0.14 ± 0.05	0.47 ± 0.09	[4]
Mules								
40.8 ± 16.8	0.098 ± 0.018	–	0.272 ± 0.037	–	0.123 ± 0.02	–	0.508 ± 0.043	[5]

Duration of waves, intervals and segments are reported as mean ± standard deviation or median value and range in brackets

[1] Bozzola et al. (2024a, b); [2] Cruz-Aleixo et al. (2023); [3] Escudero et al. (2009); [4] Kojouri et al. (2007); [5] Silva et al. (2018)

and mules. Despite their importance as working animals, the epidemiology of arrhythmias in donkeys and mules remains poorly understood due to the limited research in this area.

Existing studies suggest that donkeys and mules may be susceptible to a variety of cardiac rhythm disturbances, including atrial stand still, atrial fibrillation, atrioventricular blocks, and ventricular tachycardia, however the exact prevalence and underlying risk factors remain to be elucidated (Roberts and Dukes-McEwan 2016a, b; Ertelt et al. 2020; Lima et al. 2020).

3.1.1 Atrial Stand Still

Atrial standstill is defined as a transient or persistent arrhythmogenic condition characterized by the complete or partial absence of electrical and mechanical activity in the atria. Atrial stand still with a ventricular escape rhythm and syncope have been diagnosed in a 15-years-old donkey. The donkey was severely bradycardic (11–18 bpm), tachypneic (44 bpm), with poor pulse quality, decreased jugular refill and peripheral hypoperfusion. Cardiac auscultation showed an irregular rhythm with a heart rate ranging between 11 and 18 bpm. The ECG revealed absence of P-waves, and a variable ventricular escape rhythm at a rate of 11 bpm. The S waves were negative, with positive T waves. The RR-intervals were irregular, and ST-depression was obvious. Pericardial effusion and reduction of left and right ventricular function was evident at echocardiography. Laboratory evaluation revealed an increase of myocardial damage markers (Ertelt et al. 2020).

3.1.2 Atrial Fibrillation

Atrial fibrillation (AF) is particularly common in horses because of their large atria and high vagal tone (Reef and Marr 2010); however, it has been reported also in donkeys (Roberts and Dukes-McEwan 2016a, b; Mendoza et al. 2018). In horses AF can be paroxysmal, however, most frequently AF is permanent once it has started, and it does not convert spontaneously. At rest, it will usually not result in clinical signs, while during exercise the heart rate may become excessively high, and the decreased ventricular filling time results in a decreased stroke volume and consequently in poor performance, especially in high demanding sport disciplines such as racing, eventing or endurance (Evans and Young 2010). Atrial fibrillation (AF) is characterized by the absence of P waves, the presence of fibrillation waves or f waves and irregularly irregular RR intervals with a normal QRS morphology (Reef and Marr 2010). In the absence of underlying cardiac disease, the ventricular rate at rest is normal.

3.1.3 Third Degree AV Block and Adam-Stocks Syndrome

In third degree AV block or complete heart block, there is a complete dissociation between atria and ventricles, and none of the atrial impulses conducts through the AV node. The ventricles are left to contract according to their own intrinsic escape rhythm, which is usually slower than the normal sinus rhythm. This pathological condition, described in donkey case reports, can be caused by degenerative or inflammatory AV nodal disease and can be associated with syncope or weakness

(De Lange et al. 2021a, b; Sedlinská et al. 2021; Heun et al. 2023). The ECG shows P waves with a normal morphology and a regular PP interval. P waves have no relationship with the QRS complexes. The rate of the P waves is usually high and QRS complexes can be bizarrely shaped although a ventricular escape rhythm originating close to the AV node is usually regular and results in a close to normal QRS complex (De Lange et al. 2021a, b; Sedlinská et al. 2021; Heun et al. 2023). Third-degree AV block seems to be a common cardiac conduction disorder in donkeys, particularly in young small breeds (Decloedt 2016; Mendoza et al. 2018) and there are individual case reports of pacemaker implantation in donkeys with this kind of block (Pibarot et al. 1993; De Lange et al. 2021a, b; Sedlinská et al. 2021; Heun et al. 2023). A familial history of syncope was reported in one donkey with several siblings suffering from episodes of collapse, suggesting a hereditary syndrome (Decloedt 2016). Adams-Stock syndrome (cardiac syncope), characterized by loss of consciousness due to a sudden decrease in the cardiac output, has been also reported in donkeys (Pibarot et al. 1993). Both the bradyarrhythmia have been treated with an internal pacemaker (Pibarot et al. 1993; De Lange et al. 2021a, b).

3.1.4 Ventricular Tachycardia

Ventricular dysrhythmias are the consequence of abnormal impulses that arise somewhere in the ventricular myocardium. Ventricular tachycardia (VT) may be paroxysmal or sustained if it persists for many minutes or hours. It nearly always indicates underlying cardiac or systemic disease (Reef and Marr 2010). Ventricular tachycardia is characterized by abnormal QRS' complexes which are not related to P waves. When the QRS' complexes all have the same morphology, monomorphic VT is present, when the ventricular impulses arise from more than one location, QRS' complexes have different morphologies (polymorphic VT) (Reef and Marr 2010).

3.2 Diagnosis and Therapy of Arrhythmias

The diagnostic approach to arrhythmias in donkeys and mules is largely similar to that employed in horses, involving a combination of clinical examination, electrocardiography, and, in some cases, advanced imaging techniques, such as echocardiography (Lima et al. 2020). However, it is important to note that donkeys and mules exhibit anatomical and physiological variations from horses, which can influence the interpretation of diagnostic findings and the selection of appropriate treatment strategies (Burden and Thiemann 2015). For instance, donkeys have a larger heart, a narrower thorax, and a shorter esophagus compared to horses, all of which can impact the placement and interpretation of electrocardiographic leads (Rezakhani and Yazdanmehr 1977; Kojouri et al. 2007; Escudero et al. 2009; Burden and Thiemann 2015).

Regarding the management of arrhythmias in donkeys and mules, the available literature suggests that the general principles of treatment employed in horses can be cautiously applied to these equids. However, reports on treatments for cardiovascular pathologies are anecdotal in donkeys and mules (Mendoza et al. 2018).

Verapamil, at the dose of 0.3 mg/kg/IV, has been used in healthy nonpregnant jennies with transient secondary effect (sinus tachycardia, wandering pacemarker, first-degree AV block, and second-degree AV block). Therefore, caution should be taken when we use this drug as treatment for supraventricular arrhythmias in donkeys (Kojouri et al. 2007). The effects of intravenous infusion at 14 μg/kg of digoxin in healthy donkeys were also investigated (Tharwat and Al-Sobayil 2014) with no variations in electrocardiogram, biochemistry, acid-base and electrolytic, or cardiac troponin I plasma concentrations. The authors concluded that digoxin appears to be a reasonable treatment for arrhythmias and heart failure in donkeys, with a wider safety range compared with horses (Tharwat and Al-Sobayil 2014). Glycoside toxicity due to Oleander spp. ingestion has been described in a 9-year-old gelding miniature donkey (Smith et al. 2003).

3.3 New Devices for Heart Rate Monitoring and ECG Recordings

There are several devices available to record heart rate (HR) and ECG traces; some are designed for long-term or exercise-induced recordings, while others are intended for short-term use, such as standard electrocardiographs (Verheyen et al. 2010). However, the latter can sometimes be impractical due to their relatively high cost, bulky size, and requirement for an electrical connection, which may not always be available in field settings where donkeys are often examined. Fortunately, several portable devices have been developed and validated for both human and veterinary applications over the last few years, which can help veterinarians to overcome these problems. While the use of these portable digital devices is widely reported in horses (Vera et al. 2019; Coelho et al. 2019; Welch-Huston et al. 2020; Alberti et al. 2020; Mott et al. 2021; Vitale et al. 2021; Bindi et al. 2024), only few studies described their use in donkeys and mules (McLean et al. 2012; Guccione et al. 2014; Vezzosi et al. 2018; Kos et al. 2020; Panzera et al. 2020; Bozzola et al. 2024b).

Few studies have assessed the heart rate variability in donkeys and mules using a telemetric heart rate monitor for real-time recording (Polar®) (McLean et al. 2012; Silva et al. 2018; Panzera et al. 2020). A study assessed the cardiac rhythm of resting donkeys using a long term Holter ECG, showing a good interpretation of most ECG tracings after 24-h monitoring (Guccione et al. 2014). A case report focused their research on pacemaker implantation in miniature donkeys, mentioning the use of a portable digital ECG device (Televet, Germany) to record arrhythmias in the animals evaluated (De Lange et al. 2021a, b). Recently, a study assessed the use of a smartphone-based digital stethoscope device (Eko DUO ECG + Digital Stethoscope, Eko Devices Inc., USA) in recording ECG traces in donkeys. A good agreement between standard ECG tracings and digital ECG tracings was found, demonstrating that this device may be considered a valid, practical, and easy-to-use alternative to the standard ECG for assessing heart rhythm in donkeys in the field (Bozzola et al. 2024b).

4 Echocardiography in Donkeys and Mules

Over the last few years, echocardiography has developed into the most important diagnostic tool for assessing the equine heart. Using different echo-modalities, including two-dimensional (2D-mode), mono-dimensional (M-mode), and Doppler studies, transthoracic echocardiography allows a complete examination of the heart. It gives important information about cardiac function, vessels and chamber dimensions, blood flow patterns, heart abnormalities, and hemodynamic modifications (Reef et al. 2005). Similar to horses, an echocardiographic examination is mainly indicated in donkeys and mules with heart murmurs, to distinguish between physiological and pathological murmurs, and to evaluate the clinical relevance of the valvular regurgitation. Moreover, it can also provide useful information about other clinical situations, such as during the presence of arrhythmias, suspicions of heart failure, and so on (Reef et al. 2014). The echocardiographic examination is performed from both the left and right side of the thorax, using a phased-array transducer (2.5 MHz), and an ECG tracing is often recorded simultaneously. The echocardiographic views used in donkeys and mules are the same to those previously described for horses (Long et al. 1992).

Due to the stoic nature of donkeys, the lower prevalence of loud heart murmurs previously reported (2%), and their lower athletic requirement, the diagnosis of cardiac abnormalities and heart diseases is more difficult compared to horses (Roberts and Dukes-McEwan 2016a). The global sdonkey population is highly heterogeneous, including purebred animals and mixed-breed donkeys, therefore it is important to provide normal reference ranges for echocardiographic measurements in this species, assessing the possible influence of age, sex, breed, and bodyweight (BW) on echocardiographic parameters.

Several studies have reported reference ranges for M-mode and B-mode echocardiographic measurements in healthy donkeys (Amory et al. 2004; Hassan and Torad 2015; Roberts and Dukes-McEwan 2016b; Farag and Ibrahim 2020; Cruz-Aleixo et al. 2023). Only a few studies have reported reference ranges for Doppler echocardiographic parameters in this species (Amory et al. 2004; Marzok et al. 2022; Cruz-Aleixo et al. 2023). However, it is often challenging to compare these reference ranges, as many echocardiographic parameters are frequently measured from different echocardiographic views across the studies. Furthermore, certain studies assess different echocardiographic parameters, rendering direct comparison unfeasible. Overall, Amory et al. (2004), Cruz-Aleixo et al. (2023), Farag and Ibrahim (2020), and Roberts and Dukes-McEwan (2016a, b) found similar reference ranges for most echocardiographic parameters (Amory et al. 2004; Roberts and Dukes-McEwan 2016b; Farag and Ibrahim 2020; Cruz-Aleixo et al. 2023), whereas different reference ranges were found in assessing Baladi Egyptian working donkeys (Hassan and Torad 2015). Some representative data for selected echocardiographic measurements are reported in Table 3.

Only few studies have assessed the influence of sex and age on echocardiographic parameters, showing that these parameters do not influence cardiac measurements in donkeys (Roberts and Dukes-McEwan 2016b; Farag and Ibrahim 2020; Bozzola

Table 3 Reference values of selected echocardiographic parameters in donkeys

Measurement	Normal values		Breed	N	Bodyweight	Ref
	Mean ± SD	Interval			Min–Max	
PAD (mm)	32.2 ± 4.1	25.8–43.3	Mixed	36	130–262	[1]
	32.8 ± 4.5	22.5–43.0	Pega	50	230–260	[2]
	32 ± 6.0	24.0–45.0	Mixed	20	98–260	[3]
ABS (mm)	35.6 ± 3.9	29.5–43.2	Mixed	36	130–262	[1]
	24.6 ± 1.8	21.8–28.7	Baladi Egyptian	30	150–220	[4]
	28 ± 3.0	23.0–33.0	Mixed	20	98–260	[3]
ASV (mm)	42 ± 4.5	33.2–52.5	Mixed	36	130–262	[1]
	36.2 ± 2.8	30.4–40.0	Baladi Egyptian	30	150–220	[4]
	41 ± 6.0	34.0–55.0	Mixed	20	98–260	[3]
AJT (mm)	35.7 ± 3.8	28.4–43.6	Mixed	36	130–262	[1]
	40.4 ± 5.9	27.0–51.7	Pega	50	230–260	[2]
	24.3 ± 1.6	21.0–27.1	Baladi Egyptian	30	150–220	[4]
	33 ± 4.0	27.0–47.0	Mixed	20	98–260	[3]
LAD (mm)	63.7 ± 6.3[a]	46.8–74.3	Mixed	36	130–262	[1]
	52.7 ± 6.4[b]	40.3–65.1	Pega	50	230–260	[2]
	64 ± 8.0[a]	50–80	Mixed	20	98–260	[3]
AOD (mm)	40.8 ± 4.3	32.9–48.7	Mixed	36	130–262	[1]
	40.3 ± 5.9	26.7–51.4	Pega	50	230–260	[2]
RVIDd (mm)	20.7 ± 3.9	13.5–30.4	Mixed	36	130–262	[1]
	15.1 ± 2.2	11.0–18.9	Baladi Egyptian	30	150–220	[4]
	18.1 ± 6.1	10.0–37.4	Pega	50	230–260	[2]
	26 ± 5.0	16.0–36.0	Mixed	20	98–260	[3]
IVSd (mm)	16.2 ± 2.4	10.7–22.4	Mixed	36	130–262	[1]
	14.7 ± 2.1	11.8–19.4	Baladi Egyptian	30	150–220	[4]
	17 ± 4.0	11.0–31-0	Mixed	44	150–350	[5]
	17.4 ± 3.4	12.6–25.9	Pega	50	230–260	[2]
	16 ± 2.0	13.0–19.0	Mixed	20	98–260	[3]
LVIDd (mm)	68.1 ± 6.7	54.0–85.7	Mixed	36	130–262	[1]
	48.2 ± 5.8	38.4–58.3	Baladi Egyptian	30	150–220	[4]
	61 ± 8	38.0–91.0	Mixed	44	150–350	[5]
	68 ± 10.8	46.0–91.7	Pega	50	230–260	[2]
	65 ± 12.0	50.0–93.0	Mixed	20	98–260	[3]
LVFWd (mm)	17 ± 2.6	11.1–23.2	Mixed	36	130–262	[1]
	14.2 ± 1.5	11.6–19.8	Baladi Egyptian	30	150–220	[4]
	28 ± 5.0	15.0–59.0	Mixed	44	150–350	[5]
	17.3 ± 3.3	11.7–26.7	Pega	50	230–260	[2]
	16 ± 2.0	13.0–20.0	Mixed	20	98–260	[3]
IVSs (mm)	25.8 ± 3.5	18.8–33.4	Mixed	36	130–262	[1]
	24.4 ± 2.1	21.0–29.0	Baladi Egyptian	30	150–220	[4]
	29 ± 4.0	19.0–58.0	Mixed	44	150–350	[5]
	30.5 ± 5.5	18.3–45.9	Pega	50	230–260	[2]

(continued)

Table 3 (continued)

Measurement	Normal values		Breed	N	Bodyweight	Ref
	Mean ± SD	Interval			Min–Max	
	29 ± 4.0	23.0–38.0	Mixed	20	98–260	[3]
LVIDs (mm)	40.7 ± 5.2	29.3–51.2	Mixed	36	130–262	[1]
	27.5 ± 3.3	21.3–34.4	Baladi Egyptian	30	150–220	[4]
	37 ± 5.0	21.0–65.0	Mixed	44	150–350	[5]
	40.7 ± 7.1	24.5–54.7	Pega	50	230–260	[2]
	37 ± 8.0	27.0–58.0	Mixed	20	98–260	[3]
LVFWs (mm)	22.7 ± 3.6	15.4–30.3	Mixed	36	130–262	[1]
	21 ± 2.4	16.7–28.6	Baladi Egyptian	30	150–220	[4]
	36 ± 4.0	19.0–51.0	Mixed	44	150–350	[5]
	25.9 ± 4.6	17.7–38.4	Pega	50	230–260	[2]
	27 ± 3.0	23.0–35.0	Mixed	20	98–260	[3]
FS (%)	39.7 ± 7.3	27.0–56.0	Mixed	36	130–262	[1]
	42.7 ± 5.9	33.2–63.5	Baladi Egyptian	30	150–220	[4]
	42.4 ± 5.5	31.0–56.0	Mixed	44	150–350	[5]
	39.9 ± 6.0	27.9–55.8	Pega	50	230–260	[2]
	43 ± 5.0	38.0–54.0	Mixed	20	98–260	[3]

Abbreviations: *PAD* pulmonary artery diameter, *ABS* aortic diameter at the level of aortic valve, *AVS* aortic diameter at the level of sinus of Valsalva, *AJT* aortic diameter at the level of sinotubular junction, *LAD* left atrial diameter, *AOD* aortic root diameter, *RVID* right ventricular internal diameter, *IVS* interventricular septal thickness, *LVID* left ventricular internal diameter, *LVFW* left ventricular free-wall thickness, *FS* fractional shortening, *d* diastole, *s* systole

[a]Long axis, [b]short axis

References: [1] Roberts and Dukes-McEwan (2016b); [2] Cruz-Aleixo et al. (2023); [3] Amory et al. (2004); [4] Hassan and Torad (2015); [5] Farag and Ibrahim (2020)

et al. 2024a), however several studies have focused their attention on the influence of bodyweight on echocardiographic measurements, reporting conflicting results. Roberts and Dukes-McEwan (2016a, b) found no significant association with bodyweight on most of the M-mode variables (Roberts and Dukes-McEwan 2016b). This contrasts with significant linear regression association reported by other authors, where bodyweight was significantly associated with most of the echocardiographic parameters evaluated (Amory et al. 2004; Hassan and Torad 2015; Bozzola et al. 2024a). Since donkeys are often employed in working and sport activities, training is another important variable to consider when assessing the heart in this species (Bozzola et al. 2024a). In fact, in horses it has been demonstrated that training affects cardiac dimensions (Buhl et al. 2005). In donkeys, only one study has assessed the influence of training on echocardiographic parameters, reporting that training affects some echocardiographic measurements in healthy mixed-breed donkeys, and suggesting that this variable should be considered when assessing cardiac dimension in this species (Bozzola et al. 2024a).

Mules, like donkeys, are often employed for work purposes, such as tourism, trail rides, territorial defense, and military operations (Tucker et al. 1995; Canisso et al.

Table 4 Echocardiographic parameters (mean ± standard deviation) of mules compared to reported values in horses

	Tucker et al. (1995)	Long et al. (1992)	Lescure and Tamzali (1984)
Measurement	Mules ($n = 10$)	Thoroughbreds ($n = 27$)	Mixed sporting breeds ($n = 100$)
IVSd (mm)	28 ± 3.0	30.2 ± 4.0	30.6 ± 6.0
LVIDd (mm)	111 ± 15.0	119 ± 7.0	110.6 ± 13.0
LVIDs (mm)	67 ± 13.0	73.5 ± 7.0	61.1 ± 9.0
LVFWd (mm)	33 ± 3.0	23.9 ± 3.0	29.2 ± 5.0
FS (%)	40.4 ± 5.7	38.8 ± 4.6	44.1 ± 6.4

Abbreviation: *IVS* interventricular septal thickness, *LVID* left ventricular internal diameter, *LVFW* left ventricular free-wall thickness, *FS* fractional shortening, *d* diastole, *s* systole

2019). They have an innate ability to withstand higher workloads and there are references stating that mules have greater endurance than horses (Travis 1990). Despite their importance as a working animal, there are few studies in the literature about echocardiographic examination of this species (Tucker et al. 1995; Brown et al. 2024). Only one study has reported baseline information about selected echocardiographic measurements in 10 mules (Table 4), finding no significant difference in cardiac dimensions compared to reported values in horses of similar size (Lescure and Tamzali 1984; Long et al. 1992; Tucker et al. 1995).

5 Congenital Heart Diseases

5.1 Patent Ductus Arteriosus

The patent ductus arteriosus (PDA) is a rare congenital defect that can occur alone or combined with other malformations (Schwarzwald 2018). In the literature, PDA has been reported in a donkey foal (Lowe 1972).

The ductus arteriosus is a fetal vessel derived from the left sixth aortic arch. Its function is to permit the blood to shunt from the pulmonary artery into the descending aorta. The functional closure occurs within 72 h after birth; if it does not close, a left-to-right-shunt from aorta to pulmonary artery occurs. Premature foals with pulmonary hypertension are more susceptible to develop PDA (Schwarzwald 2018). No reports of mules are present in the literature.

Clinical signs include a continuous machinery murmur and thrill loudest over the pulmonary artery. Depending on the magnitude of the shunt, the signs can be more severe, and foals may develop pulmonary edema and cardiac heart failure.

Echocardiography will show a volume overload of the pulmonary artery, left atrium and ventricle. PDA can be better visualized from the left cranial thorax by Doppler echocardiographic assessment of the pulmonary artery. A continuous high-velocity turbulent flow directed toward the pulmonary valve is an echocardiographic sign of PDA.

5.2 Septal Defect

The ventricular septal defect (VSD) is the most important congenital heart disease in horses and a genetic basis has been reported for some breeds. In donkeys, VSD has been reported (Roberts and Dukes-McEwan 2016a) on an ante-mortem assessment of cardio-vascular diseases. In particular, the VSD was diagnosed in one donkey out of 202 and the septal defect was associated with aortic prolapse. The 10 year-old donkey was a gelding with a bilateral heart murmur (grade 2/6), basilar early to mid-systolic with no diastolic component. Echocardiographic assessment confirmed a restrictive VSD associated with aortic prolapse, and no significant aortic regurgitation was found. No case reports are present in the literature regarding mules.

The location of VSD depends on embryogenesis of the lesion and influences the clinical manifestations (Schwarzwald 2018). Usually, the VSD is perimembranous as it is located in the membranous portion of the septum in the left ventricular outflow tract immediately below the right coronary cusp of the aortic valve and the tricuspid valve (Lombard et al. 1983; Marr 2010; De Lange et al. 2021a, b), but it can be also infundibular (in fibrous continuity between the aortic valve and the pulmonary valve), or muscular (surrounded by muscular tissue (De Lange et al. 2021a, b). The shunt is usually left to right, but bidirectional shunts may be present with more severe lesions (i.e. bicuspid pulmonary valve) (Marr 2010; Schwarzwald 2018).

Clinical signs of isolated VSD are consistent with a harsh holo-or pan-systolic murmur on the right or left or both thoracic sides. Concerning the perimembranous VSD, the murmur is loudest just below the tricuspid valve region, while the infundibular VSD murmur is loudest cranially on the left side, below the pulmonary valve area. The muscular VSD may be characterized by a left- or right-sided systolic murmur with a different radiation pattern (Schwarzwald 2018; De Lange et al. 2021a, b). Finally, if an aortic regurgitation is present, a holo-diastolic decrescendo murmur can be revealed over the aortic valve.

Echocardiography is the method of choice for the diagnosis of the perimembranous and infundibular VSD, using standardized scanning planes. The muscular VSD can be difficult to identify. Identification of shunting can be diagnosed using color Doppler, while continuous-wave Doppler is needed to estimate the pressure difference between the two ventricles (Schwarzwald 2018).

5.3 Congenital Valvular Diseases

A recent study (Xue and Cavanaugh 2022) reported mitral valve dysplasia in a Nevisian donkey jenny aged 3 years old. Jenny presented for the assessment of a left-sided III/VI systolic musical heart murmur diagnosed during a pre-anesthetic exam. Echocardiographic exam revealed a single mitral regurgitation jet apparent on the right parasternal long axis view, and two mitral regurgitant jets revealed on the left parasternal apical and oblique views. Moreover, a bridge of tissue between the septal and the mural mitral leaflets was found. The heart was mildly remodeled and no clinical signs were present. A diagnosis of mitral valve dysplasia with moderate mitral valve regurgitation was presumed based on the jenny's age, history and echocardiographic findings. Early degenerative valvular disease was considered unlikely based on the literature for horses (Scansen 2019). A familial predisposition to early onset valvular degeneration has not been described in donkeys.

6 Congestive Heart Failure

(Marr 2010; Schwarzwald 2018)

This syndrome is not common in equids. It can develop in foals and mature horses as a consequence of different cardiac disorders, but the most common cause of congestive heart failure (CHF) is valvular heart disease. The onset of atrial fibrillation may be the trigger to cross from compensated to uncompensated heart failure. The signs may include lethargy, anorexia, weight loss, coughing, tachypnea, ventral edema, and colic along with tachycardia, tachypnea, and a loud third heart sound. Right-sided CHF is more frequent in equids, and it is characterized by ventral and limb edema and a pathologic jugular pulse. Left-sided heart failure is more difficult; it is characterized by pulmonary venous congestion and edema (interstitial more than alveolar type). Echocardiography is the diagnostic modality of choice to understand the underlying cause and apply the specific therapy needed for the primary cardiac disease. The aim of the therapy in CHF is to restore the cardiac output and reduce congestion and hemodynamic problems.

Loop diuretics, such as furosemide, is the treatment of choice to reduce the volume overload and edema in equids. The furosemide dose is 1–3 mg/kg IV (loading dose) followed by 0.12 mg/kg/h CRI (Redpath and Bowen 2019). No paper on the use of furosemide is present in the literature for the treatment of CHF in donkeys or mules. One recent paper reported the effects of furosemide administration at different doses on clinical variables, electrocardiographic indices and serum electrolytes in healthy miniature horses (Banihashemi et al. 2023).

Attention must be focused on renal function when administering furosemide in order to minimize prerenal azotemia. Oxygen needs to be administered (5–10 L/min) if respiratory distress is present.

Acepromazine may be indicated as a vasodilator, but systemic blood pressure must be monitored to avoid hypotension. In donkeys and mules, the acepromazine dose is 0.03–0.1 mg/kg (Mendoza et al. 2019) for sedation.

Ionotropic support may be aimed with dobutamine (acute phase) (1–5 mg/kg/min i.v.) or digoxin (chronic phase) (2.2 mg/kg i.v., bid or as required) (Redpath and Bowen 2019). Some studies are present in the literature on digoxin in donkeys (Tharwat and Al-Sobayil 2014), but not for the treatment of CHF.

Angiotensin-converting enzyme (ACE) inhibitors may be used to reduce volume overload, cause vasodilation, decrease renal sodium and water reabsorption and act as a protective agent of the myocardium. Few studies on ACE inhibitors are present in the literature for healthy horses (Gehlen et al. 2003; Luciani et al. 2007; Afonso et al. 2013; Serrano-Rodríguez et al. 2016), and horses affected by CHF (Guglielmini et al. 2002).

7 Valvular Diseases

Most studies related to cardiac diseases in donkeys have focused on the anatomical and pathological description of the lesions (Roberts and Dukes-McEwan 2016a; Morrow et al. 2011). A retrospective study on post-mortem findings in 1444 aged donkeys revealed that vascular disease (including thickening, stenosis, or arterio-sclerosis of the great vessels) was the second most common finding (60%). Valvular diseases were observed in 30% of the cases, and myocardial diseases in 11.2% (Morrow et al. 2011).

A more recent study conducted on the observation of clinical, echocardiographic, and pathological findings in a population of 202 donkeys has further highlighted the aortic valve as the valvular structure most affected by degenerative diseases and valvular insufficiency. While no mitral, tricuspid, or pulmonary regurgitation was detected, suggesting a lower prevalence of these types of valvular incompetence in this species compared to horses. It is worth noting, however, that Color Doppler in this study was only performed in the right parasternal view; in horses, it is usually performed in both the right and left parasternal views (Roberts and Dukes-McEwan 2016a). Therefore, it is possible that the presence of valvular regurgitation has been underestimated.

Moreover, a study reported the left anterior cusp of the aortic valve as the main localization of the valvular lesions in mules (Bishop et al. 1966). This type of lesion has been identified through echocardiography with greater frequency in elderly subjects, although it is rarely associated with audible heart murmurs upon ausculta-tion (Roberts and Dukes-McEwan 2016a; Decloedt 2016). The low prevalence of heart sounds is likely related to the low sensitivity of auscultation in this species, confirming echocardiography, particularly Color Doppler, as the gold standard for identifying valvular regurgitations (Roberts and Dukes-McEwan 2016a; Decloedt 2016).

Clinical information concerning structural or functional heart diseases are there-fore lacking in this species, especially in the geriatric population. Similarly, there are currently no reports regarding valvular diseases in young donkeys.

Both in horses and especially in donkeys, aortic insufficiency is generally a well-tolerated condition (Reef 1998). In aging individuals with underlying degenerative valvular disease, the progression is slow, and if the animal is not working, an annual check-up is sufficient to determine the progression of regurgitation severity (Reef 1998). However, a more guarded prognosis is given to subjects with fenestrations, discontinuities, or bacterial endocarditis as these valvular alterations can progress more rapidly. In these cases, more frequent monitoring is recommended, as it is for horses with severe aortic insufficiency (Reef 1998).

Case Study Questions

Case Study 1 During a routine examination, you note that the donkey's jugular vein remains distended even when the head is at a normal position. What could be the cause of this persistent jugular vein distension? How might this finding relate to right-sided heart function?

Answer: Persistent distension of the jugular veins indicates elevated central venous pressure or cranial vena cava obstruction. Possible causes include elevated right atrial pressures, often due to heart failure; impaired right ventricular diastolic function, which leads to blood backup; tricuspid regurgitation that increases atrial pressure; and arrhythmias affecting normal atrial contraction. Distended jugular veins reflect right-sided heart dysfunction, commonly linked to heart failure, which causes congestion and elevated central venous pressure, and right ventricular issues that result in increased venous pressure.

Case Study 2 You detect a continuous murmur in a young donkey, heard most prominently over the left hemithorax. What are the possible causes of this type of murmur, and what congenital heart conditions should you consider in this case?

Answer: A continuous murmur in a young donkey, especially when heard prominently over the left hemithorax, may suggest several potential causes. One of the most common is PDA, which occurs when the ductus arteriosus fails to close after birth, allowing blood to flow from the aorta to the pulmonary artery. Other possibilities include aortopulmonary connections and arteriovenous malformations such as truncus arteriosus, and pseudotruncus arteriosus.

Case Study 3 You are asked to analyze hematologic and biochemical data from a 1-month-old Amiata donkey foal. How might age influence the expected results, and what changes might you expect compared to an adult donkey?

Answer: Foals typically exhibit higher levels of lymphocytes and neutrophils due to their developing immune systems, along with an increased proportion of immature red blood cells. Biochemical parameters may show variations in total protein levels, often higher during lactation, and differences in enzyme levels reflecting immature organ function.

Case Study 4 An adult donkey's ECG shows a prolonged QRS complex. What might this indicate about the donkey's cardiac condition, and what further diagnostic steps would you take?

Answer: A prolonged QRS complex in an adult donkey's ECG may indicate a ventricular conduction abnormality, such as a bundle branch block or ventricular dysrhythmia, suggesting potential underlying cardiac issues. To further evaluate the condition, an echocardiogram can visualize heart structure and function, while laboratory tests can check for myocardial damage markers. Holter monitoring may assess arrhythmias over 24 h, and cardiac stress testing could reveal issues not evident at rest.

Case Study 5 A 13-year-old donkey presents with severe bradycardia (heart rate: 11–20 bpm), tachypnea, and poor pulse quality. Based on the description provided, which arrhythmia is the donkey most likely suffering from? What diagnostic tests would you prioritize to confirm the diagnosis? Considering the potential causes of this condition, what treatment options could be explored?

Answer: The 15-year-old donkey is most likely suffering from atrial stand still, characterized by absent electrical activity in the atria leading to a slow ventricular escape rhythm. To confirm this diagnosis, prioritize an ECG to identify the absence of P waves, echocardiography to assess cardiac structure and function, and blood tests to check for myocardial damage markers. Treatment options may include pacemaker implantation for severe cases, management of underlying conditions affecting the AV node, and supportive care like fluid therapy.

Case Study 6 A 10-year-old mixed-breed donkey presents with a heart murmur during a routine check-up. Using echocardiography, you are asked to distinguish between a physiological and a pathological murmur. What steps would you take during the echocardiographic examination to determine the cause of the murmur? How might the donkey's breed, age, and bodyweight influence your interpretation of the echocardiographic results?

Answer: Position the donkey comfortably and use a phased-array transducer (2.5 MHz) to examine both thoracic sides while recording an ECG. Obtain standard views, including the left parasternal long axis, to evaluate heart chambers and valves. Classify the murmur based on location, timing, quality, and intensity, using color Doppler to identify any valvular regurgitation or shunting. When interpreting results, consider the donkey's breed, age, and bodyweight, as these factors may influence echocardiographic reference ranges and the likelihood of pathological murmurs.

Case Study 7 A 8-year-old donkey has been diagnosed with a VSD and associated aortic prolapse. The animal presents with a bilateral heart murmur grade 2/6. What are the expected echocardiographic findings in this case, and how do they correlate with the clinical symptoms? How does the location of the VSD (perimembranous, infundibular, or muscular) influence the clinical presentation and prognosis?

Answer: Echocardiographic findings would likely show a left-to-right shunt from the left ventricle to the right, indicating right ventricular volume overload and potential aortic regurgitation. The bilateral heart murmur grade 2/6 results from turbulent blood flow due to the VSD. The location of the VSD affects clinical presentation and prognosis: perimembranous VSDs may cause more severe symptoms, infundibular VSDs can lead to dynamic obstruction, and muscular VSDs often have a less pronounced impact. Overall, the echocardiogram is crucial for determining the defect's significance and guiding management.

Case Study 8 A 16-year-old donkey presents with lethargy, ventral edema, and tachypnea. An echocardiogram reveals signs consistent with left-sided heart failure. What clinical signs would indicate a transition from compensated to uncompensated heart failure in this case? Describe the echocardiographic findings you would expect to see in a donkey with left-sided CHF

Answer: Signs indicating a transition from compensated to uncompensated heart failure may include worsening respiratory distress, increased ventral and limb edema, anorexia, weight loss, greater lethargy, coughing, and changes in heart sounds, such as a loud third heart sound. Echocardiographic findings would typically show pulmonary venous congestion with dilated pulmonary veins, left atrial enlargement due to increased pressure, impaired left ventricular diastolic function, possible mitral regurgitation, and signs of interstitial pulmonary edema. These findings collectively support the diagnosis of left-sided CHF and inform therapeutic management.

Case Study 9 A veterinarian is considering the use of ACE inhibitors for a donkey diagnosed with CHF. How do ACE inhibitors work to alleviate symptoms of CHF, and what are their potential benefits and risks in equids? Given the limited literature on ACE inhibitors in donkeys, what considerations should be taken into account before administering this treatment?

Answer: ACE inhibitors help alleviate CHF symptoms by promoting vasodilation, reducing blood pressure, and lowering heart workload. Benefits include improved cardiac function and reduced fluid retention, while risks involve hypotension, renal impairment, and hyperkalemia. Due to limited research in donkeys, careful monitoring of renal function and electrolytes is essential, starting with a low dose and gradually increasing it.

Case Study 10 A 5-year-old donkey presents with exercise intolerance and a subtle murmur. An echocardiogram reveals aortic insufficiency. What clinical signs would you expect to see in this donkey, and what kind of finding do you expect to find during cardiac auscultation? Discuss the role of echocardiography in managing this case and the specific parameters you would assess

Answer: You would expect exercise intolerance, lethargy, and possibly signs of heart failure. Aortic insufficiency typically presents as a diastolic murmur best heard over the left hemithorax. Echocardiography is vital for evaluating heart function and structure. Key parameters to assess include left ventricular function, aortic valve morphology, regurgitant volume, and the size of the left atrium and ventricle. This imaging technique helps confirm the diagnosis, monitor progression, and guide treatment decisions.

Case Study 11 A donkey is diagnosed with bacterial endocarditis affecting the aortic valve. What clinical signs would likely be present in this case, and what would be your immediate treatment plan? Discuss the prognosis for donkeys with bacterial endocarditis and the factors that may influence recovery and long-term management.

Answer: In a donkey with bacterial endocarditis affecting the aortic valve, you would see signs like fever, lethargy, anorexia, ventral edema, and possibly a diastolic heart murmur. Immediate treatment should include broad-spectrum intravenous antibiotics, and supportive care with anti-inflammatory medications, fluids, and oxygen therapy if respiratory distress is present. The prognosis is guarded and depends on the severity of the disease, promptness of treatment, and the donkey's overall health, necessitating regular follow-ups for optimal recovery.

References

Afonso T, Giguère S, Rapoport G, Berghaus LJ, Barton MH, Coleman AE (2013) Pharmacodynamic evaluation of 4 angiotensin-converting enzyme inhibitors in healthy adult horses. J Vet Intern Med 27(5):1185–1192. https://doi.org/10.1111/jvim.12153

Alberti E, Stucchi L, Pesce V, Stancari G, Ferro E, Ferrucci F, Zucca E (2020) Evaluation of a smartphone-based electrocardiogram device accuracy in field and in hospital conditions in horses. Vet Rec Open 7(1):e000441. https://doi.org/10.1136/vetreco-2020-000441

Amory H, Bertrand P, Delvaux V, Sandersen C (2004) Doppler echocardiographic reference values in healthy donkeys. In: Matthews NS, Taylor TS (eds) Veterinary care of donkeys. International Veterinary Information Service, Ithaca, pp 2919–0904

Banihashemi SH, Samimi AS, Samimi K, Tajik J (2023) Effects of double intravenous furosemide administration at different doses on clinical variables, electrocardiographic indices and serum electrolytes in miniature donkeys. Vet Res Forum 14(1):53–57. https://doi.org/10.30466/vrf.2020.127810.2954

Bature I, Shehu BM, Barje PP (2018) Serum biochemical parameters of donkeys (Equus asinus) as affected by age, location and sex in northwestern Nigeria. J Anim Prod Res 30:123–133

Bindi F, Vezzosi T, Sala G, Freccero F, Marmorini P, Bonelli F, Sgorbini M (2024) Evaluation of a smartphone electrocardiograph in healthy foals and comparison to standard base-apex electrocardiography. Vet Res Commun 48(2):1227–1232. https://doi.org/10.1007/s11259-023-10206-x

Bishop SP, Cole CR, Smetzer DL (1966) Functional and morphologic pathology of equine aortic insufficiency. Pathol Vet 3(2):137–158. https://doi.org/10.1177/030098586600300203

Blissitt K (2010) Auscultation. In: Marr CM, Bowen M (eds) Cardiology of the horse, 2nd edn. Saunders Elsevier, UK, pp 91–104

Bozzola C, Guffanti I, Ortolina A, Bronzo V, Zucca E (2024a) Influence of age, sex, bodyweight, and training on echocardiographic measurements in healthy mixed-breed donkeys. Vet Res Commun 48(5):3339–3347. https://doi.org/10.1007/s11259-024-10461-6. Advance online publication

Bozzola C, Ortolina A, Guffanti I, Alberti E, Bronzo V, Zucca E (2024b) Evaluation of a digital stethoscope for electrocardiographic recording in donkeys: preliminary results. J Equine Vet Sci 135:105048. https://doi.org/10.1016/j.jevs.2024.105048

Brown K, Williams Louie E, Pinn-Woodcock T, Pearson E, Pearson GB, Marr J, Hackett ES, Rath Brown L, Mitchell KJ (2024) Cardiac disease related to primary hyperthyroidism in a 20-year-old mule. Animals 14(11):1660. https://doi.org/10.3390/ani14111660

Buhl R, Ersbøll AK, Eriksen L, Koch J (2005) Changes over time in echocardiographic measurements in young Standardbred racehorses undergoing training and racing and association with racing performance. J Am Vet Med Assoc 226(11):1881–1887. https://doi.org/10.2460/javma.2005.226.1881

Burden F, Thiemann A (2015) Donkeys are different. J Equine Vet Sci 35(5):376–382. https://doi.org/10.1016/j.jevs.2015.03.005

Burden FA, Hazell-Smith E, Mulugeta G, Patrick V, Trawford R, Brooks Brownlie HW (2016) Reference intervals for biochemical and haematological parameters in mature domestic donkeys (Equus asinus) in the UK. Equine Vet Educ 28:134–139. https://doi.org/10.1111/eve.12512

Caldin M, Furlanello T, Solano-Gallego L et al (2005) Reference ranges for haematology, biochemical profile and electrophoresis in a single herd of Ragusana donkeys from Sicily (Italy). Comp Clin Pathol 14:5–12. https://doi.org/10.1007/s00580-005-0544-8

Canisso IF, Panzani D, Miró J, Ellerbrock RE (2019) Key aspects of donkey and mule reproduction. Vet Clin North Am Equine Pract 35(3):607–642. https://doi.org/10.1016/j.cveq.2019.08.014

Coelho CS, Adam GL, Agra de Omena E, Silva G, Silva de Carvalho R, Cuña de Souza VR, Fazio F (2019) Heart rate monitoring in Mangalarga Marchador horses during a field marcha test. J Equine Vet Sci 79:50–53. https://doi.org/10.1016/j.jevs.2019.05.020

Cruz-Aleixo AS, de Oliveira KC, de Oliveira Ferreira LV, Cedeo Quevedo DA, Cruz RKS, Tsunemi MH, Chiacchio SB, Lourenço MLG (2023) Electrocardiographic and echocardiographic parameters in Pega breed donkeys: a descriptive study. Animals 13(5):861. https://doi.org/10.3390/ani13050861

De Lange L, van Steenkiste G, Vernemmen I, Vera L, Cromheeke KMC, Walser U, Meert H, Decloedt A, van Loon G (2021a) Successful application of closed loop stimulation pacemakers with remote monitoring in 3 miniature donkeys with syncope. J Vet Intern Med 35(6): 2920–2925. https://doi.org/10.1111/jvim.16305

De Lange L, Vera L, Decloedt A, van Steenkiste G, Vernemmen I, van Loon G (2021b) Prevalence and characteristics of ventricular septal defects in a non-racehorse equine population (2008-2019). J Vet Intern Med 35(3):1573–1581. https://doi.org/10.1111/jvim.16106

De Palo P, Maggiolino A, Albenzio M, Casalino E, Neglia G, Centoducati G, Tateo A (2018) Survey of biochemical and oxidative profile in donkey foals suckled with one natural and one semi-artificial technique. PloS one 13(6):e0198774. https://doi.org/10.1371/journal.pone.0198774

Decloedt A (2016) Prevalence of cardiovascular disease in donkeys. Vet Rec 179(15):382–383. https://doi.org/10.1136/vr.i5547

Dezzutto D, Barbero R, Valle E, Giribaldi M, Raspa F, Biasato I, Cavallarin L., Bergagna S., McLean A., Gennero MS (2018) Observations of the hematological, hematochemical, and electrophoretic parameters in lactating donkeys (Equus asinus). J Equine Vet Sci 65:1–5. https://doi.org/10.1016/j.jevs.2018.02.013

Ertelt A, Bertram CA, Stumpff F, Kuban S, Klopfleisch R, Lübke-Becker A, Gehlen H (2020) Atrial standstill in a donkey with digitoxin intoxication. Pferdeheilkunde – Equine Medicine 36(2): 114–118.

Escudero A, González JR, Benedito JL, Prieto FR, Ayala I (2009) Electrocardiographic parameters in the clinically healthy Zamorano-leones donkey. Res Vet Sci 87(3):458–461. https://doi.org/10.1016/j.rvsc.2009.04.005
Evans DL, Young LE (2010) Cardiac responses to exercise and training. In: Marr CM, Bowen M (eds) Cardiology of the horse. Saunders Elsevier, UK, pp 35–47
Farag AMM, Ibrahim HMM (2020) Reference values and repeatability of B-mode and M-mode echocardiographic parameters in healthy donkey (Equus asinus)-the guide to assess healthy heart. J Equine Vet Sci 88:102929. https://doi.org/10.1016/j.jevs.2020.102929
Fielding D, Krause P, (1998) Donkeys. The Tropical Agriculturist. Macmillam Education Ltd, London and Brasingstoke
Fregin GF (1992) Medical evaluation of the cardiovascular system. Vet Clin North Am Equine Pract 8(2):329–346. https://doi.org/10.1016/s0749-0739(17)30455-8
Friedrichs KR, Harr KE, Freeman KP, Szladovits B, Walton RM, Barnhart KF, Blanco-Chavez J, American Society for Veterinary Clinical Pathology (2012) ASVCP reference interval guidelines: determination of de novo reference intervals in veterinary species and other related topics. Vet Clin Pathol 41(4):441–453. https://doi.org/10.1111/vcp.12006
Gay CC, Radostits OM, Mayhew I, Houston D (2000) Clinical examination of the cardiovascular system. In: Radostits OM, Mayhew IG, Houston DM (eds) Veterinary clinical examination and diagnosis. Harcourt Publishers, USA, pp 245–291
Gehlen H, Vieht JC, Stadler P (2003) Effects of the ACE inhibitor quinapril on echocardiographic variables in horses with mitral valve insufficiency. J Vet Med A Physiol Pathol Clin Med 50(9): 460–465. https://doi.org/10.1046/j.1439-0442.2003.00580.x
Girardi AM, Marques LC, de Toledo CZP et al (2015) Hematological variables of the Pêga donkey (Equus asinus) breed: influence of age and sex. Comp Clin Pathol 24, 335–342. https://doi.org/10.1007/s00580-014-1905-y
Gloria A, Veronesi MC, Carluccio R, Parrillo S, De Amicis I, Contri A (2018) Biochemical blood analysis along pregnancy in Martina Franca jennies. Theriogenology 115:84–89. https://doi.org/10.1016/j.theriogenology.2018.04.026
Goodrich EL, Webb JL (2024) Complete blood count and biochemistry reference intervals for healthy adult donkeys in the United States. Animals 14(14):2018. https://doi.org/10.3390/ani14142018
Greene HM, Hurson MJ, Wickler SJ (2006) Haematological and respiratory gas changes in horses and mules exercised at altitude (3800 m). Equine Vet J Suppl 36:551–556. https://doi.org/10.1111/j.2042-3306.2006.tb05603.x
Guccione J, Piantedosi D, Di Loria A, Veneziano V, Ciaramella P (2014) Long-term Electrocardiography Recording with Holter Monitoring in 15 Donkeys, J Equine Vet Sci 34(2):302–306, https://doi.org/10.1016/j.jevs.2013.07.001
Guglielmini C, Giuliani A, Testoni S, Corletto F, Bernardini D (2002) Use of an ACE inhibitor (ramipril) in a horse with congestive heart failure. Equine Vet Educ 14(6):297–306. https://doi.org/10.1111/j.2042-3292.2002.tb00193.x
Hassan EA, Torad FA (2015) Two-dimensional and M-mode echocardiographic measurements in the healthy donkey (Equus asinus). J Equine Vet Sci 35(4):283–289. https://doi.org/10.1016/j.jevs.2015.01.016
Heun F, Niebuhr T, Gutierrez Bautista A, Wiedmann F, Verhaar N, Kästner S, Feige K, Schmidt C (2023) Treatment of a paroxysmal atrioventricular block by implantation of a bipolar, single-chamber cardiac pacemaker in a donkey. Animals 13(17):2724. https://doi.org/10.3390/ani13172724
Kısadere I, Donmez N, Omurzakova N (2019) Serum biochemical reference values of Kyrgyz donkeys (Equus asinus). Comp Clin Pathol 28:817–823. https://doi.org/10.1007/s00580-018-2864-5
Kojouri GA, Rezakhani A, Torki E (2007) The effects of verapamil hydrochloride on electrocardiographic (ECG) parameters of domestic donkey (Equus asinus). J Equine Vet Sci 27(11): 499–503. https://doi.org/10.1016/j.jevs.2007.10.007

Kos VK, Brložnik M, Petrič AD, Avbelj V (2020) Simultaneous phonocardiography and electrocardiography using smartphone in dogs, cats and horses. In: In 2020 43rd International Convention on Information, Communication and Electronic Technology (MIPRO). IEEE, pp 327–331

Lagos J, Tadich TA (2019) Hematological and biochemical reference intervals for mules in Chile. Front Vet Sci 6:400. https://doi.org/10.3389/fvets.2019.00400

Lescure F, Tamzali Y (1984) Valeurs de reference on echocardiographic Tm ches le cheval de sport. Rev Med Vet 135:405–418

Lima TS, Silva RAF, Pereira RMF, Soares KL, Santos NTA, Sousa MS, Mendonça FS, Lucena RB (2020) Skin diseases in donkeys and mules-an update. Animals 11(1):65. https://doi.org/10.3390/ani11010065

Ljubodrag S, Vladimir D, Predrag S, Nada L, Ivica R, Ante I, Jevrosima S, Zoran S (2015) Morphological, biochemical and hematological characterization of endangered Balkan donkey breed. Acta Vet Brno 65(1):125–136. https://doi.org/10.1515/acve-2015-0010

Lombard CW, Scarratt WK, Buergelt CD (1983) Ventricular septal defects in the horse. J Am Vet Med Assoc 183(5):562–565

Long KJ, Bonagura JD, Darke PG (1992) Standardised imaging technique for guided M-mode and Doppler echocardiography in the horse. Equine Vet J 24(3):226–235. https://doi.org/10.1111/j.2042-3306.1992.tb02820.x

Lowe JS (1972) Patent ductus arteriosus in a donkey foal. N Z Vet J 20(1–2):15. https://doi.org/10.1080/00480169.1972.33992

Luciani A, Civitella C, Santori D, Sconza S, Guglielmini C (2007) Haemodynamic effects in healthy horses treated with an ACE-inhibitor (Ramipril). Vet Res Commun 31(Suppl 1):297–299. https://doi.org/10.1007/s11259-007-0051-2

Maas L, Morgan JM, Case J, Chell DD, McLean AK (2025) Pilot study: heart rate and heart rate variability indices in mules evaluated by 24-hour electrocardiogram. Animals 15:2438. https://doi.org/10.3390/ani15162438

Marr CM (2010) Cardiac murmurs: congenital heart disease. In: Cardiology of the horse. WB Saunders, pp 193–205

Marzok M, Almubarak AI, Al Mohamad Z, Salem M, Farag AM, Ibrahim HM, El-Ashker MR, El-Khodery S (2022) Reference values and repeatability of pulsed wave Doppler echocardiography parameters in normal donkeys. Animals 12(17):2296. https://doi.org/10.3390/ani12172296

McLean AK, Heleski CR, Yokoyama MT, Wang W, Doumbia A, Dembele B (2012) Improving working donkey (Equus asinus) welfare and management in Mali, West Africa. J Vet Behav 7(3):123–134. https://doi.org/10.1016/j.jveb.2011.10.004

McLean AK, Wang W, Navas-Gonzalez FJ, Rodrigues JB (2016) Reference intervals for hematological and blood biochemistry reference values in healthy mules and hinnies. Comp Clin Pathol 25:871–878. https://doi.org/10.1007/s00580-016-2276-3

Mendoza FJ, Toribio RE, Perez-Ecija A (2018) Donkey internal medicine—part II: cardiovascular, respiratory, neurologic, urinary, ophthalmic, dermatology, and musculoskeletal disorders. J Equine Vet Sci 65:86–97. https://doi.org/10.1016/j.jevs.2018.02.025

Mendoza FJ, Perez-Ecija A, Toribio RE (2019) Clinical pharmacology in donkeys and mules. Vet Clin North Am Equine Pract 35(3):589–606. https://doi.org/10.1016/j.cveq.2019.08.011

Morrow LD, Smith KC, Piercy RJ, du Toit N, Burden FA, Olmos G, Gregory NG, Verheyen KL (2011) Retrospective analysis of post-mortem findings in 1,444 aged donkeys. J Comp Pathol 144(2–3):145–156. https://doi.org/10.1016/j.jcpa.2010.08.005

Mott R, Dowell F, Evans N (2021) Use of the Polar V800 and Actiheart 5 heart rate monitors for the assessment of heart rate variability (HRV) in horses. Appl Anim Behav Sci 241:105401. https://doi.org/10.1016/j.applanim.2021.105401

Panzera M, Alberghina D, Statelli A (2020) Ethological and physiological parameters assessment in donkeys used in animal assisted interventions. Animals 10(10):1867. https://doi.org/10.3390/ani10101867

Pibarot P, Krins A, Salmon Y, Di Fruscia R (1993). Implantation of a programmable atrioventricular pacemaker in a donkey with complete atrioventricular block and syncope, Equine Veterinary Journal, 25(3):248-251. https://doi.org/10.1111/j.2042-3306.1993.tb02955.x

Redpath A, Bowen M (2019) Cardiac therapeutics in horses. Vet Clin North Am Equine Pract 35(1):217–241. https://doi.org/10.1016/j.cveq.2018.11.004

Reef VB (1998) Cardiovascular ultrasonography. In: Equine diagnostic ultrasound, 1st edn. Saunders, pp 215–272

Reef VB, Marr CM (2010) Dysrhythmias: assessment and medical management. In: Marr CM, Bowen M (eds) Cardiology of the horse. Saunders-Elsevier, pp 159–178

Reef VB, Whittier M, Allam LG (2005) Echocardiography. Clin Tech Equine Pract 3:274–283. https://doi.org/10.1053/j.ctep.2005.03.013

Reef VB, Bonagura J, Buhl R, McGurrin MK, Schwarzwald CC, van Loon G, Young LE (2014) Recommendations for management of equine athletes with cardiovascular abnormalities. J Vet Intern Med 28(3):749–761. https://doi.org/10.1111/jvim.12340

Rezakhani A, Yazdanmehr MM (1977) The normal electrocardiographic parameters of the domestic donkey (Equus asinus). Zentralblatt fur Veterinarmedizin Reihe A 24(8):672–679. https://doi.org/10.1111/j.1439-0442.1977.tb01746.x

Roberts SL, Dukes-McEwan J (2016a) Assessment of cardiovascular disease in the donkey: clinical, echocardiographic and pathological observations. Vet Rec 179(15):384. https://doi.org/10.1136/vr.103733

Roberts SL, Dukes-McEwan J (2016b) Echocardiographic reference ranges for sedentary donkeys in the UK. Vet Rec 179(13):332. https://doi.org/10.1136/vr.103734

Scansen BA (2019) Equine congenital heart disease. Vet Clin North Am Equine Pract 35(1): 103–117. https://doi.org/10.1016/j.cveq.2018.11.001

Schwarzwald CC (2018) Disorders of the cardiovascular system. In: Reed SM, Bayly WM, Sellon DC (eds) Equine internal medicine, 4th edn. Elsevier, USA, pp 387–541

Sedlinská M, Kabeš R, Novák M, Kološ F, Melková P (2021) Single-chamber cardiac pacemaker implantation in a donkey with complete AV block: a long-term follow-up. Animals 11(3):746. https://doi.org/10.3390/ani11030746

Senior JM (2013) Not small horses: improving treatments for donkeys. Vet Rec 173(12):292–293. https://doi.org/10.1136/vr.f5646

Serrano-Rodríguez JM, Gómez-Díez M, Esgueva M, Castejón-Riber C, Mena-Bravo A, Priego-Capote F, Serrano Caballero JM, Muñoz A (2016) Pharmacokinetics and pharmacodynamics of ramipril and ramiprilat after intravenous and oral doses of ramipril in healthy horses. Vet J 208: 38–43. https://doi.org/10.1016/j.tvjl.2015.10.024

Sgorbini M, Bonelli F, Rota A, Baragli P, Marchetti V, Corazza M (2013) Hematology and clinical chemistry in Amiata donkey foals from birth to 2 months of age. J Equine Vet Sci 33(1):35–39. https://doi.org/10.1016/j.jevs.2012.04010

Silva GA, Rodrigues LM, Monteiro BS, de Souza VR, Manso Filho HC, Coelho CS (2018) Effect of a Marcha field test on some blood and electrocardiographic parameters of mules. J Equine Vet Sci 70:42–47. https://doi.org/10.1016/j.jevs.2018.08.009

Smith PA, Aldridge BM, Kittleson MD (2003) Oleander toxicosis in a donkey. J Vet Intern Med 17(1):111–114. https://doi.org/10.1111/j.1939-1676.2003.tb01332.x

Svendsen ED (2008) Physiological parameters. In: Duncan J, Hadrill D (eds) The professional handbook of the donkey. Whittet Books, p 380

Tharwat M, Al-Sobayil F (2014) Influence of the cardiac glycoside digoxin on cardiac troponin I, acid-base and electrolyte balance, and haematobiochemical profiles in healthy donkeys (Equus asinus). BMC Vet Res 10:64. https://doi.org/10.1186/1746-6148-10-64

Travis L (1990) The mule. Allen JA and Co, Great Britain

Trimboli F, De Amicis I, Di Loria A, Ceniti C, Carluccio A (2020) Reference ranges for hematological and biochemical profile of Martina Franca donkeys. Front Vet Sci 7:602984. https://doi.org/10.3389/fvets.2020.602984

Tucker RL, Wickler SJ, London C, Wyle A, Gabbard M (1995) Selected echocardiographic parameters and right sided pressures of the mule. Equine Vet J 27(S18):117–119. https://doi.org/10.1111/j.2042-3306.1995.tb04902.x

Vera L, van Steenkiste G, Decloedt A, de Clercq D, van Loon G (2019) Comparison of smartphone-based and standard electrocardiography in healthy horses. In: Proceedings of the British Equine Veterinary

Verheyen T, Decloedt A, De Clercq D, Deprez P, Sys SU, van Loon G (2010) Electrocardiography in horses–part 1: how to make a good recording. Vlaams Diergeneeskundig Tijdschrift 79(5): 331–336

Vernaccini M, Bindi F, Bonelli F, Nocera I, Sgorbini M (2024) Hematology and clinical chemistry in mule foals from birth to two months of age: a preliminary study. J Equine Vet Sci 132: 104977. https://doi.org/10.1016/j.jevs.2023.104977

Vezzosi T, Sgorbini M, Bonelli F, Buralli C, Pillotti M, Meucci V, Tognetti R (2018) Evaluation of a smartphone electrocardiograph in healthy horses: comparison with standard base-apex electrocardiography. J Equine Vet Sci 67:61–65. https://doi.org/10.1016/j.jevs.2018.03.006

Vitale V, Vezzosi T, Tognetti R, Fraschetti C, Sgorbini M (2021) Evaluation of a new portable 1-lead digital cardiac monitor (eKuore) compared with standard base-apex electrocardiography in healthy horses. PLoS One 16(8):e0255247. https://doi.org/10.1371/journal.pone.0255247

Ware WA, Bonagura JD, Scansen BA (2021) Cardiovascular examination. In: Cardiovascular disease in companion animals. CRC Press, pp 33–44

Welch-Huston B, Durward-Akhurst S, Norton E, Ellingson L, Rendahl A, McCue M (2020) Comparison between smartphone electrocardiography and standard three-lead base apex electrocardiography in healthy horses. Vet Rec 187(9):e70. https://doi.org/10.1136/vr.105759

Xue C, Cavanaugh SM (2022) Echocardiographic findings of a donkey (Equus asinus) Jenny with suspected mitral valve dysplasia. J Equine Vet Sci 111:103871. https://doi.org/10.1016/j.jevs.2022.103871

Clinical Approach to the Donkey, Mule or Hinny with Skin Disease

Stephen D. White, Derek C. Knottenbelt, and Philip Wayuta Mshelia

Abstract

Donkeys (*Equus asinus*) and their hybrids (mules and hinnies) are used throughout the world primarily as beasts of burden, but occasionally for other functions, as a meat and milk source or as pets. Although closely related to other equids, they have some unique features of their own with regard to dermatologic disease. This article highlights the various dermatoses noted or reported in donkeys, as well as comparisons with horses when differing in prevalence, presentation, or treatment.

Keywords

Dermatitis · Skin · Pruritus · Cutaneous habronemiasis · Culicoides · Pigeon fever

1 Introduction and Definitions

Donkeys (*Equus asinus*) and their hybrids are species utilized throughout the world primarily as beasts of burden, but occasionally for other functions, as a meat or milk source or as pets. In this chapter unless otherwise indicated, diseases mentioning donkeys are assumed to apply to their hybrids.

S. D. White (✉)
School of Veterinary Medicine, University of California, Davis, CA, USA
e-mail: sdwhite@ucdavis.edu

D. C. Knottenbelt
Equine Medical Solutions, Carbeth Gardens Cottage, Killearn, Scotland, UK

P. W. Mshelia
Department of Veterinary Medicine, Faculty of Veterinary Medicine, Ahmadu Bello University, Zaria, Nigeria

A. K. McLean et al. (eds.), *Donkey and Mule Medicine: Clinical Medicine and Reproduction*, https://doi.org/10.1007/978-3-032-18540-2_2

Donkeys have numerous skin diseases, which may parallel their horse relatives, but also have unique features. This chapter provides an overview and methodical approach to skin disease in donkeys based upon clinical signs. The most common reasons for owners of donkeys to seek care for skin disease are:

- Pruritus
- Nodules
- Alopecia
- Crusts and scale

In one retrospective study of donkeys presenting to the teaching hospitals of four veterinary schools, the most common skin diseases were insect bite hypersensitivity, sarcoid, habronemiasis, superficial bacterial infection, and dermatophytosis (White et al. 2019).

2 Pruritus

2.1 Ectoparasites

Similar to other equids, ectoparasites are a common cause of pruritus in donkeys and their hybrids. These include *Culicoides* spp. (Yeruham et al. 1993; Lincoln et al. 2015), the biting louse *Werneckiella equi* (Veneziano et al. 2012), and the sucking louse *Haematopinus asini* (Veneziano et al. 2013).

Various species of *Culicoides* may feed on the donkey host and induce a hypersensitivity response via salivary antigens (Barbosa et al. 2024). Treatment of *Culicoides* spp. is multi-faceted. These midges are very small- the smallest being 1 mm in size. Their size enables them to easily fly through the mesh of normal fly and mosquito screens. Ultra-fine netting or screens placed in windows (60 squares to the square inch or pore size of 0.1825mm^2) can be used to keep these insects out of stables; such nets have been shown to be amongst the most effective ways of reducing numbers of flies in stables. *Culicoides* spp. are also poor fliers. Overhead and window fans that maintain air flow will keep midges from reaching the donkey. Outdoors, paddocks on a windy hill will also reduce numbers of biting flies. *Culicoides* spp. are primarily active, and feed, from dusk until dawn (sunset-sunrise). It is therefore helpful to stable the animals during this time. Fly sheets and face masks provide a mechanical barrier to protect against fly bites. These should cover as much of the surface of the animal as possible. Insect repellants containing permethrin (2% spray) can be useful but their effect is fairly short lived necessitating frequent applications.

Pruritus also may be managed with corticosteroids such as prednisolone (starting at 1 mg/kg q24h, then tapering down to a minimal effective dose) or dexamethasone (starting at 0.05–0.01 mg/kg q24h, then again tapering down to the least effective dose). Care must be taken to avoid adverse effects such as laminitis ('founder').

Hyposensitization using allergens or recombinant allergens may be administered either subcutaneously or orally. The effectiveness of this treatment is however controversial. Vaccination of donkeys against their own interleukins (IL5 or IL 31) may bring encouraging results in the future (Fettelschoss-Gabriel et al. 2018; Olomski et al. 2020).

Treatment of lice is somewhat different, as these ectoparasites live on the host. Topical treatment with insecticides such as permethrin derivatives (Veneziano et al. 2012; Veneziano et al. 2013), lime sulfur, and/or systemic products such as avermectins (at least for sucking lice) have all been shown to be effective.

Recently a case of the tropical rat mite (*Ornithonyssus bacoti*) was reported in a donkey, secondary to cutaneous lymphoma (Dumitrache et al. 2023).

2.2 Allergy

Donkeys may develop atopic dermatitis (environmental allergy), similar to that reported for horses (Stepnik et al. 2012) with presumably the same treatment options (antihistamines, hyposensitization). There is a report of successful treatment of an atopic miniature donkey with cyclosporine at 5 mg/kg q24h, then tapered eventually to q72h (Falk et al. 2015).

3 Nodules

3.1 Infections

Deep fungal infections causing nodules have been reported in donkeys. These include *Sporothrix schenkii* (Crothers et al. 2009; Irizarry-Rovira et al. 2000 [Fig. 1]), *Cryptococcosus* sp. (Khodakaram-Tafti and Dehghani 2006), *Histoplasma capsulatum* (North American histoplasmosis) (Cooper et al. 1994), *Histoplasma capsulatum* var. *farciminosum* ('farcy' –the cause of equine epizootic lymphangitis in east Africa) (Powell et al. 2006), and *Prototheca* sp.

In general, deep fungal infections present clinically with nodules and ulcers, often with a purulent exudate. Thickened ('corded') lymphatics may be noted, especially in *Histoplasma capsulatum* var. *farciminosum* infections. Sporotrichosis tends to affect the inguinal area and medial aspect of the hind legs although other areas of the body may also be affected.

Diagnosis may be achieved by impression smear cytology, tissue culture or histopathology. Treatment with oral potassium iodide (KI) and fluconazole may be helpful in controlling the disease but may not resolve them completely. Reported dosage regimens for KI are 0.5–2 mg/kg daily for 2.5 years or 40 mg/kg daily for 10 days; dose for fluconazole is 1 mg/kg daily for 2 years (Crothers et al. 2009; Powell et al. 2006). Surgical removal of well-defined fungal granulomas may be curative but if only partial removal is possible, adjunctive topical treatment with 4% tincture of iodine applied directly to the site may be helpful (Gateechew 2004;

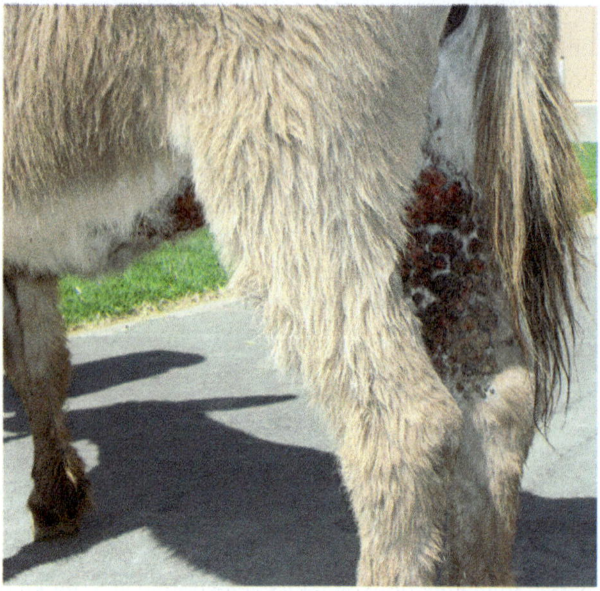

Fig. 1 Nine-year-old jenny with sporotrichosis. Note ulcerative plaques on inner thigh and ventral abdomen. (Reproduced with permission from John Wiley & Sons; from reference (White et al. 2019))

Scantlebury and Reed 2009). *Pythium insidiosum* is the organism causing a severe cutaneous and subcutaneous disease seen most commonly in tropical and subtropical areas of the world. *P insidiosum* is neither bacteria, fungus nor parasite, but is in its own phylum. It is a water-borne organism, thus most lesions are on the face, legs or ventrum, where the animal was in contact with water while feeding. Lesions are ulcerative, proliferative and exudative and are sometimes referred to as 'swamp cancer'. There is one report of a donkey with pythiosis that had the larger lesions excised whereupon the smaller lesions resolved spontaneously (Maia et al 2016). A combination of mefenoxam (an agricultural anti-fungal), prednisolone, terbinafine and minocycline have been anecdotally reported to successfully treat horses with pythiosis.

Deep pyoderma, especially when caused by *Corynebacterium pseudotuberculosis* may cause nodules, ulcers and draining tracts similar to deep fungal infections (Fig. 2). These may occur on the legs, cranial thorax or ventral abdomen.

Viral infections causing nodules in donkeys include molluscum contagiosum (Fox et al. 2012), as well as bovine papillomavirus (see equine sarcoid, below).

3.2 Infestations

Habronema spp. ('summer sore', 'jack sore') can afflict donkeys and their hybrids. Anecdotally, cutaneous habronemiasis has been thought to cause more severe lesions in donkeys and mules than in horses. The prevalence is highest in regions with a temperate/tropical climate. It is caused by any of the three species of

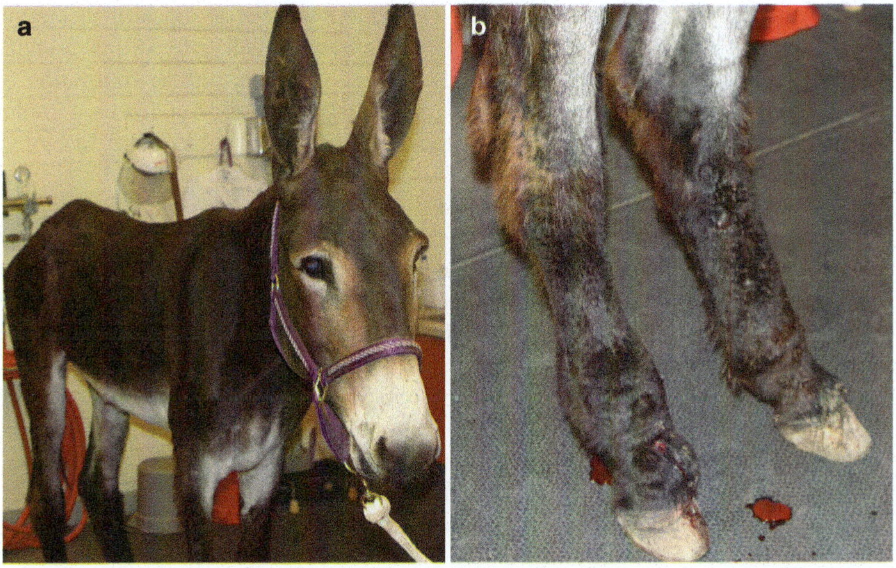

Fig. 2 Seven-year-old mammoth jenny with *Corynebacterium pseudotuberculosis* infection. Note (**a**- photo on left) thin body condition and (**b**- photo on right) nodules on distal legs. (Reproduced with permission from John Wiley & Sons; from reference (White et al. 2019))

Habronema: *H. muscae*, *H. microstoma*, and *H. megastoma*. The adult worms reside in the stomach where they cause little reaction with the exception of *H. megastoma* which produces varying sized nodules that usually occur near the margo plicatus. The females are viviparous and the larvae are passed in the feces where they are ingested by the maggots of flies which serve as the intermediate host. *H. muscae* and *H. megastoma* develop in the house fly, *Musca domestica*. *H. microstoma* develops in the stable fly, *Stomoxys calcitrans*. The normal life cycle is completed when the infective larvae are deposited by flies around the donkey's lips from where they are swallowed.

Cutaneous habronemiasis is a form of aberrant parasitism when the larvae gain access to the deeper layers of the skin in a presumably hypersensitive host. It is important to remember that infective larvae of *Habronema* spp. cannot penetrate normal healthy skin. Thus, for cutaneous habronemiasis to develop, the larvae must be deposited in previously damaged skin or mucous membranes (Pusterla et al. 2003).

The disease is seasonal, first appearing in the spring and in most cases spontaneously regressing in the winter months. As the occurrence of the disease is sporadic, with only a few equids in any given area being affected, a hypersensitivity reaction is suspected. Once the disease develops in a donkey it usually will recur each succeeding summer unless stringent preventative measures are taken. The usual locations for the lesions are the medial canthus of the eye, the conjunctivae, the male genitalia (especially the urethral process), and the distal extremities (hock, canon, pastern, fetlock, coronary band). The lesions consist of areas of ulceration with or without granulation tissue and usually containing small gritty yellow nodules ('kunkers'). Biopsy of the lesions reveal granulation tissue, diffuse infiltration of eosinophils, and foci of coagulation necrosis of eosinophils surrounded by a granulomatous response. Cross sections of dead larva can usually be found in biopsy tissue in approximately 50% of cases.

Diagnosis is often based on the history, the location and the characteristic appearance of the lesions. The major differential diagnoses include exuberant granulation tissue ('proud flesh'), fibroblastic sarcoid and squamous cell carcinoma. All three of these however may be secondarily infected with *Habronema* and comorbidity is relatively common. Thus, if the lesions do not respond to habronemiasis treatment, or if there is any reason to suspect that there is an underlying neoplasm, a biopsy should be obtained for histopathology. Scrapings of the lesions are usually unrewarding in demonstrating the larvae.

Treatment should be aimed at decreasing the size of the lesion by surgical means, reducing inflammation through the use of topical, intralesional or systemic corticosteroid, and preventing recurrence. Topical or systemic avermectin administration can also be useful. Fly control, including protective covering such as fly masks and leg wraps, as well as regular deworming with ivermectin, are recommended to reduce the incidence of habronemiasis (Pusterla et al. 2003).

A form of cutaneous onchocerciasis which causes severe ulceration in the withers and neck region is encountered in donkeys in Africa; it is the adult parasite that causes this problem. The possible causative species is *Onchocerca raillieti* (White 2013). While ivermectin is often used for the treatment of both ecto- and endoparasitism, it has been reported as having both greater and lesser gastrointestinal absorption in donkeys than in horses. This awaits more clarifying studies (Grosenbaugh et al. 2011).

Donkeys are susceptible to infestation with the protozoan parasite *Besnoitia bennetti* (Davis et al. 1997; Dubey et al. 2005; Elsheikha et al. 2005, 2008; Ness et al. 2012, 2014). This organism also afflicts cattle, and is closely related to *B tarandi* which afflicts reindeer (*Rangifer tarandus*) (Davis et al. 1997). The disease presents as cystic nodules in the skin, nares, and sclera. Scaling of the skin may be seen. Multiple donkeys in a herd may be affected (Elsheikha et al. 2005; Ness et al. 2012). Diagnosis is by histopathology, serologic testing, and PCR (Elsheikha et al. 2005; Ness et al. 2012, 2014; Tinkler et al. 2024). Unfortunately, there is no known effective treatment.

3.3 Sterile Nodular Panniculitis

This is a rare, idiopathic inflammatory condition of the subcutaneous fat of equids (Bell 2001; White et al. 2019 [Fig. 3]).

Its importance is related to differential diagnoses of other more common nodular diseases. The etiology of ANY panniculitis seen on histopathology must initially include infections and drug eruptions. Thus 'sterile nodular panniculitis' refers to

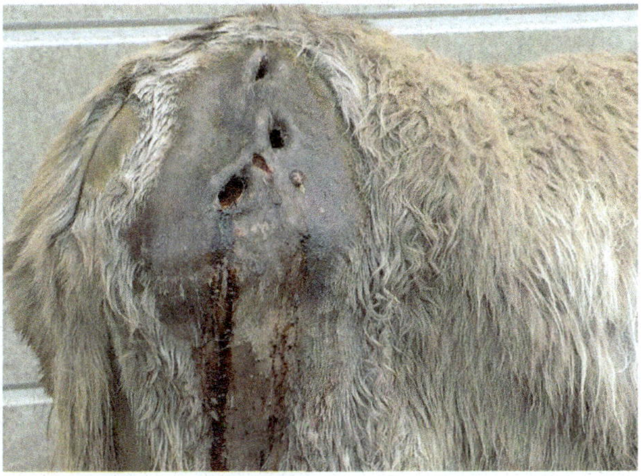

Fig. 3 Seventeen-year-old jenny with sterile panniculitis. Note large draining tracts. (Reproduced with permission from John Wiley & Sons; from reference (White et al. 2019))

sterile subcutaneous inflammatory nodules as a descriptive term of unknown etiologic factors.

The disease presents as deep-seated cutaneous nodules and plaques, which may be single, multiple or generalized, and vary in size from a few millimeters to several centimeters in diameter. Nodules may be firm and well circumscribed or soft and ill defined. They are initially subcutaneous but may become attached to the skin as they progress. The lesions may become cystic, ulcerate and frequently develop draining tracts of an oily, yellowish-brown to bloody discharge. Pain is variable and lesions may heal with depressed scars. Concurrent systemic signs of anorexia, depression, lethargy and pyrexia may be present.

Diagnosis is primarily based on skin biopsy which shows granulomatous to pyogranulomatous inflammation of the subcutaneous fat with negative results for special stains for bacteria, mycobacteria, and fungi as well as culture for these organisms. Treatment with prednisolone can be effective, given orally at 1.1–2.2 mg/kg/day, or 20–30 mg dexamethasone/500 kg/day. Lesions usually subside within 7–14 days. Relapses may occur and require long-term alternate day prednisolone maintenance therapy. There is one report of the use of sterile maggots as a treatment in an old donkey (Bell 2001).

3.4 Neoplasia

Sarcoids are the most common cutaneous tumor in donkeys (Knottenbelt 2019). Donkeys (particularly male donkeys) may have a higher incidence than horses. As in horses and zebras, a number of researchers have been able to isolate bovine papilloma virus DNA from sarcoids in donkeys (Nasir et al. 1997; Reid et al.

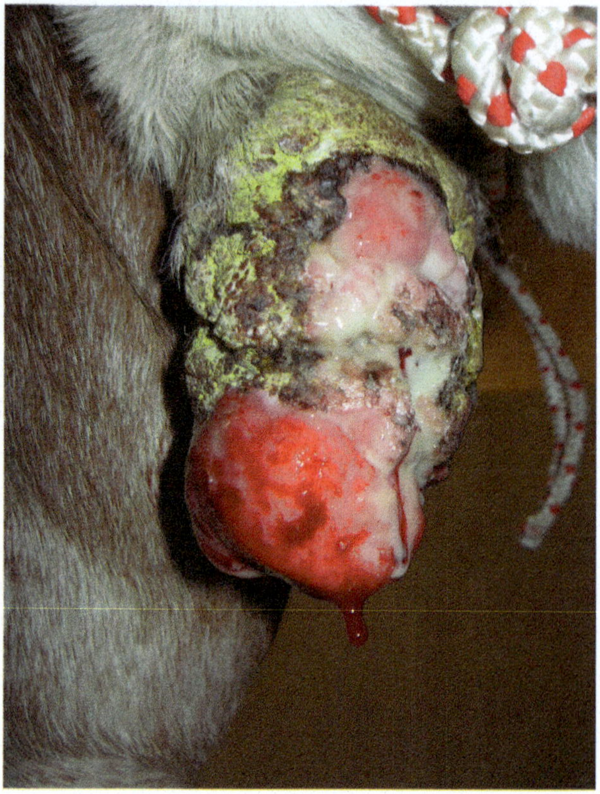

Fig. 4 Five-year-old jack with ulcerated verrucous/nodular sarcoid

1994a, 1994b). The categories of sarcoids: occult, verrucous (warty), nodular, fibroblastic, mixed (Fig. 4) and malevolent have all been reported in donkeys. One review of options for treatment noted that electrosurgical excision resulted in the highest treatment success rate in equids (Haspeslagh et al. 2016). Topical treatments with imiquimod 5% cream or *Sanguinaria canadensis* + zinc chloride can be effective (Pettersson et al. 2020).

Fibrolipomas are encountered in working donkeys in third-world countries. These tumors are especially noticeable when working conditions are difficult due to environment, humidity changes, poorly fitting tack, or trauma. Diagnosis is best established through histopathology, showing a fibrous component intermixed with lobules of adipose tissue. Treatment is aimed at correcting the initiating factor(s) and if possible, excision of the tumor.

4 Alopecia

4.1 Infections

The dermatophytes *Trichophyton verrucosum*, *T. mentagrophytes*, and *Microsporum racemosum* have been reported as causing alopecia as well as scaling in donkeys (Abdalla et al. 2005; Nardoni et al. 2016). Typical lesions include alopecia, often spreading outward in a centrifugal pattern (the 'ring' of ringworm).

Diagnosis may be made by fungal culture of hairs and/or scale, direct examination of the hair for spores or hyphae, PCR, or biopsy for histopathology (with special stains for fungi). Treatment has not been extensively studied in the donkey, but probably would include topical antifungal washes or dips of miconazole-chlorhexidine, enilconazole, or lime sulfur. While terbinafine orally at 5 mg/kg for 10–14 days has anecdotally been noted as effective for dermatophytosis in horses, its dose and safety has not been investigated in donkeys (Williams et al. 2011).

4.2 Alopecia Areata

Alopecia areata is an idiopathic disease of various species typified by areas of nonpruritic alopecia. It has been reported in one donkey (White et al. 2019 [Fig. 5]).

Histologically, a lymphocytic infiltrate surrounding the base of the hair follicle is seen; however, in longstanding cases this infiltrate may not be present. Lesions may be focal or progress to whole body alopecia (often sparing the mane and tail). Most horses regrow the hair, but this may take up to 2–3 years. Corticosteroids may hasten the hair regrowth. This disease is also often mistaken for ringworm, due to the circular nature of the alopecia. In a recent review of 15 cases, a number of horses' lesions waxed and waned often with the season (Hoolahan et al. 2013).

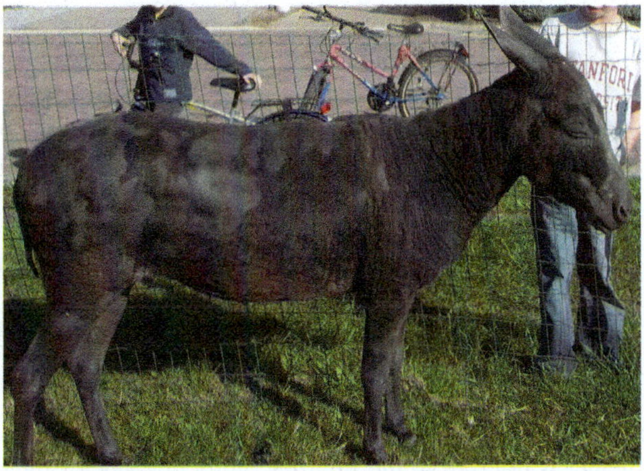

Fig. 5 Seven-year-old jenny with alopecia areata. Note hair loss over caudal half of trunk, rear legs and face. (Reproduced with permission from John Wiley & Sons; from reference (White et al. 2019))

5 Scale and Crust

Scale is defined as excessive stratum corneum. Crust is the result of scale plus dried serum, blood, inflammatory cells and/or organisms, etc.

5.1 Infections

Perhaps the most important bacterial skin infection is that caused by *Dermatophilus congolensis*. Three conditions must be present for *Dermatophilus* to manifest itself: a carrier animal, moisture and skin abrasions. Chronically affected animals are the primary source of infection, however they only become a serious source of infection when their lesions are moistened, which results in the release of zoospores, the infective stage of the organism. Mechanical transmission of the disease occurs by both biting and nonbiting flies, ticks, and possibly fomites. As normal healthy skin is impervious to infection with *D. congolensis*, some predisposing factor that results in decreased resistance of the skin is necessary for infection to occur, prolonged wetting of the skin by rain being one of the most important.

The disease is usually seen during the autumn and winter months, with the dorsal surface of the animal most commonly affected. Occasionally the lesions involve the lower extremities when animals are kept in wet pastures ("dew poisoning"). In the early stages of the disease the lesions can be felt more easily than they can be seen. Thick crusts can be palpated under the hair coat. Removing the crusts and attached hair exposes a pink, moist skin surface, with the removed hair assuming the form of a 'paintbrush'. The under surface of the crusts is usually concave with the bulbs ("roots") of the hairs protruding.

Diagnosis is by demonstrating the "railroad track" cocci on impression smears: a portion of one of the crusts should be macerated and mixed with a few drops of sterile water on a glass slide, Gram stained and examined microscopically. Alternatively, bacterial culture or histopathology may be utilized for diagnosis. A thick crust composed of alternating layers of parakeratotic stratum corneum, dried serum, and degenerating neutrophils is the most characteristic change.

Superficial folliculitis, leading to pronounced alopecia, may be a prominent feature of the disease. In biopsy sections stained with Gram stain, the branching, filamentous organisms can be observed in the crusts and in the follicles.

Treatment is removal from the wet environment, removal of crusts (with care, as this may be painful), washing with iodophors, 2–3% chlorhexidine, or lime sulfur, and antibiotics (penicillin: 22,000 mg/kg procaine pen G intramuscularly twice daily or trimethoprim sulfa orally: 30 mg/kg q12h) for 7 days (Outerbridge and Ihrke 2003).

As the crusts are important in contagion, these should be disposed of rather than brushed on to the ground.

Donkeys' metabolism of drugs may not be the same as horses (Peck et al. 2002), and some antibiotics may cause adverse reactions (Grosenbaugh et al. 2011). The reader is referred to the Pharmacology chapter for more information.

Interestingly, a recent report showed that *healthy* donkeys' skin microbiota varied both with the age of the donkey and the body site sampled (Ma et al. 2023).

5.2 Immune-Mediated Causes of Crust and Scale

Pemphigus foliaceous (PF) has been reported in donkeys. In one jenny, the disease occurred during, then regressed after, 2 of its 5 pregnancies (Bourdeau and Baudry 2005) (Fig. 6).

PF in equids can occur at any age (Vandenabeele et al. 2004). Crusts are the most common lesions, and frequently start on the face and legs then become generalized. Alternatively, crusts may be limited to the coronary band. Peripheral edema is common, and a low grade anemia combined with a leukocytosis may be seen.

Diagnosis is by biopsy, showing subcorneal pustules with acantholytic cells (epidermal cells that have lost their coherence with the surrounding cells). It is very important not to surgically prepare the area to be biopsied, as the acantholytic cells are usually in the crusts; the biopsy site can be cleaned after taking the biopsy to limit the risk of infection. As certain strains of *Trichophyton* spp. of dermatophytes may also cause acantholysis, any histology suggestive of PF should also have stains for fungi performed.

Treatment is reliant on corticosteroids at immunosuppressive doses (prednisolone 1 mg/kg q12h, dexamethasone approximately 0.05–0.1 mg/kg q24h; doses are tapered to a minimal effective alternate daily dose). There are also reports of azathioprine at 1–3 mg/kg q24–48 h being used for various auto-immune skin diseases in horses; this has not been investigated in donkeys (White et al. 2000). However, azathioprine was used in 8 horses at a dose of 1–3 mg/kg for 1 month, then q48h with no deleterious effects (White et al. 2005); this is used as a steroid-sparing drug, used with corticosteroids in an attempt to eventually decrease the steroid needed.

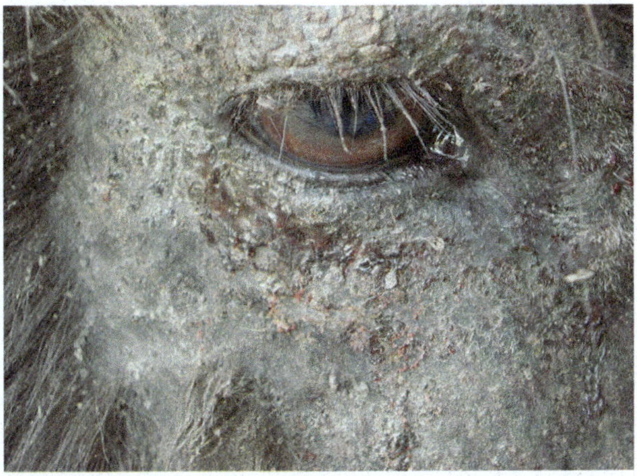

Fig. 6 Eight-year-old Normand jenny with pemphigus foliaceus, occurring only during three of five pregnancies. Note crusts and scales around eye. (Reproduced with permission from John Wiley & Sons; from reference (White et al. 2019))

Equine sarcoidosis has been seen in a donkey (Wayne Rosenkrantz, personal communication 2012). This rare disease is typified by excess scale and occasionally nodules. It should not be confused with the neoplasm equine sarcoid noted above. Equine sarcoidosis is usually generalized but may occasionally present with only limited skin areas (legs or face). It may also affect internal organ systems, particularly the gastrointestinal tract and the lungs, leading to a poorer prognosis.

Diagnosis is by biopsy, which shows non-caseating granulomas consisting of aggregates of epithelioid cells and multinucleated giant cells. Neutrophils, lymphocytes, and plasma cells are present in variable numbers. An infectious etiology has not been demonstrated in the majority of cases in the literature. Traditionally, immunosuppressive therapy as noted above for pemphigus foliaceus has been advised, but some horses may spontaneously resolve, particularly if stress or work can be reduced (Sloet van Oldruitenborgh-Oosterbaan and Grinwis 2013).

Three donkeys with fluctuating or chronic loss of weight, lethargy, exfoliative dermatitis and peripheral eosinophilia were diagnosed with the idiopathic disease multisystemic eosinophilic epitheliotropic disease (MEED), a rare condition of equids characterized by eosinophilic infiltration of multiple organs. All three donkeys were euthanized due to poor prognosis and welfare concerns (Paraschou et al. 2023).

6 Equine Pituitary Pars Intermedia Dysfunction

Also known as Equine Cushing's disease, this occurs in older equids, classically with an inappropriately long hair coat or failure to shed a winter coat, laminitis, and poor overall health (recurrent infections). This has been reported in donkeys (Koutinas et al. 2004; Rickards 2010; White et al. 2019). For more information at the diagnosis and treatment of this condition refer to donkey nutrition and endocrinology chapters.

7 Junctional Epidermalolysis Bullosa

As has been reported in horses, junctional epidermalolysis bullosa has been noted in a donkey foal (Sloet van Oldruitenborgh-Oostrbaan and Boord 2002) (Fig. 7). The disease was characterized by ulceration of large areas of the skin on the legs and ears. Interestingly, the foal was noted to be 'normal' at birth. The disease is fatal, and affected foals should be euthanized.

8 Albinism

Albinism has been recorded in donkeys, particularly those on the island of Asinara, Italy. All the breed individuals on the island have oculo-cutaneous albinism type 1. This widespread manifestation of a recessive autosomal mutation is probably the

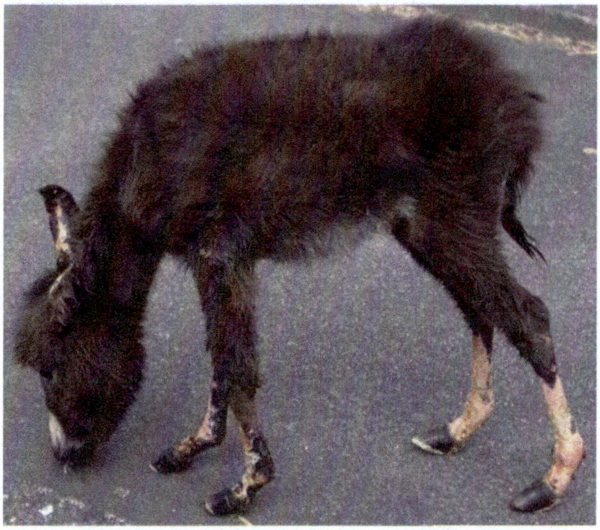

Fig. 7 One-month-old donkey colt with epidermolysis bullosa. Note loss of surface skin on distal legs and deformation of ears. (Reproduced with permission from John Wiley & Sons; from reference (White et al. 2019))

result of isolation and inbreeding (Utzeri et al. 2015). It may be associated with a difference in red blood cell values between the Asinara donkeys and those of nearby Sardinia (Cappai et al. 2024).

Questions

1. Which of the following is a common cause of pruritus in donkeys and is transmitted by midges?
   A. *Haematopinus asini*
   B. *Dermatophilus congolensis*
   C. *Culicoides* spp hypersensitivity
   D. *Habronema megastoma*
   Correct Answer: C. *Culicoides* spp. hypersensitivity
2. What distinguishes cutaneous habronemiasis from other skin conditions in donkeys?
   A. It only occurs in cold climates
   B. Larvae must penetrate intact healthy skin
   C. It is caused by mites that burrow under the skin
   D. It requires larvae to enter previously damaged skin
   Correct Answer: D. It requires larvae to enter previously damaged skin
3. Which diagnostic method is used to confirm *Dermatophilus congolensis* infection?
   A. PCR of nasal secretions
   B. Impression smear showing "railroad track" cocci

   C. ELISA for dermatophytes
   D. Serology for eosinophilic markers
   Correct Answer: B. Impression smear showing "railroad track" cocci
4. Sarcoids in donkeys are most closely associated with which viral agent?
   A. Equine herpesvirus
   B. Onchocerca raillieti
   C. Bovine papillomavirus
   D. Molluscum contagiosum virus
   Correct Answer: C. Bovine papillomavirus
5. Which of the following is a rare, fatal skin disorder in donkey foals that presents with widespread skin ulceration but normal appearance at birth?
   A. Pemphigus foliaceus
   B. Junctional epidermolysis bullosa
   C. Equine sarcoidosis
   D. Multisystemic eosinophilic epitheliotropic disease
   Correct Answer: B. Junctional epidermolysis bullosa

References

Abdalla WG, Suliman EA, Gabbar AE (2005) A report on *Trichophyton verrucosum* in donkeys in Sudan. Sudan J Vet Res 20:83–85

Barbosa JD, Sodré MHS, Barbosa CC, da Costa PSC, Oliveira CMC, Ferreira TTA, da Silveira JAS, Lamego EC, Paz MC, Caldeira RD, Matos PCM, Serruya A, Salvarani FM, da Silva N, Silveira S (2024) Allergic dermatitis in Pêga breed donkeys (*Equus asinus*) caused by *Culicoides* bites in the Amazon Biome, Pará, Brazil. Animals 14(9):1330. https://doi.org/10.3390/ani14091330

Bell NJ (2001) Use of sterile maggots to treat panniculitis in an aged donkey. Vet Rec 149:768–770

Bourdeau P, Baudry J (2005) Pemphigus-type bullous dermatosis associated with pregnancy in a female donkey. Informations Dermatol Vét 11:19–24

Cappai MG, Senes A, Pilo G (2024) Albinism and blood cell profile. The peculiar case of Asinara donkeys. Animals 14(9):1330

Cooper VL, Kennedy GA, Kruckenber SM, Vorhies MW (1994) Histoplasmosis in a miniature Sicilian burro. J Vet Diagn Invest 6:499–501

Crothers S, White SD, Ihrke PJ, Affolter VK (2009) Sporotrichosis: a retrospective evaluation of 23 cases seen in northern California (1987–2007). Vet Dermatol 20:249–259

Davis WP, Peters DF, Dunstan RW (1997) Besnoitiosis in a miniature donkey. Vet Dermatol 8: 139–143

Dubey JP, Sreekumar C, Donovan T, Rozmanec M, Rosenthal BM, Vianna MC, Davis WP, Belden JS (2005) Redescription of *Besnoitia bennetti* (Protozoa: Apicomplexa) from the donkey (*Equus asinus*). Int J Parasitol 35:659–672

Dumitrache MO, Gyorke A, Julien F, Kondratjeva J, Cadiergues MC (2023) Case report: identification of the tropical rat mite (*Ornithonyssus bacoti*) on a domestic donkey in France. Front Vet Sci 10:1141290

Elsheikha HM, Mackenzie CD, Rosenthal BM, Marteniuk JV, Steficek B, Windsor S, Saeed AM, Mansfield LS (2005) An outbreak of besnoitiosis in miniature donkeys. J Parasitol 91:877–881

Elsheikha HM, Mansfield LS, Morsy GH (2008) Studies on *Besnoitiosis bennetti* in miniature donkeys. J Egypt Soc Parasitol 38:171–184

Falk EF, Restifo M, Lam ATH, Ferrer L (2015) Use of ciclosporin modified to treat atopic dermatitis in a miniature donkey. Vet Dermatol 26:134

Fettelschoss-Gabriel A, Fettelschoss V, Thoms F (2018) Treating insect-bite hypersensitivity in horses with active vaccination against IL-5. J Allergy Clin Immunol 142:1194–1205

Fox R, Thiemann A, Everest D, Steinbach F, Dastjerdi A, Finnegan C (2012) Molluscum contagiosum in two donkeys. Vet Rec 170:649

Gateechew A (2004) Clinical trial of iodides combined with ancilliary treatment on Epizootic Lymphangitis in cart horses at Debra Zeit and Akaki towns. DVM Thesis, Faculty of Veterinary Medicine, Addis Ababa University

Grosenbaugh DA, Reinemeyerr CR, Figueiredo MD (2011) Pharmacology and therapeutics in donkeys. Equine Vet Educ 23:523–530

Haspeslagh M, Vlaminck LE, Martens AM (2016) Treatment of sarcoids in equids: 230 cases (2008–2013). J Am Vet Med Assoc 249:311–318

Hoolahan DE, White SD, Outerbridge CA, Shearer PL, Affolter VK (2013) Equine alopecia areata: a retrospective descriptive study at the University of California-Davis (1980–2011). Vet Dermatol 24:282–285

Irizarry-Rovira AR, Kaufman L, Christian JA, Reberg SR, Adams SB, DeNicola DB, Rivers W, Hawkins JF (2000) Diagnosis of sporotrichosis in a donkey using direct fluorescein-labeled antibody testing. J Vet Diagn Invest 12:180–183

Khodakaram-Tafti A, Dehghani S (2006) Cutaneous cryptococcosis in a donkey. Comp Clin Pathol 15:271–273

Knottenbelt DC (2019) Skin Disorders of the Donkey and Mule. Veterinary Clinics of North America: Equine Practice 35:493–514

Koutinas CK, Saridomichelakis MN, Mylonakis ME, Hamhougias K, Fytianou A, Diakakis N, Koutinas AF (2004) Equine hyperadrenocorticism: a report of 4 natural cases. J Hellenic Vet Med Soc 55:21–33

Lincoln VJ, Page PC, Kopp C, Mathis A, von Niederhäusern R, Burger D, Herholz C (2015) Protection of horses against *Culicoides* biting midges in different housing systems in Switzerland. Vet Parasitol 210:206–214

Ma Q, Yue Y, Kou X, Hou W, Wang M, Yang X, Liu G, Li Y, Wang C (2023) Dynamic distribution of skin microorganisms in donkeys at different ages and various sites of the body. Animals (Basel) 13:566

Maia LA, Olinda RG, Araújo TF, Firmino PR, Nakazato L, Neto EGM, Riet-Correa F, Dantas AFM (2016) Cutaneous pythiosis in a donkey (Equus asinus) in Brazil. J Vet Diagn Invest 28:436–439

Nardoni S, Rocchigiani G, Papini RA, Veneziano V, Brajon G, Martini M, Salari F, Mancianti F (2016) Dermatophytosis in donkeys (*Equus asinus*) due to *Microsporum racemosum*, an unusual geophilic agent. Med Mycol Case Rep 12:8–10

Nasir L, McFarlane ST, Torrontegui BO, Reid SW (1997) Screening for bovine papillomavirus in peripheral blood cells of donkeys with and without sarcoids. Res Vet Sci 63:89–90

Ness SL, Peters-Kennedy J, Schares G, Dubey JP, Mittel LD, Mohammed HO, Bowman DD, Felippe MJ, Wade SE, Shultz N, Divers TJ (2012) Investigation of an outbreak of besnoitiosis in donkeys in northeastern Pennsylvania. J Am Vet Med Assoc 240:1329–1337

Ness SL, Schare G, Peters-Kennedy J, Mittel LD, Dubey JP, Bowman DD, Mohammed HO, Divers TJ (2014) Serological diagnosis of *Besnoitia bennetti* infection in donkeys (*Equus asinus*). J Vet Diagn Invest 26:778–782

Olomski F, Fettelschoss V, Jonsdottir S, Birkmann K, Thoms F, Marti E, Bachmann MF, Kündig TM, Fettelschoss-Gabriel A (2020) Interleukin 31 in insect bite hypersensitivity-alleviating clinical symptoms by active vaccination against itch. Allergy 75:862–871

Outerbridge CA, Ihrke PJ (2003) Folliculitis: Staphylococcal pyoderma, dermatophilosis, dermatophytosis. In: Robinson NE (ed) Current therapy in equine medicine 5. Saunders, St. Louis, pp 197–200

Paraschou G, Vogel PE, Lee AM, Trawford RF, Priestnall SL (2023) Multisystemic eosinophilic epitheliotropic disease in three donkeys. J Comp Pathol 201:105–108

Peck KE, Matthew NS, Taylor TS, Mealey KL (2002) Pharmacokinetics of sulfamethoxazole and trimethoprim in donkeys, mules, and horses. Am J Vet Res 63:349–353
Pettersson CM, Broström H, Humblot P, Bergvall KE (2020) Topical treatment of equine sarcoids with imiquimod 5% cream or *Sanguinaria canadensis* and zinc chloride—an open prospective study. Vet Dermatol 31:471–e126
Powell RK, Bell NJ, Abreha T, Asmamaw K, Bekelle H, Dawit T, Itsay K, Feseha GA (2006) Cutaneous histoplasmosis in 13 Ethiopian donkeys. Vet Rec 158:836–837
Pusterla N, Watson JL, Wilson WD, Affolter VK, Spier SJ (2003) Cutaneous and ocular habronemiasis in horses: 63 cases (1988–2002). J Am Vet Med Assoc 222:978–982
Reid SW, Gettinby G, Fowler JN, Ikin P (1994a) Epidemiological observations on sarcoids in a population of donkeys (*Equus asinus*). Vet Rec 134:207–211
Reid SW, Smith KT, Jarrett WF (1994b) Detection, cloning and characterisation of papillomaviral DNA present in sarcoid tumours of *Equus asinus*. Vet Rec 135:430–432
Rickards K (2010) Treatment of hyperadrenocorticism in donkeys. Vet Rec 166:152
Scantlebury C, Reed K (2009) Epizootic lymphangitis. In: Mair TS, Hutchinson RE (eds) Infectious diseases of the horse. Equine Veterinary Journal, pp 397–406
Sloet van Oldruitenborgh-Oosterbaan MM, Grinwis GC (2013) Equine sarcoidosis. Vet Clin N Am Equine Pract 29:615–627
Sloet van Oldruitenborgh-Oostrbaan MM, Boord M (2002) Equine dermatology workshop. In: Thoday KL, Foil CS, Bond R (eds) Advances in veterinary dermatology, vol 4. Blackwell Science, Oxford, pp 286–290
Stepnik C, Outerbridge CA, White SD, Kass P (2012) Equine atopic skin disease and response to allergen specific immunotherapy (ASIT): a retrospective study at the University of California-Davis (1991–2008). Vet Dermatol 23:29–36
Tinkler SH, Villa L, Manfredi MT, Walshe N, Jahns H (2024) First report of *Besnoitia bennetti* in Irish donkeys: an emerging parasitic disease in Europe. Ir Vet J 77:2
Utzeri VJ, Bertolini F, Ribani A, Schiavo G, Dall'Olio S, Fontanes L (2015) The albinism of the feral Asinara white donkeys (Equus asinus) is determined by a missense mutation in a highly conserved position of the tyrosinase (Tyr) gene deduced protein. Anim Genet 47:120–124
Vandenabeele SIG, White SD, Kass P, Ihrke PJ (2004) Pemphigus foliaceus in the horse: 20 cases. Vet Dermatol 15:381–388
Veneziano V, Neglia G, Galietti A, Rufrano D, Bassini A, Mariani U, Gokbulut C (2012) Efficacy of alphacypermetrin pour-on against natural *Werneckiella equi* infestation on donkeys (*Equus asinus*). Parasitol Res 111:967–973
Veneziano V, Galietti A, Marian U, Di Loria A, Piantedosi D, Neola B, Guccione J, Gokbulut C (2013) Field efficacy of eprinomectin against the sucking louse *Haematopinus asini* on naturally infested donkeys. Vet J 197:512–514
White SD (2013) Donkey dermatology. Vet Clin N Am Equine Pract 29:703–708
White SD, Rosychuk RAW, Outerbridge CA, Fieseler KV, Spier S, Ihrke PJ, Chapman P (2000) Thiopurine methyltransferase in red blood cells of dogs, cats, and horses. J Vet Intern Med 14: 499–502
White SD, Maxwell LK, Szabo N, Hawkins JL, Kolias-Baker C (2005) Pharmacokinetics of azathioprine following single-dose intravenous and oral administration and effects of azathioprine following chronic oral administration in horses. Am J Vet Res 66:1578–1583
White SD, Bourdeau PJ, Brément T, Bruet V, Vandenabeele SI, Haspeslagh M, Sloet van Oldruitenborgh-Oosterbaan MM (2019) Skin disease in donkeys (*Equus asinus*): a retrospective study from 4 veterinary schools. Vet Dermatol 30:247–254
Williams MM, Davis EG, KuKanich B (2011) Pharmacokinetics of oral terbinafine in horses and dogs. J Vet Pharmacol Ther 34:232–237
Yeruham I, Braverman Y, Orgad U (1993) Field observations in Israel on hypersensitivity in cattle, sheep and donkeys caused by *Culicoides*. Aust Vet J 70:348–352

Endocrines Disorders of Donkeys and Mules

F. J. Mendoza, R. A. Perez-Ecija, and R. E. Toribio

Abstract

Donkey metabolic syndrome (DMS) and pituitary pars intermedia dysfunction (PPID) are the most common endocrine disorders in donkeys and mules. However, other less frequent disturbances such as calcium and phosphorus imbalances or thyroid gland disorders can also be observed. There is recent information on the diagnosis and treatment of PPID, DMS, obesity, and hyperinsulinemia-associated laminitis; however, fewer studies compared to horses. In mules, there is a lack of information on endocrine disorders. This chapter compiles endocrine information available in donkeys and mules.

Keywords

Calcium and phosphorus disturbances · Donkey metabolic syndrome · Pituitary pars intermedia dysfunction · Thyroid gland disorders

F. J. Mendoza (✉) · R. A. Perez-Ecija
Department of Animal Medicine and Surgery, College of Veterinary Medicine, University of Cordoba, Cordoba, Spain
e-mail: fjmendoza@uco.es; alejandro.perez.ecija@uco.es

R. E. Toribio
Department of Veterinary Clinical Sciences, College of Veterinary Medicine, The Ohio State University, Columbus, OH, USA
e-mail: toribio.1@osu.edu

A. K. McLean et al. (eds.), *Donkey and Mule Medicine: Clinical Medicine and Reproduction*, https://doi.org/10.1007/978-3-032-18540-2_3

1 Introduction

Endocrine disorders are common in donkeys, with donkey metabolic syndrome (DMS) and pituitary pars intermedia dysfunction (PPID), being the most common ones. Less frequently diagnosed conditions include thyroid gland abnormalities and disturbances of calcium and phosphorus regulation.

Mule-specific information is not available, although they may develop endocrine and metabolic disturbances similar to donkeys.

2 Pituitary Pars Intermedia Dysfunction (PPID)

2.1 Introduction and Epidemiology

PPID (formerly known as Cushing's disease) is a condition of older animals. Since donkeys have a longer life expectancy than horses, it is thought that PPID is more prevalent in this species, with higher risk in animals older than 15 years (Gehlen et al. 2020a).

Contrary to horses, no sex or breed predisposition have been demonstrated in donkeys (Durham et al. 2022). The effect of geography (latitude) on ACTH could be similar to horses.

No data is available for mules, although PPID should most probably show a similar prevalence to horses.

2.2 Pathophysiology

PPID is caused by the loss of hypothalamic dopaminergic inhibition of the melanotropes in the pituitary pars intermedia, resulting in increased synthesis and secretion of proopiomelanocortin (POMC) and derived cleavage peptides including adrenocorticotropic hormone (ACTH), α- and β-melanocyte-stimulating hormone (MSH), β-endorphin (β-END), and corticotropin-like intermediate lobe peptide (CLIP) (Mendoza et al. 2024).

2.3 Clinical Signs

The most common clinical sign reported by owners is hypertrichosis, but this can be challenging in donkey breeds with long hair coats. Changes of behavior, the second most common clinical sign in horses with PPID, can go unnoticed in donkeys due to their calm behavior. Other clinical signs include insulin dysregulation (ID), recurrent laminitis, weight loss, muscle wasting, and fat redistribution (difficult to detect sometimes), pendulous abdomen (potbelly), polydipsia/polyuria, predisposition to infections, parasitism, and infertility. Orthopedic problems are rarely reported, in part because donkeys are not typically used for sport events, although they can be used for work. Other musculoskeletal disorders could be a consequence of ageing.

2.4 Diagnosis

Diagnosis of PPID is based on resting plasma ACTH concentrations and dynamic challenges (Tables 1 and 2).

Healthy donkeys have higher resting plasma ACTH concentrations than horses and mules regardless of season (Gehlen et al. 2020b), thus clinicians should not use equine ACTH reference values. A seasonal effect on plasma ACTH concentrations has been also documented in donkeys and mules, with higher concentrations in autumn (Gehlen et al. 2020b), increasing from July to November (Durham et al. 2022; Humphreys et al. 2022) in donkeys, and from July to October in mules (Gehlen et al. 2020b) (Table 1). The seasonal ACTH dynamics in mules is more

Table 1 Seasonally adjusted resting plasma ACTH concentrations for donkeys and mules

		ACTH (pg/mL)				
	Analyzer	December–June	July–November	August	September–October	Reference
Donkey	Immulite 2000 xpi	5–55.4		19.5–143		Gehlen et al. (2020b)
	Immulite 1000	17.3 (5–80.5)	40.9 (12.4–214)	88.1 (46–259)	97 (35.7–319)	Humphreys et al. (2022)
	Immulite 1000	35 (22–66)	63 (40–149.5)	105 (68–187)	136 (77- > 200)	Durham et al. (2022)[a]
Mule	Immulite 2000 xpi	<5–37.1		9.8–68.7		Gehlen et al. (2020b)

Data are expressed as median (range) or minimum–maximum values

[a]Retrospective study from laboratory records (no disease status classification). *ACTH* adrenocorticotropic hormone

Table 2 Protocols for dynamic testing in donkeys suspected of pituitary pars intermedia dysfunction (PPID)

Test	Preparation	Step #1 SAMPLE 1	Step #2 ADMINISTER	Step #3 SAMPLE 2	Step #4 ID DIAGNOSIS
TRH	Feeding does not affect sampling. Do not perform if there is evidence of pain (acute laminitis), stress (illness, after exercise or sedation with α2-agonists).	Resting ACTH (frozen plasma EDTA)	0.5–1 mg/IV, protirelin or chemical-grade TRH	10 minutes	ACTH >110 pg/mL
DST		Resting cortisol (chilled serum or plasma)	40 μg/kg/IM, dexamethasone	15–19 hours	Cortisol >1 μg/mL

ACTH adrenocorticotropic hormone; *DST* dexamethasone suppression test; *TRH* thyrotropin-releasing hormone similar to horses than to donkeys. However, seasonally adjusted ACTH ranges have not been established for donkeys or mules.

A circadian rhythm in ACTH has been reported in donkeys, with higher concentrations early in the day compared to the afternoon. This study also reported no influence of cold weather on ACTH concentrations. Noteworthy, an ELISA kit was used in this study (Gao et al. 2022). It should be noted that plasma ACTH concentrations differ between different techniques and analyzers (Mendoza et al. 2024). Thus, care should be taken when comparing results between studies and laboratories.

In animals with clinical signs consistent with PPID but low or non-diagnostic resting ACTH concentrations, dynamic testing is recommended (Table 2). The thyrotropin-releasing hormone (TRH) stimulation test, following protocols published for horses, is a reliable method in donkeys (Mejia-Pereira et al. 2019), and likely in mules. A lower protirelin (synthetic TRH) dose (0.5 mg/IV) is used in small and miniature donkeys (<250 kg of bodyweight) (Table 2). The dexamethasone suppression test (DST) could be used when protirelin or chemical-grade TRH are not available (Table 2), although the DST is more likely to give equivocal results in early stages of the disease (Mejia-Pereira et al. 2019).

Blood samples for ACTH measurement should be collected in EDTA-containing tubes to avoid false negative results. It is recommended to centrifuge, separate the plasma, freeze rapidly, and ship under refrigerated conditions (See chapter "Urogen ital Disorders" for more information about diagnostic testing strategies).

Other peptides such as α-MSH or β-END could potentially be evaluated, although they have not been studied in donkeys or mules, and these tests are not commercially available.

2.5 Treatment and Prognosis

Pergolide is the treatment of choice for donkeys and mules with PPID. Recent pharmacologic studies about this drug in donkeys showed that pergolide has a faster metabolism and clearance in donkeys compared to horses (Xue et al. 2023), but also displays a cumulative pattern, thus similar dosing to horses can be used (Table 3). No pergolide pharmacologic studies have been performed in mules.

Since PPID is a chronic, progressive, and irreversible disease, life-long treatment is imperative. In most donkeys, clinical signs (behavior and physical condition: lethargy, haircoat, weight gain or loss, and lameness) improve with treatment. Follow-up ACTH measurement 3 and 6 months after treatment initiation, and then annually, is recommended. If there is evidence of relapse, it is recommended to measure ACTH.

If plasma ACTH concentration does not decrease after this time, the pergolide dose can be increased. If anorexia develops with pergolide treatment, the daily dose can be reduced, split between morning and afternoon, or stopped and resumed 4–5 days later with lower doses and increase progressively. In donkeys not responding to pergolide, combination with cyproheptadine should be considered.

Table 3 Common drugs used for treatment of metabolic and endocrine disturbances in donkeys and mules

Drugs	Dose	Route	Interval	Observations
Anti-H1 receptor, anticholinergic (M) and antiserotonergic (5-HT)				
Cyproheptadine[a]	0.2–0.6 mg/kg	PO	12–24 h	An appetite stimulant
Biguanides				
Metformin[a]	15–50 mg/kg	PO	8–12 h	Anorexia; does not induce hypoglycemia
Dipeptidyl peptidase-4 inhibitor (DPP-4)				
Sitagliptin[a]	1.5 mg/kg	PO	24 h	
Dopamine receptor agonist type 2				
Cabergoline[a]	0.005–0.01 mg/kg	IM	7 d	Anorexia, lethargy, colic, diarrhea
Pergolide	0.002–0.01 mg/kg	PO	24 h	Anorexia, lethargy and oral ulcers
Sodium-glucose transporter type 2 (SGLT-2) inhibitors				
Canagliflozin[a]	0.3–0.6 mg/kg	PO	24 h	Increases blood triglycerides
Dapaglifozin[a]	0.02 mg/kg	PO	24 h	
Ertugliflozin[a]	0.03–0.05 mg/kg	PO	24 h	
Velagliflozin[a]	0.3 mg/kg	PO	24 h	
Synthetic thyroxine (tT4)				
Sodium levothyroxine[a]	0.05–0.1 mg/kg	PO	24 h	Improves insulin sensitivity

D days; *H* hours; *IM* intramuscular; *PO* per os

[a]Dosage extrapolated from horses and based on clinical experience due to lack of donkey or mule specific pharmacokinetic information (Mendoza et al. 2024)

Although cabergoline is occasionally used in horses failing to respond to pergolide (Sundra et al. 2024a), no information regarding its use is available in donkeys.

Prognosis is good in donkeys and mules under proper pharmacologic treatment and regular ACTH monitoring. Additional geriatric husbandry care is needed, with regular deworming, vaccinations, dental care, and farriery.

3 Donkey/Asinine Metabolic Syndrome (DMS/AMS)

3.1 Introduction and Epidemiology

DMS is the most common endocrine disorder in donkeys. This syndrome has three major features: obesity (regional adiposity) (Fig. 1), ID, and hyperinsulinemia-associated laminitis (endocrinopathic recurrent laminitis) (Fig. 2).

Donkeys of any age, gender and breed can develop DMS (Mendoza et al. 2018, 2024). Although this syndrome is more frequent in developed countries (due to overfeeding and obesity), it can also occur in developing countries where donkeys have access to high-quality pastures.

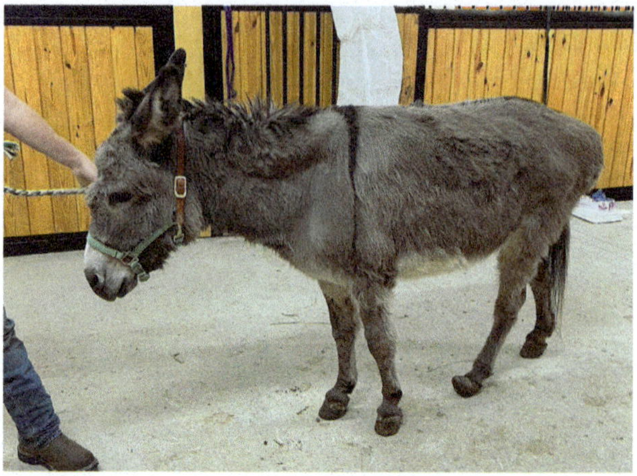

Fig. 1 Donkey with severe obesity (BCS: 9/9 or 5/5, and NS: 4/4)

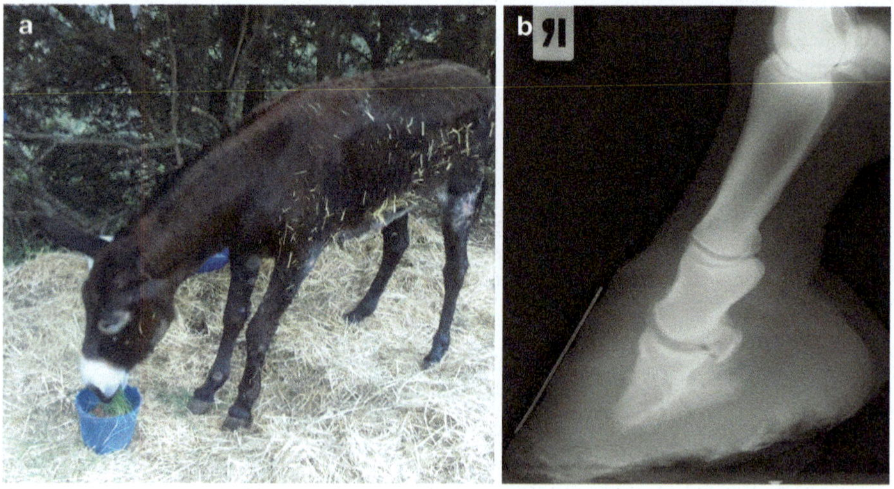

Fig. 2 (**a**) (photo on the left) Obese donkey showing typical painful standing position with forelimb under the body; (**b**) (photo on the right) Latero-medial radiography of a donkey with hyperinsulinemia-associated laminitis. Note the rotation of third phalanx (coffin bone)

Mules also develop metabolic syndrome, with features similar to donkeys and horses; however, epidemiological information is not available.

3.2 Pathophysiology

In DMS, obesity is a proinflammatory state where macrophages residing in the adipose tissue release cytokines (Mendoza et al. 2019). These mediators contribute to peripheral insulin resistance to reduce glucose uptake, reduce hepatic insulin clearance, and increase insulin production by pancreatic β-cells (Mendoza et al. 2024). All these changes result in ID and hyperinsulinemia, which favors the development of hyperinsulinemia-associated laminitis (HAL). The role of the insulin-like growth factor 1 (IGF-1) receptor and other factors on abnormal hoof growth and HAL development have not been evaluated in donkeys.

An exaggerated enteroinsular axis (EIA) response has been proposed to contribute to ID in horses (Fitzgerald et al. 2019). Although donkeys have a functional EIA, the relationship between EIA dysfunction and ID has not been studied in donkeys (Mendoza et al. 2022a).

3.3 Clinical Signs

Obesity (regional adiposity) and recurrent endocrinopathic laminitis are the main clinical features of this condition (Mendoza et al. 2024) (Fig. 1). Laminitis can also be observed in lean donkeys or mules. Infertility and hyperlipemia are frequent findings in these animals (Mendoza et al. 2019).

3.4 Diagnosis

3.4.1 Obesity

Obesity diagnosis is based on a donkey-specific BCS (>6 out of 9; or > 3 out of 5) and neck score (NS, >2 out of 4) (Pearson and Ouassat 2000; Mendoza et al. 2015b; The Donkey Sanctuary 2018) (Table 4). Ultrasonography can be useful to assess intra-abdominal and subcutaneous fat. Plasma leptin concentrations, which are positively correlated with plasma triglyceride concentrations, body condition and neck scores (Mendoza et al. 2015b), can also be used in the diagnosis. Hepatic enzyme activity can be increased in some obese donkeys with ID due to hepatic lipidosis. Obese donkeys tend to have higher insulin concentrations and lower insulin sensitivity compared to lean ones (Pritchard et al. 2019).

3.4.2 Insulin Dysregulation

The diagnosis of ID is based on resting serum/plasma insulin concentrations. Resting insulin concentrations are similar between donkeys, mules, and horses, although some studies have found higher concentrations in donkeys (Gehlen et al. 2020b). A seasonal effect has been observed in mules compared to donkeys (Gehlen et al. 2020b), but this change is small and lacks clinical relevance; thus, seasonally adjusted ranges for mules are not needed. Insulin concentrations can differ between analytical methods (radioimmunoassay versus Immulite 1000 versus Immulite 2000×pi) (Warnken et al. 2016). Caution must be used when fasting donkeys to measure resting insulin concentrations due to risk of hyperlipemia, thus, providing some hay or straw overnight is suggested. The correlation between resting insulin after feeding and laminitis has not been evaluated in donkeys.

Table 4 Body condition scores (BCS) and neck score (NS) more commonly used in donkeys

BCS	Pearson A. University of Edinburgh[a]	BCS	The Donkey Sanctuary[b]	NS	Mendoza et al., 2015[c]
1	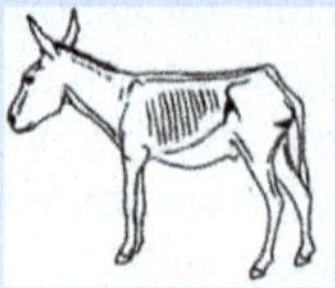	1		0	
2	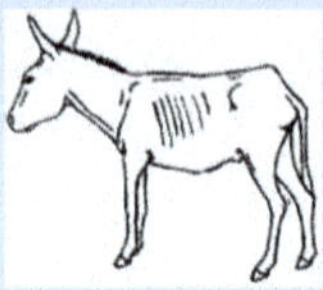	2		1	
3	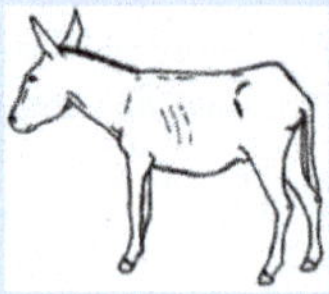	3		2	
4		4		3	
5		5		4	
6					
7					
8					

(continued)

Table 4 (continued)

BCS	Pearson A. University of Edinburgh[a]	BCS	The Donkey Sanctuary[b]	NS	Mendoza et al., 2015[c]
9					

[a]Adapted from Pearson and Ouassat (2000)
[b]Adapted from The Donkey Sanctuary (*https://www.thedonkeysanctuary.org.uk/for-owners/owners-resources/donkey-body-condition-score-chart*)
[c]Modified from Mendoza et al. (2015b)

Current equine metabolic syndrome guidelines establish an insulin cut-off concentration of 45 μIU/mL for ID diagnosis (Bertin et al. 2024). However, a donkey and mule-specific insulin cut-off values for ID diagnosis have not been established. Insulin concentrations of less than 20 μIU/mL make ID unlikely (Dugat et al. 2010). In donkeys or mules with clinical signs compatible with DMS but resting serum insulin concentration lower than <20 μIU/mL, or animals without DMS signs but increased insulin concentrations (equivocal range: 20–50 μIU/mL), further evaluation using dynamic testing should be performed (Mendoza et al. 2024).

Dynamic tests have been characterized in healthy donkeys (Mendoza et al. 2015a, 2022a) (Table 5). Glucose dynamic is right-shifted in donkeys, suggesting lower intestinal absorption (oral challenges), delayed EIA response, reduced cellular glucose uptake (hepatic or peripheral tissues) and/or decreased renal clearance. This indicates that species-specific sample timing and cut-off values should be used in donkeys. While the oral sugar test (OST) and the intravenous insulin tolerance test (IVITT) are the dynamic tests recommended for ID diagnosis in horses (Bertin et al. 2024), data on their reliability for ID diagnosis or risk of HAL in donkeys or mules are lacking. In a previous study, the intravenous glucose tolerance test (IVGTT) was more reliable than the combined glucose-insulin test (CGIT) to assess ID in donkeys (Mendoza et al. 2022b). An in-feed OST or the use of other commercial glucose mixtures have not been evaluated in donkeys or mules.

3.4.3 Laminitis

The diagnosis of laminitis is similar between donkeys, horses, and mules (Thiemann et al. 2022) (Fig. 2). However, it is important to use donkey-specific radiographic measurements (Thiemann et al. 2022). Specifics on the diagnosis are described elsewhere in this book (see chapters "Clinical Approach to the Donkey, Mule or Hinny with Skin Disease" and "Gastrointestinal Disorders of Donkeys and Mules").

Table 5 Protocols for dynamic testing in donkeys with insulin dysregulation

Test	Preparation	Step #1 SAMPLE 1	Step #2 ADMINISTER	Step #3 SAMPLE 2	Step #4 ID DIAGNOSIS
OST	Provide one flake of hay overnight or 6 h prior to sampling.	Baseline blood sample to determine resting glucose (oxalate or fluoride tubes) and insulin (serum or plasma depending on laboratory).	0.45 mL/kg/ PO of Karo Light corn syrup	60–75 minutes[a]	Insulin >60 μIU/mL[b]
OGT			Dextrose (20% solution) 1 g/ kg through a nasogastric tube	120–150 minutes[a]	Insulin >80 μIU/mL[b]
IVGTT			Bolus of 50% dextrose (300 mg/kg/ IV)	150–180 minutes	Glucose above baseline[b]
CGIT			Bolus of 50% dextrose (150 mg/kg, IV) followed by regular fast action insulin (0.1 IU/kg/IV)	60–75 minutes	Glucose above baseline[b]

CGIT combined glucose–insulin test; *ID* insulin dysregulation; *IVGTT* intravenous glucose tolerance test; *OGT* oral glucose tolerance test; *OST* oral sugar test
[a]Precise blood collection time has not been validated for ID diagnosis in donkeys and mules
[b]Insulin cut-off value has not been validated for ID diagnosis in donkeys and mules

3.5 Treatment

The management of metabolic syndrome in donkeys and mules is based on reducing body weight, increasing insulin sensitivity (decrease hyperinsulinemia), and managing laminitis.

3.5.1 Obesity

Implementation of a weight loss program based on progressive diet modifications (considering the risk of hyperlipemia) is crucial in these animals (Burden and Bell 2019). Ideal diets should have less than 10% of soluble non-structural carbohydrates (NSC), and daily dry matter forage consumption should be between 1 and 2% of body weight (Mendoza et al. 2024). Ideally, in order to fulfill these requirements, hay should be evaluated for nutritional composition and energy content. NSC can be reduced by soaking the hay in cold tap water (at least 60 minutes) or warm tap water (30 minutes), or using a hay steamer prior to feeding. It is important to take into consideration that longer soaking times promotes the leach of other nutrients and minerals. Commercial “weight-loss” diets marketed for horses are not appropriate for donkeys due to their caloric content and the high energy efficiency of this species. Straw is a good source of forage to promote weight loss, but a ration balancer supplement contributing minerals and vitamins, or a high-quality oil (linseed or corn) source (if hypertriglyceridemia is not present) should be added. It is recommended to use a weight scale for close monitoring of weight loss (See chapter "Neurologic Examination of Donkeys and Mules"). Severe caloric restriction (<1% BW) is not recommended due to the high risk of hyperlipemia.

Grazing of high-quality grass should be avoided despite the benefits of increased physical activity. In donkeys that cannot be housed, it is important to reduce the number of hours at pasture (also consider a grazing muzzle) and avoid rapidly growing pasture due to sugar accumulation. Grain and treats (e.g. apples, carrots, mints, etc.) should be avoided. The use of grazing muzzles in donkeys is controversial but should be considered on an individual basis.

Daily physical activity (walker or hand walking) improves insulin sensitivity and weight loss. In obese donkeys with laminitis (or other orthopedic problems) where exercise should be limited, or in animals refractory to these measures, levothyroxine sodium (Table 3) or resveratrol to promote insulin sensitivity could be used, although research in donkeys and mules is lacking.

3.5.2 Insulin Dysregulation

Weight loss can be sufficient to correct ID in mild to moderate cases of hyperinsulinemia and HAL. In lean ID donkeys or when diet modifications are insufficient to decrease serum insulin concentrations, metformin could be considered (Table 3) using similar protocol to horses (Sundra et al. 2024b), although no pharmacokinetic data in donkeys or mules are available (Mendoza et al. 2024). If the donkey becomes anorectic, reducing the dose for a few days and resuming with a lower dose or frequency should be considered.

Pharmacologic studies of sodium-glucose transporter type 2 (SGLT-2) inhibitors have not been performed in donkeys, although they are used at similar doses as horses (Table 3). SGLT-2 inhibitors increase blood triglyceride concentrations and hepatic enzyme activity in horses, ponies, and donkeys, and they should be measured periodically in these animals due to the increased risk of hyperlipemia (Sundra et al. 2024b).

Other drugs, such as dipeptidyl peptidase-4 inhibitor (e.g. sitagliptin) have been studied in horses with ID but not in donkeys or mules.

3.5.3 Laminitis

Equine analgesic protocols to manage laminitis apply to donkeys (non-steroidal anti-inflammatory drugs [NSAIDs], acetaminophen, gabapentin, etc.). NSAIDs have faster clearance in donkeys and dose or dose frequency may need to be increased. It is also important to consider hoof anatomical differences for corrective trimming and support (Thiemann et al. 2022). See further information in the podiatry and farrier chapters.

3.6 Prognosis

Donkeys with ID can have a good prognosis as long as weight loss is reached, laminitis pain controlled, and hoof measurements remain in place. In contrast, donkeys with severe laminitis (sinking or rotation), where hyperinsulinemia is not controlled with weight loss or are refractory to medical treatment, the prognosis is poor.

4 Thyroid Gland Diseases

There is scarce information about thyroid gland disorders in donkeys (Mendoza et al. 2024) and mules. Healthy donkeys have higher plasma thyroid hormone (TH) concentrations (fT3, tT3, rT3, fT4 and tT4) than horses, and younger donkeys have higher TH concentrations than older ones (Mendoza et al. 2013) (Table 6). No information is available in mules at this moment.

Some drugs (phenylbutazone, dexamethasone), transport, and stressful situations (e.g., weaning, isolation, restraint) affect blood TH concentrations. Thus, donkey-specific reference ranges should be used to avoid misdiagnoses, also taking into consideration the analytical method used (ELISA versus radioimmunoassay versus chemiluminescence immunoassay). No TH data are available in mules.

Dynamic tests (TRH stimulation test, TSH stimulation test, and T3 suppression test) to assess thyroid function have not been evaluated in donkeys. Thus, based on available information, equine protocols can be used (Kritchevsky et al. 2023).

The prevalence of thyroid gland adenomas or malignant neoplasia (adenocarcinomas) in donkeys seems to be lower than in horses, despite long-life

Table 6 Reference ranges for hormones and mineral metabolism parameters in donkeys

Parameter	Units	Donkeys
fT3	pg/mL	1.4–2.1
tT3	ng/dL	52–78
rT3	ng/mL	0.5–0.8
fT4	ng/dL	0.4–0.5
tT4	mg/dL	2.6–4.1
Ionized calcium	mmol/L	1.7 ± 0.04
Ionized magnesium	mmol/L	0.6 ± 0.04
Total calcium	mmol/L	3.4 ± 0.2
Total magnesium	mmol/L	1.0 ± 0.08
Calcitriol	pg/mL	13.4 ± 5.9
Phosphorus	mmol/L	1.1 ± 0.3
PTH	pg/mL	20.4 ± 21.2

Data are expressed as range or mean ± standard deviation. *fT3* free triiodothyronine; *fT4* free thyroxine; *PTH* parathyroid hormone; *rT3* reverse triiodothyronine; *tT3* total triiodothyronine; *tT4* total thyroxine. Based on data from Lopez et al. (2006), and Mendoza et al. (2013)

expectancy in this species. Hyperthyroidism secondary to a bilateral thyroid gland adenocarcinoma was described in a geriatric mule with concurrent PPID (Brown et al. 2024).

Hypothyroidism (congenital or secondary to iodine deficient diet) can be observed in donkey foals.

5 Calcium and Phosphorus Homeostasis Disorders

Donkeys have higher ionized calcium and calcitriol, but lower parathyroid hormone (PTH) concentrations than horses (Mendoza et al. 2019, 2024) (Table 6); thus, donkey-specific reference ranges should be used to avoid misdiagnoses. No information is available in mules.

Oxalate-rich plants or diets with high phosphorus and low calcium content can cause nutritional secondary hyperparathyroidism in donkeys (Mendoza et al. 2017).

Donkeys may develop hypercalcemia, hypophosphatemia and normal-to-low PTH concentrations due to tumors producing PTH-related protein, a paraneoplastic condition known as humoral hypercalcemia of malignancy (Perez-Ecija et al. 2018). Creatinine concentrations in these animals are usually normal (unless the cause is a renal neoplasia). Donkeys and mules with chronic renal failure have azotemia, hypercalcemia, hypophosphatemia, and hypermagnesemia.

6 Conclusion

Metabolic syndrome is a common endocrine disorder in both donkeys and mules. PPID seems to have a higher prevalence in donkeys compared to mules and horses. Further studies are needed in donkeys and mules to improve the diagnostic and treatment approaches in this species.

Study Questions (5 or so)

1. What of the following risk factors does not predispose to insulin resistance?
   (a) Breed
   (b) Genetic
   (c) Exercise
   (d) Diet
   (e) **Geography**
2. What of the following statements is correct regarding PPID diagnosis?
   (a) Baseline plasma ACTH measurement is inaccurate and it does not depend on analyzers.
   (b) A seasonal effect of ACTH concentration has not been observed in donkeys.
   (c) The cut-off of plasma ACTH concentration is similar among horses, donkeys, and mules.
   (d) **The dynamic challenge recommended for PPID diagnosis in mules is the TRH test.**

3. What of the following is incorrect concerning ID diagnosis?
   (a) Plasma insulin concentration cut-off is similar to horses.
   (b) **CGIT is the dynamic test recommended for ID diagnosis in mules**.
   (c) OST is the dynamic test recommended for ID diagnosis in donkeys.
   (d) Glucose disposal is right-shifted in donkeys.
4. What of the following statements is incorrect concerning ID and PPID treatment?
   (a) **Metformin administration is the first step for DMS treatment before of exercise and diet control**.
   (b) Ertugliflozin administration can induce an increase of serum triglycerides and liver enzymes concentrations.
   (c) Cabergoline can be used in PPID donkeys when they do not respond to pergolide.
   (d) High quality straw is the best source of forage for donkeys.
5. What of the following statements is incorrect?
   (a) Donkeys have lower serum PTH and higher calcitriol concentrations than horses.
   (b) **Primary hypothyroidism is commonly observed in obese jennies with reproductive disorders**.
   (c) Donkeys have lower serum thyroid hormones concentrations than horses.
   (d) Nutritional secondary hyperparathyroidism could be suspected in donkeys feeding diets with high phosphorus and low calcium content.

References

Bertin FR, Bailey S, Burns T, De Laat M, Durham A, Frank N, Kritchevsky J, Menzies-Gow N (2024) Recommendations for the diagnosis and management of equine metabolic syndrome (EMS) and insulin dysregulation. The Equine Endocrinology Group. https://equineendocrinologygroup.org/

Brown K, Williams Louie E, Pinn-Woodcock T, Pearson E, Pearson GB, Marr J, Hackett ES, Rath Brown L, Mitchell KJ (2024) Cardiac disease related to primary hyperthyroidism in a 20-year-old mule. Animals 14:1660. https://doi.org/10.3390/ani14111660

Burden FA, Bell N (2019) Donkey nutrition and malnutrition. Vet Clin North Am Equine Pract 35: 469–479. https://doi.org/10.1016/j.cveq.2019.08.004

Dugat SL, Taylor TS, Matthews NS, Gold JR (2010) Values for triglycerides, insulin, cortisol, and ACTH in a herd of normal donkeys. J Equine Vet Sci 30:141–144. https://doi.org/10.1016/j.jevs.2010.01.054

Durham AE, Potier JF, Huber L (2022) The effect of month and breed on plasma adrenocorticotropic hormone concentrations in equids. Vet J 286:105857. https://doi.org/10.1016/j.tvjl.2022.105857

Fitzgerald DM, Walsh DM, Sillence MN, Pollitt CC, De Laat MA (2019) Insulin and incretin responses to grazing in insulin-dysregulated and healthy ponies. J Vet Intern Med 33:225–232. https://doi.org/10.1111/jvim.15363

Gao R, Shi L, Guo W, Xu Y, Jin X, Yan S, Shi B (2022) Effects of housing and management systems on the growth, immunity, antioxidation, and related physiological and biochemical indicators of donkeys in cold weather. Animals 12:2405. https://doi.org/10.3390/ani12182405

Gehlen H, Schwarz B, Bartmann C, Gernhardt J, Stöckle SD (2020a) Pituitary pars intermedia dysfunction and metabolic syndrome in donkeys. Animals 10:2335. https://doi.org/10.3390/ani10122335

Gehlen H, Twickel S, Stockle S, Weber C, Bartmann CP (2020b) Diagnostic orientation values for ACTH and other parameters for clinically healthy donkeys and mules (insulin, triglycerides, glucose, fructosamines, and γ-GT). J Anim Physiol Anim Nutr 104:679–689. https://doi.org/10.1111/jpn.13279

Humphreys S, Kass PH, Magdesian KG, Goodrich E, Berryhill E (2022) Seasonal variation of endogenous adrenocorticotropic hormone concentrations in healthy non-geriatric donkeys in Northern California. Front Vet Sci 9:981920. https://doi.org/10.3389/fvets.2022.981920

Kritchevsky J, Bertin FR, Frank N, Schott HC, Breuhaus B (2023) Recommendations on diagnosis and management of thyroid disease and the use of thyroxine in horses. Equine Endocrinology Group. https://equineendocrinologygroup.org/

Lopez I, Estepa JC, Mendoza FJ, Rodriguez M, Aguilera-Tejero E (2006) Serum concentrations of calcium, phosphorus, magnesium and calciotropic hormones in donkeys. Am J Vet Res 67:1333–1336. https://doi.org/10.2460/ajvr.67.8.1333

Mejia-Pereira S, Perez-Ecija A, Buchanan BR, Toribio RE, Mendoza FJ (2019) Evaluation of dynamic testing for pituitary pars intermedia dysfunction diagnosis in donkeys. Equine Vet J 51:481–488. https://doi.org/10.1111/evj.13034

Mendoza FJ, Perez-Ecija RA, Toribio RE, Estepa JC (2013) Thyroid hormone concentrations differ between donkeys and horses. Equine Vet J 45:214–218. https://doi.org/10.1111/j.2042-3306.2012.00622.x

Mendoza FJ, Aguilera-Aguilera R, Gonzalez-De Cara CA, Toribio RE, Estepa JC, Perez-Ecija A (2015a) Characterization of the intravenous glucose tolerance test and the combined glucose-insulin test in donkeys. Vet J 206:371–376. https://doi.org/10.1016/j.tvjl.2015.08.015

Mendoza FJ, Estepa JC, Gonzalez-De Cara CA, Aguilera-Aguilera R, Toribio RE, Perez-Ecija A (2015b) Energy-related parameters and their association with age, gender, and morphometric measurements in healthy donkeys. Vet J 204:201–207. https://doi.org/10.1016/j.tvjl.2015.03.004

Mendoza FJ, Toribio RE, Perez-Ecija A (2017) Nutritional secondary hyperparathyroidism in equids: overview and new insights. Equine Vet Educ 29:558–563. https://doi.org/10.1111/eve.12681

Mendoza FJ, Toribio RE, Perez-Ecija A (2018) Donkey internal medicine - part I: metabolic, endocrine, and alimentary tract disturbances. J Equine Vet Sci 65:66–74. https://doi.org/10.1016/j.jevs.2018.02.001

Mendoza FJ, Toribio RE, Perez-Ecija A (2019) Metabolic and endocrine disorders in donkeys. Vet Clin North Am Equine Pract 35:399–417. https://doi.org/10.1016/j.cveq.2019.07.001

Mendoza FJ, Buzon-Cuevas A, Toribio RE, Perez-Ecija A (2022a) Characterisation of the oral glucose and sugar tolerance tests and the enteroinsular axis response in healthy adult donkeys. Equine Vet J 54:1123–1132. https://doi.org/10.1111/evj.13544

Mendoza FJ, Mejia-Moreira S, Buchanan BR, Toribio RE, Perez-Ecija A (2022b) Evaluation of the combined glucose-insulin and intravenous glucose tolerance tests for insulin dysregulation diagnosis in donkeys. Equine Vet J 54:531–540. https://doi.org/10.1111/evj.13482

Mendoza FJ, Toribio RE, Perez-Ecija A (2024) Metabolic and endocrine insights in donkeys. Animals 14:590. https://doi.org/10.3390/ani14040590

Pearson RA, Ouassat M (2000) Estimation of live weight. In: Pearson RA, Ouassat MA (eds) Guide to live weight estimaton and body condition scoring of donkeys, 1st edn. Thomson Colour Printers, Glasgow

Perez-Ecija A, Toribio RE, Mendoza FJ (2018) Endocrine tumours in equids. Equine Vet Educ 30:664–670. https://doi.org/10.1111/eve.12803

Pritchard A, Nielsen B, Mclean A, Robison C, Yokoyama M, Hengemuehle S, Bailey S, Harris P (2019) Insulin resistance as a result of body condition categorized as thin, moderate, and obese in domesticated U.S. donkeys (Equus asinus). J Equine Vet Sci 77:31–35. https://doi.org/10.1016/j.jevs.2019.02.011

Sundra T, Kelty E, Rossi G, Rendle D (2024a) Retrospective assessment of the use of extended-release cabergoline in the management of equine pituitary pars intermedia dysfunction. Front Vet Sci 11:1332337. https://doi.org/10.3389/fvets.2024.1332337

Sundra T, Rossi G, Rendle D, Lester G (2024b) A practical approach to hyperinsulinaemia in horses with equine metabolic syndrome. Equine Vet Educ 36:325–336. https://doi.org/10.1111/eve.13938

The Donkey Sanctuary (2018) In: Evans L, Crane M (eds) The clinical companion of the donkey, 1st edn. Matador, Leicestershire

Thiemann AK, Buil J, Rickards K, Sullivan RJ (2022) A review of laminitis in the donkey. Equine Vet Educ 34:553–560. https://doi.org/10.1111/eve.13533

Warnken T, Huber K, Feige K (2016) Comparison of three different methods for the quantification of equine insulin. Vet Res 12:196. https://doi.org/10.1186/s12917-016-0828-z

Xue C, Davis J, Berghaus LJ, Hanafi A, Vaughn SA, Hart KA (2023) Pharmacokinetic properties of pergolide mesylate following single and multiple-dose administration in donkeys (Equus asinus). Equine Vet J 55:1078–1085. https://doi.org/10.1111/evj.13917

Gastrointestinal Disorders of Donkeys and Mules

F. J. Mendoza, R. E. Toribio, and R. A. Perez-Ecija

Abstract

Although physiological differences between donkeys, mules and horses have been described regarding the gastrointestinal tract, these animals are affected by similar gastrointestinal disorders. Diagnostic and treatment approaches for donkeys and mules are similar to horses. Pharmacokinetic and pharmacodynamic studies for drugs commonly used in horses with gastrointestinal diseases have not been performed in donkeys and mules. Thus, doses reported for horses can be used with caution in donkeys and mules.

Keywords

Colic · Diarrhea · Endoparasitism · Enterolith · Gastric and large colon impactions · Inflammatory bowel disease

1 Introduction

While there are differences between donkeys, mules, and horses for some gastrointestinal diseases, these animals are affected by the same conditions. The lack of epidemiological information for some disorders is likely the result of less veterinary care in donkeys and mules compared to horses.

F. J. Mendoza (✉) · R. A. Perez-Ecija
Department of Animal Medicine and Surgery, College of Veterinary Medicine, University of Cordoba, Cordoba, Spain
e-mail: fjmendoza@uco.es; alejandro.perez.ecija@uco.es

R. E. Toribio
Department of Veterinary Clinical Sciences, College of Veterinary Medicine, The Ohio State University, Columbus, OH, USA
e-mail: toribio.1@osu.edu

A. K. McLean et al. (eds.), *Donkey and Mule Medicine: Clinical Medicine and Reproduction*, https://doi.org/10.1007/978-3-032-18540-2_4

Anatomic and physiological differences between horses, donkeys, and mules, warrants species-specific approaches for gastrointestinal problems. For example, the large intestine of donkeys has a greater ability to retain water, which is related to their evolutionary adaptation to limited water access in arid and warm regions (Thiemann 2024). Intestinal transit is slower in donkeys than horses (Thiemann 2024). Donkeys and mules are better adapted to digest poor-quality forage and can thrive better than horses in rough environments. Thus, when donkeys have access to high-quality forage and concentrated grain, they are prone to obesity and colic. In addition, donkeys and mules have unique feeding behaviors, since under natural conditions they spend long periods of time grazing (up to 12–18 h per day). Restrictions to grazing or reduced number of daily meals can predispose to gastro-intestinal disorders such as gastric ulcers, impaction, and colic.

Treatment for most gastrointestinal disorders in donkeys and mules is similar to horses, however, there is scarce species-specific pharmacologic information for most drugs. Thus, clinicians must extrapolate doses and therapeutic protocols from horses, which works in most instances, but also entails potential risks.

2 Dental and Oral Cavity Disorders

Dental disorders are frequent in donkeys due to poor dental care in some situations and a long-life expectancy (Rodrigues and Lilly 2019). Sharp points, diastemata, and periodontal disease are common, although neoplasm can be also observed in donkeys. Donkeys have a narrower arcade compared to horses. Protocols used in horses for oral cavity examination can be applied to donkeys and mules. Caution must be taken in older donkeys where temporomandibular joint disease (arthrosis) or reduced mandibular ossification are relatively common, and incorrectly placed mouth-speculums or abrupt handling could result in fractures or additional problems. Similarly to horses, caries of maxillary teeth or oronasal fistula due to foreign bodies (poor quality forage) can lead to secondary maxillary sinusitis. Chronic weight loss, recurrent choke and colic, or presence of non-digested fiber or grain in feces are markers of dental or oral disturbances. Halitosis can indicate maxillary sinusitis, but also impacted food material in diastemata, periodontal disease or oral wounds.

Sialolithiasis, formation of salivary calculi or “stones” within the ducts of the major salivary glands and commonly leading to their partial or total obstruction, is common in donkeys. Clinical signs such as pain on palpation, anorexia, dysphagia, and parotid gland and duct distension can be observed. Other differential diagnosis affecting these structures should be taken into account, such as abscessation, neo-plasia, and sialadenitis. Diagnosis is based on clinical signs and radiograph. Surgery under general anesthesia or standing (oral or percutaneous approach) can be performed (Rodrigues et al. 2013).

Although donkeys have a similar dentition formula as horses, special considerations should be taken for dental floating compared to horses (See Xue 2026). Rest periods are recommended during floating, mainly in aged donkeys. Ideally, oral cavity examination and dental care should be performed at least annually in all donkeys and mules, since oral problems could predispose to malnutrition, weight loss, colic, hyperlipemia, and wellness related problems.

3 Esophageal Disturbances

The most common esophageal disorder in donkeys and mules is obstruction/impaction (choke). Other disorders reported in horses (e.g. megaesophagus, neoplasia, diverticulum) are rarely observed in donkeys and mules (Mendoza et al. 2018b). Pigmentation of the proximal esophageal mucosa can be observed in grey donkeys without functional impairment.

Risk factors for choke include poor dental condition, poor quality forage, low water consumption (dehydration), age, and stressful situations.

Clinical signs are similar to those described in horses, including nasal discharge with feed material, sialorrhea, extended neck posture, respiratory noise, cough, and respiratory distress. Secondary complications include esophageal rupture, diverticula, stenosis, post-obstruction megaesophagus, and aspiration pneumonia.

Diagnosis is based on clinical signs, nasogastric intubation, and endoscopy. Esophageal lavage with warm water under deep sedation (α2-agonists such as xylazine) and scopolamine administration (Table 1) is recommended to relax esophageal muscle fibers and allow dilation. Mineral oil administration must be avoided due to the high risk of aspiration pneumonia. Endoscopy after resolution is recommended to assess the integrity of the esophageal mucosa, to discard post-obstruction damage, and to detect the primary cause of dysfunction or impaction such as neoplasia, diverticula, strictures, or abscesses. In animals with esophageal injury, there is a higher risk of secondary complications (strictures), which are handled similar to horses. Broad spectrum antimicrobials are recommended. External esophageal obstruction secondary to abscesses (*Streptococcus equi* ssp. *equi*) or neoplasia can result in intermittent choke.

4 Gastric Problems

4.1 Gastric Impaction

Gastric impaction is a common disorder in donkeys and mules. Although it may occur in any breed, gender and age, elderly donkeys are more susceptible, mainly because of poor dental care. Other risk factors are poor quality roughage, low water consumption, gastric parasites (*Gasterophilus* spp.), foreign bodies (e.g. plastic bags and ropes), and liver disease (Mendoza et al. 2018a). Gastric ulcers and gastritis may also predispose to gastric impaction due to pyloric dysfunction or delay in gastric emptying (which seems to be longer than in horses).

Since clinical signs may go unnoticed for a long time, this condition should be discarded in any colicky donkey, or even in those that are lethargic, anorexic, with behavioral changes (slow feeding), or reluctance to move.

Table 1 Common drugs used in gastrointestinal disturbances in donkeys and mules

Drugs	Dose	Route	Interval	Dosage regimen	Observations	Donkey reference
Antimicrobials						
Azithromycin[a]	10 mg/kg	PO	24 h		Foals <6 months of age only	
Chloramphenicol[a]	50–60 mg/kg	PO	6 h		Where legally available	
Doxycycline	10 mg/kg	PO	12 h			Chapuis et al. (2021)
Enrofloxacin	5–7.5 mg/kg	IV-PO	24		Overdosing: lethargy or seizures	Sekkin et al. (2012)
Penicillin	20,000 IU/kg	IV	6–8–12			Oukessou et al. (1994)
Oxytetracycline	5–10 mg/kg	IV	12–24 h	7 days	Diluted in 1 L of saline solution and administered slowly in 30–45 min	Horspool and McKellar (1990)
Polymyxin B[a]	5000–6000 IU/kg	IV	8–12 h		Nephrotoxic, not used in azotemic animals. Diluted in 500 mL of saline solution and administered in 20 min	
Sulfonamide/ Trimethoprim	12.5	IV	12			Peck et al. (2002)
	20–30 mg/kg[a]	PO	12		Likely this dose is not adjusted for trimethoprim	
Anticholinergics						
Hyoscine (scopolamine N-butyl bromide)[a]	0.3 mg/kg	IV	20–60 min	As needed		
Antipyretic and antispasmodic						
Metamizole (dipyrone)	25 mg/kg	IV	12 h			Arcoverde et al. (2023)

Analgesic, antipyretic, anti-inflammatory						
Flunixin meglumine	1.1 mg/kg	IV	8–12–24 h			Cheng et al. (1996)
Paracetamol (acetaminophen)	10 mg/kg[a]	IV	8–12			
	20–30 mg/kg	PO				Sullivan and Vogel (2023)
Pentoxifylline[a]	10–15 mg/kg	IV-PO	8–12 h			
Gastric ulcers						
Misoprostol[a]	1–5 µg/kg	PO	8–12 h			
Omeprazole[a]	4 mg/kg	PO	24 h			
Sucralfate[a]	20–40 mg/kg	PO	6–8 h			
Cathartics, absorbents, chelators and laxatives						
Activated charcoal[a]	0.5–1 g/kg	IG	24 h			
Coca-Cola®[a]	1 L/200 kg	IG	4 h		Without caffeine and sugar and remove gas prior to administration	
Dimercaprol[a]	2.5–5 mg/kg	IM	24 h		Arsenic, lead and mercury toxicosis	
Magnesium sulfate[a]	1 g/kg	IG	24 h			
Mineral oil[a]	1 L/200 kg	IG	24 h			
Sodium sulfate[a]	1 g/kg	IG	24 h			
Smectite (Biosponge®)[a]	0.5–1 g/kg	IG-PO	12–24 h			
Corticosteroids *(inflammatory bowel disease)*						
Dexamethasone[a]	0.05–0.2 mg/kg	IV-IM-PO	24			

(continued)

Table 1 (continued)

Drugs	Dose	Route	Interval	Dosage regimen	Observations	Donkey reference
Prednisolone[a]	0.5–1 mg/kg	PO	12–24			
Prokinetics						
Lidocaine	Loading dose: 1.3 mg/kg	IV	20 min bolus	Some clinicians prefer to use without bolus	Extrapyramidal signs (ataxia, circling, obnubilation, etc.) if toxicosis (commonly fast administration).	Wood and Loomes (2022)
	Maintenance: 0.025–0.05 mg/kg/min	IV	CRI			
Metoclopramide	0.1–0.5 mg/kg[a]	IV-IM-SC-PO	4–8 h		Extrapyramidal signs (ataxia, circling, obnubilation, etc.) if toxicosis.	
	0.04 mg/kg	IV	CRI			Wood and Loomes (2022)
Neostigmine[a]	0.005–0.02 mg/kg	IV-IM-SC	4–6 h		Sialorrhea, tachycardia, abdominal discomfort	Wood and Loomes (2022)
Anti-parasitic drugs *(against parasites causing gastrointestinal disease)*						
Fenbendazole	5–10 mg/kg	PO	24 h	5 days	Roundworm. Cyathostomins resistance has been observed. Caution in early pregnant jennies due to potential teratogenicity	Grosenbaugh et al. (2011)
Ivermectin	0.2 mg/kg	PO	Once		Overdosing: blindness, staggering and seizures. No use in younger than 6 months. Resistance observed in some countries.	Gokbulut et al. (2005)

Moxidectin	0.2 mg/kg	PO	Once		Roundworm, *D. arnfieldi*, and arthropods. Overdosing: staggering and seizures	Trawford et al. (2005)
	0.4 mg/kg	PO	Once		Roundworm, *D. arnfieldi*, arthropods, and encysted cyathostomin larvae. Overdosing: staggering and seizures	The Donkey Sanctuary (2018)
Praziquantel	1–2.5 mg/kg	PO	Once		*A. perfoliata*	
Pyrantel pamoate	19 mg/kg	PO	Once		Roundworm	Gokbulut et al. (2014)
	38 mg/kg	PO	Once		*A. perfoliata*	The Donkey Sanctuary (2018)

IG intragastric, *IM* intramuscular, *IV* intravenous, *PO* per os

[a]Dosages with asterisk are extrapolated from horses or doses from clinical experience (no reference) both for donkeys and mules, since no donkey- or mule-specific studies are available.

Diagnosis can be suspected if nasogastric reflux contains a large amount of feed material or has a fermented odor. In some cases, the tube can be hard to pass into the stomach or is obstructed with feed material. Abdominal ultrasound can be helpful if the stomach is visualized beyond the 11–12th intercostal space (gastric distention).

Treatment is similar to horses, with gastric lavage with warm water being the cornerstone. Dioctyl sodium succinate or Coca-Cola (without sugar and caffeine) can also be administered through the nasogastric tube to facilitate the dissolution of the impaction. Gastroscopy is important to make a definitive diagnosis, to assess resolution, and to identify predisposing causes such as gastric ulcers, pyloric dysfunction, or neoplasia (squamous cell carcinoma, adenocarcinoma). Gastric mucosa biopsy can be helpful to achieve a diagnosis.

4.2 Gastric Ulcers

Donkey gastric ulcer syndrome (DGUS) is common (Thiemann and Sullivan 2019), although prevalence appears to be lower (40–50%) compared to horses (60–90%).

Donkeys are usually fed diets with low sugar content and rarely face stressful situations such as frequent transport, training, or competition, which could explain this difference. However, this could also be due to few studies on gastric ulcers in this species. Similar to horses, both glandular (DGGD) or squamous (DSGD) portions can be affected (Fig. 1), but DSGD seems to be more common.

Since poor performance is not a common clinical feature in this species and mild colic signs can go unnoticed due to their stoic behavior, gastroscopy is recommended in donkeys or mules with reduced appetite or chronic weight loss to rule out DGUS.

Gastroscopy is performed like horses, although shorter fasting time is recommended in donkeys due to their higher risk of hyperlipemia. The presence of feed material in the stomach after fasting suggests delayed gastric emptying or

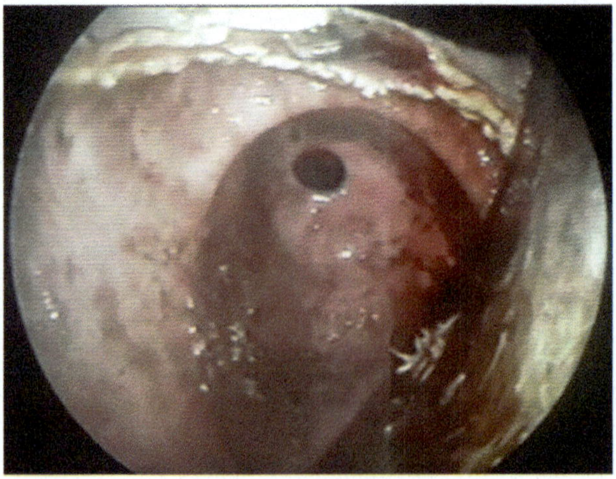

Fig. 1 Gastroscopy showing severe hemorrhagic gastric disease of the glandular portion of the stomach, and a deep and hemorrhagic ulcer in the lesser curvature of the stomach

impaction, although it will also depend on fasting duration. Lesions are scored using the equine grading system (Sykes et al. 2015).

Information for proton pump inhibitors is lacking in donkeys, but they are used according to equine dosing protocols, in particular oral omeprazole (Table 1). Since the bioavailability of different omeprazole formulations is unknown in donkeys and mules, drugs marketed for horses are used. Omeprazole is recommended to be administered between 30–60 min prior to feeding despite information not being available in these species. Parenteral administration of proton-pump inhibitors such as esomeprazole, pantoprazole, or lansoprazole have not been evaluated in these species, although they are occasionally used in donkeys and mules. Additional drugs such as misoprostol and sucralfate can be administered as for horses despite the lack of pharmacological studies (Table 1).

Feeding strategies, including diets with low sugar content and rich in omega-3 fatty acids, prolonged grazing time (12–18 h/day), and frequent feedings (3–4 times/day) are recommended. High quality forage is preferred, and grain or concentrates should be avoided. Pectin has not been evaluated in donkeys and mules, but beet pulp seems to be beneficial. Antimicrobials are not recommended to treat gastric ulcers and the role of certain pathogens (e.g. *Helicobacter pylori*) have not been elucidated in equids. Any stressful situation and long-term administration of nonsteroidal anti-inflammatory drugs (NSAIDs; mainly phenylbutazone) should be avoided.

5 Small and Large Intestine Colic

5.1 Diagnosis

Signs of abdominal pain are usually milder in donkeys compared to horses due to their stoic behavior and high pain tolerance (Burden and Thiemann 2015). This often results in a delayed diagnosis and subsequent treatment, which can compromise the prognosis. Clinical signs commonly used in horses to discern between medical versus surgical colic are less useful in donkeys, since often donkeys with surgical lesions may display mild signs (Thiemann 2024). Several pain scales for donkeys have been developed, such as Donkey Pain Scale (DOPS), and Donkey Composite Pain Assessment (Equus-Donkey-Compass), Equine Utrecht University Scale for Donkey Facial assessment of Pain (Equus-Donkey-FAP) (Van Dierendonck et al. 2020; Oliveira et al. 2021).

Signs of colic observed in donkeys include dullness, poor appetite, separation from companions, low head carriage, reluctance to move, and immobile backwards ears. Due to their stoic nature, heart rate must be interpreted with caution in donkeys and mules, since it is influenced by several factors (e.g. hypovolemia/dehydration, endotoxemia, fever, anaemia). Due to their high tolerance to dehydration/hypovolemia, peripheral pulse quality and capillary refill time should be interpreted with caution when assessing hemodynamic derangements.

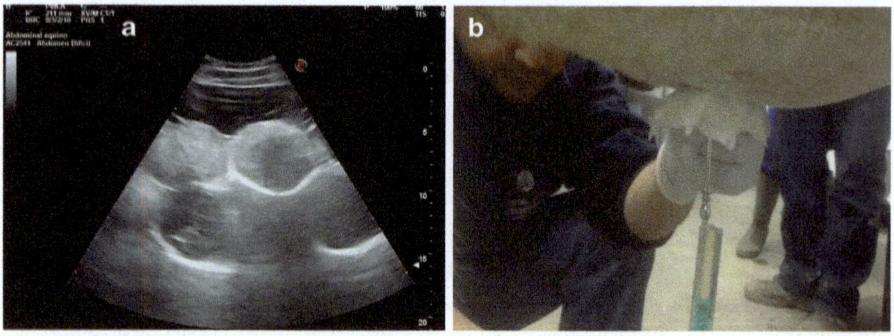

Fig. 2 (**a**) Abdominal ultrasound in an adult donkey with colic displaying severe small intestine distention and ileus (sedimented feed material in the jejunum). (**b**) Abdominocentesis technique and body landmark in an adult donkey with colic using a teat cannula. Note the normal gross appearance of the peritoneal fluid

Donkeys have narrow nasal passages, so a smaller bore nasogastric tube should be used for gastric intubation in comparison with horses. Epistaxis is commonly observed if a wide-bore tube is used or if the tube is passed or retrieved gently. The pharyngeal recess is deeper in donkeys compared to horses, which can make intubation challenging (See Alonso 2026, for more information about the anatomy of donkeys and mules).

Rectal palpation is crucial to diagnose abnormalities such as large colon impactions (most common cause of colic in donkeys), large colon displacement, and small intestinal distension; however, this may not be feasible in small size breeds. Restraint, sedation, rectal local anesthetics, or intravenous scopolamine may promote rectal relaxation and facilitate examination, in addition to good lubrication.

Blood profile may not be useful to discern medical from a surgical colic, but it is helpful to assess the systemic status of the animal, including endotoxemia, hypovolemia, acid-base and electrolyte imbalances, as well as renal function.

Abdominal ultrasound is a useful tool to evaluate intestinal wall thickness, intestinal content, gastric dilation, gut motility, small intestinal distention (Fig. 2a), presence and characteristic of free peritoneal fluid, colon displacement, and abdominal masses (abscess, neoplasia). The FLASH protocol previously described in horses can be used in donkeys (Busoni et al. 2011). Specific information on abdominal viscera measurements and ultrasonographic organ location have been reported in healthy standard size jennies (Ibrahim and El-Ashker 2020; Accorroni et al. 2025).

Abdominocentesis is valuable in the diagnosis of medical versus surgical colic (Fig. 2b). Reference ranges available for horses can be extrapolatable to donkeys and mules and interpretation is similar (Table 2). Thick ventral subcutaneous and peritoneal fat deposits can make abdominocentesis challenging, needing a longer teat cannula to cross the fat and reach the peritoneal cavity. Anatomical landmarks are similar to horses. Donkeys with peritoneal fluid lactate concentrations higher

Table 2 Characteristics and reference ranges of common parameters measured in the peritoneal fluid

Parameter	Unit	Donkeys	Mules
Macroscopic aspect	–	Transparent yellow pale	Transparent yellow pale
WBC	× 10^3/μL	<3	<3
RBC	× 10^6/μL	<0.5	<0.5
PCV	%	<1	<1
TP	mg/dL	<2.5	<2.5
Lactate	mmol/L	<2	<2
Glucose	mg/dL	>80% serum glucose	>80% serum glucose
Cellular distribution	%	Neutrophils: 50–80% Mononuclear cells (macrophages and mesothelial cells): 50–20% Lymphocytes: 0–10% Eosinophils: Not present	Neutrophils: 50–80% Mononuclear cells (macrophages and mesothelial cells): 50–20% Lymphocytes: 0–10% Eosinophils: Not present

PCV packed cell volume, TP total protein, *RBC* red blood cells, *WBC* white blood cells

than blood lactate concentrations, and high leukocyte counts, and total protein concentrations are more likely to require surgery. If after intravenous fluid therapy the blood lactate concentration decreases but peritoneal lactate concentration remains elevated or increases, these animals will likely require exploratory laparotomy.

Abdominal radiographs can be performed to diagnose sand accumulation in the large colon and enteroliths. Field radiography machines can be used in small size donkey breeds to obtain a diagnostic image, but a more powerful hospital machine will be necessary in larger breeds.

5.2 Medical Colic

Large intestine (cecum and colon) impactions are the most common medical colic in donkeys (prevalence 55%), frequently located in the pelvic flexure (Mendoza et al. 2018b). Small colon impactions can also be observed, with foreign bodies such as ropes or plastic bags being a frequent cause in developing countries. Predisposing factors for impaction are similar to those described for gastric impaction (low-quality forage, poor dental care, low water consumption, and endoparasitism). Enteral isotonic fluids (or warm plain water) together with cathartics (magnesium sulfate or sodium sulfate) and laxatives (mineral oil or paraffin) are the cornerstones of treatment (Table 1), as long as no gastric reflux is present. Two to three liters every

1–2 h can be administered in a donkey of 150–200 kg. Mortality rate for large intestine impaction can be high (50%) in animals diagnosed late. Caution must be taken in the management of these animals, in particular concerning complete feed withholding and the risk of hyperlipemia and secondary ileus. In this sense, enteral fluids can be supplemented with glucose (caution with laminitis), or partial parenteral nutrition can be initiated (see Parkes and Bukhari 2026, for further information regarding treatment).

Rectal prolapse is common in donkeys secondary to impaction or endoparasitism (Mendoza et al. 2018b). Rectal prolapse is treated similar to horses (hyperosmotic solution and purse-string suture), and prognosis will depend on prolapse time, intestine portion prolapsed, and underlying cause.

Regarding other small intestine disturbances, it is worth mentioning that spasmodic colic can go unnoticed in donkeys and mules due to their stoic behavior. Proximal enteritis and ileal impaction appear to be less frequent in donkeys and mules than in horses, although can be observed in animals with high parasite burdens. Since ivermectin and fenbendazole resistance is an emerging problem in some countries, high parasitism can be seen in donkeys dewormed with these drugs. Enteritis secondary to equine coronavirus infection has been described in donkeys, displaying leukopenia, neutropenia, and lymphopenia. Similarly, *Salmonella* spp. can cause severe gastrointestinal disease. Diagnosis is reached by fecal PCR or culture (see Chapter "Urogenital Disorders" for more information about testing strategies in cases of gastrointestinal disease). Treatment is supportive (fluids, NSAIDs, management of secondary complications). Broad-spectrum antimicrobials can be administered in case of severe leukopenia and risk of sepsis. Isolation is important until 2–3 consecutive negative fecal PCRs or cultures 1 week apart are obtained. Pharmacologic studies about prokinetic drugs (lidocaine, metoclopramide, neostigmine) have not been performed in donkeys and mules, but equine dosing protocols are used (Table 1).

Grass sickness can be considered in donkeys and mules with signs compatible with this condition in countries where this disease is present, with the final diagnosis depending upon intestinal biopsies.

Similar to horses, donkeys develop endotoxemia and secondary complications such as laminitis, hyperlipemia, disseminated intravascular coagulopathy, and organ failure. Special caution must be taken in obese donkeys and those with endocrine disorders (e.g. insulin dysregulation) due to their higher predisposition to laminitis and hyperlipemia. Endotoxemia is managed like horses, although pharmacological studies have not been performed in donkeys and mules for the most commonly used drugs. Further information about laminitis and hyperlipemia can be found elsewhere in this book (Danese et al. 2026 and chapter "Endocrines Disorders of Donkeys and Mules").

5.3 Surgical Colic

The degree of abdominal pain in donkeys and mules often underscore the severity of intestinal lesions. Surgical colic (volvulus, hernias, intussusceptions, colon displacement) often have higher mortality rates in donkeys than in horses (Merridale-Punter et al. 2023). In addition, complications such as hyperlipemia worsen the prognosis (Thiemann 2024). Despite obesity being common in donkeys, pedunculated lipomas causing small intestine strangulation are rare.

Surgical procedures (both laparotomy and laparoscopy) are similar to horses. However, special considerations must be taken with general anaesthesia (see Chapter "Veterinary Diagnostic Sample Collection, Laboratory Submission and Testing Strategies (Antemortem and Postmortem Considerations) for Donkeys and Mules").

Post-surgery management in donkeys and mules is similar to horses. However, special care must be taken considering the high risk of hyperlipemia in this species. If partial parenteral nutrition is implemented, hyperglycemia can potentially result in hyperinsulinemia, increasing the risk of laminitis (mainly in animals with obesity and insulin dysregulation). When possible, it is recommended to house donkeys with another mate to reduce stress and facilitate recovery.

6 Diarrhea in Donkeys

6.1 Acute Diarrhea

The cecum and large colon in donkeys have a great capacity to retain and absorb water, thus watery diarrhea is infrequent in this species compared to horses (Du Toit et al. 2010), even if evidence of colitis is observed by ultrasound. Clinicians and owners must be careful not to confuse colitis with moderately loose feces, which can be observed frequently in these animals after a stressful situation.

Infectious and non-infectious causes of acute diarrhea in donkeys are similar to those described in horses (Sullivan 2024). However, epidemiological information about prevalence, risk factors, outcome and mortality are unavailable.

6.1.1 Infectious diarrhea

Diagnosis of acute bacterial diarrhea (*Salmonella* spp., *Clostridioides difficile*, *Clostridium perfringens*, *Neorickettsia risticii* [Potomac horse fever, PHF], and *Lawsonia intracellularis* [LI]) is achieved by fecal PCR, serologic methods, and culture. The same diagnostic principles (sample collection, number of samples and conservation) for horses apply to donkeys and mules (Mendoza et al. 2018b) (see Chapter "Urogenital Disorders" for more information about diagnostic testing strategies). PCR, mostly after pre-incubation in enrichment broth, is the method with the highest specificity and sensitivity, and has a good agreement with fecal culture. However, fecal cultures of consecutive samples (3–5) remains the gold standard diagnostic technique for *Salmonella* spp. Intestinal parasites, including

Cyathostomins, *Parascaris equorum*, and *Strongyloides* spp. cause diarrhea in donkeys (Sullivan 2024). Fecal egg count is the most common, rapid, and cheap diagnostic tool to assess intestinal endoparasitism.

Equine coronavirus can cause enterocolitis in donkeys and fecal PCR is a reliable test to detect this virus.

Intravenous oxytetracycline administration is the preferred treatment against PHF and LI, although oral doxycycline is also effective (Table 1). For other bacterial pathogens, antibiotics administration is controversial and at least for horses, there is no evidence that antimicrobials improve outcome. The therapeutic cornerstone for acute diarrhea includes administration of isotonic intravenous fluids, correction of electrolyte and acid-base imbalances, control of systemic inflammation (pentoxifylline, polymyxin B, NSAIDs), and prevention of laminitis (cryotherapy) (Sullivan 2024). Plasma transfusion should be considered in animals with severe hypoproteinemia. There are no commercial sources of donkey plasma; thus, if a healthy compatible gelding donkey is not available, equine plasma could be considered. When equine plasma is used, it is important to perform serial hemograms in the next days to detect immune mediated disorders such as hemolytic anaemia and thrombocytopenia. Oral smectite can be administered to donkeys following equine dosing protocols (Table 1). Oral fenbendazole or macrocyclic lactones (ivermectin or moxidectin) in combination with praziquantel are effective against most intestinal parasites, including tapeworms (Bell et al. 2021). Ivermectin and fenbendazole resistance has been described in some countries. There are not anthelmintics specifically approved for donkeys or mules and products used are formulated for horses.

Implementation of biosecurity measures (isolation, disinfection) is important for animals with diarrhea, mainly in contagious causes.

6.1.2 Non-infectious diarrhea

Dysbiosis secondary to antimicrobial administration (antibiotic-associated diarrhea) or alimentary changes are common causes of diarrhea. Although microbiome studies have not been performed in diarrheic donkeys receiving antibiotics, it is assumed that *Clostridioides difficile* can be involved in severe cases. Diagnosis is based on history of antimicrobial administration or diet modifications. Prognosis is usually good without further treatment, except with *C. difficile* proliferation, where additional therapeutic measurements are commonly needed as described for infectious diarrhea.

Right dorsal colitis due to long-term or high doses of NSAIDs can also be observed in donkeys and mules (Mendoza et al. 2018b). Caution must be taken when NSAIDs are given to dehydrated animals, even at recommended doses. Similar to horses, diagnosis is based on previous history, clinical signs (although watery diarrhea is rarely observed in donkeys), hematology (leukopenia and neutropenia), biochemistry (hypoalbuminemia) and abdominal ultrasound (increased intestinal wall thickness, increased luminal fluid content). The prognosis is guarded and depends on the presence and severity of secondary complications. Fluid therapy and correction of acid-base and electrolytes imbalances are the most important aspects of treatment. Plasma transfusion can be beneficial but large volumes are required and can be costly. Synthetic colloids can be used to increase the oncotic pressure. Additional treatments such as pentoxifylline, misoprostol, paracetamol (acetaminophen), metamizole (dipyrone), and smectite can be administered (Table 1). High quality hay should be provided, but concentrates, grain and high sugar feedstuffs should be avoided. If anorexia is present, intravenous glucose infusion should be started to prevent hyperlipemia (see Parkes and Bukhari 2026, for further information).

Other disturbances causing diarrhea such as hyperlipemia and pancreatitis should also be considered. Usually, pancreatitis is secondary to fatty infiltration in hyperlipemic donkeys. Clinical signs are non-specific, and diagnosis is based on plasma amylase and lipase concentrations.

6.2 Chronic Diarrhea

6.2.1 Infectious diarrhea

Chronic parasitosis or chronic salmonellosis are common causes of chronic infectious diarrhea in donkeys and mules. Fecal egg count is recommended. Cyathostomin eggs may not be present in feces despite being the causative agent. Eosinophilia is not a typical feature in intestinal parasitosis. An ELISA test is available for cyathostomins in horses but its reliability in donkeys and mules is unknown. *Fasciola hepatica* (flukes) must be considered in some countries (e.g. UK, Spain), with increasing incidence in recent years (Bell et al. 2021). These donkeys can have increased plasma liver enzyme activities. Similar treatments as described in parasitic acute diarrhea can be effective.

Donkeys and mules can be asymptomatic carriers of *Salmonella* spp., representing a risk for horses, and other donkeys and mules. Serial fecal PCR can be valuable to identify these animals. Isolation is recommended until at least two consecutive negative PCR are obtained.

6.2.2 Non-infectious diarrhea

Sand accumulation in the large colon and cecum is a common cause of chronic diarrhea in animals fed on sandy areas. Similar to horses, abdominal radiography is helpful to reach a final diagnosis and to evaluate the response to the treatment (Mendoza et al. 2018b). In small donkeys and mules, abdominal radiographs can be done in the field with a portable machine. Abdominal ultrasound could help in severe cases. In milder or moderate cases, the administration of laxatives (mineral oil or paraffin) and cathartics (magnesium sulfate) can be useful, but in severe cases surgery to evacuate the large colon will be needed. In animals fed on the ground or in sandy areas, daily oral administration of psyllium or plantago ovata can be preventive.

Donkeys and mules can develop inflammatory bowel disease (IBD), but epidemiological data are unavailable. A number of disorders can be included under IBD depending on the type of cellular infiltrate, including granulomatous, lymphocytic-plasmocytic, and eosinophilic enteritis. Diagnosis is based on clinical signs (weight loss, diarrhea), biochemistry (hypoalbuminemia), abdominal ultrasound (increased intestinal wall thickness), and rectal and/or duodenal biopsy. An alimentary allergy panel should be performed. Treatment is similar to horses and includes long-term oral glucocorticoids (dexamethasone or prednisolone), low carbohydrate and high fat diets, regular deworming, and even fecal microbial transplantation from a healthy heard mate. Prognosis is variable, depending on the IBD type, but is usually poor. Alimentary lymphoma can have a similar presentation as IBD and in some cases it can be difficult to differentiate from lymphocytic-plasmocytic enteritis. Donkeys can develop multisystemic eosinophilic epitheliotropic disease (MEED), which in addition to intestinal disease have epidermal lesions (Timko 2024). MEED is treated as IBD, and the prognosis is poor.

Abdominal neoplasia, abscesses, and peritonitis must also be considered as differential diagnoses in cases of chronic diarrhea.

7 Peritonitis and Abdominal Masses

Primary gastrointestinal neoplasia, neoplasia in other systems (reproductive or urinary tract), and metastases in the liver or spleen can cause colic signs and diarrhea (Mendoza et al. 2018b). Clinical signs are commonly subtle and vague, such as lethargy, anorexia, chronic weight loss, poor clinical appearance, recurrent colic, or chronic diarrhea. Diagnosis is similar to horses. Although epidemiological information is lacking in donkeys and mules, the prevalence could be high in donkeys due to their long-life expectancy. Intestinal (alimentary) or abdominal lymphoma is the most common abdominal neoplasia in donkeys, usually with multicentric involvement (Paraschou et al. 2024). Chemotherapy has been described in donkeys, following similar multimodal protocols as described previously in horses (Luethy et al. 2019). Long-term outcome is poor in donkeys and mules with abdominal neoplasia.

Donkeys with abdominal abscesses display non-specific clinical signs, including lethargy, anorexia, fever, colic, or diarrhea. Laboratory findings are not specific, such as leucocytosis, lymphocytosis, monocytosis, and hyperglobulinemia. Rectal palpation (depending on body size), abdominal ultrasound and abdominocentesis can be useful diagnostic tools, but abscesses cannot be discarded if abnormalities are not found. Oral long-term administration of antibiotics with good tissue penetration (e.g. doxycycline, chloramphenicol, sulfonamides, rifampicin, or enrofloxacin) is the recommended treatment (Table 1). Depending on location, drainage or surgery may be necessary.

Peritonitis is diagnosed and managed like horses, with variable prognosis. Primary peritonitis has a good prognosis with proper treatment while the prognosis is grave with gut perforation. Abdominal drainage and lavage as described for horses may be required for some cases.

8 Toxicology

Information about toxins causing gastrointestinal signs in donkeys or mules is scarce (Bates 2024); thus, it is assumed that these species are sensitive to similar toxins as horses (Stegelmeier and Davis 2018), although magnitude of clinical signs can be different. Due to curious behaviour of the donkey, poisoning could be more common from plants and pesticides (Bates 2024). Cantharidin, arsenic, mercury, acorns (tannin metabolites), and *Solanum* spp. are examples of toxins that cause gastrointestinal disturbances in donkeys and mules. Clinical signs include hypersalivation, gastritis, colic, and diarrhea that can vary from watery to hemorrhagic (arsenic or mercury) depending on the toxin. Removing the toxic source, gastric lavage, and administration of activated charcoal, mineral oil or paraffin are the recommended treatments, followed by systemic support (correct fluid, acid-base, and electrolytes abnormalities). Intravenous dimercaprol could be administered in cases of arsenic poisoning as a chelating agent.

Organophosphates and plants containing cardiac glycosides (e.g. Nerium oleander) can also produce diarrhea, but neurological signs or cardiac arrhythmias, respectively, are usually more predominant than gastrointestinal signs.

9 Conclusions

Gastrointestinal disorders are very common in both in donkeys and mules. While there are anatomic and physiological differences between these species and horses (e.g. gastric emptying, large intestine ability to retain water, better adaptability to digest poor-quality forage, tendency to obesity and colic when fed high-quality forage and concentrated grain), the diagnosis and treatment for most diseases is similar to horses. Special care must be taken in donkeys and mules when extrapolating drug doses, since most lack species-specific pharmacological studies.

Questions

1. Which of the following statement is true in donkeys and mules with diarrhea due to cyathostomin infestation?
   (a) **Eggs can be absent in feces, even when these parasites are the cause of the diarrhea.**
   (b) Eosinophilia is a common finding in donkeys affected with these parasites.
   (c) There is a specific ELISA test available, highly reliable for donkeys and mules.
   (d) All of the aforementioned statements are true.
2. During the treatment of a large intestine impaction in a donkey, we must consider…
   (a) Complete feed withholding is advised in these patients.
   (b) 10–20 liters of enteral warm plain water with magnesium sulfate and paraffin is a common therapeutic approach.

   (c) This type of impaction is common in donkeys and linked to high mortality rates.
   (d) **All of them are true.**
3. When assessing a colicky donkey, we must remember that…
   (a) Signs are usually more obvious and marked compared to horses.
   (b) Heart rate is a reliable marker of pain in these species.
   (c) Peripheral pulse quality and capillary refill time are the more reliable signs of hemodynamic derangement in donkeys.
   (d) **Subtle signs, such as dullness, separation from companions or backwards ears are the most common signs in colicky donkeys**.
4. Concerning the utility of laboratory parameters in gastrointestinal disorders of mules, it is true that…
   (a) Blood profile can differentiate between a medical and a surgical colic.
   (b) A mule with a high blood lactate/peritoneal lactate rate, leukopenia and hypoproteinemia is more likely to need a surgery.
   (c) **An exploratory laparotomy should be performed in a donkey maintaining higher peritoneal lactate compared to plasmatic lactate concentrations after intravenous fluid therapy**.
   (d) Fasciolosis (fluke) can be discarded if liver enzyme activities are not increased or eosinophilia is not observed.

References

Accorroni L, Bazzano M, Marchegiani A, Spaterna A, Laus F (2025) Abdominal ultrasonography in healthy female standard donkeys. Animals 15:129. https://doi.org/10.3390/ani15020129

Alonso D (2026) Saddle fit in mules and donkeys. In: McLean AK et al. (eds) Donkey and mule medicine: foundations, breeds, and management, vol 1. Springer, Cham

Arcoverde KN, Alves LSA, Cavalcante JM, Maranhao ACPM, Jurema AP, Araújo-Silva G, JTP U, De Paula VV (2023) Pharmacokinetic monitoring of dypirone in Northeastern Brazilian donkeys (Equus asinus). Res Vet Sci 164:105034. https://doi.org/10.1016/j.rvsc.2023.105034

Bates N (2024) Poisoning in donkeys. UK-Vet Equine 7:10–20. https://doi.org/10.12968/ukve.2023.7.1.10

Bell N, Thiemann A, Sullivan R (2021) Endoparasite control for donkeys in the UK. UK-Vet Equine 5:84. https://doi.org/10.12968/ukve.2021.5.2.84

Burden F, Thiemann A (2015) Donkeys are different. J Equine Vet Sci 35:376–382. https://doi.org/10.1016/j.jevs.2015.03.005

Busoni V, De Busscher V, Lopez D, Verwilghen D, Cassart D (2011) Evaluation of a protocol for fast localised abdominal sonography of horses (FLASH) admitted for colic. Vet J 188:77–82. https://doi.org/10.1016/j.tvjl.2010.02.017

Chapuis RJJ, Smith JS, French HM, Toka FN, Peterson EW, Little EL (2021) Nonlinear mixed-effect pharmacokinetic modeling and distribution of doxycycline in healthy female donkeys after multiple intragastric dosing-preliminary investigation. Animals 11:2047. https://doi.org/10.3390/ani11072047

Cheng Z, McKellar Q, Nolan A, Lees P (1996) Preliminary pharmacokinetic and pharmacodynamic studies on flunixin meglumine in donkeys. Vet Res Commun 20:469–472

Danese T, Raspa F, Valle E, Cavallarin L, Giribaldi M (2026) Donkey nutrition and feeding strategies. In: McLean AK et al. (eds) Donkey and mule medicine: foundations, breeds, and management, vol 1. Springer, Cham

Du Toit N, Burden FA, Getachew M, Trawford AF (2010) Idiopathic typhlocolitis in 40 aged donkeys. Equine Vet Educ 22:53–57. https://doi.org/10.2746/095777309x480894

Gokbulut C, Boyacioglu M, Karademir U (2005) Plasma pharmacokinetics and faecal excretion of ivermectin (Eqvalan paste) and doramectin (Dectomax, 1%) following oral administration in donkeys. Res Vet Sci 79:233–238. https://doi.org/10.1016/j.rvsc.2005.01.002

Gokbulut C, Aksit D, Smaldone G, Mariani U, Veneziano V (2014) Plasma pharmacokinetics, faecal excretion and efficacy of pyrantel pamoate paste and granule formulations following per os administration in donkeys naturally infected with intestinal strongylidae. Vet Parasitol 205: 186–192. https://doi.org/10.1016/j.vetpar.2014.06.026

Grosenbaugh DA, Reinemeyer CR, Figueiredo MD (2011) Pharmacology and therapeutics in donkeys. Equine Vet Educ 23:523–530. https://doi.org/10.1111/j.2042-3292.2011.00291.x

Horspool LJI, McKellar QA (1990) Disposition of oxytetracycline in horses, ponies and donkeys after intravenous administration. Equine Vet J 22:284–285

Ibrahim HMM, El-Ashker MR (2020) Reference values and repeatability of transabdominal ultrasonographic gastrointestinal tract thickness and motility in healthy donkeys (Equus asinus). J Equine Vet Sci 92:103153. https://doi.org/10.1016/j.jevs.2020.103153

Luethy D, Frimberger AE, Bedenice D, Byrne BS, Groover ES, Gardner RB, Lewis T, Macdonald VS, Proctor-Brown L, Tomlinson JE, Rassnick KM, Johnson AL (2019) Retrospective evaluation of clinical outcome after chemotherapy for lymphoma in 15 equids (1991–2017). J Vet Intern Med 33:953–960. https://doi.org/10.1111/jvim.15411

Mendoza FJ, Toribio RE, Perez-Ecija A (2018a) Aspects of clinical relevance in donkeys. In: Reed SM, Bayly WM, Sellon DC (eds) Equine internal medicine, 4th edn. ElSevier, Amsterdam, pp 1513–1524

Mendoza FJ, Toribio RE, Perez-Ecija A (2018b) Donkey internal medicine—part I: metabolic, endocrine, and alimentary tract disturbances. J Equine Vet Sci 65:66–74. https://doi.org/10.1016/j.jevs.2018.02.001

Merridale-Punter MS, Prutton JSW, Stefanovski D, Worsman FCF, Payne RJ, Wylie CE, Archer DC, Menzies-Gow NJ, Coles L, Mcgovern KF, Parker RA (2023) Outcome following emergency laparotomy in 33 UK donkeys: a retrospective multicentre study. Equine Vet J 55:222–229. https://doi.org/10.1111/evj.13578

Oliveira MGCD, Paula VVD, Mouta AN, Lima IDO, Macêdo LBD, Nunes TL, Trindade PHE, Luna SPL (2021) Validation of the Donkey Pain Scale (DOPS) for assessing postoperative pain in donkeys. Front Vet Sci 8:671330. https://doi.org/10.3389/fvets.2021.671330

Oukessou M, Aissa M, Hmidouch A (1994) Pharmacokinetics study of benzylpenicillin and amoxicillin in the donkey (Equus asinus). Indian Vet J 71:1077–1081

Paraschou G, Xue C, Egan R, Bolfa P (2024) Multicentric lymphoma in a donkey with intestinal and bone marrow involvement. Vet Res 20:57. https://doi.org/10.1186/s12917-024-03903-5

Parkes RSV, Bukhari SSUH (2026) Biomechanics in donkeys and mules. In: McLean AK et al. (eds) Donkey and mule medicine: foundations, breeds, and management, vol 1. Springer, Cham

Peck KE, Matthews NS, Taylor TS, Mealey KL (2002) Pharmacokinetics of sulfamethoxazole and trimethoprim in donkeys, mules, and horses. Am J Vet Res 63:349–353. https://doi.org/10.2460/ajvr.2002.63.349

Rodrigues JB, Lilly G (2019) Dental disorders of donkeys. Vet Clin North Am Equine Pract 35: 529–544. https://doi.org/10.1016/j.cveq.2019.08.008

Rodrigues JB, Mora S, Bastos E, Viegas C, San Roman F (2013) Percutaneous approach for sialolith removal in a donkey. J Vet Dent 30:30–33. https://doi.org/10.1177/089875641303000104

Sekkin S, Gokbulut C, Kum C, Karademir U (2012) Plasma disposition of enrofloxacin following intravenous and intramuscular administration in donkeys. Vet Rec 171:447. https://doi.org/10.1136/vr.100653

Stegelmeier BL, Davis TZ (2018) Toxic causes of intestinal disease in horses. Vet Clin North Am Equine Pract 34:127–139. https://doi.org/10.1016/j.cveq.2017.11.008

Sullivan R (2024) Colitis in donkeys. UK-Vet Equine 8:134. https://doi.org/10.12968/ukve.2024.8.3.134

Sullivan R, Vogel P (2023) Therapeutics in veterinary care of the donkey. UK-Vet Equine 7:32. https://doi.org/10.12968/ukve.2023.7.1.32

Sykes BW, Hewetson M, Hepburn RJ, Luthersson N, Tamzali Y (2015) Equine gastric ulcer syndrome in adult horses. J Vet Intern Med 29:1288–1299. https://doi.org/10.1111/jvim.13578

The Donkey Sanctuary (2018) In: Evans L, Crane M (eds) The clinical companion of the donkey, 1st edn. Matador, Leicestershire

Thiemann A (2024) Gastrointestinal disorders in donkeys. UK-Vet Equine 8:66. https://doi.org/10.12968/ukve.2024.8.2.66

Thiemann AK, Sullivan RJE (2019) Gastrointestinal disorders of donkeys and mules. Vet Clin North Am Equine Pract 35:419–432. https://doi.org/10.1016/j.cveq.2019.08.001

Timko K (2024) Approach to inflammatory bowel disease. Vet Clin North Am Equine Pract 40: 287–306. https://doi.org/10.1016/j.cveq.2024.04.004

Trawford AF, Burden F, Hodgkinson JE (2005) Suspected moxidectin resistance in cyathostomin in two donkey herds at the donkey santuary, UK. In: 20th International Conference of the World Association for the Advancement of Veterinary Parasitology. New Zealand

Van Dierendonck MC, Burden FA, Rickards K, Van Loon J (2020) Monitoring acute pain in donkeys with the Equine Utrecht University Scale for Donkeys Composite Pain Assessment (EQUUS-DONKEY-COMPASS) and the Equine Utrecht University Scale for Donkey Facial Assessment of Pain (EQUUS-DONKEY-FAP). Animals 10:354. https://doi.org/10.3390/ani10020354

Wood F, Loomes K (2022) Suspected lidocaine toxicity and the use of metoclopramide and neostigmine to treat post-operative small intestinal ileus in a donkey. Vet Rec Case Rep 10: e481. https://doi.org/10.1002/vrc2.481

Xue C (2026) Physical examination of the donkey and their hybrid crosses. In: McLean AK et al. (eds) Donkey and mule medicine: foundations, breeds, and management, vol 1. Springer, Cham

Donkey and Mule Immunity: Genomic Basis and Functional Mechanisms

Francisco Javier Navas González

Abstract

Donkeys (*Equus asinus*) and their hybrids, particularly mules, represent globally significant equids whose immune systems remain underexplored compared to those of horses. This chapter synthesizes current knowledge on the genomic and functional foundations of donkey and mule immunity, integrating comparative, molecular, and evolutionary perspectives. Genomic studies reveal extensive immune gene expansion, signatures of positive selection, and regulatory roles for microRNAs, underscoring adaptive responses shaped by unique ecological and domestication histories. Diversity within the Major Histocompatibility Complex (MHC), along with species-specific polymorphisms in serum proteins and esterases, further highlights immunogenetic specialization. At the cellular and molecular levels, donkeys possess a complete complement of innate and adaptive immune mechanisms, encompassing granulocytes, monocytes, dendritic cells, interferons, natural killer cells, and lymphoid populations. Adaptive immunity features a structurally complex immunoglobulin locus, distinctive antibody repertoires, and functional T-cell subsets, including regulatory T cells and γδ T cells, with implications for disease resistance, vaccination, and maternal–neonatal immunity. Mules and hinnies, as genomic mosaics, present unique immunogenetic configurations that influence pathogen recognition and immune regulation. Finally, the chapter contextualizes donkey immune responses across the phases of activation, effector function, homeostasis, and memory, emphasizing the need for species-specific veterinary tools. Collectively, these insights position donkey and mule immunology as a critical frontier for advancing equid health, comparative immunology, and One Health frameworks.

F. J. Navas González (✉)
Department of Genetics, University of Córdoba, Córdoba, Spain

The Worldwide Donkey Breeds Project, University of Córdoba, Córdoba, Spain
e-mail: fjnavas@uco.es

A. K. McLean et al. (eds.), *Donkey and Mule Medicine: Clinical Medicine and Reproduction*, https://doi.org/10.1007/978-3-032-18540-2_5

Keywords

Regulatory T cells · Pattern recognition receptors · Immunoglobulins · IgG subclasses · Toll-like receptors · V(D)J recombination · Mucosal immunity · Antibody diversity

1 Introduction

Donkeys (*Equus asinus*) are globally important but historically understudied equids. With an estimated population exceeding 50 million animals worldwide, donkeys play critical roles in agricultural labor, transportation, dairy and meat production, and as companion animals—especially in developing regions (Wang et al. 2021). Despite their significance, donkeys have received considerably less scientific attention than horses, particularly in the field of immunology. This research disparity has led to a reliance on horse-based immunological data for donkey veterinary care, even though donkeys exhibit distinct physiological and immunological characteristics that necessitate species-specific understanding (Perzyna et al. 2025).

At the molecular level, donkeys possess the full suite of innate and adaptive immune machinery identified in humans and rodents, including pattern recognition receptors such as Toll-like receptors (TLRs), complement proteins, phagocytic cells, and lymphocyte populations (Qiu et al. 2025; Perzyna et al. 2025). Genomic and transcriptomic analyses confirm the presence of twelve conserved TLR genes in donkeys—TLR3, TLR4, TLR5, TLR9 among them—with approximately 99.7% sequence homology across the Equidae family (Qiu et al. 2025). These TLRs allow key innate immune cells—macrophages, dendritic cells, and neutrophils—to detect pathogen-associated molecular patterns (PAMPs) and initiate inflammatory responses fundamental to host defense.

Comparative genomics highlight evolutionary adaptations unique to the donkey immune system. Genes associated with inflammatory responses and innate immune regulation, such as ITIH4, show signatures of positive selection, suggesting functional optimization in response to species-specific environmental pressures (Perzyna et al. 2025). Additional selective pressures on metabolic genes underscore the interconnected nature of immunological and physiological evolution in donkeys (Perzyna et al. 2025).

The adaptive immune repertoire of donkeys follows the classic mammalian architecture. B and T cell repertoires are shaped by somatic recombination, enabling antigen specificity and immune memory. The donkey immunoglobulin heavy-chain (IGH) locus, mapped to chromosome 7, spans approximately 1.19 megabases and contains 117 variable (VH) segments (23 functional), 44 diversity (DH), and 8 joining (JH) segments (Qiu et al. 2025). While broadly similar to the equine IGH locus, subtle structural differences—such as shorter, less cysteine-rich CDR3H regions in donkeys—suggest functional diversity in antibody configuration and epitope recognition (Qiu et al. 2025).

Processes driving antibody diversification, including V(D)J recombination, N-nucleotide addition, and somatic hypermutation with A → G and G → A transitions, are conserved between donkeys and horses, supporting a robust and adaptable antibody repertoire (Qiu et al. 2025). Donkeys produce all five major immunoglobulin classes—IgM, IgD, IgG, IgA, and IgE—with glycoproteomic analyses revealing IgG subclasses 1, 2, 3, 5, 6, and 7. Notably, IgG4 is absent in donkeys, distinguishing their IgG profile from other mammals (Marihonnaiah and Belur Shivappa 2024). Each IgG subclass exhibits unique glycosylation patterns, influencing effector function, complement activation, and Fc receptor binding.

Recent comparative genetic and genomic studies between donkeys and horses have uncovered key adaptations in the donkey immune system that likely arose in response to distinct environmental pressures and domestication history. Donkeys exhibit expanded immune-related gene families, accelerated evolution of immune pathways, unique microRNA repertoires for post-transcriptional regulation, and remarkable diversity within their Major Histocompatibility Complex (MHC). Together, these features illustrate how non-model livestock species can evolve robust disease resistance mechanisms (Huang et al. 2015).

The present chapter will explore the genomic foundations, molecular mechanisms, and evolutionary drivers of donkey and mule immunity, highlighting the genetic architecture that shapes their species-specific immune responses.

2 Genetic and Genomic Insights into Donkey and Mule Immune Evolution

Recent comparative genetic and genomic studies between donkeys (*Equus asinus*) and horses have uncovered key adaptations in the donkey immune system that likely arose in response to distinct environmental pressures and domestication history. Donkeys exhibit expanded immune-related gene families, accelerated evolution of immune pathways, unique microRNA repertoires for post-transcriptional regulation, and remarkable diversity within their Major Histocompatibility Complex (MHC). Together, these features illustrate how non-model livestock species can evolve robust disease resistance mechanisms (Huang et al. 2015).

Mules (offspring of a mare and a jack donkey) and hinnies (offspring of a jenny donkey and a stallion) represent unique equid hybrids whose genomes integrate both horse and donkey genetic backgrounds. This hybrid genomic architecture inevitably extends to immune system encoding, where differences in parental contribution may influence immune gene expression, allelic dominance, and regulatory balance. The inheritance of immune-related loci such as the Major Histocompatibility Complex (MHC), Toll-like receptor (TLR) repertoires, and immunoglobulin gene clusters can create novel combinations of functional alleles, potentially altering antigen presentation, pathogen recognition, and antibody diversity relative to either parental species. Moreover, genomic imprinting and epigenetic interactions between horse- and donkey-derived immune genes may affect the magnitude of inflammatory responses and vaccine reactivity in these hybrids. The asymmetry between mules and hinnies, particularly due to maternal cytoplasmic inheritance, gonosome linked and or related traits and mitochondrial DNA (maternal/paternal horse versus donkey origin), further suggests that immune metabolism and cellular energetics may differ between the two hybrids, contributing to subtle variations in their immunological performance and disease resistance profiles.

2.1 Inbreeding Thresholds and Immunological Consequences in Donkeys and Mules

In donkeys, as in other domestic equids, rising inbreeding coefficients (F) are linked to measurable declines in immune function and overall fitness. Pedigree analysis of the critically endangered Catalonian donkey (Equus asinus asinus) demonstrated that when individual inbreeding exceeded $F = 0.0625$—the level equivalent to first-cousin matings—there was a significant drop in foal survival and colostral IgG quality (Gutiérrez et al. 2005). Below this threshold, colostral IgG concentrations averaged over 20 mg/mL, but once F surpassed 0.0625, mean IgG fell by nearly 30%, correlating with a threefold increase in early-life mortality (Gutiérrez et al. 2005).

Similar conclusions arise from a recent genomic survey of Hetian Qing donkeys (former Gola donkey) in China, where the average inbreeding coefficient was low ($F \approx 0.003$) and heterozygosity remained high. However, individuals whose F approached 0.05 displayed an uptick in long runs of homozygosity (ROH) >5 Mb, a genomic signature linked in other species to compromised immune gene diversity (Liu et al. 2024). Conservation geneticists generally recommend maintaining F below 0.05 in small livestock populations to avoid inbreeding depression in immunity, reproduction, and growth.

Altogether, empirical and theoretical work converge on critical thresholds: • $F \approx 0.05$ marks the onset of detectable losses in heterozygosity and immune gene diversity (Liu et al. 2024). • $F \approx 0.0625$ signals a sharp decline in colostral antibody transfer and neonatal survivorship in donkeys (Gutiérrez et al. 2005).

These data underline the need for breeders to monitor pedigree-based inbreeding coefficients closely and to aim for $F < 0.05$ in mating schemes, thereby preserving colostral quality and foal viability.

In contrast, the mule and the hinny, by their very hybrid nature, are not the result of progressive inbreeding but rather a cross between genetically divergent species, which prevents further hybrid propagation due to sterility (although they carry the genomic consequences derived from inbreeding that their parent already presented, and which can hinder their genetic and as a result overall health). This reproductive barrier acts as a natural safeguard against inbreeding depression in hybrids but also underscores the fundamental difference between donkeys and mules and hinnies in terms of genetic management: whereas donkeys require systematic monitoring to avoid immunological erosion through inbreeding, mules and hinnies are evolutionary "dead ends," whose immune system is shaped by the unique, one-generation combination of horse and donkey genomes.

2.2 Immune Gene Expansion and Rapid Evolution

Genome-wide analyses identified 283 gene families significantly expanded in donkeys relative to horses, notably including those involved in protein digestion, olfactory transduction, and immune regulation (Huang et al. 2015). In parallel, 1292 genes in the donkey lineage display signatures of accelerated evolution compared to their equine orthologs. Many of these rapidly evolving genes participate in lymphocyte differentiation (GO:0045619), telomere maintenance, forebrain development, and aerobic respiration. The expansion and rapid diversification of immune-regulatory gene families may underlie donkeys' empirically observed resilience to infectious diseases and their strong vaccine-induced antibody responses (Huang et al. 2015).

2.3 microRNAs and Epigenetic Regulation

Beyond protein-coding genes, the donkey genome harbors over 1190 annotated microRNAs, several of which are uniquely expanded compared to horses. These miRNAs are predicted to fine-tune immune cell development, cytokine secretion, and inflammatory cascades through post-transcriptional silencing of key transcripts (Huang et al. 2015). Although functional validation remains forthcoming, this epigenetic layer likely contributes to the precise calibration of donkey immune responses, preventing both under-reaction to pathogens and harmful hyperinflammation.

2.4 Major Histocompatibility Complex Diversity and Function

The MHC locus is pivotal for antigen presentation and adaptive immunity. Donkeys display distinct MHC characteristics—both in class I and class II regions—reflecting their evolutionary trajectory and pathogen encounters.

2.4.1 MHC Class I Polymorphism and Conservation

Pioneering work by Campbell (2007) employed RT-PCR on peripheral lymphocyte cDNA to characterize 35 unique donkey MHC class I sequences, including 12 classical and 7 non-classical alleles. Classical alleles featured highly polymorphic peptide-binding domains with shorter transmembrane segments (~105 bp), whereas non-classical alleles showed longer transmembrane regions (~120 bp) and limited variability (Birch et al. 2006). Phylogenetic analyses revealed that some donkey alleles are identical to horse MHC variants—such as equine allele 7–6 at locus 19—implying trans-species polymorphism preserved by balancing selection mechanisms (Adams and Parham 2001; Hughes and Nei 1988). Furthermore, haplotypic gene-content variation within donkey populations parallels observations in horses, cattle, and macaques, underscoring the dynamic complexity of MHC class I gene organization (Otting et al. 2005).

2.4.2 MHC Class I Alloantigen Recognition and Interspecies Compatibility

Serological and functional assays have confirmed extensive immunogenetic compatibility between donkeys and horses. Baker et al. (2001) identified five donkey leukocyte antigens (DoLA) using antisera from both intraspecies and interspecies pregnancies, demonstrating that these polymorphic class I molecules elicit cytotoxic T-lymphocyte responses across species barriers. Full-length cDNA sequencing showed conservation of peptide-binding grooves and T-cell receptor contact residues, explaining the maternal–fetal tolerance observed in mule and hinny gestations and validating donkeys as valuable comparative models in equid immunogenetics (Baker et al. 2001).

2.4.3 MHC Class II Diversity: DRA, DRB, and DQB Loci

Class II MHC genes present extracellular antigens to $CD4^+$ helper T cells and exhibit notable diversity in donkeys. Contrary to most mammals, the DRA locus—typically conserved—displays high polymorphism in exon 2 (the peptide-binding domain) across donkey populations from Italy, Jordan, and Kenya, with the greatest allelic richness in Kenyan donkeys (Vranova et al. 2011). DRB and DQB loci also show extensive allelic variation: three DRB and three DQB genes were characterized, with DRB1 and DQB1 being most polymorphic. Many alleles appear species- or population-specific, while others demonstrate ancient trans-species polymorphisms maintained by positive selection at antigen-binding sites (Klumplerova et al. 2020). Intriguingly, DQB2 was absent in some wild ass lineages, suggesting lineage-specific gene loss or divergence, yet the overall class II MHC architecture remains conserved across Equidae.

2.5 Serum Esterases

Kaminski and Gajos (1964) employed histochemical and immunoelectrophoretic techniques to compare donkey and equine serum esterases. She identified a non-lipoprotein "α_1-esterase" present in horse serum but entirely absent in donkey serum. Conversely, the lipoprotein-associated α-ω (E_4) esterase was conserved across both species, forming identical precipitin lines in immunodiffusion assays. This species-specific absence of α_1-esterase in *Equus asinus* underscores its evolutionary divergence and provides a reliable biochemical marker for donkey-equid phylogenetics (Kaminski and Gajos 1964; Kaminski 1969).

2.6 Acidic Prealbumin Polymorphisms

Braend and Romagnoli (1980) characterized three acidic prealbumin (Pr) phenotypes—Pr M, Pr T, and Pr MT—in Italian donkey populations via starch-gel electrophoresis. The system follows simple Mendelian codominance with allele frequencies of 0.87 (PrM) and 0.13 (PrT), matching Hardy–Weinberg expectations. These prealbumins, homologous to α_1-antitrypsin, likely modulate protease activity during inflammation, suggesting roles in acute-phase responses analogous to those in human immunology (Braend and Romagnoli 1980; Ek 1979).

3 Components of Donkey Immune System

Donkeys (*Equus asinus*) possess a fully developed immune system that mirrors that of other mammals in both innate and adaptive branches. Their innate immunity features classical cellular components—neutrophils, eosinophils, basophils, mast cells, monocytes, dendritic cells, and type I interferons—as well as a repertoire of pattern-recognition receptors (such as TLRs), complement proteins, and chemokine and cytokine networks that orchestrate early defense responses. This rapid response machinery not only wards off pathogens but also shapes downstream adaptive responses. On the adaptive side, donkeys exhibit the same fundamental mechanisms of humoral and cellular immunity found across mammals. A complex immunoglobulin (Ig) gene locus gives rise to diverse Ig isotypes (including IgM, IgD, IgG, IgA, IgE), and IgG subclasses with specific glycosylation patterns. B-cell–mediated antibody responses and T-cell–mediated responses—including helper, cytotoxic, and regulatory T cells—reflect a robust capability for immunological memory and pathogen clearance. This chapter revisits the major cellular and molecular pillars of donkey immunity, highlighting both established parallels with equine models and emerging donkey-specific specializations that merit deeper investigation.

3.1 Innate Immunity

3.1.1 Neutrophils, Eosinophils, Basophils, and Mast Cells

Donkeys possess the same core granulocytic immune cells seen in other mammals—neutrophils, eosinophils, basophils—as well as tissue-resident mast cells. These cells are integral to pathogen defense, allergic responses, and tissue regulation. Neutrophils act as first-line phagocytes, deploying respiratory bursts, degranulation, and NETosis to eliminate bacteria and debris. For instance, extensive neutrophil recruitment occurs in jenny uteri following artificial insemination, where seminal plasma triggers polymorphonuclear neutrophil (PMN) influx and NET formation, supporting sperm selection and endometrial clearance (Radar-Chafirovitch et al. 2025). In systemic tick-borne infections (e.g., *Theileria, Ehrlichia*), donkeys show marked neutropenia with concurrent anemia and thrombocytopenia, highlighting intense acute-phase inflammatory responses. In healthy lungs, bronchoalveolar lavage fluid features about 3% neutrophils (0–22%), a percentage comparable to or lower than that in horses. Eosinophils contribute heavily to anti-helminth defenses. Donkeys infected with *Parascaris equorum* or *Strongylus* species exhibit pronounced eosinophilia, associated with reactive oxygen and degranulation activity. Rare conditions such as multisystemic eosinophilic epitheliotropic disease (MEED) in donkeys are marked by widespread eosinophilic organ infiltration

(Paraschou et al. 2024). In respiratory samples, eosinophil proportions average 2% but can range dramatically (up to 60%), indicating subclinical lung infestation or airway inflammation. Basophils are key mediators of Th2-type inflammation and allergic responses, releasing histamine, heparin, and eosinophil-attracting factors. While direct donkey-specific research is sparse, basophil levels rise significantly during parasitic and tick-borne infections, suggesting their active role in systemic eosinophilic and allergic responses. Mast cells, found in connective tissue, skin, and mucous membranes, release histamine, heparin, and proteases. They play central roles in allergic inflammation, neovascularization, wound repair, and pathogen defense. Although detailed studies in donkeys are limited, mast cell function is presumed analogous to that in horses and other mammals. However, standard cytology methods (e.g., BALF staining) often underreport mast cell presence, calling for refined techniques to fully assess their distribution and role.

Reference ranges for neutrophils, eosinophils, basophils, and mast cells in donkeys can vary depending on multiple factors such as age, breed, and health status, but general values provide a useful framework for interpretation. Neutrophil counts typically range from 2.4 to 6.3 × 10^9/L, comprising approximately 29.4–69.3% of total white blood cells. Eosinophils are usually found in lower quantities, with absolute counts between 0.3 and 2.8 × 10^9/L and percentages ranging from 2.7% to 19.8%. Basophils are even less frequent, accounting for 0.2–0.6% of leukocytes; however, they show a high coefficient of variation, indicating notable inter-individual variability. Mast cells are not routinely included in standard hematological panels, as noted by the National Institutes of Health (NIH), but they play an important immunological role during parasitic infestations and inflammatory responses. For instance, during tick infestations, mast cells infiltrate affected tissues and degranulate, releasing mediators that contribute to local immune defense.

These values, however, are not static and may be influenced by several physiological and environmental factors. Age plays a particularly significant role: in Martina Franca donkeys, eosinophil counts were observed to increase with age, likely reflecting cumulative antigenic exposure and parasitic challenge, while basophil and monocyte counts tended to decrease in older animals. Similarly, breed differences have been documented, with Ragusana donkeys showing higher lymphocyte and neutrophil counts than crossbred donkeys (Caldin et al. 2005). Physiological states such as pregnancy, lactation, or recent illness can also alter leukocyte distribution. Moreover, variations in diet, housing, management, and even sampling time due to circadian rhythm may subtly affect these parameters. Consequently, while general reference ranges serve as valuable benchmarks, accurate interpretation of leukocyte data in donkeys must always consider individual factors such as age, breed, physiological status, and clinical context (Fig. 1).

3.1.2 Monocytes, Dendritic Cells, and Type I Interferons

In donkeys, as in other equids, monocytes are a critical component of innate immunity and are readily identifiable within the peripheral blood mononuclear cell (PBMC) fraction by the expression of CD14. In horses, $CD14^+$ monocytes are well

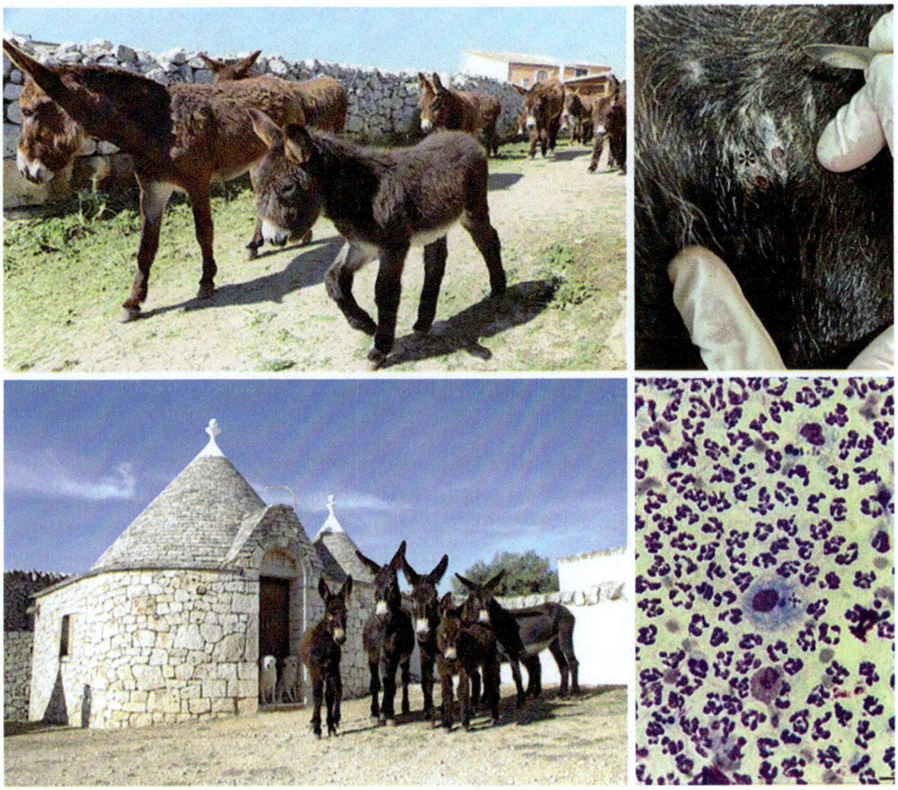

Fig. 1 Ragusana (upper left) and Martina Franca (lower left) Donkey Breeds; Cytological sampling and processing (Upper right) Donkey presenting two alopecic lesions with epithelial discontinuity and Apposition cytology made on the skin lesion (*), with an intense mixed inflammatory infiltrate composed predominantly by neutrophils, activated macrophages phagocytosing neutrophils and apoptotic bodies, and eosinophils. Absence of amastigote forms of *Leishmania* spp. (Diff Quik stain, 100X) (lower right). (Carbonara et al. 2024)

characterized, representing the classical monocyte subset and comprising the majority of PBMC monocytes (Perzyna et al. 2025). While similar detailed subtype classifications (e.g., classical, intermediate, non-classical) are not yet fully defined in donkeys, CD14 expression in donkey monocytes suggests a preserved pattern of lineage differentiation and function.

Dendritic cells (DCs), essential for bridging innate and adaptive immune responses, have not been as extensively studied in donkeys. However, insights from equine research indicate that monocyte-derived dendritic cells (mo-DCs) can be generated in vitro by stimulating $CD14^+$ PBMCs with granulocyte–macrophage colony-stimulating factor (GM-CSF) and interleukin-4 (IL-4). In horses, these mo-DCs adopt an MHC-II^high $CD86^+$ $CD83^+$ CD14^low phenotype (Ziegler-Heitbrock 2007). Although comparable studies have yet to be reported in donkeys,

Fig. 2 Pêga (right) and Kyrgyz (left) donkey breeds

it is likely that donkey mo-DCs exhibit phenotypic and functional profiles like those of horses, including the potential to mature and activate antigen-specific T cells.

Beyond mo-DCs, the presence of plasmacytoid DCs (pDCs) in horses is inferred through functional studies showing robust type I interferon (IFN-α) responses following in vitro viral stimulation. Equine PBMCs secrete high levels of IFN-α when exposed to equine influenza virus and equine herpesvirus-1, implicating pDC-like cells as the source. Donkey immunity likely follows suit, with conserved innate viral sensing pathways activating pDCs to secrete IFN-α and other type I IFNs.

The equine genome encodes multiple type I interferons—including at least six IFN-α, four IFN-β, eight IFN-τ, three IFN-δ, one IFN-κ, and one IFN-ε—all of which have functional genes and are expressed across various cell types in response to viral infection (Cocco et al. 2023). Donkeys, sharing high genomic conservation with horses, are presumed to express a similar breadth of type I IFNs, enabling them to mount strong antiviral defenses.

Breed differences may also affect monocyte counts, as noted in comparative studies across Ragusana, Pêga, and Kyrgyz donkeys (Fig. 2) (Caldin et al. 2005; Girardi et al. 2017; Kısadere et al. 2019). In Martina Franca donkeys, monocyte counts range from 0.3 to 1.1×10^9/L (cells/μL), representing 2.6–8.5% of circulating leukocytes. Younger donkeys (1–3 years) tend to exhibit slightly higher monocyte levels compared to adults, potentially reflecting greater innate immune activation during developmental stages. In contrast, dendritic cells (DCs)—although functionally critical for antigen presentation and initiation of adaptive responses—are not routinely quantified in clinical settings. Their presence in peripheral blood is minimal, typically comprising less than 1% of mononuclear cells, and identification requires flow cytometric analysis using markers such as MHC-II, CD11c, and CD123. Due to these technical limitations, no formal reference intervals for DCs have been established in donkeys. Similarly, type I interferons (IFN-α and IFN-β), which are key antiviral cytokines predominantly produced by plasmacytoid dendritic cells, are generally undetectable in the circulation of healthy animals. Their expression is transient and stimulus-dependent, becoming measurable only in the context of viral infections or experimental immunostimulation. In donkeys, data on circulating type I interferons are lacking, and extrapolations are typically made from equine models using equine-specific ELISA kits or PCR-based assays targeting interferon-stimulated genes. These findings underscore the importance of contextual interpretation and highlight the need for further immunological research in donkeys, particularly for undercharacterized cell types and cytokine responses.

3.1.3 Innate Lymphoid Cells, Other Innate Immune Cells and Pattern Recognition Receptors

Donkeys host a diverse repertoire of innate lymphoid and pattern recognition elements that closely parallel those found in horses and other mammals. Natural killer (NK) cells are key innate lymphocytes involved in immediate defense against virus-infected and tumorigenic cells. These cells do not require prior sensitization and play an essential role in the surveillance of infected or abnormal cells. In horses, NK cells have been well characterized by a $CD3^-$, $CD16^+$, $NKp46^+$ phenotype (Kydd et al. 1982), and given the close genetic relationship between donkeys and horses, it is highly plausible that donkeys harbor analogous NK cell populations, although direct functional studies in Equus asinus remain limited. Cross-reactivity of anti-equine and anti-bovine monoclonal antibodies has demonstrated that donkey NK cells are large granular lymphocytes identifiable by the absence of CD3 and the presence of activating receptors such as CD8α, CD16 (FcγRIII), NKp46, and NKG2D (Dissen et al. 2008). These cells typically comprise 2–10% of circulating lymphocytes and reside in blood, spleen, lymph nodes, and mucosal tissues, homing rapidly to sites of inflammation via chemokine receptors such as CXCR3 and CCR5 (Birch et al. 2006). Upon activation, donkey NK cells employ conserved effector mechanisms—including perforin/granzyme-mediated cytotoxicity, Fas ligand–induced apoptosis, and secretion of interferon-γ and tumor necrosis factor-α—to orchestrate and amplify downstream adaptive immune responses (Wu et al. 2017). Recent evidence indicates that NK cells express multiple Toll-like receptors (e.g., TLR1, TLR2, TLR3, TLR5–9), enabling direct activation by microbial ligands such as lipopolysaccharide and flagellin and further enhancing their cytotoxic and cytokine responses (Noh et al. 2020). Although donkey-specific profiling of NK-cell TLR expression is not yet complete, the high degree of conservation suggests similar innate sensing capacities in donkeys. Table 1 summarizes the expression profiles of key immunological markers observed in donkey natural killer (NK)-like cells, highlighting conserved features and species-specific variations in surface receptor patterns.

Natural killer T (NKT) cells are a subset of T lymphocytes that co-express NK surface receptors and an αβ T-cell receptor (TCR) to recognize lipid antigens presented by the non-classical MHC molecule CD1d (Kumar et al. 2017). CD1D gene expression has been identified in donkeys, implying that donkey NKT cells may perform immune surveillance and rapid cytokine production analogous to equine iNKT responses to α-galactosylceramide. In horses, invariant NKT (iNKT) cells are detected using CD1d tetramers loaded with α-GalCer and express markers such as CD3, CD56, and CD161, rapidly releasing interleukin-4 and interferon-γ

Table 1 Marker expression in donkey NK-like cells

Marker	Expression in Donkey NK-like Cells	Notes
CD3	–	Absence of pan-T cell marker
CD8α	+	Often used to enrich cytotoxic lymphocytes
CD16 (FcγRIII)	+	Mediates antibody-dependent cellular cytotoxicity (ADCC)
CD2, CD5	+/–	Species-dependent; CD2 common on equine NK subsets
NKp46	+ (likely)	Conserved activating receptor across mammals
NKG2D	+ (likely)	Recognizes stress-induced ligands on target cells

Table 2 Marker expression in donkey NKT-like cells

Marker	Expression in Donkey NKT-like Cells	Notes
CD3	+	Pan-T cell marker
CD56	+ (presumed)	NK-associated marker; cross-reactive with human clone
CD161 (NKR-P1A)	+ (likely)	Common to T/NK hybrids
TCR Vα24-Jα18	?	Invariant chain in humans; equine ortholog unconfirmed
CD1d tetramer	+ (experimental)	Loaded with α-GalCer analog to identify iNKT

upon activation (Lebedinskaya et al. 2006). Preliminary application of equine CD1d–α-GalCer tetramers to donkey peripheral blood mononuclear cells suggests the presence of analogous NKT populations in *Equus asinus* (Kumar et al. 2017). Table 2 outlines the phenotypic signature of donkey NKT-like cells, detailing surface marker expression that reflects both T cell lineage and innate-like features, with noted experimental approaches and species-specific uncertainties.

γδ T cells, characterized by their γδ TCR, reside preferentially in mucosal sites—skin, gut, and lungs—where they participate in both defense against infections and regulation of tissue homeostasis. Well-documented in equine blood and mucosal tissues, γδ T cells have also been detected in donkey PBMCs by flow cytometry, supporting their role in systemic and mucosal immunity in donkeys.

A crucial arm of innate recognition involves pattern recognition receptors (PRRs), which detect pathogen-associated molecular patterns (PAMPs) and damage-associated molecular patterns (DAMPs). Donkeys express orthologs of Toll-like receptors (TLRs 1–10), NOD-like receptors (NOD1, NOD2), RIG-I-like receptors (DDX58), and C-type lectin receptors (e.g., Dectin-1 [CLEC7A], DC-SIGN [CD209])—enabling detection of bacterial lipopolysaccharide (via TLR4), viral RNA (via TLR3, TLR7/8), flagellin (via TLR5), and fungal carbohydrates. In horses, TLRs 2–10 have been documented across PBMCs, lung, skin, and reproductive tissues (Battista et al. 2014), and comparable profiles are presumed in donkeys given their high genomic similarity.

Together, NK, NKT, and γδ T lymphocytes form an "innate–adaptive interface" capable of immediate effector functions—cytotoxicity and rapid cytokine production (e.g., IFN-γ, IL-17)—while shaping downstream adaptive responses. PRR engagement, particularly via TLRs, triggers inflammatory signaling cascades such as NF-κB, MAPK, and interferon regulatory factors (IRFs) to coordinate these cellular responses (Noh et al. 2020). Despite these conserved parallels, functional studies in donkeys remain sparse. Species-specific functional data on donkey NK, NKT, and γδ T cells; the tissue distribution of PRRs; and downstream signaling pathways—especially cytokine output and transcriptional activation—are needed. Investment in donkey-specific immunological tools—monoclonal antibodies, validated primers, and standardized functional assays—will be essential to fully elucidate the mechanisms of donkey innate immunity and to inform improved diagnostic, preventive, and therapeutic strategies.

3.2 Adaptative Immunity

Adaptive immunity in donkeys, as in other mammals, is a highly specialized and dynamic defense system that enables long-lasting protection against specific pathogens. This branch of the immune system is characterized by its ability to recognize, remember, and respond more effectively to antigens upon repeated exposure. Despite their close genetic relationship to horses, donkeys possess unique physiological and immunological traits that warrant independent study. In recent years, increasing attention has been directed toward understanding the components and functionality of the donkey adaptive immune response, including the development and diversity of T and B lymphocytes, immunoglobulin gene expression, and antibody-mediated protection. These insights are critical not only for improving disease prevention and treatment in donkeys but also for advancing comparative immunology among equids.

3.2.1 Adaptive Cellular Immunity

Donkeys (*Equus asinus*) possess a complete and functional adaptive T-cell compartment analogous to that of horses and other mammals (Perzyna et al. 2025). Tetanus and influenza vaccinations in donkeys have elicited robust expansions of both $CD4^+$ helper (Th) and $CD8^+$ cytotoxic T lymphocytes, as well as transient increases in $CD4^+FoxP3^+$ regulatory T cells (Tregs), confirming the presence of canonical T-cell subsets including Th, CTL, and Treg populations (Hamza et al. 2011; Perzyna et al. 2025). These findings suggest phenotypic and functional conservation of adaptive cellular immunity between donkeys, horses, and other species.

3.2.1.1 $CD4^+$ Helper and $CD8^+$ Cytotoxic T Cells

Donkey $CD4^+$ helper T cells are presumed to differentiate into Th1, Th2, and Th17 subsets similarly to equines. Th1 cells produce IFN-γ to combat intracellular pathogens; Th2 cells secrete IL-4, IL-5, and IL-13 to support humoral immunity; Th17 cells release IL-17 to drive inflammation and contribute to pathogen clearance. $CD8^+$ cytotoxic T cells (CTLs) are thought to eliminate infected cells through perforin-mediated cytolysis and additional IFN-γ production. Though donkey-specific data remain limited, vaccination trials have demonstrated strong Th1/Th17 and regulatory responses, indicating that donkey T cells are functionally competent (Perzyna et al. 2025). By analogy to horses, donkey Th1 cells likely express IFN-γ, Th2 cells produce IL-4, IL-5, and IL-13, and Th17 cells secrete IL-17; donkey CTLs exhibit strong proliferative and cytokine responses following stimulation, yet gaps remain in understanding lineage-specific markers and subset dynamics in this species (Perzyna et al. 2025).

3.2.1.2 Regulatory T Cells (Tregs)

Regulatory T cells are a specialized subset of $CD4^+$ lymphocytes that maintain immune homeostasis and self-tolerance (Hamza et al. 2011). Classical markers include high expression of CD25 (IL-2 receptor α chain) and the transcription factor FoxP3, essential for lineage specification (Garg et al. 2014; Hamza et al. 2011). Natural Tregs (nTregs) develop in the thymus as $CD4^+CD25^{high}FoxP3^+$ cells, while peripherally induced Tregs (iTregs) differentiate from $CD4^+CD25^-$ precursors under the influence of IL-2 and TGF-β1 (Hamza et al. 2011). Tregs exert suppressive functions via high CD25 expression that limits IL-2 availability, constitutive CTLA-4 expression, and secretion of IL-10 and TGF-β1 (Hamza et al. 2011). In donkeys, $CD4^+FoxP3^+$ Treg expansion from ~6.8% to ~11.7% of $CD4^+$ T cells 1 month after tetanus–influenza immunization supports the presence of functional, detectable regulatory populations (Perzyna et al. 2025). Although specific surface markers for donkey Tregs are not fully characterized, parallels with horses suggest comparable regulatory mechanisms, reinforced by increased IL-10 production in donkey B cells and monocytes following immunization (Perzyna et al. 2025).

3.2.1.3 Age-Dependent T-Cell Responses to EHV-1

Equine herpesvirus-1 (EHV-1) infects donkeys, as confirmed by serological surveys, but direct studies of their T-cell responses are lacking. In horses, neonatal foals exhibit minimal IFN-γ in response to EHV-1, with early cytokine production attributed to $CD8^+$ CTLs, whereas adults develop strong $CD4^+$Th1-biased memory (Perzyna et al. 2025). Assuming similar immunodynamics, donkey foals likely rely on CTL-derived IFN-γ early in life, while adults transition to Th1-dominant $CD4^+$ responses—a hypothesis that remains to be empirically validated in donkeys (Perzyna et al. 2025).

3.2.1.4 Maternal T-Lymphocyte Transfer via Colostrum

Jennies, like mares, transfer maternal leukocytes, including $CD4^+$ and $CD8^+$ T cells, via colostrum. Although specific characterization in donkeys is minimal, equine studies reveal that colostral T cells are enriched for Th1/Th17 phenotypes—producing IFN-γ and IL-17 upon stimulation—thereby seeding the neonatal gut with pre-activated lymphocytes that enhance early mucosal defenses (Perzyna et al.

2025). The functional role and kinetics of these maternal cells in neonatal donkeys remain to be studied.

3.2.1.5 Structural Complexity and Evolution of the T Cell Receptor Gamma Locus

The donkey T-cell receptor gamma (TRG) locus, crucial for γδ T-cell function, spans approximately 860 kb and contains 34 variable (TRGV), 21 joining (TRGJ), and 14 constant (TRGC) genes arranged into 14 tandem V–J–C cassettes (Massari et al. 2023). This modular configuration permits both intra-cassette and frequent trans-cassette V–J recombination, substantially diversifying the complementarity-determining region 3 (CDR3) and enhancing antigen-specific recognition. Despite close taxonomic proximity, the donkey TRG locus surpasses that of the horse in size and segment number, reflecting rapid evolutionary adaptation within Equus and underscoring the immunogenetic versatility of donkeys (Massari et al. 2023).

3.2.2 Adaptive Systemic–Humoral Immunity in Donkeys (Antibodies)

The adaptive humoral response in donkeys relies on a diverse array of immunoglobulins (Igs) generated through somatic recombination, hypermutation, and class switching. Although donkeys share many immunological features with horses, species-specific studies have revealed distinctive aspects of their antibody repertoire, innate humoral factors, and clinical applications.

3.2.2.1 Historical Foundations

3.2.2.1.1 Early Serum Protein Electrophoresis

Pioneering work in the 1960s and 1970s applied starch-gel and polyacrylamide electrophoresis to characterize equid serum proteins, including albumins, globulins, and transferrins (Braend 1965, 1967, 1970; Gahne 1966). These studies in horses provided the methodological framework later adopted for donkeys (Braend and Romagnoli 1980), enabling the detection of species-specific immune markers and informing modern immunoassays.

3.2.2.1.2 Origins of Equid Serum Therapy

The concept of heterologous antiserum began with Behring and Kitasato's tetanus work in 1890 and extended to donkey serum against tuberculosis in 1894 (Glatman-Freedman and Casadevall 1998) and cobra venom in 1895 (Pucca et al. 2019). Serum sickness—an immune-complex (type III) hypersensitivity—led to deeper insights into antibody structure and function (Allen and Dalton 1975).

3.2.2.2 Innate Humoral Factors

3.2.2.2.1 Lysozyme

Donkey serum and secretions contain lysozyme, an enzyme hydrolyzing bacterial peptidoglycans. Average serum concentrations (1.86 ± 0.12 mg/L) lie between those of horses and mules and correlate with management conditions; older donkeys

(>20 years) show significant declines, suggesting age-related immunosenescence (Sotirov 2004).

3.2.2.2.2 Complement Activity

The alternative pathway of complement activation (APCA) operates independently of antibodies. Donkeys exhibit mean APCA values of 62.42 ± 2.32 CH_{50}, intermediate between horses (68.92 ± 2.32 CH_{50}) and mules (59.11 ± 1.49 CH_{50}). Both lysozyme and APCA are modulated by environmental and physiological factors (Sotirov 2004).

3.2.2.3 Immunoglobulin Classes and Subclasses

3.2.2.3.1 Ig Isotypes in Donkeys

Donkeys express the full spectrum of mammalian Igs—IgM, IgD, IgG, IgE, and IgA—alongside κ and λ light chains (Allen and Dalton 1975). Three major IgG subclasses (IgGa, IgGb, IgGc) cross-react with equine reagents, and at least six donkey IgG subtypes (analogous to equine IgG1, IgG2, IgG3, IgG5, IgG6, IgG7) have been confirmed proteomically (Marihonnaiah and Belur Shivappa 2024). Donkey IgM forms pentamers for complement activation, and secretory IgA in colostrum and milk provides mucosal immunity (Gnanesh Kumar and Rawal 2020). Table 3 below summarizes each heavy-chain isotype in donkey, indicating gene name, isotype, subclass (for IgG), confirmation status, known structural features, and key literature.

3.2.2.3.2 Genomic Architecture of Ig Loci

By synteny with the horse, the donkey IGH locus on chromosome 7 spans ~1.19 Mb, containing 117 VH (23 functional), 44 DH, and 8 JH segments (Qiu et al. 2025). The horse IGH on chromosome 24 harbors 104 VH (21 functional; 74 pseudogenes), 44 DH (16 functional), and 9 JH (6 functional) segments. Light-chain loci (IGK on chr 12 and IGL on chr 17) are inferred to mirror equine counts of ~60 Vκ/5 Jκ/1 Cκ and 144 Vλ/7 Jλ–Cλ clusters, respectively (Sun et al. 2010). The equine immunoglobulin repertoire is dominated by lambda light chains (≈92% of transcripts), contrasting with other mammals such as mice (≈95% κ) and humans (≈34% λ). Table 4 presents a comparison of the immunoglobulin gene segment counts between the donkey (*Equus asinus*) and the horse (*Equus caballus*), highlighting similarities and differences in their immune system gene compositions.

3.2.2.4 Antibody Repertoire Generation

3.2.2.4.1 V(D)J Recombination and Junctional Diversity

Donkeys and horses share preferential usage of IGHV4–IGHD2–IGHJ6 combinations, yet donkeys incorporate shorter nontemplated (N) nucleotide additions. This results in briefer CDR3H loops with fewer cysteines compared to the horse's extensive N-additions and cysteine-rich "knob" structures (Qiu et al. 2025).

Table 3 Heavy-chain isotypes in donkey: gene name, isotype, subclass, confirmation status, structural features, and references

Gene(s)	Isotype	Status	Subclass	Structural Features	References
IGHM	IgM	Inferred from horse	–	Four Cμ domains; secreted pentamer and membrane forms via alternative splicing; multiple N-linked glycans	Wagner et al. (2004, 2006)
IGHD	IgD	Inferred from horse	–	Three Cδ domains; extended hinge	Wagner et al. (2004, 2006)
IGHG1–IGHG7	IgG	Partially confirmed (IgG1,2,3,5,6,7)	IgG1–IgG7[a]	Three Cγ domains + hinge; CH2 N-glycosylation; IgG5 extra CH1 glycan; subclass-specific glycan motifs	Allen and Dalton (1975) and Marihonnaiah and Belur Shivappa (2024)
IGHE	IgE	Inferred from horse	–	Four Cε domains + hinge; single Cε4 glycosylation	Wagner (2006)
IGHA	IgA	Confirmed (serum & milk)	–	Three Cα domains + hinge; J-chain and secretory component; two N-glycosylation sites	Gnanesh Kumar and Rawal (2020)

[a]IgG4 has not been detected in donkey proteomes and may be rare or absent

Table 4 Comparative immunoglobulin gene segment counts in donkey (*Equus asinus*) and horse (*Equus caballus*)

Locus	Segment Type	Donkey (*Equus asinus*)	Horse (*Equus caballus*)
IGH	VH	117 (23 functional)	104 (21 functional; 74 pseudogenes; 9 ORFs)
	DH	44	44 (16 functional)
	JH	8	9 (6 functional)
	CH	Inferred (M, D, G_1–G_7, E, A)	11 (IgM, IgD, seven IgG, IgE, IgA)
IGK	Vκ	NR	60
	Jκ	NR	5
	Cκ	NR	1
IGL	Vλ	NR	144
	Jλ	NR	7
	Cλ	NR	7

NR not reported for donkey (extrapolated from equine data) (Walther et al. 2015)

3.2.2.4.2 Somatic Hypermutation

Both species exhibit A↔G transition biases during somatic hypermutation, refining antigen specificity post-activation. Mutation hotspots and frequencies differ subtly, reflecting breed and environmental influences (Qiu et al. 2025).

3.2.2.5 Functional Roles of Antibody Isotypes

3.2.2.5.1 IgM and IgG

Pentameric IgM initiates classical complement activation. Donkey IgG subclasses effect neutralization, opsonization, complement fixation, and Fcγ receptor–mediated clearance. Equine studies suggest IgG1, IgG4, and IgG7 dominate antiviral and intracellular pathogen responses, while IgG3 and IgG5 associate with extracellular proteins and Th2-type immunity (Perzyna et al. 2025; Marihonnaiah and Belur Shivappa 2024).

3.2.2.5.2 IgA and IgE

Secretory IgA enforces mucosal immune exclusion and shapes microbiota; in donkeys, fecal IgA titers correlate inversely with parasite egg counts (Tombak et al. 2020). IgE binds FcεRI on mast cells/basophils to mediate anti-parasite defenses and allergic responses, such as insect-bite hypersensitivity (Barbosa et al. 2024).

3.2.2.6 Clinical and Comparative Context

3.2.2.6.1 Neonatal Passive Transfer

Donkey foals ingest colostral IgG (~6.8–75 mg/mL) and IgA (0.5–2.4 mg/mL) to bridge immunologic naivety; serum IgG reaches 8–14 mg/mL within 12–24 h (Veronesi et al. 2014; Turini et al. 2021).

3.2.2.6.2 Hypoallergenic Donkey Milk

Low casein content and distinct Ig profiles render donkey milk a tolerable alternative for cow's milk-allergic individuals, with reduced IgE reactivity to milk proteins (Derdak et al. 2020).

3.2.2.6.3 Equid Comparisons

Donkeys and horses share most Ig characteristics, yet donkeys display shorter CDR3H loops, unique glycosylation (e.g., IgG5 extra CH1 glycan), and species-specific miRNA regulation of immune genes (Huang et al. 2015; Marihonnaiah and Belur Shivappa 2024).

3.2.3 Mucosal Adaptive Immunity and Gastrointestinal Nematode Infections

The gastrointestinal tract of donkeys is a major portal of entry for parasitic nematodes, and its mucosal immune system provides the first line of defense. Although detailed studies in donkeys remain scarce, investigations in horses and other livestock offer a framework for understanding how the organized lymphoid tissues, effector cells, and secreted immunoglobulins coordinate to limit parasite establishment and survival.

3.2.3.1 Adaptive Immune Mechanisms against Gastrointestinal Nematodes

Upon infection with strongyles (e.g., *Cyathostominae* spp., *Strongylus vulgaris*) or ascarids (*Parascaris equorum*), donkeys mount a predominantly Th2-biased response. Activated $CD4^+$ T cells secrete interleukin-4 (IL-4), IL-5, and IL-13, which drive eosinophil proliferation, mast cell hyperplasia, and goblet-cell–mediated mucus production. Eosinophils release major basic protein and eosinophil peroxidase during degranulation, damaging larval cuticles, while mast cells amplify local inflammation through histamine and protease release.

Secretory immunoglobulin A (IgA) is particularly important in the gut lumen. Parasite-specific IgA can prevent larval exsheathment, block worm adhesion to epithelial surfaces, and promote antigen clearance via immune exclusion. Concurrently, locally produced IgE binds FcεRI on mucosal mast cells and basophils, enabling rapid degranulation upon antigen re-encounter. Over successive infections, many donkeys develop partial immunity characterized by diminished worm burdens and reduced fecal egg counts, although this resistance is influenced by age, nutrition, and concurrent disease.

3.2.3.2 Acquired Resistance and Genetic Variation in Parasite Control

Equids vary in their ability to acquire protective immunity. Field observations in endemic regions have identified "low egg-shedding" donkeys that maintain stable, low fecal egg counts across seasons. These individuals exhibit stronger mucosal Th2 cytokine profiles, elevated parasite-specific IgA, and upregulated genes involved in eosinophil chemotaxis and mucin synthesis. In contrast, high egg-shedders display signs of immune tolerance or insufficient effector activation, resulting in higher parasite loads.

Genetic studies in horses suggest moderate heritability of nematode resistance traits, and preliminary data indicate similar trends in donkeys. Selective breeding for low egg-shedding phenotypes could reduce reliance on anthelmintics and slow the emergence of drug resistance. However, robust immunogenetic analyses and identification of molecular markers are urgently needed to implement such strategies in donkey populations.

3.2.3.3 Host Response: The Example of *Setaria equina*

Setaria equina, a filarial nematode inhabiting the peritoneal cavity, is often considered nonpathogenic but can provoke localized inflammation under heavy burden or aberrant migration. In donkeys, microfilariae interact with pattern recognition receptors on dendritic cells and macrophages, leading to antigen presentation and Th2 polarization (Abdel Rahman 2020).

Peritoneal eosinophils and basophils migrate to sites of larval deposition, where degranulation releases cytotoxic mediators. Circulating levels of parasite-specific IgG and IgE correlate with both systemic and mucosal responses, although chronic

S. equina infection appears to downregulate host immunity via excretory–secretory products that inhibit antigen-presenting cell activation. Understanding this modulation has implications for both equine health and comparative filarial immunology (Abdel Rahman 2020).

3.2.3.4 Immunological Architecture of the Donkey Hindgut

The donkey's hindgut—comprising the cecum and colon—harbors a dense network of lymphoid follicles, intraepithelial lymphocytes (IELs), and specialized secretory cells. Goblet cells secrete mucins that entrap pathogens, while Paneth-like cells release antimicrobial peptides. In response to nematode invasion, the hindgut mucosa is infiltrated by Th2 cells and associated cytokines, resulting in epithelial hyperplasia and strengthened mucus barriers (Li et al. 2022).

High numbers of eosinophils and mast cells within the lamina propria underscore the region's readiness to mount anti-parasite defenses. Despite these insights, detailed phenotyping of donkey gut lymphocyte subsets, immunoglobulin-secreting plasma cells, and cytokine expression remains incomplete. Comparative studies with horses and ruminants can guide targeted research, ultimately informing vaccine design and novel control strategies for equid nematodiasis.

4 Immune Response Phases

Donkeys possess a competent and nuanced immune system that follows the classical mammalian immune response trajectory: acute, recognition, activation, effector, homeostasis, and memory phases. However, their immunophysiology is shaped by species-specific traits, including metabolic adaptations, stoic clinical behavior, and unique inflammatory profiles. Understanding these phases in donkeys is essential for improving diagnostic accuracy, tailoring vaccination protocols, and managing diseases such as endotoxemia and hyperlipemia.

4.1 Acute Phase

The acute phase response (APR) is the first line of systemic defense in donkeys (Equus asinus) and other mammals, rapidly mobilizing innate immunity in response to infection, trauma or tissue damage. Activated macrophages and dendritic cells recognize pathogen- or damage-associated molecular patterns and release pro-inflammatory cytokines—interleukin-1 (IL-1), interleukin-6 (IL-6) and tumor necrosis factor-alpha (TNF-α). These cytokines circulate to the liver, where they engage hepatocyte receptors and activate transcription factors such as STAT3, shifting hepatic protein synthesis toward positive acute phase proteins (APPs) and away from constitutive proteins like albumin (Gabay and Kushner 1999; Cray et al. 2009).

4.1.1 Mechanisms and Triggers

Upon binding of pattern recognition receptors on innate immune cells to PAMPs or DAMPs, intracellular signalling cascades including NF-κB and JAK-STAT pathways are activated. The resulting surge in IL-1, IL-6 and TNF-α induces hepatocellular transcriptional reprogramming, upregulating genes encoding positive APPs while downregulating genes for negative APPs and other homeostatic proteins (Gabay and Kushner 1999).

4.1.2 Key Acute Phase Proteins in Donkeys

Donkey hepatocytes synthesize a spectrum of positive APPs that fulfil antimicrobial, antioxidant, coagulation and immunomodulatory roles. Serum Amyloid A (SAA) binds bacterial membranes and recruits phagocytes, Haptoglobin (Hp) sequesters free hemoglobin to limit oxidative damage, Fibrinogen (Fb) supports clot formation and tissue repair, C-reactive Protein (CRP) opsonizes pathogens for clearance, and α_1-Acid Glycoprotein (AGP) modulates leukocyte trafficking and cytokine production. Albumin serves as a negative APP, decreasing during the APR to conserve amino acids for synthesis of positive APPs (Eckersall and Bell 2010).

4.1.3 Kinetics and Species-Specific Dynamics

In experimentally induced endotoxemia, donkeys exhibit a rapid elevation of SAA and Hp within 2 h of lipopolysaccharide challenge, peaking earlier but to lower magnitudes than horses. Baseline SAA levels in healthy Andalusian donkeys range from 0.1 to 0.6 mg/L, while Hp spans 75–2261 mg/L; both markers typically return to baseline within 48–72 h if inflammation resolves (Perez-Ecija et al. 2021). Fibrinogen increases more slowly and is therefore more informative in subacute or chronic inflammatory contexts, whereas albumin concentrations decline proportionally to the APR's intensity (Perez-Ecija et al. 2021).

4.1.4 Metabolic and Endocrine Modulation

Insulin has emerged as a key modulator of APR intensity in donkeys. Compared to horses, donkeys maintain lower basal insulin and triglyceride levels, which appear to attenuate cytokine-driven APP production, particularly during hyperlipemia. In this metabolic disorder, elevated triglycerides correlate strongly with disease severity and may outperform SAA as prognostic indicators. Sex hormones also influence the APR: male donkeys (jacks) display higher SAA responses than jennies or geldings, implicating testosterone-mediated upregulation of TNF-α pathways.

4.1.5 Pharmacological Modulation

Non-steroidal anti-inflammatory drugs such as flunixin meglumine effectively suppress the APR in donkeys by reducing circulating TNF-α, Hp and SAA levels in experimental endotoxemia models. Similarly, exogenous insulin therapy in hyperlipemic or critically ill equids enhances survival by combining anti-inflammatory effects with metabolic regulation of hepatic APP synthesis (Samimi et al. 2019).

4.1.6 Clinical and Diagnostic Considerations

Because donkeys often mask clinical signs, measurement of APPs like SAA and Hp offers sensitive detection of subclinical inflammation. However, direct application of equine reference ranges risks misclassification; validated donkey-specific intervals and immunoassays are essential to avoid underdiagnosis or overtreatment. Serial monitoring of APP trends, rather than reliance on single measurements, provides the most reliable assessment of therapeutic response and disease resolution in this stoic species (Thiemann and Poore 2019).

4.2 Recognition and Activation

Following the acute phase, the immune system enters the recognition phase, where pattern recognition receptors (PRRs) and antigen-presenting cells (APCs) detect pathogen-associated molecular patterns (PAMPs). This leads to the activation phase, characterized by the proliferation and differentiation of lymphocytes.

In donkeys, vaccination studies using tetanus toxoid and equine influenza antigens have demonstrated robust activation of CD4+ helper T cells, CD8+ cytotoxic T cells, and regulatory T cells (CD4+ FoxP3+). CD4+ T cells increased from 25.1% to 37.3% 1 month post-vaccination, while CD8+ T cells rose from 20.6% to 32.2% by the second month. Regulatory T cells peaked at 11.7% before returning to baseline, indicating a transient but effective modulation of immune intensity (Perzyna et al. 2025).

4.3 Effector Phase

The effector phase involves the deployment of immune mechanisms to eliminate pathogens. In donkeys, this includes a shift in monocyte subsets, with an increase in CD14− MHCII+ cells that enhance antigen presentation. Concurrently, PanB/CD21+ B lymphocytes decrease, along with reductions in IL-10+, IL-17+, and Ki67+ subsets, reflecting a tightly regulated humoral response (Perzyna et al. 2025).

This phase is critical for pathogen clearance and is influenced by both innate and adaptive immune components. The balance between pro-inflammatory and regulatory signals ensures effective defense without excessive tissue damage.

4.4 Homeostasis

Once the antigen is cleared, the immune system transitions into homeostasis. Effector cells undergo apoptosis, and regulatory mechanisms restore equilibrium. APP levels such as SAA and Hp normalize within 48–72 h if inflammation resolves, providing biochemical confirmation of recovery (Perez-Ecija et al. 2021).

This phase is essential for preventing chronic inflammation and maintaining immune readiness. In donkeys, the resolution of the APR is often subtle, requiring laboratory confirmation due to their stoic demeanor.

4.5 Memory

The final phase is memory, where memory B and T cells persist to ensure rapid and effective responses upon re-exposure. This phase underpins the success of vaccination protocols and long-term immunity in donkeys. Despite limited species-specific vaccine formulations, studies confirm that donkeys retain immunological memory comparable to horses (Perzyna et al. 2025).

Memory responses are particularly relevant in herd health management and outbreak prevention. They also inform booster schedules and vaccine efficacy assessments.

5 Conclusions

The immune system of donkeys and mules is shaped by a combination of evolutionary legacy, genomic architecture, and functional specialization. While sharing many core mechanisms with horses, these equids exhibit distinctive immunological features that influence disease resistance, pathogen tolerance, and adaptive capacity. Key elements such as immunoglobulin diversity, IgG subclasses, and V(D)J recombination highlight the molecular sophistication of adaptive immunity, while innate components including Toll-like receptors and pattern recognition receptors provide essential first-line defense. The role of regulatory T cells, mucosal immunity, and neonatal protection through colostrum further illustrate the balance between effective immune activation and regulation in these species. Comparative immunogenetics underscores that donkeys and their hybrids cannot be fully understood through the lens of horse immunology alone. Their differential responses to parasitic infections, tick-borne diseases, and conditions such as besnoitiosis emphasize the need for species-specific diagnostics and management strategies. Moreover, hybrid animals such as mules serve as natural models for studying the immunogenetic consequences of interspecies reproduction, broadening our understanding of equid immune diversity. As genomic tools expand and functional studies deepen, the integration of molecular insights with applied veterinary practice will be central to improving the health and welfare of donkeys and mules. This knowledge not only supports conservation and productivity but also advances One Health approaches by clarifying how these often-overlooked equids interact with pathogens in shared ecosystems. Future progress will depend on sustained investment in comparative immunology, targeted diagnostic development, and the incorporation of genetic diversity into breeding and conservation programs.

Self-Evaluation Questionnaire

1. Which immune mechanism ensures donkeys generate a wide repertoire of antibodies against pathogens?
   A. Somatic hypermutation
   B. V(D)J recombination
   C. Passive immunity
   D. Epitope masking
   Answer: B
2. Why is reliance on horse-based immunological diagnostics problematic for donkeys?
   A. Donkeys lack lymphoid tissues
   B. Donkeys express distinct IgG subclasses
   C. Donkeys have no adaptive immunity
   D. Horse immunology is already incomplete
   Answer: B
3. Which feature of neonatal donkey immunity most depends on colostrum intake?
   A. Expression of Toll-like receptors
   B. Passive transfer of maternal antibodies
   C. Expansion of regulatory T cells
   D. Development of eosinophilia
   Answer: B
4. What role do pattern recognition receptors (PRRs) play in donkey immunity?
   A. Suppressing adaptive responses
   B. Generating antibody subclasses
   C. Detecting conserved microbial motifs
   D. Inducing immunoglobulin recombination
   Answer: C
5. In which disease do donkeys show distinctive immunological tolerance compared to horses?
   A. Equine influenza
   B. Besnoitiosis
   C. Strangles
   D. African horse sickness
   Answer: B
6. Why are regulatory T cells considered central to donkey immune balance?
   A. They enhance V(D)J recombination
   B. They promote eosinophilia
   C. They prevent excessive inflammation
   D. They increase colostrum absorption
   Answer: C

References

Abdel Rahman MMI (2020) Morphological and molecular characterization of *Setaria equina* in donkeys. Beni-Suef Univ J Basic Appl Sci 9(1):17

Adams EJ, Parham P (2001) Species-specific evolution of MHC class I genes in the higher primates. Immunol Rev 183:41–64

Allen PZ, Dalton EJ (1975) Studies on equine immunoglobulins. IV. Immunoglobulins of the donkey. Immunology 28(1):187–197. PMID: 1116869; PMCID: PMC1445747

Baker JM, Stidworthy M, Gull T, Novak J, Miller JM, Antczak DF (2001) Conservation of recognition of antibody and T-cell-defined alloantigens between species of equids. Reprod Fertil Dev 13(7):635–645. https://doi.org/10.1071/RD01081

Barbosa JD, Sodré MHS, Barbosa CC, da Costa PSC, Oliveira CMC, Ferreira TTA, da Silveira JAS, Lamego EC, Paz MC, Caldeira RD, Magalhães Matos PC, Serruya A, Salvarani FM, Silveira N d S e S (2024) Allergic dermatitis in Pêga breed donkeys (Equus asinus) caused by culicoides bites in the Amazon biome, Pará, Brazil. Animals 14(9):1330. https://www.mdpi.com/2076-2615/14/9/1330

Battista JM, Tallmadge RL, Stokol T, Felippe MJ (2014) Hematopoiesis in the equine fetal liver suggests immune preparedness. Immunogenetics 66(11):635–649

Birch J, Murphy L, MacHugh ND, Ellis SA (2006) Generation and maintenance of diversity in the cattle MHC class I region. Immunogenetics 58(8):670–679

Braend M (1965) Nomenclature of polymorphic protein systems. Nature 206:1067

Braend M (1967) Variation of horse prealbumins in acidic starch gels. Acta Veterinaria Scandinavica 8:193–194

Braend M (1970) Genetics of horse acidic prealbumins. Genetics, 65, 495–503

Braend M, Romagnoli A (1980) Variation of acidic prealbumins in the donkey (Equus asinus). Anim Blood Groups Biochem Genet 11:77–80

Caldin M, Furlanello T, Solano-Gallego L et al (2005) Reference ranges for haematology, biochemical profile and electrophoresis in a single herd of Ragusana donkeys from Sicily (Italy). Comp Clin Pathol 14:5–12. https://doi.org/10.1007/s00580-005-0544-8

Campbell J (2007) MHC class I sequence polymorphism in horses and donkeys. Master's thesis, Cornell University. https://hdl.handle.net/1813/2931

Carbonara M, Mendoza-Roldan JA, Bezerra-Santos MA, de Abreu Teles PP, Lia RP, Locantore F et al (2024) Leishmania spp. in equids and their potential vectors in endemic areas of canine leishmaniasis. PLoS Negl Trop Dis 18(7):e0012290

Cocco R, Rizzo M, Carta C, Arfuso F, Piccione G, Luridiana S, Crovace A, Passino ES, Sechi S (2023) Immune and inflammatory response of donkeys (Equus asinus) tested positive to tick-borne pathogens. J Equine Vet Sci 120:104190. https://doi.org/10.1016/j.jevs.2022.104190. Epub 2022 Dec 9

Cray C, Zaias J, Altman NH (2009) Acute phase response in animals: a review. Comp Med 59(6): 517–526

Derdak R, Sakoui S, Pop OL, Mureşan C, Vodnar D, Boutaina A, Vulturar R, Adina C, Suharoschi R, Soukri A, El Khalfi B (2020) Insights on health and food applications of Equus asinus (donkey) milk bioactive proteins and peptides—an overview. Foods 9:1302. https://doi.org/10.3390/foods9091302

Dissen E, Fossum S, Hoelsbrekken SE, Saether PC (2008) NK cell receptors in rodents and cattle. Semin Immunol 20(6):369–375. https://doi.org/10.1016/j.smim.2008.09.007. Epub 2008 Nov 1

Eckersall PD, Bell R (2010) Acute phase proteins: biomarkers of infection and inflammation in veterinary medicine. Vet J 185(1):23–27

Ek N (1979) The characterization of equine prealbumin (Pr) protein by antigen-antibody crossed electrophoresis. Acta Vet Scand 20:180–190

Gabay C, Kushner I (1999) Acute-phase proteins and other systemic responses to inflammation. N Engl J Med 340(6):448–454

Gahne B (1966) Studies on the inheritance of electrophoretic forms of transferrins, albumins, prealbumins and plasma esterases of horses. Genetics 53:681–694

Garg SK, Delaney C, Toubai T, Ghosh A, Reddy P, Banerjee R, Yung R (2014) Aging is associated with increased regulatory T-cell function. Aging Cell 13(3):441–448. https://doi.org/10.1111/acel.12191

Girardi A, Toledo C, Silva P, Marques L (2017) The effect of age and sex on serum proteins in the Pega donkey (Equus asinus). Vet Med 62:10–15. https://doi.org/10.17221/397/2014-VETMED

Glatman-Freedman A, Casadevall A (1998) Serum therapy for tuberculosis revisited: reappraisal of the role of antibody-mediated immunity against mycobacterium tuberculosis. Clin Microbiol Rev 11(3):514–532. https://doi.org/10.1128/CMR.11.3.514. PMID: 9665981; PMCID: PMC88894

Gnanesh Kumar BS, Rawal A (2020) Sequence characterization and N-glycoproteomics of secretory immunoglobulin A from donkey milk. Int J Biol Macromol 155:605–613. https://doi.org/10.1016/j.ijbiomac.2020.03.253

Gutiérrez JP, Marmi J, Goyache F, Jordana J (2005) Pedigree information reveals moderate to high levels of inbreeding and a weak population structure in the endangered Catalonian donkey breed. J Anim Breed Genet 122(5):378–386. https://doi.org/10.1111/j.1439-0388.2005.00590.x

Hamza E, Gerber V, Steinbach F, Marti E (2011) Equine CD4(+) CD25(high) T cells exhibit regulatory activity by close contact and cytokine-dependent mechanisms in vitro. Immunology 134(3):292–304. https://doi.org/10.1111/j.1365-2567.2011.03489.x

Huang J, Zhao Y, Bai D, Shiraigol W, Li B, Yang L et al (2015) Donkey genome and insight into the imprinting of fast karyotype evolution. Sci Rep 5:14106. https://doi.org/10.1038/srep14106

Hughes AL, Nei M (1988) Pattern of nucleotide substitution at major histocompatibility complex class I loci reveals overdominant selection. Nature 335(6186):167–170

Kaminski M (1969) Common and species-specific serum esterases of Equidae I. Horse and donkey. Biochim Biophys Acta 191:611–620. https://doi.org/10.1016/0005-2744(69)90354-4

Kaminski M, Gajos E (1964) Comparative examination of carboxylic esterases in sera of horse, donkey and their hybrids. Nature 201(4920):716–718

Kısadere I, Donmez N, Omurzakova, N (2019) Serum biochemical reference values of Kyrgyz donkeys (Equus asinus). Comparative Clinical Pathology 28(3):817–823

Klumplerova M, Splichalova P, Oppelt J, Futas J, Kohutova A, Musilova P, Kubickova S, Vodicka R, Orlando L, Horin P (2020) Genetic diversity, evolution and selection in the major histocompatibility complex DRB and DQB loci in the family Equidae. BMC Genomics 21:677. https://doi.org/10.1186/s12864-020-07089-6

Kumar A, Suryadevara N, Hill TM, Bezbradica JS, Van Kaer L, Joyce S (2017) Natural killer T cells: an ecological evolutionary developmental biology perspective. Front Immunol 8:1858

Kydd J, Miller J, Antczak DF, Allen WR (1982) Maternal anti-foetal cytotoxic antibody responses of equids during pregnancy. J Reprod Fertil Suppl 32:361–369

Lebedinskaya OV, Lebedinskaya EA, Donenko FV, Kuzovlev EN (2006) Cytotoxic activity of natural killer T-cells in mice with tumor-affected liver. Bull Exp Biol Med 141(1):76–79

Li Y, Ma Q, Shi X, Liu G, Wang C (2022) Integrated multi-omics reveals novel microbe-host lipid metabolism and immune interactions in the donkey hindgut. Front Immunol 13:1003247

Liu B, Gong S, Tulafu H, Zhang R, Tao W, Adili A et al (2024) Analysis of the genetic relationship and inbreeding coefficient of the Hetian Qing donkey through a simplified genome sequencing technology. Genes 15(5):570. https://doi.org/10.3390/genes15050570

Marihonnaiah S, Belur Shivappa GK (2024) Site-specific N-Glycoprofiling of immunoglobulin G subtypes from donkey Milk. J Agric Food Chem 72(47):26547–26557. https://doi.org/10.1021/acs.jafc.4c07168

Massari S, Giannico F, Paolillo NV, Pala A, Jambrenghi AC, Antonacci R (2023) Genomic and comparative analysis of the T cell receptor gamma locus in two Equus species. Front Immunol 14:1264949. https://doi.org/10.3389/fimmu.2023.1264949

Noh JY, Yoon SR, Kim TD, Choi I, Jung H (2020) Toll-like receptors in natural killer cells and their application for immunotherapy. J Immunol Res 2020:2045860. https://doi.org/10.1155/2020/2045860. PMID: 32377528; PMCID: PMC7199539

Otting N, Heijmans CM, Noort RC, de Groot NG, Doxiadis GG, van Rood JJ, Watkins DI, Bontrop RE (2005) Unparalleled complexity of the MHC class I region in rhesus macaques. Proc Natl Acad Sci 102(5):1626–1631

Paraschou G, Xue C, Egan R, Bolfa P (2024) Multicentric lymphoma in a donkey with intestinal and bone marrow involvement. BMC Veterinary Research 20(57). https://doi.org/10.1186/s12917-024-03903-5

Perez-Ecija A, Buzon-Cuevas A, Aguilera-Aguilera R, Gonzalez-De Cara C, Mendoza Garcia FJ (2021) Reference intervals of acute phase proteins in healthy Andalusian donkeys and response to experimentally induced endotoxemia. J Vet Intern Med 35(1):580–589

Perzyna M, Grzędzicka J, Milczek-Haduch D, Dąbrowska I, Trela M, Pawliński B, Witkowska-Piłaszewicz O (2025) Immunological responses to tetanus and influenza vaccination in donkeys. J Vet Intern Med 39(2):e70137. https://doi.org/10.1111/jvim.70137

Pucca MB, Cerni FA, Janke R, Bermúdez-Méndez E, Ledsgaard L, Barbosa JE, Laustsen AH (2019) History of envenoming therapy and current perspectives. Front Immunol 10:1598. https://doi.org/10.3389/fimmu.2019.01598

Qiu Y, Jiang J, Yi X, Wang S, Sun X (2025) Exploration of the differential expression patterns of immunoglobulin heavy chain genes in horses and donkeys. Dev Comp Immunol 166:105360. https://doi.org/10.1016/j.dci.2025.105360

Radar-Chafirovitch A, Quaresma M, Yánez-Ortiz I, Leiva B, Ferreira-Dias G, Payan-Carreira R, Miro J, Pires M d A (2025) Donkey endometrium: characterization of resident immune cells. Res Vet Sci 184:105516. https://doi.org/10.1016/j.rvsc.2024.105516

Samimi AS, Samimi K, Karimiafshar M, Tajik J (2019) Comparative anti-inflammatory effects of insulin and flunixin on acute-phase responses and cardiovascular biomarkers during inflammatory phase in miniature donkeys. J Equine Vet Sci 81:102788

Sotirov L (2004) Lysozyme and complement concentrations in horses, donkeys and mules. Rev Med Vet 155(4):221–225

Sun Y, Wang C, Wang Y, Zhang T, Ren L, hu X, Zhang R, Meng Q, Guo Y, Fei J, Li N, Zhao Y (2010) A comprehensive analysis of germline and expressed immunoglobulin repertoire in the horse. Dev Comp Immunol 34:1009–1020. https://doi.org/10.1016/j.dci.2010.05.003

Thiemann AK, Poore LA (2019) Hoof disorders and farriery in the donkey. Vet Clin N Am Equine Pract 35(3):643–658. https://doi.org/10.1016/j.cveq.2019.08.012

Tombak KJ, Budischak SA, Hauck S, Martinez LA, Rubenstein DI (2020) The non-invasive measurement of faecal immunoglobulin in African equids. Int J Parasitol Parasites Wildl 12: 105–112. https://doi.org/10.1016/j.ijppaw.2020.05.005

Turini L, Bonelli F, Nocera I, Meucci V, Conte G, Sgorbini M (2021) Evaluation of different methods to estimate the transfer of immunity in donkey foals fed with colostrum of good IgG quality: a preliminary study. Animals 11(2):507. https://doi.org/10.3390/ani11020507

Veronesi F, Morganti G, Ravagnan S et al (2014) Molecular and serological detection of tick-borne pathogens in donkeys (Equus asinus) in Italy. Vet Microbiol 173(3–4):348–354. https://doi.org/10.1016/j.vetmic.2014.08.017

Vranova M, Alloggio I, Qablan M et al (2011) Genetic diversity of the class II major histocompatibility DRA locus in European, Asiatic and African domestic donkeys. Infect Genet Evol 11(5): 1136–1141. https://doi.org/10.1016/j.meegid.2011.04.010

Wagner B (2006) Immunoglobulins and immunoglobulin genes of the horse. Dev Comp Immunol 30(1–2):155–164. https://doi.org/10.1016/j.dci.2005.06.008

Walther S, Rusitzka TV, Diesterbeck US, Czerny CP (2015) Equine immunoglobulins and organization of immunoglobulin genes. Dev Comp Immunol 53(2):303–319. https://doi.org/10.1016/j.dci.2015.07.017

Wang Y, Wang T, Zhang Z, Wang C (2021) Survival rate of donkey foals: status quo and improvement methods. Afr J Biotechnol 20(8):340–348. https://doi.org/10.5897/AJB2021.17376

Wu Y, Tian Z, Wei H (2017) Developmental and functional control of natural killer cells by cytokines. Front Immunol 8:930

Ziegler-Heitbrock L (2007) The CD14+ CD16+ blood monocytes: their role in infection and inflammation. J Leukoc Biol 81(3):584–592. https://doi.org/10.1189/jlb.0806510. S2CID 31534841

Donkey and Mule Immunology: Clinical Insights and Comparative Applications

Francisco Javier Navas González

Abstract

Donkeys and mules represent underexplored yet clinically and scientifically valuable models for understanding equid immunology. This chapter synthesizes current knowledge on the clinical applications of donkey and mule immune systems, highlighting both parallels and distinctions with horses. Neonatal immunity is characterized by agammaglobulinemia at birth and complete reliance on colostral immunoglobulins for passive protection, while species-specific thresholds and diagnostics for passive transfer are emerging. Immunophenotyping has confirmed cross-reactivity with equine antibodies, enabling characterization of lymphocyte subsets and functional adaptive responses to vaccination. Comparative studies reveal that donkeys mount effective innate and adaptive responses to pathogens and vaccines, though with notable differences such as lower viral loads and distinct hematological reference values. Clinical sections address key immune-mediated conditions including neonatal isoerythrolysis, immunosenescence, hypersensitivity disorders, autoimmune syndromes, anaphylaxis, and immunometabolic challenges. The chapter further explores donkey milk as both a hypoallergenic food and an immunomodulatory agent, as well as unique immunogenetic and endocrinological interactions shaping disease susceptibility. From a One Health perspective, donkeys' roles in zoonotic pathogen reservoirs, animal-assisted interventions, and production systems underscore the importance of welfare-centered immune health management. Advances in genomics and immunophenotyping are now bridging long-standing gaps, offering tools for species-specific diagnostics, vaccination strategies, and therapeutic interventions. Collectively, donkey and mule

F. J. Navas González (✉)
Department of Genetics, The Worldwide Donkey Breeds Project, University of Córdoba, Córdoba, Spain
e-mail: fjnavas@uco.es

A. K. McLean et al. (eds.), *Donkey and Mule Medicine: Clinical Medicine and Reproduction*, https://doi.org/10.1007/978-3-032-18540-2_6

immunology provides not only improved veterinary care for these animals but also comparative insights into mammalian immune evolution and translational biomedical research.

Keywords

Passive immunity · Colostrum · Neonatal immunity · Pattern recognition receptors · Toll-like receptors · Mucosal immunity · Eosinophilia · Tick-borne diseases · Virus · Bacteria · Comparative immunogenetics · Species-specific diagnostics · Immunosenescence · Hypersensitivity · Autoimmune disorders · Immunometabolism · Donkey milk

1 Introduction

From a clinical standpoint, donkeys and mules provide critical models for understanding equid immune responses under natural disease, vaccination, and management conditions. Peripheral blood immunophenotyping has revealed that donkeys express monocytic ($CD14^+$) and lymphoid (MHC II^+, $FoxP3^+$) markers nearly identical to those found in horses. Cross-reactivity of antibodies developed for equine CD4, CD8, and CD21 enables precise immunological characterization of donkey leukocyte subsets (Perzyna et al. 2025).

Upon infection with tick-borne pathogens such as Theileria, Ehrlichia, and Rickettsia, donkeys mount classical innate responses—characterized by declines in red blood cells, hemoglobin, platelets, and lymphocytes, as well as increases in eosinophils, basophils, and serum globulins—reflecting acute-phase inflammation and immune activation (Qiu et al. 2025; Perzyna et al. 2025). During parasitic infections such as Parascaris equorum, donkeys express elevated levels of proinflammatory cytokines including interleukin-1β (IL-1β), tumor necrosis factor-alpha (TNF-α), and oxidative stress markers (Attia et al. 2022).

Functional studies affirm that donkeys mount effective adaptive responses to vaccination. Tetanus and influenza vaccines elicit expansion of $CD4^+$ helper and $CD8^+$ cytotoxic T cell populations, along with transient elevations in regulatory T cells ($CD4^+FoxP3^+$ and $CD8^+FoxP3^+$), mirroring the phases of mammalian adaptive immunity (Perzyna et al. 2025). Post-vaccination, $CD4^+$ T cell frequencies increase from 25% to 37%, $CD8^+$ T cells from 20% to 30%, and regulatory subsets temporarily rise before returning to baseline, indicating coordinated immune activation.

Donkey neonates are born agammaglobulinemic due to epitheliochorial placentation, which prevents prenatal antibody transfer. Thus, passive immunity relies entirely on colostral IgG uptake. Colostrum from vaccinated jennies typically contains 8–9 g/dL of IgG, sufficient to establish protective serum titers in foals within the first 24 hours postpartum (Turini et al. 2021). Radial immunodiffusion assays provide a practical method for field monitoring of passive transfer. Failure of adequate IgG uptake—termed failure of passive transfer (FPT)—is a major risk factor for neonatal sepsis, underscoring the importance of maternal vaccination and early colostrum intake (Turini et al. 2021; Di Francesco et al. 2020).

Importantly, donkeys are not immunologically identical to horses. While fundamental processes—such as B cell activation, class switching, plasma cell differentiation, and T cell effector functions—are conserved, detailed immunokinetic studies reveal subtle distinctions. For example, EIAV-infected donkeys display viral loads approximately 1000 times lower than infected horses, with minimal clinical symptoms, suggesting enhanced subclinical resistance (Perzyna et al. 2025). Additionally, donkey immune responses to bacterial and viral vaccines include secretion of interferon-gamma (IFN-γ) and interleukin-4 (IL-4), although comprehensive cytokine profiles remain incomplete.

From a hematological perspective, donkeys possess lower RBC counts, reduced hemoglobin concentrations, and higher mean corpuscular volume (MCV) compared to horses. These differences must be considered in both research and clinical diagnostics (Perzyna et al. 2025). To date, no congenital immunodeficiencies have been definitively characterized in donkeys, though analogous conditions such as severe combined immunodeficiency (SCID) have been observed in horses and may occasionally present in donkeys.

Recent studies underscore the urgency of species-specific research. For instance, the first in vivo study of innate inflammatory responses in Sardinian donkeys exposed to tick-borne pathogens revealed leukocyte shifts and hyperglobulinemia, suggesting robust inflammation but also highlighting the absence of deeper data on cytokine signaling or macrophage behavior (Cocco et al. 2023). Likewise, research on adaptive responses shows strong post-vaccination T-cell activation in donkeys, but more nuanced studies are needed to refine vaccine schedules and dosing (Perzyna et al. 2025; Di Francesco et al. 2020).

Genomic advancements have further enriched clinical applications. The donkey genome is now sequenced to high resolution, revealing divergence from horses and identifying rapidly evolving gene families related to lymphocyte regulation (Huang et al. 2015). Such findings provide molecular frameworks for understanding species-specific immune behavior. Comparative studies have already shown that certain equine reference values—such as serum proteins and acute-phase reactants—may not translate accurately to donkeys, necessitating recalibration for clinical use.

From a comparative immunology perspective, donkeys offer a unique lens into mammalian immune evolution. Sharing a common ancestor with horses approximately 7–15 million years ago, donkeys retain broad immunologic homology while also exhibiting lineage-specific adaptations (Huang et al. 2015). Their inclusion in immunological research not only improves veterinary care for this species but also expands the comparative models available to biomedical science.

Donkey immunology thus encompasses the full spectrum of mammalian immune defenses, integrating ancient innate mechanisms and adaptive cellular complexity. Despite many parallels and foundational similarities with horses, donkeys exhibit unique immunogenetic, molecular, and functional characteristics which not only justify but also compel researchers to perform species-specific studies. In these regards, the historical neglect of donkey immunology is being remedied by recent genomic, immunophenotypic and functional studies (Qiu et al. 2025), which are beginning to close the knowledge gap. Such advances offer new tools to optimize veterinary health care, vaccination, and disease management, while also enriching comparative and evolutionary mammalian immunology in this often-overlooked species.

This chapter will examine clinical, diagnostic, and comparative applications of donkey and mule immunology, focusing on vaccination, neonatal immunity, infectious disease responses, and their broader implications for veterinary and biomedical research.

2 The Immune System of the Neonatal Donkey

The neonatal period represents a critical window in the immunological development of donkeys, during which survival depends heavily on the successful transfer of maternal immunity. Unlike some species, foals are born agammaglobulinemic, and in donkeys this vulnerability is further shaped by species-specific physiological and immunological characteristics that differ from horses. The reliance on colostrum for passive immunity, the timing of gut permeability closure, and the thresholds for adequate immunoglobulin absorption all highlight the importance of tailored clinical approaches in neonatal donkey care. Understanding these early immune dynamics is essential for improving disease prevention, diagnostic accuracy, and management strategies in young donkeys, while also providing comparative insights into equid immunology.

2.1 Ontogeny of the Donkey Immune System

Donkey foals are born agammaglobulinemic; intestinal FcRn-mediated uptake of colostral IgG peaks within 6 hours postpartum and declines by 24 hours. Successful passive transfer (>4 g/L serum IgG) provides immediate protection against enteric and respiratory pathogens (EIV, Salmonella, rotavirus) while shaping neonatal immune ontogeny. Early complement components and neonatal NK cells complement humoral defenses until foal-derived IgG synthesis commences at 4–6 weeks (Turini et al. 2021).

2.1.1 Prenatal Lymphoid Development

The donkey thymus is the first lymphoid organ to form, supporting T-cell maturation by approximately day 100 of gestation, while detectable B cells appear around day 200 (Giguère and Polkes 2005). By birth, foals are immunocompetent—possessing all the cellular machinery for adaptive responses—but remain immunologically naïve, lacking antigenic memory and relying on passive immunity until their own systems mature.

2.1.2 Postnatal Maturation

Complement activity and lymphocyte function are initially weak in neonatal donkeys but improve rapidly over the first weeks of life. Functional assays demonstrate that by 2–3 months of age, many innate and adaptive parameters approach adult levels, provided passive antibodies have bridged the early immunologic gap (Giguère and Polkes 2005).

2.2 Passive Immunity Transfer: Colostrum and Immunoglobulins

2.2.1 Colostral Immunoglobulin Dynamics

Because the donkey's epitheliochorial placenta prevents in utero immunoglobulin transfer, foals must ingest colostrum to acquire maternal IgG (Veronesi et al. 2014; Turini et al. 2021). Peak absorption occurs within 6–12 hours post-partum, with intestinal "closure" by 24 hours sharply limiting further uptake (Turini et al. 2021). Studies in Italian breeds report colostral IgG concentrations of 16–75 mg/mL and resultant foal serum IgG of 8–14 mg/mL at 12–24 hours (Veronesi et al. 2014; Turini et al. 2021).

IgA, although present at lower colostral levels, contributes to mucosal defense, and postnatal rises in serum lysozyme underscore its role in early innate protection (Veronesi et al. 2014).

2.2.2 Diagnostic Thresholds and Methods

Adequate passive transfer is defined as serum IgG $\geq$ 8 mg/mL, partial failure as 4–8 mg/mL, and complete failure below 4 mg/mL—thresholds extrapolated from equine studies pending donkey-specific validation (Turini et al. 2021). Single radial immunodiffusion (SRID) remains the gold standard, but ELISA, refractometry, and the Biuret method correlate strongly ($r > 0.89$) and allow rapid field assessment (Turini et al. 2021).

2.2.3 Influencing Factors

Unlike in horses, maternal age, parity, and season appear to have minimal impact on donkey colostrum quality (Nervo et al. 2021). Nonetheless, nutritional status, breed variation, and environmental stressors can subtly affect IgG concentration and absorption, highlighting the importance of timely and adequate colostrum intake for each neonate.

2.2.4 Testing Colostrum

Assessing colostrum quality in donkeys is essential for ensuring successful passive transfer of immunity to newborn foals, primarily through immunoglobulin G (IgG). Donkey colostrum, while similar in protein content to that of horses, tends to have lower fat and is produced in smaller volumes, making early and accurate evaluation critical. Initial visual inspection—looking for a thick, creamy consistency and monitoring volume within the first 6–12 hours postpartum—can offer quick insights. For IgG quantification, tools like colostrometers (which measure specific gravity) and refractometers (which use the Brix scale, with ≥23–24% indicating good quality) provide rapid field assessments, though temperature and species-specific calibration may affect accuracy. Laboratory immunoassays such as radial immunodiffusion (RID) and ELISA offer precise IgG measurements and subclass differentiation, ideal for clinical or research settings. Microbiological testing is also important, as bacterial contamination can impair IgG absorption and foal health; hygienic collection and culture tests are recommended if mastitis or poor neonatal health is suspected. Timing is crucial—colostrum should be collected within the first 2 hours after foaling, as IgG levels decline significantly within 12–24 hours. Based on IgG concentration, colostrum is classified as high (>20–25 g/L or Brix ≥23–24%), moderate (15–20 g/L), or low (<15 g/L or Brix <20%), guiding decisions on whether supplementation with banked colostrum or commercial IgG products is necessary to prevent failure of passive transfer.

2.3 Neonatal Immunodeficiency and Emergency Management

2.3.1 Failure of Passive Transfer: Case Study

A donkey foal with complete failure of passive transfer (FPT), presented at 36 hours old with hypoglycemia and signs of neonatal maladjustment syndrome. Prompt diagnosis via refractometry led to plasma transfusion, broad-spectrum antibiotics, and supportive nutrition. Valentino's rapid clinical improvement underscores the critical need for early IgG assessment and intervention (Turini et al. 2021).

2.3.2 Neonatal Maladjustment Syndrome

Although rarely documented in donkeys (Peart 2022), neonatal maladjustment syndrome (NMS) in horses—characterized by dysregulated hypothalamic-pituitary-adrenal axis activity, aberrant neurosteroid levels, and inflammatory cytokine release—provides a likely parallel. Preliminary observations suggest that donkey foals may tolerate systemic inflammation and viral challenges better than horse foals, but controlled studies are required to confirm species differences (Gömer et al. 2022).

2.4 Neonatal Isoerythrolysis (NI)

2.4.1 Pathophysiology and Risk Factors

Neonatal isoerythrolysis results when maternal alloantibodies in colostrum target foal erythrocytes bearing paternally inherited antigens. The so-called "donkey factor," a red-cell antigen common to all donkeys but absent in horses, predisposes mule foals to high NI risk (McClure et al. 1994). Sensitization typically follows prior fetomaternal hemorrhage or transfusion events.

2.4.2 Diagnosis, Prevention, and Treatment

Screening late-gestation jennies for anti-erythrocyte antibodies via hemolytic assays or the jaundiced foal agglutination test is recommended. When positive, withholding colostrum and providing alternative IgG sources can prevent NI. Affected foals may require washed red-cell transfusions and intensive supportive care (Traub-Dargatz et al. 1995; Felippe 2018).

2.4.3 Special Considerations in Mule Pregnancies

Because donkey-mare/donkey-sire crosses invariably express the donkey factor, approximately 10% of mule foals develop NI. Additionally, maternal platelet-targeting alloantibodies can cause neonatal thrombocytopenia and hemorrhage, necessitating vigilant monitoring and possible platelet transfusions (Felippe 2018; UC Davis 2018; Ramirez et al. 1999).

2.5 Maternal–Fetal Immune Interactions

2.5.1 Mechanisms of Gestational Tolerance

Donkey pregnancies exhibit attenuated maternal immune activation compared to equine gestations. Jennies produce lower equine chorionic gonadotropin (eCG) levels, form smaller endometrial cups, and generate fewer cytotoxic alloantibodies, promoting fetal tolerance (Allen 1982; Kydd et al. 1982).

2.5.2 Immunology of Hybrid Pregnancies

In mule and hinny gestations, immunogenetic interplay becomes more complex. Mules can develop robust antibody responses against paternal donkey antigens, yet they also maintain endometrial cups and cytotoxic titers longer than mares carrying horse fetuses. These hybrid models offer unique insights into alloimmunization, maternal-fetal tolerance mechanisms, and perinatal immune pathologies (Antczak et al. 1985).

3 Immunological Aging and Immunosenescence in Older Donkeys

Immunosenescence—the gradual deterioration of immune function with age—affects both innate and adaptive compartments. In donkeys, as in other mammals, this process likely involves a skewing toward immunosuppressive regulatory T cells, shifts in serum immunoglobulins, and altered acute-phase protein profiles.

3.1 Inflammaging and T-Cell Dysregulation

Regulatory T-Cell Expansion Aging in mice and humans is marked by an increased proportion of $CD4^{+}FoxP3^{+}$ regulatory T cells (Tregs) that express higher FoxP3 levels and exhibit greater suppressive activity—including elevated IL-10 production and enhanced down-regulation of dendritic cell CD86—than young-mouse Tregs (Garg et al. 2014). Although direct data in donkeys are lacking, geriatric donkeys likely mirror this pattern, harboring more abundant, more suppressive Tregs that consume IL-2 and secrete IL-10, thereby dampening effector T-cell responses.

Naïve versus Memory T-Cell Balance In aged humans and mice, thymic involution and peripheral homeostatic shifts reduce naïve T-cell pools while expanding memory and regulatory subsets. Older horses similarly show reduced vaccine responsiveness, decreased naïve T cells, and altered CD4/CD8 ratios (Horohov et al. 2010). By analogy, aged donkeys probably exhibit fewer naïve T cells and blunted primary responses to novel antigens, increasing susceptibility to infections and reducing vaccine efficacy.

3.2 Age- and Sex-Dependent Shifts in Serum Proteins

Girardi et al. (2017) used SDS-PAGE to profile ten serum proteins in 110 clinically healthy Pega donkeys, revealing key age- and sex-related trends.

3.2.1 Immunoglobulin A and G Dynamics

IgA concentrations declined significantly with age ($p < 0.001$), reflecting a waning reliance on mucosal immunity as antigen exposure history accumulates (Eckersall 2008; Tizard 2018). Conversely, serum IgG increased with age ($p < 0.001$), consistent with a shift toward systemic humoral memory.

3.2.2 Sex Differences in Humoral Markers

Female donkeys had higher IgG and transferrin (Tf) levels than males ($p < 0.001$), suggesting enhanced iron-sequestering and antibody readiness. Males exhibited elevated acute-phase proteins—ceruloplasmin (Cp), α_1-acid glycoprotein (α_1AG), and an uncharacterized 138 kDa protein (MWP138)—indicative of heightened basal inflammatory tone (Girardi et al. 2017). Hormonal modulation of innate immunity may underlie these differences (Eckersall and Bell 2010).

3.2.3 Acute-Phase Protein Stability and Novel Biomarkers

Although Cp, α_1AG, and haptoglobin did not vary with age, their sex-dependent abundance highlights their utility as non-specific markers of welfare and subclinical inflammation. The identification of three additional age- and sex-variable proteins (MWP23, MWP33, MWP138) points to potentially novel donkey-specific immune factors warranting proteomic and functional characterization.

3.3 Implications and Future Directions

3.3.1 T-Cell Profiling

Direct quantification of $CD4^{+}FoxP3^{+}$ Tregs, naïve/memory ratios, and cytokine outputs in aged donkeys will clarify inflammaging mechanisms.

3.3.2 Reference Ranges

Establishing donkey-specific norms for IgA, IgG, Tf, APPs, and novel MWP bands is essential, as horse values do not reliably translate.

3.3.3 Biomarker Validation

Functional studies of MWP proteins may reveal new indicators of age-related immune decline or disease susceptibility.

3.3.4 Vaccine Strategies

Understanding immunosenescence will inform adjuvant selection and vaccination schedules optimized for geriatric donkeys.

4 Endocrine–Immune Disorders

Endocrine glands and the immune system engage in a dynamic cross-talk that influences metabolism, growth, reproduction, and host defense. In donkeys, endocrine dysfunctions are infrequently reported, yet emerging studies highlight conditions in which hormonal imbalance and immune regulation intersect. Two illustrative examples are idiopathic thyroid follicular atrophy in miniature donkeys and the unique biochemical and immunological properties of pituitary gonadotropins.

4.1 Idiopathic Thyroid Atrophy in Miniature Donkeys

4.1.1 Clinical Presentation and Hormonal Deficiency

Miniature donkeys typically maintain robust thyroid function, yet Mozaffari et al. (2025) recently identified three aged miniature jennies with advanced follicular atrophy of the thyroid. Clinically, these animals exhibited persistent weight loss, poor hair coat, and lethargy, without classic signs of hypothyroidism such as cold intolerance or bradycardia. Laboratory analysis revealed significantly decreased serum triiodothyronine (T_3) and thyroxine (T_4) concentrations, confirming a state of primary hormone deficiency despite the absence of overt inflammatory markers (Mozaffari et al. 2025) (Fig. 1).

4.1.2 Histopathological Characteristics

Histological examination of thyroid tissue demonstrated profound architectural collapse of thyroid follicles, nearly complete replacement of parenchyma by adipose

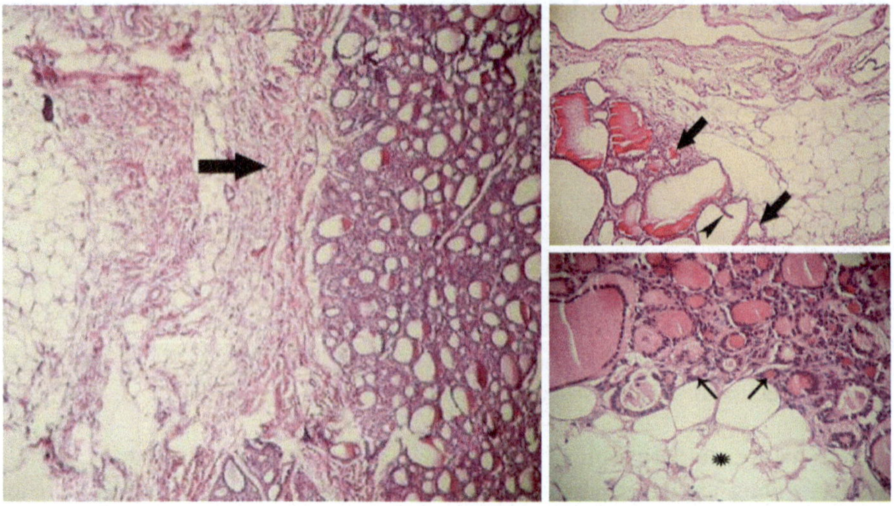

Fig. 1 Adipose tissue is seen next to clusters of follicles in irregular shape and varying size. The follicles are shrunken with empty and/or mild amount colloid (arrows) (left), Papillary projections (arrowhead) have extended into the follicular space (×100, H&E), Mature fat cells (asterisk) are in vicinity of degenerated and atrophic follicles (arrows). The mild increase of connective tissue in the interstitium of glands is visible (×200, H&E) (upper right) and A part of the thyroid gland is replaced by fat cells (asterisk) and connective tissue (arrow). The most remained thyroid follicles have small size with little eosinophilic colloid (×40, H&E) (lower right)

cells, and extensive fibrosis within the gland stroma. These degenerative changes occurred without notable lymphocytic infiltration, distinguishing this syndrome from autoimmune thyroiditis commonly observed in dogs and humans. The pattern of chronic, non-inflammatory degeneration suggests a pathogenesis rooted in atrophy rather than immune-mediated destruction (Mozaffari et al. 2025).

4.1.3 Etiology, Diagnosis, and Management

The underlying cause of this idiopathic follicular atrophy remains uncertain. Hypotheses include age-related follicular senescence, environmental exposure to goitrogenic compounds, nutritional iodine insufficiency, or mitochondrial dysfunction in follicular cells. Given the subtle clinical presentation, routine screening for serum T_3, T_4, and thyroid-stimulating hormone (TSH) in geriatric miniature donkeys is advisable to detect early dysfunction. Management strategies may encompass dietary iodine supplementation, environmental assessment for thyroid disruptors, and, if available, levothyroxine replacement therapy to restore metabolic balance (Mozaffari et al. 2025).

4.2 Endocrine Immunology: Pituitary Gonadotropins in Donkeys

4.2.1 Molecular Glycosylation Patterns

Pituitary-derived luteinizing hormone (LH) and follicle-stimulating hormone (FSH) regulate follicular development, ovulation, and steroidogenesis. Khan et al. (1984) purified donkey LH (dLH) and FSH (dFSH) and reported distinctive glycosylation profiles that diverge markedly from equine counterparts. Specifically, dLH contains only 13.5% carbohydrate and 1.9% sialic acid, compared with 26.7% and 5.8% in horse LH. Two dFSH isoforms—dFSH-A and dFSH-B—were differentiated by their oligosaccharide composition, suggesting variant half-lives and receptor affinities in vivo (Khan et al. 1984).

4.2.2 Biological Activity and Immune Modulation

Functional assays revealed that although dLH exhibits reduced in vitro receptor potency relative to equine LH, its lower sialylation confers an extended circulatory half-life, prolonging its endocrine effect. The two dFSH isoforms similarly differ in their capacity to stimulate follicular growth and estradiol production, implying that glycoform composition fine-tunes gonadal response. Beyond reproductive roles, gonadotropins can modulate immune cell function: in other species, LH and FSH receptors on lymphocytes influence cytokine secretion and cell proliferation. The unique glycosylation of donkey gonadotropins may therefore shape immunoendocrine interactions, potentially affecting seasonal immune competence (Khan et al. 1984).

4.2.3 Reproductive and Immunological Implications

Understanding the biochemical diversity of dLH and dFSH is essential for optimizing assisted-reproduction protocols in donkeys, including superovulation and embryo transfer. Moreover, exogenous administration of these hormones could have unintended immunomodulatory consequences, altering lymphocyte subsets or cytokine milieus. Future research should explore how variations in gonadotropin glycoforms affect both fertility outcomes and immune parameters, paving the way for integrated reproductive and immune health management in donkey populations.

5 Immunology Implications on Donkey Breeding

Reproductive immunology in donkeys (*Equus asinus*) is a rapidly advancing discipline at the interface of reproductive physiology, molecular immunology, and veterinary biotechnology. With donkeys regaining importance in agriculture, therapy, and biodiversity conservation, a deep understanding of their reproductive immune mechanisms is essential. These insights support both fertility optimization and the development of assisted reproductive technologies (ART), such as artificial insemination (AI) with frozen-thawed semen, while also guiding strategies for population management in feral or endangered contexts.

5.1 Uterine Immune Response Post-Insemination

Artificial insemination in donkeys elicits a strong but transient innate immune response within the uterus, characterized by the rapid recruitment of polymorphonuclear neutrophils (PMNs) into the uterine lumen in response to sperm antigens and seminal plasma (SP) components. This acute inflammatory reaction represents a physiological clearance mechanism, aimed at eliminating excess spermatozoa and reducing the risk of pathogen colonization. Interestingly, donkey spermatozoa are not typically engulfed by PMNs; instead, they either adhere to the cell surface or remain within protective halos, thereby preserving motility and viability for up to 3 hours at 37 °C, particularly when present at higher concentrations ($>500 \times 10^6$/mL). However, repeated uterine lavages to study this response can compromise endometrial integrity and reduce fertility, which has led to the development of alternative in vitro models using peripheral blood-derived PMNs. These systems provide a non-invasive and controlled approach for examining sperm–immune interactions while avoiding damage to reproductive tissues.

5.2 Seminal Plasma as an Immunological Modulator

Seminal plasma (SP) plays a crucial role in regulating both sperm survival and uterine immune tolerance in donkeys. It induces a controlled immunoactivation by stimulating polymorphonuclear neutrophils (PMNs) in a way that balances pathogen defense with the preservation of spermatozoa. Molecular studies have fractionated SP into different protein size groups, revealing that the 30–50 kDa and 50–100 kDa fractions exert the most significant effects on sperm motility and immune modulation. These active fractions are thought to contain key molecules such as cytokines, glycoproteins, and TGF-β–related factors, which together orchestrate an environment conducive to successful fertilization while maintaining uterine immune homeostasis (Table 1).

Table 1 Functional effects of donkey seminal plasma fractions on sperm and uterine immune modulation, with likely molecular components

Seminal Plasma Fraction (kDa)	Observed Effects	Likely Components
<3	Minimal impact	Small peptides
3–10	Mild activation	Antimicrobial peptides
10–30	Moderate support	Enzymes, cytokines
30–50	Strong motility preservation	Glycoproteins, TGF-β, ILs
50–100	Optimal sperm viability	Fertility-related proteins
>100	Variable	Immunoglobulins, structural proteins

5.3 Cytokine Signaling and Immune Tolerance

Seminal plasma promotes a tolerogenic uterine environment by delivering key immunomodulatory cytokines. Interleukins IL-6 and IL-10 help restrain excessive polymorphonuclear neutrophil (PMN) activation, thereby supporting sperm survival. Meanwhile, TGF-β enhances the activity of regulatory T cells (Tregs), reducing inflammation and facilitating embryo implantation. These coordinated mechanisms are conserved across equid species, highlighting the evolutionary stability of interactions between sperm and the female immune system.

5.4 Peripheral PMN Models

To bypass invasive uterine sampling, donkey reproductive immunology now applies in vitro PMN assays. In these systems, polymorphonuclear neutrophils (PMNs) are isolated from peripheral blood and co-incubated with spermatozoa and seminal plasma (SP) fractions at 37 °C. Experimental findings show that SP-treated PMNs display a reduced oxidative burst, which preserves sperm motility while simultaneously minimizing phagocytosis. This reproducible setup serves as a reliable proxy for uterine immune dynamics.

5.5 Endometrial Immune Landscape

Histological investigations of the jenny endometrium have mapped immune cell distributions. T lymphocytes are the predominant population, followed by macrophages, whereas eosinophils, neutrophils, and B cells occur only in low numbers. Importantly, in contrast to mares, jennies exhibit no significant variation in immune infiltration across different grades of endometrosis, suggesting a species-specific pattern of greater immunological resilience.

5.6 Hormonal Immunomodulation Via Inhibin Immunization

Seasonal patterns strongly influence male donkey fertility, with semen quality showing marked decline in winter. Immunization against inhibin has been shown to elevate circulating levels of follicle-stimulating hormone (FSH), luteinizing hormone (LH), anti-Müllerian hormone (AMH), and Activin A. These endocrine shifts stimulate spermatogenic activity and improve testicular histoarchitecture. The approach represents a promising tool to counteract seasonal infertility, thereby strengthening the success of artificial insemination (AI) programs in both conservation and production contexts.

5.7 Immunocontraceptive Research

For feral donkey population control, immunocontraception has emerged as a humane and non-lethal option. Zona pellucida (ZP) vaccines, prepared with either recombinant or native ZP antigens, elicit anti-ZP antibodies that interfere with sperm binding and ovulation. Reported outcomes include delayed pregnancies and reversible ovarian suppression, with only minimal adverse reactions. This strategy offers effective ecological management in areas where donkeys pose environmental pressures.

5.8 Perspectives and Applications

The study of donkey reproductive immunology provides translational insights across multiple domains. Assisted reproductive technologies (ART), including cryopreservation, artificial insemination, and embryo transfer, can benefit from integration of immunological modulators such as SP fractions and cytokines. Conservation breeding programs may capitalize on the species' apparent immunogenetic resilience to improve adaptability in situ and ex situ. At the same time, immunocontraceptive approaches furnish practical and welfare-oriented solutions for managing feral donkey populations while reducing their ecological impact.

6 Donkey Immunology in Therapeutic and Social Settings

Donkeys have become valued partners in animal-assisted interventions (AAIs), where their gentle nature supports human therapy, education, and recreation. Yet the very environments that make AAIs effective—frequent human contact, exposure to novel settings, and irregular routines—can challenge donkey immune competence. Understanding how social roles interact with immunology is vital for ensuring donkey welfare and maximizing the benefits they provide.

6.1 Immune Challenges of Animal-Assisted Interventions

In AAIs, donkeys often work in busy clinical or educational spaces, coming into contact with diverse human microbiota and environmental antigens. Chronic or acute stressors in these settings can elevate glucocorticoid release, notably cortisol, which suppresses lymphocyte proliferation, diminishes antibody production, and impairs both cell- and humoral-mediated responses (Sobrero et al. 2024). Donkeys' stoic behavior further complicates detection: they seldom display overt signs of illness, so immunological compromise may go unnoticed until disease manifests.

6.2 Management Practices and Immune Health

A survey of six AAI facilities in Northern Italy revealed inconsistent veterinary oversight and significant gaps in husbandry protocols. Only one site offered regular environmental enrichment, despite its documented role in reducing stress and bolstering immune function. Inadequate deworming schedules and suboptimal nutrition were common, further undermining immunological resilience (Sobrero et al. 2024). To counter these deficits, programs should implement routine health screenings—including complete blood counts and immunoglobulin titers—alongside pathogen surveillance, targeted anthelmintic regimens, and diets formulated to support immune cell metabolism.

6.3 Integrating a One-Welfare Approach

The One Welfare framework unites animal well-being, human health, and environmental stewardship. For donkeys in AAIs, this means designing programs that respect their behavioral needs—ample turnout, social companions, and mental stimulation—while minimizing stressors. Training handlers in low-stress techniques, establishing predictable routines, and enriching enclosures with foraging opportunities can mitigate HPA-axis activation and preserve immune homeostasis. Ultimately, healthy donkeys deliver more reliable therapeutic support, reinforcing the reciprocal benefits at the heart of One Welfare.

7 Stress and Immunomodulation in Donkeys

The nexus between stress and immunity is well-established across species, yet in donkeys, physiological stress responses often remain hidden behind their placid demeanor. Objective biomarkers are therefore indispensable for revealing how stressors—whether environmental, procedural, or immunological—reshape donkey immune function.

7.1 Hidden Stress and Reliable Biomarkers

Donkeys' innate stoicism masks pain and discomfort, making behavioral cues unreliable. Instead, glucocorticoids such as cortisol serve as quantifiable indicators of HPA-axis activation. Noninvasive sampling—particularly of feces—allows longitudinal monitoring without additional stress (Peterson et al. 2022).

7.2 Immunocontraceptive Vaccines and Biphasic Stress Responses

Peterson et al. (2022) evaluated jennies receiving two formulations of zona pellucida vaccines (recombinant ZP and native porcine ZP with Freund's adjuvants). Both preparations triggered acute elevations in fecal cortisol within 4 days, especially in animals developing injection-site abscesses. By day 35, cortisol levels fell below baseline, suggesting HPA-axis downregulation under chronic immunological challenge. This biphasic pattern—initial cortisol surge followed by a compensatory decline—mirrors findings in wildlife and laboratory models, where prolonged inflammation dampens HPA responsiveness (McEwen and Wingfield 2003).

7.3 Optimizing Protocols and Welfare Assessments

These insights demand refinement of vaccine protocols: selecting adjuvants that minimize local inflammation, reducing injection volume, or exploring alternative routes (e.g., intranasal) could temper stress. Moreover, comprehensive welfare assessments should integrate physiological metrics—fecal cortisol, salivary immunoglobulin A, heart-rate variability—with behavioral observations. As donkeys assume roles in conservation grazing and AAIs, aligning immunization strategies with stress-mitigation practices will safeguard both health and program efficacy.

8 Anaphylaxis in Donkeys

Though rare, anaphylaxis represents a life-threatening immune overreaction in donkeys. Rapid recognition and treatment are essential, yet the condition's subtle onset and the donkey's reserved behavior can delay critical interventions.

8.1 Pathophysiology of IgE-Mediated

Shock Anaphylaxis is triggered when specific IgE on mast cells and basophils binds multivalent antigens—such as Hymenoptera venom—prompting massive degranulation. Africanized honeybee stings deliver phospholipase A_2, melittin, hyaluronidase, and biogenic amines, each amplifying mast-cell activation. The resulting mediator surge increases vascular permeability, induces bronchoconstriction, and precipitates hypotension, culminating in anaphylactic shock (Tizard 2018).

8.2 Clinical Signs and Diagnosis

Symptoms typically arise within minutes of exposure. Early restlessness may give way to lateral recumbency and profound facial, tongue, and submandibular edema. Tachycardia with a weak pulse, inspiratory dyspnea, stridor, and pale or congested mucous membranes herald cardiovascular collapse. In field settings, diagnosis rests on acute onset following a known allergen exposure and recognition of hallmark signs, as laboratory assays for histamine or tryptase are rarely accessible in donkeys.

8.3 Emergency Management and Prevention

Immediate treatment follows the airway–breathing–circulation protocol. Intramuscular or intravenous epinephrine (0.01 mg/kg) reverses vasodilation and bronchospasm. Rapid isotonic crystalloid boluses restore intravascular volume, while intravenous dexamethasone (0.1 mg/kg) attenuates prolonged inflammation. H_1 and H_2 antihistamines may support vascular integrity, and oxygen supplementation or advanced airway interventions are critical if respiratory compromise persists. Removal of residual stingers prevents ongoing venom absorption (Tizard 2018).

Prognosis hinges on the speed of intervention: prompt therapy can result in full recovery but delays often lead to irreversible shock and multi-organ failure. Preventive strategies include habitat management to reduce insect encounters, pre-emptive antihistamine administration in high-risk situations, and staff training in early anaphylaxis recognition and treatment.

9 Hypersensitivity and Allergy in Donkeys

9.1 Overview of Hypersensitivity in Donkeys

Hypersensitivity reactions in donkeys span four immunopathologic types—Type I (immediate, IgE-mediated), Type II (antibody-dependent cytotoxic), Type III (immune-complex mediated), and Type IV (delayed, T-cell mediated)—mirroring the classification in other species but with donkey-specific clinical nuances. Species differences in skin thickness, hair coat density, and husbandry influence both the prevalence and presentation of allergic conditions in donkeys compared to horses.

9.2 Type I Hypersensitivity

9.2.1 Insect Bite Hypersensitivity

Insect bite hypersensitivity—analogous to "sweet itch" in horses—is the most frequent allergic disorder in donkeys. Saliva of Culicoides midges contains potent allergens that, upon cutaneous deposition, trigger an IgE-driven cascade. Clinically, affected donkeys develop intense pruritus, papular dermatitis, and regional alopecia

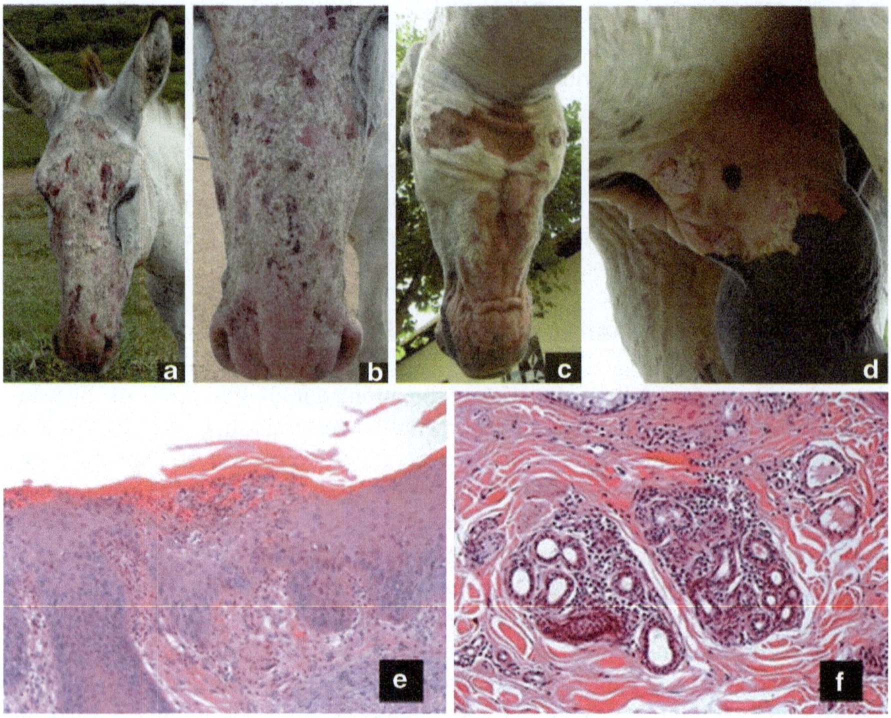

Fig. 2 Allergic dermatitis caused by *Culicoides* spp. in adult male Pêga donkeys. (**a**, **b**) Clinical presentation: hyperemia and skin thickening affecting the facial region, including the lips, extending from the nostrils to the frontal sinus, and the submandibular region, which shows alopecia. (**c**, **d**) Further clinical signs: alopecic, hyperemic, and thickened skin on the cheek (**c**), and scrotal skin exhibiting hyperemia and crust formation (**d**). (**e**, **f**) Histopathological findings: (**e**) Haired skin showing a focal intraepidermal pustule containing moderate numbers of intact and degenerated neutrophils, along with swollen keratinocytes displaying cytoplasmic vacuolation (indicative of edema). The pustule is overlaid by moderate keratin accumulation, cellular debris, and occasional erythrocytes forming a crust. The superficial dermis exhibits multifocal infiltrates of neutrophils, lymphocytes, plasma cells, and eosinophils. Severe, diffuse, irregular acanthosis and moderate, diffuse orthokeratotic hyperkeratosis are evident in the epidermis. Stained with hematoxylin and eosin (H&E). (**f**) Haired skin presenting severe inflammatory infiltration, consisting of lymphocytes, plasma cells, neutrophils, and eosinophils, distributed multifocally in the mid-dermis and around adnexal structures. H&E stain.

(mane, tail, ventrum). Chronic self-trauma leads to lichenification and predisposes to secondary bacterial or fungal infections (Fig. 2).

9.2.2 Atopic Dermatitis

Atopic dermatitis arises from sensitivity to inhaled or contact allergens—pollens, molds, and dust mites—and overlaps symptomatically with insect bite reactions. Seasonal urticaria, facial swelling, and ocular pruritus are common. Diagnosis relies on intradermal testing or allergen-specific IgE assays, which guide tailored immunotherapy via escalating subcutaneous or sublingual allergen exposure.

9.3 Type II and Type III Hypersensitivity

Although less commonly reported, donkeys may develop Type II and Type III reactions. Immune-mediated hemolytic anemia and glomerulonephritis can follow drug exposure or infections. Penicillin-induced serum sickness—a Type III reaction—manifests as fever, urticarial rash, and arthralgia, and typically resolves with drug withdrawal and short-term corticosteroids.

9.4 Parasitic Hypersensitivity: Habronematidosis

Habronematidosis ("summer sores") exemplifies a parasitic Type I/IV mixed hypersensitivity. Flies (Musca, Stomoxys) deposit Habronema larvae onto wounds or mucocutaneous junctions instead of the gastrointestinal tract. Larval death in skin elicits eosinophil-rich granulomas, ulceration, proud flesh, intense pruritus, and serosanguinous discharge. A certain limited heritable component has been observed in some breeds such as the Andalusian Donkey (Navas González et al. 2018) Histopathology confirms antigen–antibody complexes and mixed inflammatory infiltrates.

9.5 Diagnosis

A thorough history and clinical examination are essential. Intradermal skin testing remains the gold standard for Type I reactions, whereas biopsy and histopathology characterize Type IV and complex lesions such as habronematidosis. Allergen-specific IgE assays supplement skin testing for inhalant or food allergens.

9.6 Treatment and Management

Principles across hypersensitivity types include:

9.6.1 Antigen Avoidance

Stable management during peak insect activity and environmental allergen reduction.

9.6.2 Symptomatic Relief

Topical anti-pruritics, antihistamines, and non-steroidal or corticosteroid anti-inflammatories.

9.6.3 Immunomodulation

Allergen immunotherapy for atopic individuals.

9.6.4 Parasitic Control

Systemic macrocyclic lactones (ivermectin, moxidectin) to eradicate Habronema larvae, plus surgical debridement and topical corticosteroids to resolve granulomatous lesions.

9.7 Adverse Drug Reactions– Insights from Pharmacovigilance

Adverse drug reactions (ADRs) in donkeys follow classifications established in pharmacology: Type A (predictable, dose-dependent) and Type B (idiosyncratic, non–dose-related). Equine ADR data offer guidance, but species-specific pharmacokinetics require caution (Sumano López et al. 2020).

Gentamicin, at standard equine doses, can induce nephrotoxicity in donkeys due to variable drug clearance. Alpha-2 agonists (xylazine, detomidine) risk pronounced cardiovascular depression and respiratory compromise when administered intravenously. Donkeys' prolonged drug half-lives and altered receptor sensitivities mean NSAIDs and sedatives may exhibit both exaggerated effects and atypical side-effect profiles.

To mitigate ADRs, clinicians should adopt individualized dosing regimens, monitor renal and cardiovascular parameters vigilantly, and document all reactions. Establishing donkey-specific pharmacovigilance databases, conducting targeted pharmacokinetic studies, and developing species-tailored formularies will enhance therapeutic safety and welfare outcomes (Sumano López et al. 2020).

10 Hypoallergenicity, Hypersensitivity and Allergy to Donkey Products

10.1 Donkey Milk as a Hypoallergenic Food—Immunological Perspectives

Donkey milk has garnered increasing interest as both a nutritional and therapeutic substitute for cow's milk, largely because its macronutrient and protein composition more closely mirror human breast milk. In particular, donkey milk is characterized by markedly reduced levels of αs1-casein and κ-casein—the primary bovine allergens—and elevated concentrations of lysozyme and β-lactoglobulin (Vincenzetti et al. 2014). These attributes not only confer a low allergenic profile but also underpin a range of immunomodulatory benefits. This chapter first examines the immunological basis for donkey milk's hypoallergenicity, then reviews reported cases of donkey-milk allergy (including occupational and percutaneous sensitization), and finally explores its broader immune-supportive properties—from cytokine modulation to exosome-mediated signaling—before discussing practical implications and future research priorities.

10.2 Hypoallergenic Profile and Immunological Diagnostics

Compared with bovine and caprine milks, donkey milk contains minimal total casein and negligible αs1- and κ-casein, while retaining high levels of lysozyme and β-lactoglobulin (Vincenzetti et al. 2014). In children with confirmed cow's milk protein allergy (CMPA), sera showed significantly lower IgE binding to donkey milk proteins than to cow or goat proteins, confirming its reduced capacity to trigger IgE-mediated hypersensitivity (Vincenzetti et al. 2014).

Clinically, donkey milk overcomes limitations of soy-based or extensively hydrolyzed formulas, which up to 20% of CMPA infants still cannot tolerate. Its intact protein fractions facilitate non-invasive CMPA diagnostics: skin-prick tests and specific IgE assays can be performed using donkey milk antigens without risking cross-reactivity to bovine proteins (Vincenzetti et al. 2014).

10.3 Donkey Milk Allergy: Occupational and Percutaneous Sensitization

Although rare, IgE-mediated allergy to donkey milk has been documented in both occupational and non-occupational settings. While serum albumin and beta-lactoglobulin have been identified as potential allergens in donkey milk (www.allergome.com), their direct implication in donkey milk allergy has not been demonstrated. Recent findings provide strong evidence that donkey milk lysozyme is a key factor in triggering allergic reactions in patients who do not exhibit cross-reactivity with cow's milk.

10.3.1 Occupational Exposure

A 35-year-old laboratory technician with atopic dermatitis developed skin and respiratory symptoms after repeated handling of raw and processed donkey milk in a food-analysis environment. Subsequent ingestion of UHT-treated donkey milk precipitated anaphylaxis. Diagnostic work-up—including skin-prick testing (positive for donkey milk, negative for cow's milk), basophil activation tests, and immunoblotting—identified ten donkey-specific allergens (β-lactoglobulin, lactotransferrin, lysozyme, β-casein) with no cross-reactivity to bovine proteins (Giorgis et al. 2018) (Fig. 3).

10.3.2 Percutaneous Sensitization

Peeters et al. (2017) reported a 25-year-old woman who experienced facial angioedema, dyspnea, and collapse minutes after ingesting donkey milk. She had used donkey-milk-based emollients for childhood atopic dermatitis, implicating epicutaneous exposure as the priming event. Positive skin-prick and patch tests to

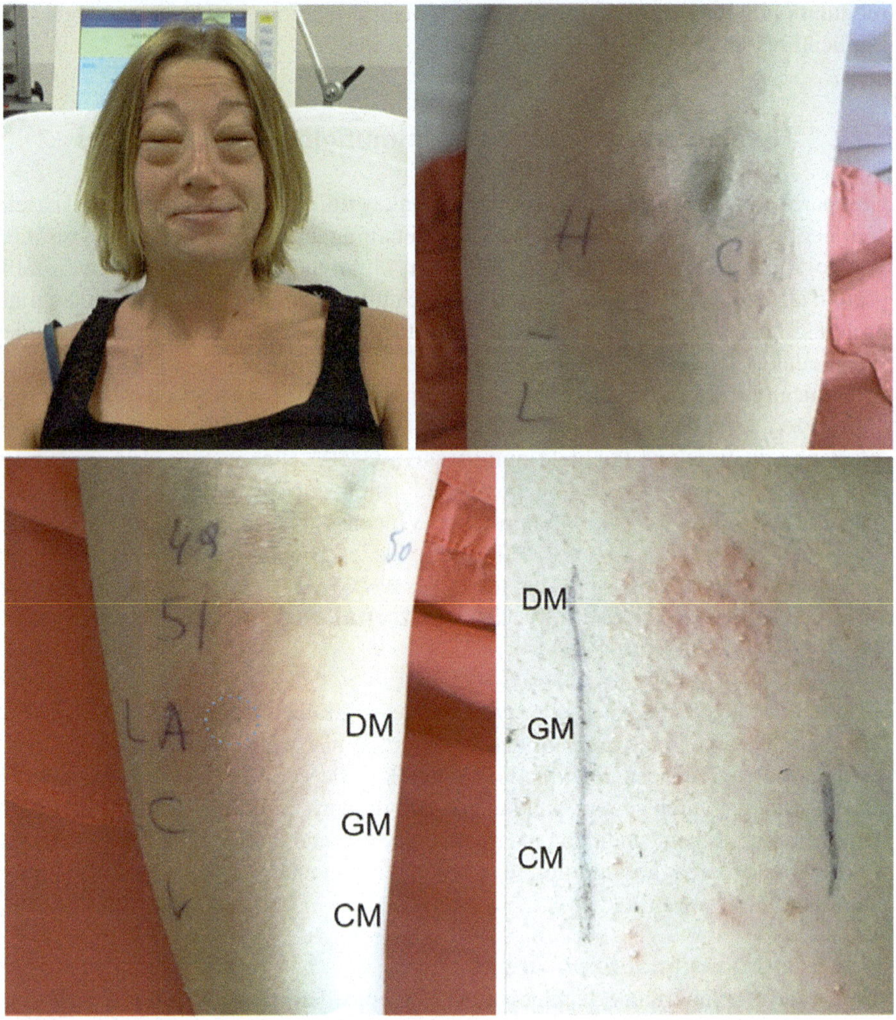

Fig. 3 Facial angioedema (upper left); Positive skin prick test with donkey's milk (DM) (++) (lower left) and negative skin prick tests with goat's milk (GM) and cow's milk (CM); results compared with histamine (H) and glycerosaline solution (upper right); positive patch test with donkey's milk (DM) at day 4 (++) (lower right)

donkey milk—and negative tests to cow's and goat's milk—demonstrated specific sensitization, likely involving both IgE-mediated (type I) and T cell–mediated (type IV) hypersensitivity (Peeters et al. 2017; Strid et al. 2005).

These cases underline that even "hypoallergenic" foods can induce sensitization when exposure is repeated via inhalation, skin contact, or ingestion. Laboratories and production facilities should implement protective measures—gloves, masks, localized ventilation—and educate staff on safe handling of donkey-derived products (Giorgis et al. 2018).

10.4 Immunomodulatory and Anti-Allergic Properties

Beyond its low allergenicity, donkey milk actively supports immune homeostasis through multiple mechanisms.

10.4.1 Cytokine Balance

In vitro studies on human peripheral blood mononuclear cells show that donkey milk stimulates both pro-inflammatory cytokines (TNF-α, IL-1β) necessary for pathogen clearance and the anti-inflammatory mediator IL-10 to prevent excessive immune activation (Jirillo and Magrone 2014). This dual modulation is particularly beneficial for individuals with chronic inflammatory conditions or immunosenescence.

10.4.2 Antimicrobial and Cardioprotective Factors

Donkey milk's high lysozyme concentration exerts direct bactericidal effects, while its promotion of nitric oxide synthesis provides cardioprotective, antibacterial, and anti-atherogenic benefits. Lactoferrin—another abundant whey protein—binds bacterial lipopolysaccharide (LPS), neutralizing endotoxins and lowering the risk of septic complications (Jirillo and Magrone 2014).

10.4.3 Clinical Outcomes in CMPA

In pediatric cohorts, more than 80% of CMPA (Cow's Milk Protein Allergy) children tolerate donkey milk without adverse events, often showing improved growth trajectories and reduced allergy symptom scores (Jirillo and Magrone 2014).

10.5 Donkey Milk Exosomes: Emerging Immune Modulators

Exosomes—30–150 nm nanovesicles loaded with proteins, RNAs, and lipids—mediate intercellular communication and immune programming. Four-dimensional label-free proteomic analysis identified 2293 proteins in donkey milk exosomes, with colostrum-derived vesicles enriched in immune-development factors such as complement C3, ACTR2, GPx3, and ANGPTL4 (upregulated >160-fold vs. mature milk). These proteins enhance leukocyte migration, epithelial barrier integrity, antioxidant defenses, and vascular maturation in the neonatal gut (Liu et al. 2025).

As lactation proceeds, exosomal cargo shifts toward metabolic regulators (ACSL1, ACSL4, PLIN2), reflecting the foal's transition from immune priming to nutritional support. The lipid bilayer protects exosomal contents from gastric degradation, improving bioavailability and highlighting their potential as natural nutraceuticals or adjunctive immunotherapies in both veterinary and human medicine (Liu et al. 2025).

11 Immune-Mediated Cytopenias

Immune-mediated cytopenias in donkeys encompass disorders in which the animal's own immune system targets and destroys circulating blood cells. The most commonly encountered of these syndromes is immune-mediated thrombocytopenia (IMT), characterized by rapid platelet destruction and a high risk of hemorrhage. Although literature on donkey IMT is sparse, case reports and extrapolations from equine medicine provide a framework for understanding its pathogenesis, diagnosis, and management.

11.1 Immune-Mediated Thrombocytopenia in Donkeys

11.1.1 Etiology and Pathogenesis

IMT arises when platelets become coated with antiplatelet immunoglobulins (IgG or IgM), leading to their removal by splenic macrophages or direct complement-mediated lysis. In donkeys, two principal triggers have been identified:

11.1.1.1 Allogeneic Plasma Transfusion

Exposure to foreign plasma proteins can sensitize the recipient, prompting antiplatelet antibody formation within 24–72 hours. Case reports in miniature donkeys illustrate how even a single transfusion may suffice to induce fulminant thrombocytopenia.

11.1.1.2 Drug-Induced Hapten Sensitization

Certain antibiotics, notably ceftiofur, act as haptens that bind platelet surfaces and elicit a cytotoxic response upon reexposure. A unique immunogenetic consideration is the "donkey factor," a redcell antigen absent in horses. When equine plasma is administered without crossmatching, donkey recipients or mule foals nursing from sensitized mares may develop alloimmune thrombocytopenia akin to neonatal isoerythrolysis.

11.1.2 Clinical Presentation and Diagnosis

Affected donkeys typically exhibit sudden spontaneous bleeding—petechiae, ecchymoses, mucosal hemorrhages, and oozing at venipuncture or surgical sites—despite normal coagulation times. A complete blood count reveals isolated thrombocytopenia (often <20,000/μL). Blood smears rule out pseudothrombocytopenia and parasites, while bone marrow aspiration, in refractory cases, shows megakaryocytic hyperplasia. Although specialized antiplatelet assays exist, diagnosis usually rests on clinical context, exclusion of other causes, and rapid platelet decline following suspected exposure.

11.1.3 Treatment and Supportive Care

Immediate withdrawal of the inciting agent—cessation of plasma transfusions or discontinuation of ceftiofur—is paramount. High-dose corticosteroids (e.g., dexamethasone 0.05–0.1 mg/kg IV every 12 hours) suppress antibody production, downregulate Fcγ receptors, and prolong platelet survival. In steroid-refractory cases, azathioprine (1–2 mg/kg PO daily) or cyclosporine (5 mg/kg PO twice daily) may be considered, although donkey-specific pharmacokinetics are lacking. When hemorrhage is life-threatening, transfusion of cross-matched whole blood or platelet-rich plasma offers hemostatic support; equine blood products require careful compatibility testing to minimize further sensitization.

Adjunctive measures include strict stall rest, fluid support to maintain hemodynamics, gastroprotection with omeprazole during corticosteroid therapy, and regular monitoring of platelet counts every 24–48 hours to guide immunosuppressive tapering.

11.1.4 Prognosis and Prevention

Donkeys whose platelet counts rebound above 50,000/μL within 48 hours of steroid initiation carry a favorable outlook. Conversely, persistent thrombocytopenia, recurrent bleeding, or complications such as pancreatitis and systemic inflammatory response syndrome portend a poor prognosis.

Preventive strategies emphasize cross-matching before all transfusions, selecting true donkey donors for neonatal plasma, and avoiding known haptenic drugs when alternatives exist. Vigilant hematologic surveillance for at least 72 hours after transfusion or high-risk drug administration enables early detection and intervention, reducing morbidity and mortality.

12 Autoimmune and Bullous Skin Disorders: Cutaneous Lupus Erythematosus in a Donkey Case

12.1 Background

Autoimmune skin diseases are exceptionally rare in donkeys, leaving much of their clinical spectrum unexplored. Cutaneous lupus erythematosus (CLE), a prototype interface dermatitis in humans and dogs, involves autoreactive immune cells attacking basal keratinocytes and dermal–epidermal junction components. Prior to 2020, no bona fide cases of donkey CLE had been documented, underscoring the novelty of recent findings (Mosca et al. 2020).

12.2 Case Presentation

12.2.1 Signalment and History

A 5-year-old neutered male donkey presented with an eight-month history of symmetrical, non-pruritic skin lesions. The owner reported progressive alopecia, erythema, scaling, and crust formation on the head, trunk, and distal limbs, unresponsive to empirical antimicrobial and antiparasitic treatments (Mosca et al. 2020).

12.2.2 Clinical Examination

Physical examination confirmed well-demarcated areas of depigmentation, fine scaling, and adherent crusts over the muzzle, periocular regions, dorsolateral trunk, and hocks. The donkey was in good body condition, afebrile, and exhibited no systemic signs such as lethargy or joint pain.

12.2.3 Diagnostic

Workup Routine hematology and biochemistry were unremarkable. Antinuclear antibody (ANA) titers fell within reference ranges, arguing against systemic lupus erythematosus. Skin biopsies underwent histopathology and immunohistochemistry:

12.2.3.1 Histopathology

Histopathology revealed a lichenoid lymphocytic interface dermatitis with vacuolar degeneration of basal keratinocytes and pigmentary incontinence.

12.2.3.2 Immunohistochemical Staining

Immunohistochemical staining demonstrated a predominance of $CD8^+$ T lymphocytes at the dermoepidermal junction, consistent with CLE's cytotoxic T-cell–mediated pathology.

12.2.3.3 Direct Immunofluorescence

Direct immunofluorescence for IgG, IgM, C3, and fibrinogen was negative, reflecting a non–immune complex–driven cutaneous lupus variant (Mosca et al. 2020).

12.3 Treatment and Outcome

12.3.1 Prednisolone Initiation

Oral prednisolone was started at 1 mg/kg daily for 2 weeks. Partial lesion regression occurred, but flares recurred during tapering, indicating insufficient monotherapy.

12.3.2 Methotrexate Maintenance

Methotrexate (MTX) was added at 0.3 mg/kg intramuscularly once weekly, with folinic acid rescue 24 hours later. Within 4 weeks, erythema and crusting markedly diminished; by 3 months, alopecic areas had repigmented and scaliness resolved entirely.

12.3.3 Monitoring and Safety

Hematologic (complete blood count) and hepatic (ALT, AST) panels were monitored monthly. No cytopenias or transaminase elevations emerged over 1 year of combined prednisolone and MTX therapy, supporting MTX's tolerability in donkeys (Mosca et al. 2020).

12.4 Discussion

This first donkey CLE case highlights several key points:

CLE in donkeys manifests clinically like canine or human disease but may lack detectable antibody deposition on immunofluorescence. A $CD8^+$ T-cell–driven interface dermatitis underlies lesion pathology, reinforcing the role of cytotoxic lymphocytes in skin autoimmunity. Methotrexate, a folate-antagonist that curtails DNA synthesis in rapidly dividing autoreactive cells, proved safe and effective as a steroid-sparing agent, mirroring its use in human and canine CLE.

Future studies should explore donkey-specific immune checkpoints, the role of environmental triggers in disease onset, and pharmacokinetics of immunosuppressants. Integrating immunohistochemistry into equine dermatology will sharpen diagnostic accuracy and broaden therapeutic horizons for donkey autoimmune disorders.

13 Serum Sickness and Passive Antibody Reactions

13.1 Pathogenesis

Serum sickness in donkeys represents a classical type III hypersensitivity arising after administration of heterologous immunoglobulins—most commonly equine-derived antitoxins or antisera. When non-donkey proteins enter circulation, the donkey immune system generates antibodies that form circulating immune complexes with the foreign immunoglobulins. Deposition of these complexes in vessel walls, synovial membranes, and renal glomeruli activates complement cascades and recruits inflammatory cells, driving the clinical manifestations of serum sickness (Tizard 2018).

13.2 Clinical Presentation

Although specific reports in donkeys are limited, extrapolation from equine data suggests onset between 7 and 14 days post-administration of passive immunotherapy. Affected donkeys may develop fever, urticarial skin lesions, polyarthritis with joint effusion, and—in severe cases—protein-losing glomerulonephritis manifesting as peripheral edema and proteinuria. Subtle signs such as decreased appetite or mild lameness can easily be missed in the stoic donkey, underscoring the need for high clinical suspicion when using heterologous sera.

13.3 Prevention and Management

To minimize risk, clinicians should preferentially use species-matched or recombinant antitoxins. When equine sera are unavoidable, premedication with non-steroidal anti-inflammatories and close post-administration monitoring for pyrexia, cutaneous reactions, and joint swelling are recommended. Future research must characterize donkey-specific immunogenic epitopes and develop homologous immunobiologics to replace cross-species products (Allen et al. 1985; McKenzie 2012).

14 Vasculitis and Vascular Immune-Mediated Disease

14.1 Mechanisms of Equid Vasculitis

Vasculitis in donkeys involves immune-mediated inflammation of vessel walls, typically through immune complex deposition (type III) or anti-endothelial autoantibodies (type II). Complement activation and leukocyte infiltration lead to fibrinoid necrosis, endothelial damage, and thrombus formation. Although documented vasculitic syndromes in horses include equine viral arteritis and purpura hemorrhagica, specific studies in donkeys are conspicuously absent (Tizard 2018).

14.2 Clinical Manifestations

Donkeys with vasculitis may exhibit localized or generalized signs. Limb or ventral edema, petechiae, mucosal bleeding, and diminished capillary refill indicate compromised vascular integrity. Systemic involvement can produce colic signs, renal impairment, or neurologic deficits if cerebral vessels are affected. Stress and immunosuppression—common in therapy or therapy-adjunct donkeys—may exacerbate vulnerability to vascular immune injury (Sobrero et al. 2024).

14.3 Diagnosis and Treatment

Diagnosis relies on clinical assessment, elevated acute-phase proteins (e.g., fibrinogen, globulins), and histopathology demonstrating leukocytoclastic vasculitis with immune complex deposition. Management parallels equine protocols: high-dose corticosteroids to quell inflammation, supportive care to maintain perfusion, and targeted antimicrobial therapy if an infectious trigger is identified. Development of donkey-specific reference intervals for vascular biomarkers would enhance early detection and treatment efficacy (McKenzie 2012).

15 Purpura Hemorrhagica-Like Syndromes in Donkeys

15.1 Etiology and Pathogenesis

Purpura hemorrhagica in horses follows Streptococcus equi infection or vaccination, whereby anti-M protein antibodies cross-react with vascular endothelium, forming immune complexes that activate complement and neutrophils. Given the similarity of equid immune systems, donkeys exposed to repeated streptococcal antigens or adjuvant-packed vaccines may experience analogous vascular reactions, although clinical evidence remains anecdotal (McKenzie 2012; Tizard 2018).

15.2 Clinical Features

Affected donkeys develop rapid, symmetrical edema of limbs, head, and ventrum, accompanied by petechiae and mucosal hemorrhages. Hypoproteinemia and proteinuria may ensue from increased vascular permeability. Because donkeys often understate pain and discomfort, these hallmarks may go unrecognized until advanced.

15.3 Prevention

Preventive measures include judicious vaccination—spacing immunizations appropriately and using purified antigen formulations—and screening for streptococcal carriage. Monitoring of serum proteins and renal function following high-risk immunizations could enable early intervention before full-blown vasculitic crises develop.

16 Immune-Associated Glomerulonephritis

16.1 Pathophysiology

Immune-associated glomerulonephritis arises when circulating or in situ immune complexes lodge in glomerular capillary walls, activating complement and recruiting inflammatory cells. Chronic deposition leads to mesangial proliferation, basement membrane thickening, and loss of filtration integrity (Tizard 2018).

16.2 Clinical Presentation in Donkeys

Although rarely reported, donkeys with glomerulonephritis may exhibit subtle signs: weight loss, intermittent edema, and mild proteinuria. Hypoalbuminemia contributes to ventral edema, while azotemia may appear in advanced disease.

16.3 Diagnosis and Monitoring

Routine serum chemistry and urinalysis—measuring protein:creatinine ratios—should be incorporated into donkey health protocols, particularly for animals receiving frequent immunotherapies. Definitive diagnosis requires renal biopsy with immunofluorescence to demonstrate immunoglobulin and complement deposition.

16.4 Future Directions

Establishing baseline renal immune parameters in donkeys will improve recognition of early glomerular disease. Longitudinal studies of donkeys undergoing chronic antigenic exposure—such as repeated serum therapies or persistent infections—are needed to delineate risk factors and refine preventive care.

17 Immune-Mediated Myopathies in Donkeys

17.1 Pathogenesis of Autoimmune Myositis

Under normal conditions, donkey myofibers do not express MHC class I or II on their sarcolemma, preserving self-tolerance. In immune-mediated polymyositis, inflammatory stimuli (e.g., vaccine adjuvants or parasitic antigens) induce MHC I/II upregulation, enabling presentation of endogenous peptides to $CD8^+$ cytotoxic and $CD4^+$ helper T lymphocytes. $CD8^+$ cells then release perforin and granzyme, causing focal myofiber necrosis, while $CD4^+$ cells secrete interferon-γ to amplify antigen presentation and recruit macrophages.

17.2 Immune Cell Infiltrates and Cytokine Profiles

Muscle biopsies in equine polymyositis reveal dense endomysial infiltrates of $CD8^+$ T cells at necrotic fibers, perimysial $CD4^+$ cuffs, and MHC II-positive macrophages. A similar pattern is anticipated in donkey lesions, accompanied by local production of TNF-α, IL-1β and interferon-γ, which both drive myofiber apoptosis and impair satellite-cell–mediated repair.

17.3 Immunodiagnostic Biomarkers

Acute episodes of donkey myositis would manifest as marked elevations in serum creatine kinase (often 10- to 50-fold) and lactate dehydrogenase, while electromyography shows short-duration, low-amplitude motor-unit potentials consistent with myopathy. Definitive diagnosis relies on histopathology and immunohistochemistry for CD3, CD4, CD8, MHC I/II and complement membrane attack complex (C5b-9) deposition on myofibers (Valberg and Mckenzie 2025).

17.4 Immunotherapeutic Strategies

Donkeys are particularly sensitive to corticosteroids; an initial prednisolone dose of 1–1.5 mg/kg/day, tapering over weeks, combined with cyclosporine (4–5 mg/kg/day) as a steroid-sparing agent, is recommended. Controlled, low-impact exercise under immunosuppression supports regeneration of healthy muscle fibers while minimizing relapse risk (Valberg and Mackenzie 2025).

18 Granulomatous and Pyogranulomatous Disorders in Donkeys

18.1 Immunobiology of Granuloma Formation

Persistent antigens—such as mycobacterial or fungal components—engage pattern-recognition receptors (e.g., TLR2, dectin-1) on donkey macrophages, triggering Th1 polarization. Interferon-γ from $CD4^+$ T helper-1 cells promotes macrophage differentiation into epithelioid cells and multinucleated giant cells, which cluster into granulomas to sequester the offending agent (Wynn et al. 2013).

18.2 Pyogranulomatous Inflammation

When acute neutrophilic and chronic macrophage responses coexist—exemplified by pythiosis—IL-8 and leukotriene B4 recruit neutrophils that release myeloperoxidase and elastase, while adjacent macrophages maintain granulomatous architecture via IL-12 and TNF-α. This results in necrotic, pus-filled nodules encased by epithelioid cells (Tizard 2018).

18.3 Immunodiagnostic Strategies

Deep tissue biopsy followed by histopathology with Ziehl–Neelsen or periodic acid–Schiff staining identifies acid-fast bacteria or fungi. Immunohistochemistry for CD68 (macrophages), myeloperoxidase (neutrophils) and CD3 (T cells), plus PCR or culture of biopsy material, confirms the specific pathogen (Tizard 2018).

18.4 Immunomodulatory Therapy

In infectious granulomas, long-term, pathogen-directed antimicrobials or antifungals are combined with anti-TNF agents (e.g., pentoxifylline at 20 mg/kg PO BID) to reduce macrophage cytokine release. Idiopathic cases may respond to low-dose corticosteroids (0.5–1 mg/kg/day) to dampen chronic macrophage activation (Tizard 2018).

19 Chronic Inflammatory Bowel Disease in Donkeys

19.1 Mucosal Immune Tolerance and Dysregulation

Donkey intestinal dendritic cells sample luminal antigens and present them via MHC II to naïve $CD4^+$ T cells. Under homeostasis, regulatory T cells (Tregs) secrete IL-10 and TGF-β to suppress effector responses. In CIBD, decreased Treg function and excessive IL-12 and IL-6 skew $CD4^+$ differentiation toward Th1 and Th17 phenotypes, driving mucosal inflammation.

19.2 Epithelial Innate Responses and Barrier Dysfunction

Pattern-recognition receptors on donkey enterocytes (TLR4 for LPS, TLR5 for flagellin) activate MyD88-dependent NF-κB signaling, inducing secretion of IL-1β, IL-8 and TNF-α. Chronic activation disrupts tight-junction proteins, increasing permeability and enabling microbial translocation that perpetuates inflammation (McKenzie 2012).

19.3 Adaptive Immune Imbalance

An imbalance favoring Th17 over Treg responses—marked by elevated IL-17A and IL-23 alongside reduced IL-10—sustains neutrophil recruitment and plasma cell infiltration in the lamina propria. Donkeys may exhibit higher baseline IL-10, delaying overt clinical signs despite ongoing subclinical mucosal injury (Thiemann and Sullivan 2019).

19.4 Immunodiagnostics

Endoscopic or full-thickness biopsy with immunohistochemical staining for CD3 (T cells), CD20 (B cells) and myeloperoxidase (neutrophils) defines infiltrate composition. Quantitative PCR for mucosal cytokine transcripts (IFN-γ, IL-17A, IL-10) provides objective measures of immune activation, while fecal calprotectin assays noninvasively assess neutrophilic inflammation (Thiemann and Sullivan 2019).

19.5 Immunotherapeutic and Prophylactic Measures

Dietary antigen elimination—transitioning to novel or hydrolyzed protein forages—reduces antigenic drive. Probiotics containing Lactobacillus and Bifidobacterium species support barrier integrity. In moderate to severe cases, corticosteroids (prednisolone 1 mg/kg/day) combined with azathioprine (1 mg/kg every other day) restore Treg/Th17 balance. Emerging biologics—anti-IL-23 antibodies or low-dose IL-10 analogs—warrant donkey-specific pharmacokinetic studies (Thiemann and Sullivan 2019).

20 Immunological Processes Underlying Donkey Hoof Inflammation

Inflammatory hoof disorders in donkeys—particularly laminitis, white line disease, and subsolar abscessation—result from a coordinated cascade of innate and adaptive immune events. Unraveling these pathways offers avenues for earlier diagnosis and more precise therapies.

20.1 Innate Immune Activation in Laminitis

Endotoxemia, often secondary to gastrointestinal dysbiosis or grain overload, initiates donkey laminitis by engaging toll-like receptor 4 on circulating monocytes and tissue macrophages. TLR4 ligation triggers NF-κB translocation into the nucleus, driving release of tumor necrosis factor-α, interleukin-1β and interleukin-6. These cytokines markedly increase vascular permeability within the lamellar microvasculature and prime endothelial cells for leukocyte adhesion and extravasation.

20.2 Neutrophil-Mediated Tissue Injury

Recruited neutrophils are central to early laminar damage. In donkeys, activated neutrophils generate reactive oxygen species and release neutrophil extracellular traps—DNA–histone webs decorated with myeloperoxidase and elastase. Although NETs immobilize pathogens, they indiscriminately degrade the basement membrane of secondary epidermal lamellae. This enzymatic assault directly contributes to keratinocyte detachment and mechanical failure of the hoof's suspensory apparatus.

20.3 Matrix Metalloproteinases and Extracellular Matrix Degradation

Pro-inflammatory cytokines amplify destruction by upregulating matrix metalloproteinases, especially MMP-2 and MMP-9, in lamellar tissue. These proteases degrade type IV collagen within the basal laminar membrane. In donkeys, an imbalance between MMPs and their tissue inhibitors (TIMPs) permits unrestrained matrix degradation, hastening laminar separation and exacerbating hoof capsule distortion.

20.4 Endothelial Dysfunction and Oxidative Stress

Cytokines and endotoxin impair endothelial nitric oxide synthase activity, reducing nitric oxide bioavailability while elevating endothelin-1 expression. The resulting vasoconstriction and capillary stasis precipitate local ischemia. Ischemic hypoxia then intensifies oxidative stress, creating a feed-forward loop in which reactive oxygen species inflict further damage on vascular and laminar cells.

20.5 Macrophage Polarization and Chronic Remodeling

As acute inflammation wanes, macrophage phenotypes shift. M1 macrophages sustain pro-inflammatory signaling, whereas M2 macrophages release transforming growth factor-β1, promoting fibroplasia. In chronic or recurrent laminitis, persistent M2 activity drives fibrotic remodeling of the laminar basement membrane. This fibrosis underlies the irreversible hoof distortion and loss of biomechanical function seen in many long-term donkey cases.

20.6 Immunopathogenesis of White Line Disease and Subsolar Abscesses

When hoof-horn integrity is compromised, bacteria invade through the white line or sole, triggering a localized inflammatory response. Monocytes, macrophages and dendritic cells recognize bacterial pathogen-associated molecular patterns, releasing cytokines that recruit additional leukocytes. The ensuing purulent inflammation and enzymatic tissue destruction culminate in abscess formation. Donkeys' stoic pain responses often delay detection, allowing lesions to extend deeper before clinical lameness manifests.

20.7 Immunological Diagnostics

Objective biomarkers can augment clinical evaluation of hoof inflammation in donkeys. Measuring acute-phase proteins—serum amyloid A and fibrinogen—provides a window into systemic inflammation, with elevated concentrations correlating with laminar pathology. At the local level, lamellar biopsies analyzed for MMP-9 activity and myeloperoxidase content yield direct indices of matrix degradation and neutrophil activation. Cytokine profiling in serum or hoof exudate—quantifying TNF-α, IL-1β and IL-6 via ELISA—further characterizes the inflammatory milieu. Establishing donkey-specific reference intervals for these parameters will improve early detection and monitoring of therapeutic responses.

20.8 Immunomodulatory Therapeutic Targets

Translating immunopathogenic insights into interventions opens new opportunities for preserving donkey hoof health. Neutralizing circulating endotoxin or blocking TLR4 early in the disease cascade can attenuate cytokine storms and reduce leukocyte infiltration. Degrading neutrophil extracellular traps with recombinant DNase offers a strategy to limit basement membrane damage. Modulating MMP activity—either through synthetic inhibitors or by enhancing TIMP expression—helps maintain laminar matrix integrity. Concurrent antioxidant therapy with vitamins E and C can quench reactive oxygen species, protecting vascular and laminar cells. Finally, therapies that guide macrophage polarization toward a balanced M1/M2 profile may prevent excessive fibrosis in chronic cases. Tailoring these approaches to Equus asinus promises to halt inflammatory cascades before irreversible hoof distortion occurs.

21 Immune-Associated Ocular Diseases in Donkeys

Immune-mediated eye disorders such as recurrent uveitis inflict significant welfare and functional losses in donkeys, yet their subtler clinical signs and scarce species-specific data hinder timely recognition. Drawing on field surveys, ultrasonographic studies and clinical experience, this section delineates epidemiology, immunopathogenesis, diagnosis, management and prevention of immune-driven ocular disease in Equus asinus.

21.1 Epidemiology and Clinical Presentation

Although equine recurrent uveitis affects up to 25% of horses in some regions, only sparse reports detail its occurrence in donkeys. A UK sanctuary study encompassing over 1400 post-mortem examinations found 2.9% of donkeys with lesions consistent with quiescent uveitis, and nearly one in five working donkeys in northern India exhibited chronic ocular inflammation—episcleral injection, aqueous flare or irregular pupil shape—despite no overt history of trauma (Bradley 2022; Grove 2008). Unlike horses, donkeys' stoic nature and retained vision in early attacks often delay presentation until secondary cataract formation, posterior synechiae or complications such as glaucoma become apparent on slit-lamp examination.

21.2 Immunopathogenesis of Recurrent Uveitis

Donkey recurrent uveitis likely mirrors the autoimmune cascade characterized in horses: initial breach of the blood–ocular barrier—frequently following systemic leptospiral exposure—permits uveal antigen presentation by iris and ciliary-body dendritic cells to naïve $CD4^+$ T lymphocytes in regional lymph nodes (Malalana et al. 2015). Activated T helper-1 and T helper-17 subsets home back to the eye, releasing interferon-γ and interleukin-17, which amplify local cytokine networks and upregulate matrix metalloproteinases that degrade basement membranes. Concurrent B-cell infiltration generates anti-uveal autoantibodies and immune complexes that activate complement and recruit neutrophils, perpetuating tissue injury (Bradley 2022; Grove 2008). Over successive flares, tissue remodeling—driven by M2 macrophage-derived transforming growth factor-β1—leads to fibrosis of the iris, lens capsule opacification and trabecular meshwork obstruction (Fig. 4).

21.3 Diagnostic Strategies

Early identification of immune-associated uveitis in donkeys hinges on thorough ophthalmic examination adapted to their behavioral traits. After initial history and

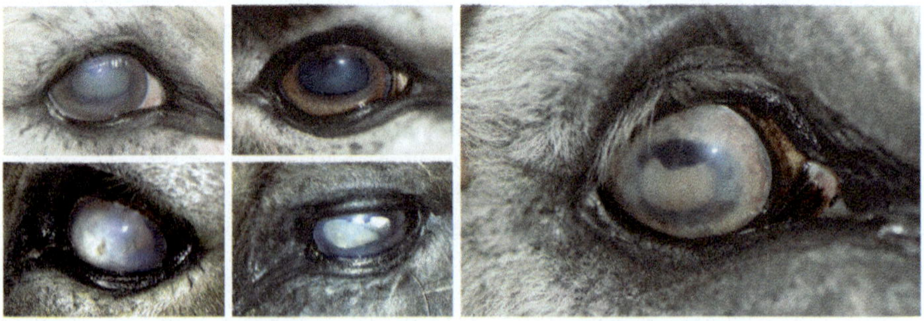

Fig. 4 Active uveitis in the right eye of a donkey, with diffuse corneal oedema, aqueous flare and keratic precipitates. There is also corneal vascularisation present (Upper left), Quiescent uveitis in the right eye of a donkey, showing iris rests and early cataract formation (upper middle), Anterior lens luxation in the left eye of a donkey with chronic uveitis (lower left), Posterior lens luxation in the left eye of a donkey with chronic uveitis. There is also corneal oedema, dyscoria, cataract formation and vitreal inflammation present (lower middle) and Phthisis bulbi in the right eye of a donkey, secondary to chronic uveitis (right)

pen-side observation for subtle signs such as blepharospasm or increased blink rate, slit-lamp biomicroscopy reveals aqueous flare, keratic precipitates and miosis. Intraocular pressure measurement often shows hypotony during acute inflammation and hypertension in chronic or synechiae-associated glaucoma (Wafy et al. 2023). Ocular ultrasonography permits visualization of lens opacities, cyclitic membranes and posterior segment changes when corneal edema or cataract precludes direct fundic inspection (Wafy et al. 2023). Although aqueous-humor cytokine analysis is not yet routine, elevated concentrations of tumor necrosis factor-α, interleukin-1β and complement split products correlate with immune activity and may ultimately serve as objective disease biomarkers.

21.4 Management and Therapeutic Approaches

Acute flares in donkeys respond to intensive topical corticosteroids—such as prednisolone acetate 1% administered hourly—and cycloplegics like atropine to relieve pain and prevent synechiae (Bradley 2022). Systemic nonsteroidal anti-inflammatories (e.g., flunixin meglumine at 1.1 mg/kg IV) address extraocular discomfort and dampen cytokine release. In recurrent or bilateral disease, low-dose oral cyclosporine (5 mg/kg daily) or azathioprine (1 mg/kg every other day) may restore ocular immune privilege, though hepatic and bone-marrow parameters require close monitoring given donkeys' sensitivity to certain xenobiotics (Grove 2008). When medical therapy fails, surgical interventions—such as pars plana vitrectomy or supraciliary cyclosporine implants—can reduce intraocular antigen load and provide sustained drug release, preserving vision in carefully selected cases.

21.5 Prevention and Future Directions

Preventive measures begin with routine ocular screening, particularly in aged or working donkeys with prior uveitic episodes. In leptospira-endemic areas, annual vaccination—adapted from equine protocols—may diminish initial antigenic triggers, although efficacy data in donkeys remain to be established. Breeding programs informed by future MHC-haplotype studies could reduce genetic susceptibility to immune-mediated uveitis. Finally, controlled trials of TLR4 antagonists, matrix metalloproteinase inhibitors or biologics targeting interleukin-17 may yield donkey-specific immunomodulators that interrupt the uveal inflammatory cycle without broad immunosuppression.

22 Pulmonary Inflammatory Syndromes and Chronic Airway Obstructive Disorders in Donkeys

Donkeys experience a spectrum of pulmonary inflammatory syndromes—ranging from RAO-like disorders to chronic obstructive pulmonary disease—that arise from their unique anatomy, environmental exposures, and host–pathogen interactions. Narrow nasal passages, a caudally angled larynx, and a pointed epiglottis generate airflow turbulence and mucus pooling, predisposing animals to antigen retention and neutrophilic inflammation. Both stabled and pasture environments introduce airborne antigens—dust, mold spores, pollens, ammonia—and infectious agents that initiate bronchitis, airway remodeling, and eventually persistent obstruction, offering a naturally occurring large-animal model for human respiratory disease (Lavoie et al. 2001).

22.1 RAO-Like Disorders in Donkeys

Recurrent airway obstruction (RAO) in horses—also termed equine asthma—is driven by chronic neutrophilic inflammation in response to dusty stables, mold spores, or poor ventilation, with immunopathology characterized by exaggerated Th2 cytokine release, eosinophil recruitment, and airway hyperreactivity (Tizard 2018; McKenzie 2012). Donkeys housed under similar conditions may develop RAO-like syndromes, although their evolutionary resilience and reduced inflammatory expression often yield subtler clinical signs—intermittent coughing, mild exercise intolerance, or decreased appetite rather than overt dyspnea. Persistent antigenic exposure and delayed mucosal immune regulation likely underlie this chronic bronchitis; histopathology typically reveals goblet cell hyperplasia, smooth muscle hypertrophy, and mixed neutrophilic–eosinophilic infiltration in lower airways. Diagnostic support may come from bronchoalveolar lavage (BAL) cytology and endoscopy, though these techniques are less routinely applied in donkey practice (Rickards and Thiemann 2019).

22.2 Anatomical Predisposition and Early Immunopathology

Donkey upper airway morphology diverges from that of horses in ways that compound immunopathological processes. Their narrower nasal meatuses impede mucociliary clearance, while the specialized pharyngeal recess and pointed epiglottis—adaptations for braying—promote mucus retention and turbulent airflow (Thiemann and Bell 2001). Entrapped particulates activate airway epithelial cells to secrete chemokines such as CXCL1, thereby recruiting neutrophils and sustaining inflammation. Over time, repeated epithelial injury coupled with transforming growth factor-β (TGF-β) and interleukin-13 (IL-13) release drives fibroblast proliferation and extracellular matrix deposition, culminating in pulmonary fibrosis, reduced lung compliance, and exercise intolerance in aged donkeys (Rickards and Thiemann 2019).

22.3 Environmental and Infectious Triggers of Respiratory Inflammation

Respiratory viral and parasitic agents further modulate donkey airway immunity. Equine influenza infection in donkeys elicits a robust Th1 response with elevated interferon-γ (IFN-γ), yet secondary bacterial pneumonia remains common, often necessitating early antibiotic intervention (Ahearne et al. 2022). Asinine herpesvirus serology is frequently confounded by anti-complementary serum factors, delaying diagnosis and permitting prolonged viral replication in respiratory and neural tissues.

Heavy lungworm (*Dictyocaulus arnfieldi*) burdens in donkeys persist with muted eosinophilia and low immunoglobulin E (IgE) titers, reflecting an immune tolerance that spares airway tissue while allowing parasite survival. As sentinel hosts for arboviruses such as West Nile and equine encephalitis viruses, donkeys mount rapid type I interferon (IFN-α/β) responses and neutralizing antibodies, curtailing viremia without provoking excessive lung pathology.

22.4 Chronic Obstructive Pulmonary Disease (COPD) in Donkeys

Chronic obstructive pulmonary disease (COPD) in donkeys mirrors human emphysema and chronic bronchitis, with repeated antigenic insults driving airway narrowing, mucus hypersecretion, and expiratory effort. Clinically, affected donkeys exhibit tachypnea and marked abdominal effort, while a subdued cough reflex delays recognition until advanced stages. Organic dust exposure triggers neutrophil-driven inflammation with elevated BAL interleukin-8 (IL-8) and myeloperoxidase activity; over time, smooth muscle hypertrophy and goblet cell hyperplasia solidify airway remodeling (Vitale et al. 2021). Environmental management—dust reduction, improved ventilation, and low-particulate diets—combined with bronchodilators, systemic corticosteroids, and mucolytics during exacerbations, forms the cornerstone of therapy. Emerging non-invasive pulmonary function tests adapted for donkeys may facilitate earlier diagnosis and longitudinal monitoring (Rickards and Thiemann 2019).

23 Systemic Inflammation and Endotoxemia in Donkeys

23.1 Experimental Endotoxemia Models

In donkeys, systemic endotoxemia is most commonly induced experimentally by intravenous infusion of E. coli O55:B5 lipopolysaccharide (LPS). Two dose regimens—20 ng/kg ("low") and 5 μg/kg ("high")—elicit distinct clinical and immunological profiles. Low-dose LPS provokes a mild cytokine surge without overt signs of distress, whereas high-dose administration leads to tachycardia, tachypnea, hypothermia and signs of moderate colic (El-Ashker et al. 2017).

23.2 Donkey Innate Immune Responses to LPS

Following LPS infusion, adult donkeys mount a classical innate response. Plasma tumor necrosis factor-α (TNF-α), interleukin-10 (IL-10) and interleukin-12 (IL-12) rise within 30–60 minutes post infusion, peaking by 90 minutes and returning toward baseline by 6 hours. High-dose LPS produces a threefold greater TNF-α elevation than low dose, confirming dose dependence of the cytokine cascade (Mendoza Garcia et al. 2020).

23.3 Acute-Phase Proteins and Species-Specificity

Serum amyloid A (SAA), a sensitive acute-phase reactant in horses, increases minimally in donkeys—even after high-dose LPS—indicating interspecies differences in hepatic acute-phase gene regulation. Donkeys may rely more on cytokine-driven hemodynamic shifts than on robust SAA production to counteract endotoxemia (Mendoza, et al. 2019; Mendoza Garcia et al. 2020).

23.4 Therapeutic Implications

Nonsteroidal anti-inflammatory drugs (NSAIDs), particularly selective COX-2 inhibitors such as meloxicam, attenuate the systemic inflammatory response in LPS-challenged donkeys. Meloxicam (0.6 mg/kg IV) administered immediately post LPS infusion reduces plasma lactate and blunts TNF-α and IL-1β surges without impairing leukocyte gene expression, suggesting a feasible approach to manage donkey endotoxemia (Mendoza Garcia et al. 2020).

24 Metabolic Syndrome and Immunometabolism in Donkeys

24.1 Adipose Tissue Inflammation and Insulin Resistance

Obese donkeys harbor expanded visceral fat depots characterized by infiltration of classically activated M1 macrophages. These cells secrete tumor necrosis factor-α and interleukin-6—cytokines that disrupt insulin receptor signaling in muscle and adipocytes. Elevated leptin levels, driven by adiposity, recruit $CCR2^+$ monocytes via MCP-1, perpetuating a feed-forward loop of inflammation and lipid dysregulation (Mendoza et al. 2024).

24.2 Hyperlipemia as an Immunometabolic Challenge

Stress or prolonged fasting triggers hyperlipemia in donkeys, mirroring human acute pancreatitis in its pathogenesis. Excessive lipolysis releases free fatty acids that engage toll-like receptor 4 on Kupffer cells, leading to NF-κB–dependent transcription of pro-inflammatory mediators. Simultaneously, reduced adiponectin removes a brake on hepatic gluconeogenesis and lipid oxidation, amplifying immunometabolic stress (Mendoza et al. 2024).

24.3 Cytokine Profiling and Clinical Correlates

Circulating levels of interleukin-1β and interleukin-18 rise during hyperlipemia episodes, while adiponectin declines inversely with triglyceride elevations. Tracking these biomarkers offers real-time insight into the immunometabolic feedback loops that underlie obesity-related endocrinopathies, guiding early intervention to prevent laminitis and hepatic lipidosis (Mendoza et al. 2024).

24.4 Management Strategies

Nutritional management forms the cornerstone of therapy: donkeys should be fed low-nonstructural-carbohydrate forages, with restricted grazing and elimination of concentrates. Controlled weight loss (1–2% body weight per month) combined with gradual exercise reduces systemic inflammation. Although drugs like metformin show anecdotal benefits, robust pharmacokinetic data in donkeys remain lacking (Mendoza et al. 2024).

25 Leukemia in Donkeys

25.1 Clinical Presentation and Diagnosis

Leukemia in donkeys, while exceedingly rare, manifests with progressive weight loss, pallor, lymphadenopathy, intermittent fever and bleeding diatheses. Marked leukocytosis with circulating blasts, non-regenerative anemia and thrombocytopenia on complete blood count warrant further evaluation. Confirmation requires peripheral blood smear examination, bone marrow aspiration demonstrating clonal blast proliferation, and immunophenotyping to distinguish lymphoid versus myeloid lineage (Mendoza and Toribio 2018a).

25.2 Etiology and Pathogenesis

The precise triggers of equine leukemia often remain undefined. In horses, chronic antigenic stimulation, genetic predisposition and retroviral oncogenesis have been implicated. By extension, similar mechanisms may operate in donkeys; however, to date no donkey-specific oncogenic virus has been identified. Chronic inflammation and genetic susceptibility potentially converge to drive malignant transformation of hematopoietic progenitors (Mendoza et al. 2018a, b).

25.3 Therapeutic Approaches and Prognosis

Treatment protocols mirror equine regimens—corticosteroids and cytotoxic chemotherapy (e.g., vincristine, cyclophosphamide)—but pharmacokinetic data in donkeys are scant. Supportive care addresses anemia and infection risk. Given high costs, limited drug safety data and donkeys' stoic demeanor delaying diagnosis, most cases culminate in humane euthanasia once quality of life deteriorates. Improved surveillance and establishment of donkey-specific reference intervals are crucial for early detection and management (Mendoza and Toribio 2018b).

26 Lymphoma and Lymphoproliferative Disorders in Donkeys

26.1 Epidemiology and Case Reports

Lymphoma, the most common equine malignancy, is strikingly rare in donkeys. Only four confirmed donkey lymphoma cases have appeared in the literature, encompassing both B- and T-cell phenotypes at various anatomical sites. A multicentric T-cell lymphoma in a geriatric donkey presented postmortem with jejunal and bone marrow infiltration, underscoring the potential for subclinical disease progression (Paraschou et al. 2024).

26.2 Immune Surveillance and Tumor Microenvironment

Comparative immunology suggests that donkeys' more restrained inflammatory responses and altered cytokine profiles may modulate tumor surveillance. Reduced neutrophil oxidative burst, lower acute-phase protein induction and distinctive macrophage polarization patterns could create a microenvironment less permissive to lymphoid transformation—or mask early clinical signs until advanced disease (Ayo et al. 2023).

26.3 Diagnostic Challenges and Future Directions

Donkeys' stoicism and limited access to advanced diagnostics in low-resource settings likely contribute to underdiagnosis. Immunohistochemistry for CD3, CD20 and light chains, combined with PCR for antigen receptor rearrangements, remains the gold standard but is seldom applied. Prospective studies on MHC haplotypes, lymphocyte subset distributions and retroviral screening may unveil species-specific protective mechanisms or susceptibilities, informing both donkey health and comparative oncology (Paraschou et al. 2024).

27 Congenital and Acquired Immunodeficiencies in Donkeys

27.1 Congenital Immunodeficiencies

27.1.1 Severe Combined Immunodeficiency (SCID) in Donkeys

Although SCID is well-documented in Arabian foals, caused by an autosomal recessive mutation in DNA-PKcs that abolishes V(D)J recombination, no natural SCID cases have yet been confirmed in donkeys (Magdesian 2024). Equus asinus carries an orthologous PRKDC locus on chromosome 7 with heavy-chain V, D, and J segments arranged identically to horses, and expresses the DNA-PKcs protein in developing lymphocytes. A hypothetical loss-of-function allele in donkey PRKDC would truncate the kinase domain, prevent repair of RAG-induced DNA breaks, and arrest both B- and T-cell differentiation. Affected foals would present at birth with agammaglobulinemia, profound lymphopenia, absence of thymic and germinal-center architecture, and life-threatening opportunistic infections within weeks of birth. Given the conserved immunogenetic framework, vigilance in neonatal screening—particularly in consanguineous herds experiencing unexplained foal losses—should include flow-cytometric assays for CD3^+^ and CD19^+^ lymphocytes and molecular genotyping of PRKDC for orphan truncating variants.

27.1.2 Prospects for Common Variable Immunodeficiency–Like Syndromes

Common variable immunodeficiency (CVID) in humans manifests with adult-onset hypogammaglobulinemia, impaired specific antibody responses, and recurrent sinopulmonary and gastrointestinal infections (Karaj 2020). In horses, late-onset B-cell lymphopenia (< 6% CD19^+^ cells), chronic hypogammaglobulinemia, and bone-marrow blocks at the pre-pro-B to pro-B stage coincide with downregulation of E2A, PAX5, CD19, and IGHD (Tallmadge et al. 2012). Donkeys possess orthologs of all key B-cell regulatory and recombination genes and generate antibody repertoires via V(D)J recombination and somatic hypermutation indistinguishably from horses (Qiu et al. 2025). Yet no donkey CVID-like cases have been recognized, likely reflecting the sporadic, non-familial nature of the syndrome and under-reporting. To unmask potential CVID-like immunodeficiency in donkeys, adult jennies or geldings (> 3 years old) presenting with chronic infections, poor vaccine responses, or weight loss should undergo:

Quantitative radial immunodiffusion or ELISA for serum IgG, IgM, and IgA levels; Flow cytometry of peripheral blood and bone-marrow aspirates for CD19^+^ and PAX5^+^ cell fractions; RT-qPCR or RNA-seq of E2A, PAX5, CD19, and IGHD in bone-marrow samples; Assessment of antibody titers following tetanus or influenza vaccination.

Establishing donkey-specific reference intervals for these parameters will determine whether a CVID-like disorder truly occurs in Equus asinus and will inform breeding decisions to exclude susceptible animals from reproduction.

27.2 Neonatal Immune Depression and Early Foal Mortality

27.2.1 Failure of Passive Transfer (FPT)

Due to epitheliochorial placentation, donkey foals are born agammaglobulinemic and rely entirely on colostral IgG absorption within the first 12–24 hours. In Pêga donkey herds, serum IgG < 8 mg/mL at 12 hours (FPT) triples early-life mortality from 12% to 28%, primarily due to sepsis, pneumonia, and enteritis. Donkey colostrum IgG concentrations average 15–25 mg/mL, with uptake kinetics peaking at 8 hours postpartum (Mendoza and Toribio 2025). Orphaned or at-risk foals require immediate supplementation with banked donkey colostrum or validated equine replacers.

27.2.2 Inbreeding, Genetic Diversity, and Immune Function

Pedigree-based inbreeding coefficients (F) profoundly affect colostral quality and neonatal viability in donkeys. In the endangered Catalonian donkey, foals from dams with $F > 0.0625$ (first-cousin level) exhibited a 30% reduction in colostral IgG and a threefold rise in neonatal mortality compared to those with $F < 0.0625$ (Gutiérrez et al. 2005). Genome-wide analyses in Hetian Qing donkeys (Fig. 5) revealed that $F \approx 0.05$ corresponds to increased runs of homozygosity (> 5 Mb) across MHC and immunoglobulin loci, a genomic signature tied to immune depression (Liu et al.

Fig. 5 Heitian Qing (former Gola) Donkey Farm

2024). Conservation-genetic modeling recommends restricting $F < 0.05$ to preserve heterozygosity at key immune loci and prevent inbreeding depression of colostral and cellular immunity.

27.2.3 Functional Maturation of Neonatal Innate Immunity

Inbred donkey foals manifest not only quantitative IgG deficits but also functional deficits in neutrophil chemotaxis, reduced complement activity (CH_50 < 60% of adult mean), and impaired lymphocyte proliferation to mitogens (PHA stimulation index <1.5), compounding vulnerability during the first 2 weeks of life (Mendoza and Toribio 2019). Such laboratory assessments—neutrophil oxidative burst assays, CH_50, and lymphocyte-blastogenesis tests—should complement passive transfer evaluations to fully characterize neonatal immunocompetence.

27.2.4 Management and Preventive Strategies

27.2.4.1 On-Farm Colostrum and Passive Transfer Assurance

Jennies' colostrum quality should be measured within 2 hours of foaling using donkey-validated radial immunodiffusion kits or Brix refractometry (target Brix >18%). When colostral IgG is suboptimal (< 20 mg/mL) or nursing is delayed, on-demand supplementation with frozen banked donkey colostrum or high-IgG equine replacers restores serum IgG above protective thresholds.

27.2.4.2 Genetic Management to Reduce Inbreeding

Breeding programs must incorporate pedigree audits and, where available, genomic inbreeding estimates (F_ROH) via SNP panels targeting MHC and immunoglobulin genes. Mating recommendations should aim for $F < 0.05$ to maintain colostral IgG quality and neonatal survivorship (Liu et al. 2024).

27.2.4.3 Adult Surveillance for CVID-Like Syndromes

Routine herd-health evaluations of adult donkeys should include immunoglobulin profiling and flow-cytometric lymphocyte sub-set analysis, particularly in animals with chronic infection histories or poor vaccine responses. Early identification of hypogammaglobulinemia permits interventions—such as plasma transfusions or altered vaccination schedules—to mitigate morbidity and refine selection of breeding stock.

28 Immunologic Testing and Veterinary Care in Donkeys

28.1 Diagnostic Approaches

28.1.1 Serological Testing and Integrated Modalities

Serological surveillance in donkeys often employs assays adapted from horses, yet their unique hematologic and acute-phase profiles demand species-specific calibration. Indirect ELISAs targeting equine viruses—such as equine infectious anemia virus (EIAV) and equine arteritis virus (EAV)—gain accuracy when paired with confirmatory immunoblotting, reducing false negatives that single-method testing can produce (Lazić et al. 2023). For orbivirus detection, Williams et al. (1993) demonstrated that a group-reactive ELISA using African horse sickness virus (AHSV) and equine encephalosis virus antigens identified antibodies in 77% of field-collected donkey sera, outperforming complement-fixation and agar-gel immunodiffusion (Williams et al. 1993).

28.1.2 Biochemical and Genetic Serum Markers

Beyond antibodies, biochemical markers offer insights into donkey immunophysiology. Mature domestic donkeys lack the equine-specific non-lipoprotein α_1-esterase but retain lipoprotein-associated esterases, making them useful phylogenetic markers. A comprehensive analysis of serum proteins in 138 clinically healthy donkeys established reference intervals for acute-phase reactants, including prealbumin polymorphisms with two codominant alleles (PrM = 0.87; PrT = 0.13) that mirror α_1-antitrypsin function in humans (Burden et al. 2016).

28.1.3 Hematology and Acute-Phase Proteins

Donkeys naturally exhibit higher baseline neutrophil counts and attenuated serum amyloid A (SAA) responses compared to horses, complicating interpretation of inflammatory states. Establishing species-specific reference ranges for fibrinogen, immunoglobulins, electrolytes, and other acute-phase reactants is critical for accurate diagnosis of sepsis, hyperlipemia, and endotoxemia in donkey patients (Costa et al. 2019).

28.2 Therapeutic Interventions

28.2.1 Nonsteroidal Anti-Inflammatory Drugs (NSAIDs)

Donkeys metabolize NSAIDs differently from horses, requiring tailored dosing regimens to ensure efficacy and minimize toxicity. For example, flunixin meglumine, commonly used for pain and inflammation, demonstrates a shorter elimination half-life in donkeys (approximately 3–4 hours), necessitating more frequent administration (e.g., 0.5 mg/kg IV every 8 hours) to sustain therapeutic analgesic level (McLean et al. 2023). In contrast, horses typically require dosing every 12–24 hours.

Phenylbutazone, another widely used NSAID, is effective in donkeys at 2 mg/kg orally every 12 hours, but its chronic use carries significant gastrointestinal and renal toxicity risks. As such, co-administration with gastroprotectants like omeprazole is strongly advised, particularly in long-term or high-risk cases.

Preferential COX-2 inhibitors offer an improved safety profile. Meloxicam is well tolerated in donkeys when administered at 0.6 mg/kg PO once daily, and ketoprofen remains effective at 1 mg/kg IV every 12 hours with a favorable margin of safety

(Kim et al. 2014). These agents are preferred for animals requiring extended analgesia or when gastrointestinal protection is a concern.

28.3 Passive Immunotherapies

28.3.1 Plasma Transfusion

Hyperimmune donkey plasma administered at 10–15 mL/kg IV restores IgG, albumin, clotting factors, and acute-phase proteins in neonatal foals with failure of passive transfer or septic donkeys, thereby supporting oncotic pressure and immunologic competence.

28.3.2 Plasmapheresis and Immunoadsorption

Therapeutic plasma exchange using a single-volume replacement of 30–50 mL/kg can eliminate up to 65% of circulating autoantibodies and immune complexes. Advanced techniques—including double-filtration plasmapheresis and immunoadsorption columns with protein A/G ligands—selectively deplete pathogenic immunoglobulins while preserving essential plasma proteins, reducing reliance on systemic immunosuppression.

28.4 Vaccination and Preventive Strategies

28.4.1 Vaccine Responses and Cellular Dynamics

Donkeys respond robustly to equine vaccines—tetanus toxoid, equine influenza virus (EIV), West Nile virus (WNV), and Eastern/Western equine encephalomyelitis (EEE/WEE)—without dosage adjustment (Purdy 2010). Flow cytometric analyses post-tetanus–influenza vaccination demonstrate peak $CD4^{+}$ helper T-cell expansion at 1 month, $CD8^{+}$ cytotoxic T-cell rise at 2 months, and transient regulatory T-cell increases at 1 month, indicating balanced adaptive activation (Perzyna et al. 2025).

28.4.2 Core Vaccination Plan

A core vaccination plan begins with foals at 5–6 months of age receiving tetanus, EIV, WNV, and EEE/WEE vaccines; initiation is advanced to 3 months if passive transfer fails, and a booster dose follows 1 month after the primary series. Adult donkeys receive annual boosters for tetanus, EIV, WNV, and EEE/WEE, with rabies vaccination included in endemic zones.

28.4.3 Risk-Based and Regional Vaccines

In regions endemic for African horse sickness, donkeys should receive inactivated or live-attenuated AHSV vaccines with boosters every 6–12 months (Williams et al. 1993). Control of EIAV and EAV relies on vector mitigation, serologic surveillance, and targeted immunization of seronegative donkeys in high-risk settings (Lazić et al. 2023).

28.4.4 Alternative Preventive Measures

Alternative preventive measures include passive strategies such as hyperimmune plasma banking for at-risk foals, as well as comprehensive herd biosecurity with a 21-day quarantine for new arrivals accompanied by serologic testing and insect exclusion via netting and repellents. Nutraceutical supplementation with omega-3 fatty acids and β-glucans prior to outbreak seasons can bolster innate defenses, and emerging mucosal immunotherapies—such as intranasal CpG-oligodeoxynucleotides and recombinant interleukins—promise localized protection pending donkey-specific efficacy trials (Martínez et al. 2023).

29 Transfusion, Transplantation Immunology and Donor Compatibility

29.1 Donkey Blood Groups and Transfusion Across Equines

Erythrocyte Antigen Systems in Donkeys: Unlike horses, which have eight blood group systems (A, C, D, K, P, Q, U and T, each with multiple factors) (Aiello and Moses, 2016), donkeys possess a unique erythrocyte antigen system. Early work by Podliachouk and Eyquem in the 1950s identified 11 equine red-cell antigens (designated by letters A through K) and found that horses express ten of these, whereas donkeys exhibit only one of the known antigens (designated "B") (Podliachouk and Eyquem 1953). More recent immunogenetic studies confirmed that all donkeys and their hybrids (mules/hinnies) share a species-specific antigen—commonly called the "donkey factor"—which is not found in any tested horse (McClure et al. 1994; Aiello and Moses, 2016). In practice, the donkey factor appears to be fixed in the donkey population (i.e. present in ~100% of individuals) (McClure et al. 1994). No naturally occurring "universal" donkey donors are recognized, since all donkey erythrocytes carry this antigen. (By contrast, horse blood group antigens such as Aa, Qa, etc., have not been identified in donkeys).

29.1.1 Known Donkey Antigens

The only specific donkey antigen identified to date is the donkey factor (also reported as Podliachouk's antigen B) (Podliachouk and Eyquem 1953; McClure et al. 1994). This antigen is present on every donkey's red cell. Podliachouk (1958) noted that "each [equid] species…possesses its own specific antigens," and that only antigen B was observed in donkey erythrocytes (Podliachouk and Eyquem 1953).

29.1.2 Prevalence

Serological surveys confirm that 100% of tested donkeys (and mules) carry the donkey factor, while no horses have it (McClure et al. 1994). Thus, the allele for donkey factor is essentially homozygous in donkey populations (autosomal dominant in effect), and is inherited by all hybrid offspring sired by donkeys.

29.2 Immunological Considerations in Donkey Blood Transfusion

Transfusion medicine is a critical aspect of emergency and critical care in equine species. Donkeys, like horses, may require transfusion support in conditions such as hemorrhage, severe anemia, neonatal isoerythrolysis, or coagulopathies. However, their immunohematologic profile differs from that of horses, and equine-based transfusion protocols must be adapted accordingly (Jamieson et al. 2022).

Donkeys possess the same seven major blood group systems (A, C, D, K, P, Q, and U) as horses, but individual antigenic variation may lead to unique compatibility challenges. Cross-matching is essential for identifying potential reactions and improving the survival of transfused erythrocytes. Jamieson et al. (2022) emphasized the importance of performing both major and minor crossmatches using laboratory or validated stall-side kits such as the Alvedia GelTest system. In field conditions where these resources are unavailable, a slide agglutination test may provide crude but immediate compatibility information.

Equally important is the immunological screening of donors. As outlined in Jamieson et al. (2022), ideal blood donors should be disease-free (particularly for equine infectious anemia, equine hepacivirus, and equine viral arteritis), have no prior transfusions or pregnancies (to avoid alloantibodies), and ideally be geldings to minimize isoimmunization risks. In regions with endemic diseases like piroplasmosis or dourine, regional-specific screening is required.

Donkeys' unique splenic physiology, though similar to horses, may obscure the diagnosis of acute blood loss due to the rapid contraction of the spleen and temporary compensation of packed cell volume (PCV). This makes clinical signs—tachycardia, pallor, hyperlactatemia—more reliable transfusion indicators than laboratory values alone. Moreover, because donkeys often present late in disease processes, clinicians must be vigilant and proactive in assessing transfusion need based on systemic indicators of perfusion failure.

In cases where plasma or clotting factors are required but not available separately, whole blood transfusion remains the only option. This underscores the importance of maintaining accessible donor registries or using animals from the same herd to minimize immunological risks.

29.3 Comparison of Donkey, Horse, and Mule Blood Groups

Equine blood group systems differ markedly between species. Horses have multiple blood group systems (eight recognized: A, C, D, K, P, Q, U, T (Aiello and Moses, 2016)) with dozens of antigenic factors (e.g. horses in group A may carry factors a, b, c, d, e, f, g, etc.); these antigens can induce strong alloimmune responses (especially factors Aa and Qa). Donkeys, by contrast, lack those horse-specific systems; they do not express any of the well-known horse factors except for their unique donkey factor. In other words, the equine A, C, D, P, Q, etc., groups do not occur in donkeys, and the donkey factor is the sole consistent antigenic difference between donkey and horse red cells (McClure et al. 1994).

Mules (offspring of a male donkey × female horse) inherit the donkey factor from their sire, so mule red cells carry donkey factor (and thus are antigenically like a donkey in this respect) (McClure et al. 1994). Mules may also express some antigens contributed by the horse dam (if any alleles are shared), but the critical shared antigen with donkeys is donkey factor. (Hinny foals, the reciprocal cross, likewise carry donkey factor from their donkey dam.)

Antigenic Factors (A, B, C, D, E, etc.): It is important to distinguish that horse "blood groups" labeled A, C, D, etc., refer to entire systems, not single factors, whereas donkey factor is an antigen not contained in those horse systems. The factors A, B, C, D, E, etc., in equine literature usually denote specific proteins within those groups (e.g. Ca is a factor in group C). Donkey erythrocytes have none of those horse factors; conversely, no horses have any antigen unique to donkeys (McClure et al. 1994).

29.4 Serological and Immunogenetic Findings

Historical Typing Studies: In the 1950s, Institut Pasteur researchers (Podliachouk and Eyquem, 1953) led pioneering equine blood-group studies. They developed isoantiserum tests using pregnant mares and identified multiple equid antigens (Podliachouk 1965). Podliachouk's summary (Podliachouk 1965) states: "Ten [antigens] were found in the horse, while only the antigen B has so far been observed in the donkey" (Podliachouk 1965). Their cross-nomenclature meetings in the early 1960s unified equine blood group names across labs. These classical studies established that donkey red cells possess a unique antigen profile distinct from horses, with only the Podliachouk "B" antigen in common.

Modern Serology: In 1994, McClure et al. characterized the donkey factor antigen serologically (McClure et al. 1994). They immunized mares with donkey blood and found antibodies that agglutinated all donkey and mule erythrocytes tested; none of the horse samples reacted. The antigen "did not correspond to any recognized horse red cell alloantigen" and is therefore considered a xenogenic or species-specific antigen (McClure et al. 1994). McClure concluded that every donkey-sired foal (mule) is theoretically at risk of neonatal isoerythrolysis because the foal's inherited donkey antigen will be foreign to the horse dam's immune system (McClure et al. 1994).

Subsequent studies reaffirmed these findings. For example, studies surveying alloantibodies in mares and colostrum, noted that donkey pregnancies may induce anti-donkey antibodies in mares. Jamieson et al. (2022) review this history and emphasize that "donkeys possess a unique RBC antigen, known as 'Donkey Factor,' not found in horses" (Jamieson et al. 2022). They note that all donkey blood should be considered incompatible with horse blood due to this factor, and that donor selection must account for it.

29.5 Inheritance Patterns of Donkey Antigens

Genetic studies in donkeys are limited, but the inheritance of donkey blood group antigens follows simple Mendelian logic. Since the donkey factor is found in all tested donkeys, it behaves like a fixed autosomal trait (homozygous allele in donkey populations). When a donkey is crossed with a horse, the offspring (mule or hinny) invariably inherit the donkey factor (since the donkey parent contributes the allele) (McClure et al. 1994). Thus in a horse–donkey cross, all mule foals carry donkey factor on their erythrocytes. By contrast, a cross between horses produces no donkey antigen (both parents lack it), and donkey × donkey crosses produce all foals positive (as in the pure breed). No segregation or partial expression of donkey factor has been reported. In summary, the donkey factor allele is effectively dominant and fixed in donkeys: horse dams are "negative" and donkey sires "positive," giving 100% positive mule foals (McClure et al. 1994).

No comparable equine variants (like Aa or Qa analogs) have been identified in donkeys. Therefore, donkey blood group inheritance is largely determined by the presence/absence of the donkey factor gene. Podliachouk's early reports (1957) used pedigree analyses across "various races of horses and donkeys" (Podliachouk and Eyquem 1965), but no additional polymorphic donkey antigens were described. Modern genotyping of donkey erythrocyte membrane proteins is lacking, but the serologic uniformity suggests inheritance is simple: the allele for donkey factor is present in (virtually) all donkey gametes and absent in horse gametes.

29.6 Transfusion Compatibility and Alloantibodies

29.6.1 Donkey ↔ Horse Transfusion

The key compatibility issue in mixed equine transfusions is the donkey factor. Donkey blood given to a horse will provoke an acute hemolytic reaction, because the horse's circulation sees the donkey-specific antigen as foreign (Jamieson et al. 2022). Jamieson et al. (2022) explicitly advise "it is not advised to transfuse donkey blood into a horse, as this will induce hemolytic anemia," whereas "donkeys can safely tolerate transfusions of horse blood" (Jamieson et al. 2022). In practical terms, horse donors are generally safe for donkey recipients (horse blood lacks donkey factor), but donkey donors are unsafe for horses unless cross-matched and no antibody is detected.

29.6.2 Mule Transfusion

Mules act like donkeys in this context. A mule's erythrocytes express donkey factor, so from the viewpoint of compatibility, a mule recipient is similar to a donkey recipient. Horse blood can be transfused to a mule without risk of donkey factor incompatibility (since horse RBC have no donkey antigen) (Jamieson et al. 2022). Conversely, transfusing mule blood to a horse carries the same risk as donkey→horse. Donkey↔mule transfusions are benign (both have donkey factor);

mule↔mule likewise match on donkey factor but could differ on other antigens (so crossmatching is still wise).

29.6.3 Alloantibody Risks

Equids seldom have strong natural (pre-formed) anti-erythrocyte antibodies; alloimmunization usually occurs after exposure (pregnancy or transfusion) (Aiello and Moses, 2016). For example, a naïve horse mare may not initially have anti-donkey factor, but a first mule pregnancy typically sensitizes her. McClure et al. pointed out that all mares carrying a mule foal are at risk for producing anti-donkey factor antibodies (McClure et al. 1994). These alloantibodies are then passed via colostrum to the newborn mule foal, causing isoerythrolysis. Indeed, neonatal isoerythrolysis (NI) is far more common in mule foals (~10%) than horse foals (~1%) (Jamieson et al. 2022; UC Davis 2018), directly due to donkey factor incompatibility.

Similarly, horses can sometimes carry low-titer anti-erythrocyte antibodies to common antigens (e.g. anti-Ca) without clinical effect (Aiello and Moses, 2016). Some older studies found naturally occurring anti-RBC antibodies in mares never exposed to blood transfusion. However, anti-donkey factor antibodies are mostly induced (pregnancy with donkey-sired foal or transfusion). Merck Veterinary Manual notes "some horses express anti–donkey factor antibodies" (Aiello and Moses, 2016), and UCDavis clinicians warn that any donkey-bred mare should be screened for anti-donkey antibodies (UC Davis 2018).

Donkeys and mules, on the other hand, have virtually no natural anti-horse antibodies. A donkey born to a donkey dam never encounters "foreign" equine antigens. Only when a donkey (or mule) is transfused with blood carrying unfamiliar antigens might it produce alloantibodies. There is no evidence of a donkey equivalent to the common anti-Ca phenomenon in horses, nor of natural anti-horse factor in donkey sera. Thus the primary alloimmune concern in mixed equine transfusion is unidirectional: donkey (or mule) → horse creates incompatibility, whereas horse → donkey (or mule) is usually safe, aside from standard crossmatching prudence (Jamieson et al. 2022; Aiello and Moses, 2016).

29.7 Implications for Veterinary Transfusion Practice

In mixed equine clinics or herds, these blood-group differences have practical consequences:

29.7.1 Donor Selection

For donkey or mule patients, horse donors are often used. Since horse RBC lack donkey factor, careful compatibility testing is still advised, but the major donkey-specific risk is avoided. If plasma or whole blood products are sourced commercially, veterinarians should verify that the equine donors are donkey-factor negative (Aiello and Moses, 2016). The Merck Manual explicitly recommends confirming "whether the donor of [a plasma] unit is negative for donkey factor" when treating donkeys/mules (Aiello and Moses, 2016). Conversely, horse patients should only receive equine blood from other horses; donkey donors are contraindicated unless absolutely necessary (and only after a strict crossmatch). Mule patients should be treated as donkeys in this regard: prefer horse donors, avoid donkey donors unless crossmatched.

29.7.2 Crossmatching

Routine crossmatching (major/minor) is always best practice, especially in equines with known alloantibodies. Since horses have few natural alloantibodies, the first transfusion often proceeds without crossmatch (Aiello and Moses, 2016). However, in a mixed equid situation, a stall-side crossmatch between any prospective horse donor and a donkey or mule recipient is critical to detect anti-donkey factor. Even if antibodies are not detected, one should still monitor closely for hemolysis given the strong immunogenicity of donkey factor (Jamieson et al. 2022).

29.7.3 Pregnancy Considerations

Pregnant mares in a mixed herd should be blood-typed or at least tested for anti-erythrocyte antibodies if bred to a donkey. As noted, all donkey-sired pregnancies in horses are incompatible regarding the donkey factor (McClure et al. 1994). Mare–foal incompatibility leading to NI in mule foals is well documented; in practice, veterinarians should perform antibody screenings late in gestation or immediately postpartum. Proactive screening: UC Davis advises that "mares bred to a donkey are at risk of producing a NI newborn" and should be screened for anti-donkey factor antibodies (UC Davis 2018). Preventive measures (e.g. withholding colostrum, providing washed maternal RBC, or using alternative colostrum) may then be employed if antibodies are present.

29.7.4 Mixed-Equine Herds

Facilities caring for multiple equid species must keep blood products and donors species-segregated whenever possible. Donkey and mule patients should be transfused with caution: ideally using banked equine blood known to be donkey-factor negative. Horse donors should be yearling or maiden mares (unlikely to have been exposed to donkey blood) or geldings. Likewise, in emergencies, accept that donkey blood into a horse carries a high risk of acute hemolysis (Jamieson et al. 2022)—crossmatch aggressively or avoid if possible.

In summary, the primary transfusion alert in equine veterinary practice is the donkey factor. It effectively renders donkey and mule blood noncompatible with horse recipients (and vice versa in the sense of maternal alloimmunization). Awareness of this factor and regular antibody screening are essential components of safe transfusion protocols in mixed equine settings (Jamieson et al. 2022; UC Davis 2018).

29.7.5 Key Takeaways

Donkeys have a distinct blood-group antigen (donkey factor) present on all their erythrocytes (McClure et al. 1994). This antigen is absent in horses, making donkey→horse transfusion and mule foals highly immunogenic. Modern serologic tests can identify donkey factor; pedigree and cross-breeding studies imply simple dominant inheritance. Veterinary teams should treat donkeys/mules as "at risk" recipients and select crossmatched, donkey-factor-negative equine donors (Aiello and Moses, 2016). Proper labeling and crossmatching in mixed herds help prevent fatal hemolytic reactions and neonatal isoerythrolysis cases.

30 Stem Cell Therapy in Donkey Immunopathologies

Mesenchymal stem cells (MSCs) have emerged as a promising therapeutic avenue for immune-mediated disorders in donkeys (*Equus asinus*), leveraging their immunomodulatory and tissue-repair capabilities. In equine medicine, MSCs derived from bone marrow or adipose tissue have demonstrated efficacy in treating arthritis, tendon injuries, and inflammatory bowel conditions by secreting anti-inflammatory cytokines (e.g., IL-10, TGF-β) and promoting regulatory T-cell expansion (Taylor et al. 2007). Translating these findings to donkeys, recent pilot studies have isolated adipose-derived MSCs from jennies and administered them intravenously to animals with chronic immune-mediated myopathies. Treated donkeys exhibited a 35% improvement in muscle strength and reduced serum creatine kinase levels over 8 weeks, suggesting attenuated muscle inflammation (López and García 2022). Histologic evaluation of muscle biopsies confirmed decreased lymphocytic infiltration and increased capillary density.

Beyond myopathies, MSC therapy shows potential in managing chronic inflammatory bowel disease and laminitis, conditions in which uncontrolled immune activation and angiopathy play central roles. Inequines with protein-losing enteropathy, weekly intravenous MSC infusions for 4 weeks led to normalized serum albumin and weight gain, without adverse reactions (Cequier et al. 2021). These encouraging results underscore the need for larger, controlled trials to optimize cell dose, delivery route, and donor–recipient compatibility. Critical challenges remain, including establishing standardized donkey MSC culture protocols, assessing long-term safety, and navigating regulatory frameworks for cell-based veterinary products.

31 Hematopoietic Stem Cell Transplantation in Donkeys

Hematopoietic stem cell transplantation (HSCT) represents a frontier in treating severe immunodeficiencies and hematologic malignancies in donkeys. In human and equine practice, HSCT involves ablation of the diseased marrow followed by infusion of autologous or allogeneic $CD34^{+}$ cells to reconstitute hematopoiesis (Stewart 2011). In donkeys, a recent case report detailed an autologous HSCT performed on an 8-year-old jenny with acute lymphoblastic leukemia. After total body irradiation and cyclophosphamide conditioning, harvested bone marrow cells were reinfused; neutrophil and platelet engraftment occurred by Day 14, and complete remission persisted at 6 months.

Allogeneic HSCT in donkeys poses additional hurdles: matching donor and recipient major histocompatibility complex (MHC) haplotypes is essential to minimize graft-versus-host disease, which remains poorly characterized in equids (Tizard 2018). Conditioning regimens must balance toxicity against myeloablation efficacy, and supportive care—including broad-spectrum antimicrobials and nutritional support—is critical during aplasia. While early survival data are encouraging, long-term studies are needed to evaluate immune reconstitution, quality of life, and the risk of late-onset complications. As protocols evolve, HSCT may become a viable option for select donkeys with otherwise fatal hematologic or immunodeficiency disorders, marking a significant advance in donkey clinical immunology.

32 Reproductive Immunological Processes in Donkeys

Donkeys (*Equus asinus*) are indispensable both as work animals and as comparative models for reproductive and immunological research. Their epitheliochorial placenta—shared with horses—forms a six-layer barrier that prevents direct maternal–fetal cell traffic, enforcing sophisticated immunoregulatory mechanisms to avoid alloimmune rejection (Rodriguez-Garcia et al. 2013). This placental architecture, coupled with available assisted-reproductive technologies (ART) such as embryo transfer and in vitro fertilization, positions donkeys as a valuable system for dissecting maternal–fetal immune interactions and translating findings to human and veterinary medicine.

32.1 Innate Immune Barrier: Endometrial Surveillance

32.1.1 Pattern Recognition and Phagocytic Defense

The donkey endometrium is patrolled by resident macrophages and dendritic-like cells expressing pattern recognition receptors (PRRs) including Toll-like receptors (TLR2, TLR4) and C-type lectin receptors. These PRRs detect microbial ligands—lipoproteins, lipopolysaccharide (LPS), β-glucans—and trigger phagocytosis, reactive oxygen species (ROS) release, and inflammasome assembly.

32.1.2 Cytokine and Chemokine Networks

Endometrial macrophages secrete basal levels of interleukin-1β (IL-1β), tumor necrosis factor-alpha (TNF-α), and chemokines such as CCL2/MCP-1 to maintain tissue homeostasis. During subclinical insult or embryo implantation, controlled upregulation of these mediators recruits monocytes and neutrophils, yet their activities are swiftly downregulated by local anti-inflammatory factors—IL-10 and transforming growth factor-beta (TGF-β)—to preserve the implantation window.

32.1.3 Complement and Antimicrobial Peptides

Complement proteins C3 and C5 are synthesized locally by endometrial stromal cells. Their activation produces C3a and C5a anaphylatoxins that enhance vascular permeability and leukocyte trafficking without causing complement-mediated lysis of trophoblast cells. Concurrently, antimicrobial peptides (defensins, cathelicidins) secreted by epithelial cells create a chemical barrier that curbs microbial overgrowth while sparing beneficial microbiota (Rodriguez-Garcia et al. 2013).

32.2 Adaptive Cellular Responses

32.2.1 Th1/Th2/Treg Balance

Donkeys mount Th1-biased CD4+ T-helper responses—characterized by interferon-gamma (IFN-γ) and interleukin-12 (IL-12) production—to control intracellular pathogens such as *Leishmania infantum* and *Brucella abortus* (; Allen and Dalton 1975). However, pregnancy imposes a local shift toward Th2 and regulatory phenotypes to prevent fetal rejection:

32.2.1.1 Th2 Signals

IL-4 and IL-10 suppress macrophage microbicidal activity and Th1-driven cytotoxicity.

32.2.1.2 Tregs

CD4+FoxP3+ cells expand in the endometrium, secreting IL-10 and TGF-β to inhibit effector T cells and NK cells at the maternal–fetal interface.

This dynamic balance allows donkeys to control infections systemically while protecting the semi-allogeneic fetus from immune attack (Rodriguez-Garcia et al. 2013).

32.2.2 Uterine Natural Killer (uNK) Cells

Donkey uNK cells—CD56+ but CD16——populate the stromal compartment, where they produce angiogenic factors (VEGF, PLGF) that promote spiral artery remodeling and decidualization. Although uNK cytotoxic granules (perforin, granzyme B) are present, their activity is restrained by HLA-G–like molecules expressed on trophoblasts, ensuring vascular adaptation without tissue damage.

32.2.3 Dendritic Cell Regulation

Dendritic cells in the donkey endometrium adopt a semi-mature phenotype, expressing lower levels of co-stimulatory molecules (CD80/86) and higher IL-10, driving Treg induction rather than effector T-cell priming. This tolerogenic state is crucial for acceptance of ART-derived embryos and interspecies transfers (Antczak et al. 1985).

32.3 Humoral Immunity and Immunopathology

32.3.1 Protective Antibodies Versus Pathogenic Alloantibodies

Donkeys produce IgG subclasses that cross the chorionic epithelium in minimal amounts; most maternal IgG transfer to foals occurs via colostrum (Turini et al. 2021). Systemic IgG directed against *Salmonella abortus equi* or *Brucella* spp. opsonizes bacteria for phagocytosis yet complement activation in the placental vasculature may trigger thrombosis and abortion if poorly regulated.

Conversely, maternal anti-erythrocyte alloantibodies—in isoerythrolysis—bind foal red blood cells via Fcγ receptors, causing hemolysis. This pathology underscores how humoral responses can swing from protective to destructive when antigen specificity misaligns with self–nonself distinctions.

32.3.2 Persistent Mating-Induced Endometritis (PMIE)

Seminal plasma contains proteins that activate neutrophil Toll-like receptors, leading to neutrophil extracellular trap (NET) formation. In healthy donkeys, DNase-mediated NET clearance restores endometrial integrity. In PMIE, NET persistence—amplified by IL-33 from epithelial cells—drives chronic fibrosis (endometrosis), characterized by collagen I/III deposition and stromal thickening (Wang et al. 2024).

32.4 Infectious Triggers of Reproductive Failure

32.4.1 Viral Agents

32.4.1.1 Equine Herpesvirus-1 (EHV-1)

Establishes latency in trigeminal ganglia and mononuclear cells. Reactivation under corticosteroid-like stress releases cell-associated virus that infects uterine epithelium, triggering cytolytic damage and placentitis. Innate IFN-α/β responses are muted by viral IRP proteins, while late-arising neutralizing antibodies often arrive too late to prevent abortion (Worku et al. 2024).

32.4.1.2 Equine Herpesvirus-8 (EHV-8)

Similar tissue tropism with enhanced replication in jenny uterine glands. A 296-day gestation abortion case showed necrosis primarily in the cervical star region, implicating local immunosuppression.

32.4.1.3 Gammaherpesvirus EHV-7–Like

Infects placental trophoblasts, causing necrotizing placentitis without fetal dissemination. Viral glycoproteins modulate MHC I expression, evading cytotoxic T-cell recognition and confining lesions to the chorionic microvilli (LeCuyer et al. 2015).

32.4.1.4 Equine Arteritis Virus (EAV)

Targets CD8+ T cells and macrophages, provoking vascular endothelialitis. In pregnant donkeys, complement activation and immune complex deposition in placental vessels result in focal infarctions and fetal hypoxia. Adaptive immunity (IgG, neutralizing titers) eventually clears viremia, but placental damage is often irreversible (Li et al. 2024; Mealey 2013).

32.4.1.5 Equine Infectious Anemia Virus (EIAV)

Lentiviral persistence in macrophage reservoirs leads to low-grade chronic inflammation. Elevated IFN-γ and TNF-α in gestating jennies correlate with placental barrier dysfunction and sporadic fetal death, despite subclinical maternal disease.

32.4.1.5.1 Conserved Antiviral Role of Tetherin Across Equid Species

Tetherin (BST-2/CD317) is a key antiviral protein in mammals that restricts the release of enveloped viruses and primes the immune response. In donkeys and horses, Yin et al. (2014) found tetherin orthologues differ slightly in structure—three amino acid deletions and substitutions in donkeys—but both retain equivalent antiviral efficacy against EIAV and HIV-1 particles, activate NF-κB signaling, localize to the plasma membrane, and are vulnerable to viral antagonism. Their shared ability to stimulate cytokines like IL-6 and IFN-β suggests that tetherin's antiviral function is conserved across equids, and differences in EIAV symptoms are likely shaped by other immune factors, positioning tetherin as a foundational player in cross-species viral defense.

32.4.2 Bacterial Abortion

Brucella abortus equinus survives in macrophage phagosomes by inhibiting phagosome–lysosome fusion. Donkeys mount granulomatous responses with epithelioid macrophages and T-cell rims, but fail to sterilize tissues. Chronic carrier states in uterine lymphoid aggregates perpetuate endemic abortion—detected only by repeated serology or isolation (Allen and Dalton 1975).

Salmonella enterica serotypes invade trophoblasts via Type III secretion, stir acute-phase cytokines (IL-6, SAA), and attract neutrophils that release proteases damaging placental architecture.

32.5 Maternal–Fetal Tolerance and Immunoprophylaxis

To overcome interspecific barrier failures, early passive immunization with pregnant mare serum transfers protective alloantibodies and soluble factors—IL-4, IL-10—that skew immune responses toward tolerance. Active immunization with paternal lymphocytes before day 20 of gestation educates maternal dendritic cells to present donor antigens in a non-immunogenic context, fostering long-lived Tregs that protect the conceptus (Allen et al. 1985).

32.6 Immunological Strategies for Reproductive Control

32.6.1 Immunocontraception

Vaccines targeting zona pellucida glycoproteins (ZP2, ZP3) conjugated to carrier proteins elicit high-titer IgG that sterically hinders sperm binding. Booster intervals (6–12 months) maintain antibody levels above protective thresholds. Alternative adjuvants—saponins, aluminum hydroxide—are under investigation to reduce injection-site reactions observed with Freund's formulations (French et al. 2020).

32.6.2 Immunocastration

GnRH-conjugate vaccines (e.g., Bopriva®) induce anti-GnRH antibodies that neutralize endogenous GnRH, suppress luteinizing hormone (LH) release, and cause reversible testicular atrophy. Donkeys show sustained antibody titers and azoospermia for 4–6 months post-boost, enabling non-surgical male fertility control in extensive systems (Rocha et al. 2018).

33 Infectious Diseases Reservoirs and Zoonotic Immunity: Donkeys and the One Health Paradigm

Donkeys play a vital immuno-epidemiological role in the transmission and maintenance of zoonotic and equine-specific pathogens. While often clinically asymptomatic, donkeys demonstrate unique immunological profiles that allow them to control or limit infections without eliminating pathogens. Their immune system, characterized by robust innate responses and finely modulated adaptive mechanisms, supports the One Health concept by linking environmental, animal, and human health.

33.1 Immune Responses to Protozoal and Bacterial Pathogens

Donkey immunity to protozoal infections, notably *Leishmania infantum* and *Besnoitia bennetti,* is dominated by a Th1-skewed cell-mediated response. This involves IFN-γ production, macrophage activation, and nitric oxide release to restrict intracellular parasite replication. Seropositive donkeys rarely show clinical leishmaniasis, indicating immune control rather than pathogen clearance. In besnoitiosis, infected donkeys generate antigen-specific IgG and show cellular immune activation despite the frequent absence of lesions. Chronic carrier states are maintained by incomplete sterilizing immunity, suggesting a balance between parasite suppression and immune evasion (Fig. 6).

Against hemoprotozoa like *Theileria equi* and *Babesia caballi*, donkeys mount early innate responses including complement activation and NK cell–mediated cytotoxicity, followed by splenic phagocytic clearance of parasitized erythrocytes. These mechanisms contribute to low parasitemia and minimal pathology, despite chronic infection.

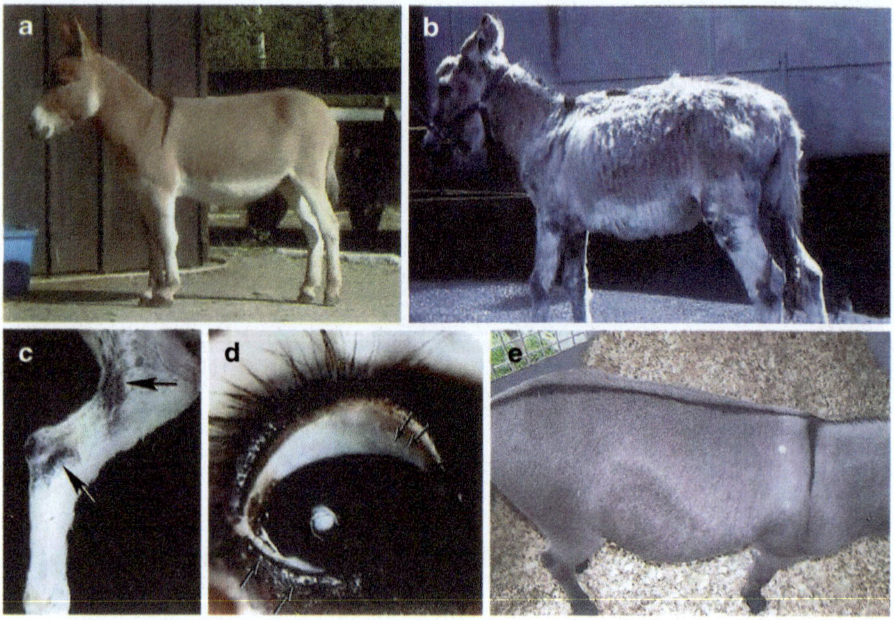

Fig. 6 Clinical progression and tissue cysts of *Besnoitia bennetti* in a miniature donkey (**a**) Photograph taken at approximately 1 year of age, prior to the onset of clinical signs. (**b**) Eight months after symptom onset and shortly before euthanasia. The animal exhibits severe cachexia, a deteriorated hair coat, and widespread dermatologic lesions including alopecia, hypotrichosis, hyperpigmentation, skin thickening, and crusting. Affected regions include the lateral neck, shoulders, carpi, stifles, hocks, periocular areas, ears, and muzzle. (**c**) Lateral view of the hind limb showing alopecia and hypotrichosis, with irregular, thickened, crusty, hyperpigmented skin over the caudal stifle and lateral hock (indicated by arrows). (**d**) Multifocal pinpoint white granular lesions (indicative of *Besnoitia* tissue cysts) visible within the sclera and conjunctiva of the eye (arrows). (**e**) Lateral view of donkey no. 1 following treatment with sulfamethoxazole (SMZ) and trimethoprim (TM), showing notable improvement in hair coat condition

Bacterial pathogens such as *Rhodococcus equi* induce granulomatous inflammation marked by macrophage and multinucleated giant cell accumulation. Adult donkeys display efficient clearance, likely due to mature neutrophil activity and a balanced Th1/Th2 cytokine response. In contrast, Salmonella abortus equi activates strong acute-phase responses, including elevated haptoglobin and serum amyloid A, which are counterbalanced by complement-mediated bacterial lysis in the placenta. Nevertheless, immune failure at the maternal–fetal interface permits abortion, underscoring the limitations of systemic immunity in reproductive tissues (Corpas et al. 1999).

33.1.1 Brucellosis in Donkeys—Immunological Implications

Brucellosis, caused by Brucella spp., is a chronic intracellular infection that evades immune clearance by suppressing antigen presentation and inhibiting macrophage killing. In donkeys, the immune response remains poorly characterized, though seroprevalence studies suggest widespread exposure. Brucella organisms exploit the host's phagocytic cells, persisting within macrophages and subverting lysosomal fusion. Donkey immune control likely depends on sustained Th1-type responses, particularly interferon-gamma (IFN-γ) secretion, though the limited clinical signs suggest either successful containment or immunological tolerance. The absence of detailed cytokine profiling in infected donkeys reflects a critical research gap.

33.1.2 Immunological Aspects of Clostridium Infection

Clostridium infections, particularly those caused by *Clostridium difficile*, pose a serious threat to neonatal miniature donkey foals due to their immature immune systems and frequent failure of passive transfer (FPT) (Giguère and Polkes 2005; Turini et al. 2021). In the absence of sufficient maternal immunoglobulin G (IgG) from colostrum, foals lack the antibodies needed to neutralize bacterial toxins and mount an effective immune response (Veronesi et al. 2014). This immunological gap leaves them vulnerable to opportunistic pathogens and systemic complications such as sepsis (Cray et al. 2009). *C. difficile* produces potent toxins that disrupt gut epithelial integrity and trigger excessive cytokine release, overwhelming the underdeveloped innate defenses of neonatal donkeys, including neutrophil and macrophage activity (Gabay and Kushner 1999; Watt et al. 2015). The adaptive immune system, still in early development, cannot compensate for this deficit, and mucosal immunity—especially secretory IgA—is insufficient to protect the gastrointestinal tract (Jirillo and Magrone 2014). Plasma transfusion is often required to provide immediate passive immunity, while supportive care and antimicrobials help control infection (Attia et al. 2022). Ultimately, early recognition and immunological support are critical to survival, and recovery depends on the foal's ability to gradually build endogenous immunity and restore mucosal balance (Mendoza and Toribio 2025) (Fig. 7).

Tetanus in donkeys, caused by the neurotoxin of *Clostridium tetani*, triggers a potent immune response when animals are vaccinated with tetanus toxoid. Upon

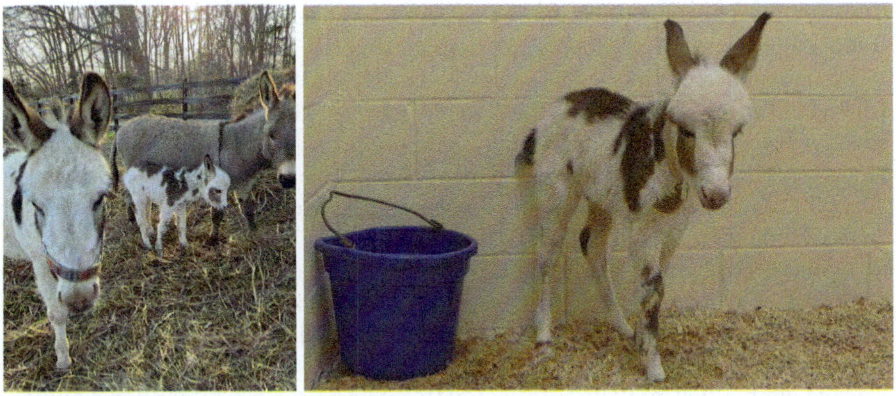

Fig. 7 *Clostridium difficile* case in Miniature Donkeys

vaccination, the donkey's immune system activates both innate and adaptive pathways. Dendritic cells and monocytes present the toxoid antigen to T cells, initiating a cascade of cytokine signaling and lymphocyte activation. Studies show that vaccinated donkeys exhibit a significant increase in CD4+ and CD8+ T cells, indicating strong cell-mediated immunity (Perzyna et al. 2025). Regulatory T cells (CD4+ FoxP3+) also peak shortly after vaccination, suggesting a balanced immune modulation. B cells (PanB/CD21+) produce specific antibodies against the tetanus toxoid, although their numbers decline post-vaccination as effector functions take over (Perzyna et al. 2025). These responses confirm that donkeys, despite being immunologically distinct from horses, can mount robust and protective immunity when vaccinated appropriately. This supports the continued use of equine-based vaccination protocols in donkeys, while highlighting the need for species-specific immunological data (Kahn 2010) (Fig. 8).

33.2 Donkey Immunity to Viral Infections

Donkeys exhibit an immunological equilibrium in response to viral infections, often controlling viral replication while allowing persistent low-level infection. For Equine Infectious Anemia Virus (EIAV), donkeys show extremely low viral loads and delayed seroconversion. Their subdued humoral response likely reflects a modulated innate antiviral state, possibly involving type I interferons and APOBEC3-mediated restriction.

Similarly, Equine Herpesviruses (EHV-1/-4) are controlled by cytotoxic T cells and NK cells, although immunological stress—especially in pregnant or lactating donkeys—can trigger viral reactivation. The humoral response is measurable but often insufficient for full protection, highlighting the importance of mucosal immunity and latent reservoir dynamics.

In Equine Viral Arteritis (EAV), viral tropism for $CD14^{+}$ monocytes and $CD3^{+}$ T cells activates transient viremia and mucosal inflammation without abortion. The lack of severe outcomes suggests that donkeys may exhibit a more regulated T-cell susceptibility phenotype than horses, possibly linked to differential expression of cellular receptors or T-regulatory cell activity (Li et al. 2024).

Equine Influenza Virus (EIV) elicits robust innate responses, particularly in foals, where failure of passive transfer leads to uncontrolled viral replication. In adults, the adaptive response includes virus-neutralizing antibodies and $CD8^{+}$ T-cell responses, though disease severity remains age-dependent. The high mortality observed in young unvaccinated donkeys highlights the critical role of maternal antibodies and colostral immunity in antiviral protection.

Equine Hepacivirus (EqHV) infection in donkeys results in low-grade viremia and minimal hepatic inflammation. Transcriptomic studies suggest individual variation in interferon-stimulated gene (ISG) activation, with some donkeys exhibiting efficient antiviral responses and others a muted profile. The general absence of liver damage implies that donkeys resolve EqHV infection via tightly regulated immune pathways that limit immunopathology.

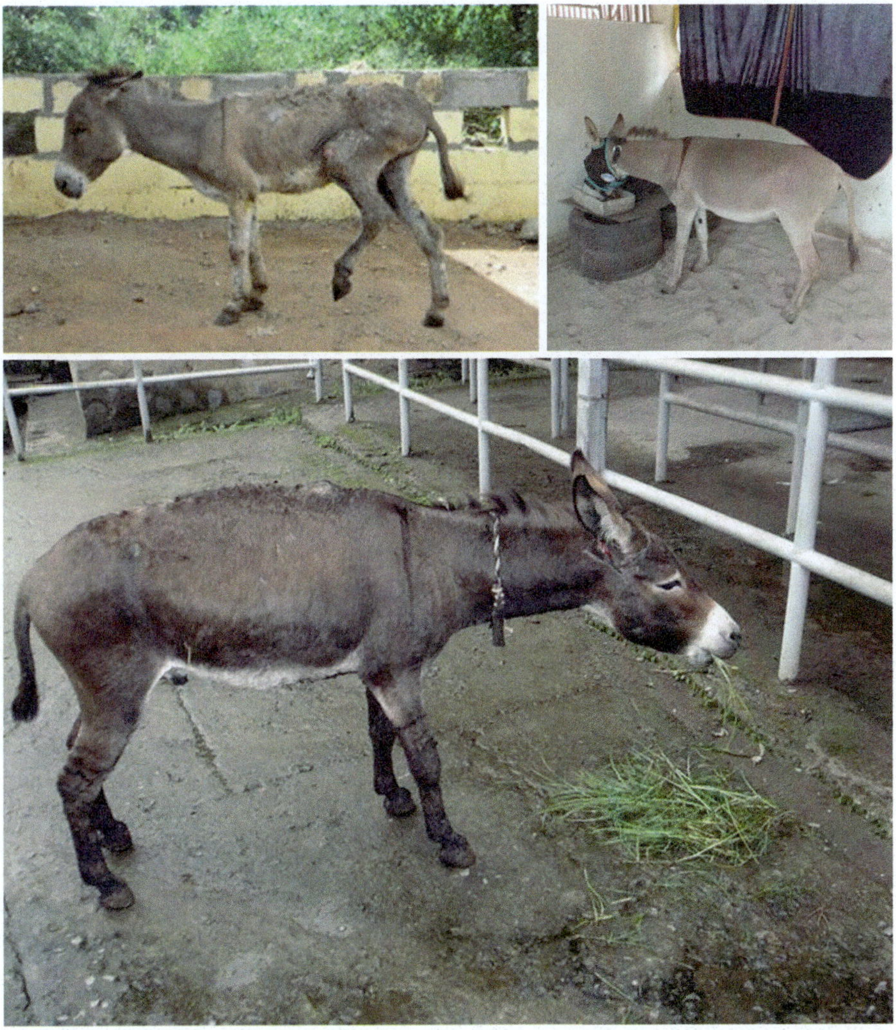

Fig. 8 Tetanus in donkeys

Upon natural Culicoides-borne exposure, African horse sickness virus (AHSV) initially replicates in regional lymphoid tissues and endothelial cells, where toll-like receptor recognition of viral double-stranded RNA triggers type I interferon release and MHC I upregulation, restraining viral spread and priming natural killer activity. Low-level viraemia emerges by days 3–5 and is typically cleared by day 17 in immunocompetent donkeys as robust adaptive responses develop: by days 7–10, neutralizing antibodies against outer-capsid proteins VP2 and VP5—and group-specific antibodies to VP7—peak and confer sterilizing immunity, while $CD8^+$ cytotoxic T lymphocytes eliminate infected cells and $CD4^+$ helper T cells secrete

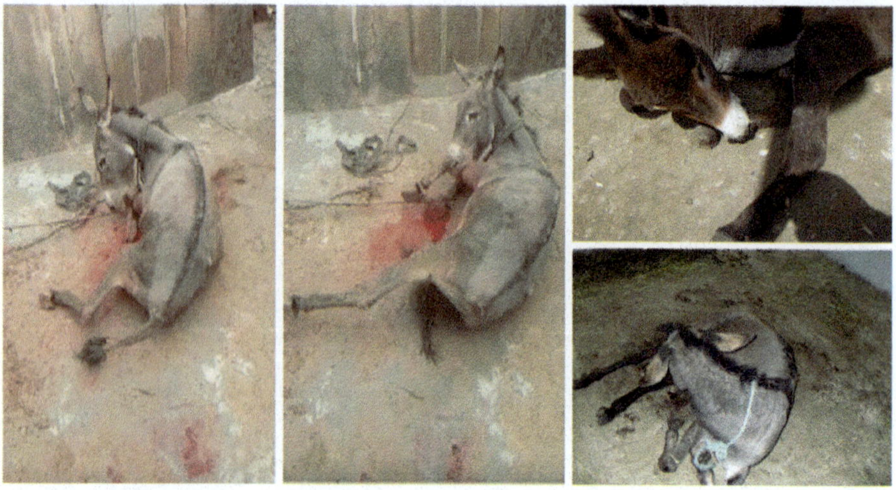

Fig. 9 Rabies in donkeys

IFN-γ and IL-2 to drive B-cell class switching and affinity maturation. Memory B cells targeting conserved VP7 epitopes mount rapid secondary responses to heterologous serotypes without detectable viraemia, underscoring the efficiency of donkey antiviral immunity and explaining their subclinical course and potential role as short-term reservoirs (Williams et al. 1993).

When a donkey is exposed to the rabies virus, its immune system initiates a complex response aimed at neutralizing the pathogen, though the virus is highly adept at evading detection (Fig. 9). Initially, the innate immune system produces type I interferons to slow viral replication, but the virus's replication in muscle tissue and rapid entry into peripheral nerves allows it to bypass early immune defenses (Katz et al. 2017; Banyard and Tordo 2018). As the virus travels toward the central nervous system (CNS), the adaptive immune system begins producing neutralizing antibodies—primarily targeting the viral glycoprotein—through B cell activation, supported by CD4+ T helper cells (World Health Organization [WHO] 2017). However, once the virus reaches the CNS, immune clearance becomes extremely difficult due to the brain's immune-privileged status, which limits antigen presentation and immune cell access (Ozkaraca et al. 2022). Cytotoxic T cells have limited efficacy in this environment, and the virus suppresses apoptosis and cytokine signaling to further evade destruction (Ozkaraca et al. 2022). Vaccination plays a critical role in rabies prevention, as it primes the immune system to produce protective antibodies before the virus can reach the CNS, offering donkeys a reliable defense against this otherwise fatal disease (Kirui et al. 2011; WHO 2017) (Fig. 9).

33.3 Tick-Borne Pathogens and Donkey Immunity

Donkeys exposed to tick-borne pathogens such as *T. equi, B. caballi*, and *Anaplasma phagocytophilum* exhibit signs of systemic inflammation and chronic antigenic stimulation. Their immune response includes elevated eosinophils and hyperglobulinemia, reflecting prolonged engagement of both innate and humoral arms (Villa et al. 2025). Despite this, overt disease is rare, indicating that immune mechanisms effectively suppress pathogen expansion. Cytokines like IL-1, IL-6, and TNF-α likely contribute to the inflammatory milieu, although formal profiling is limited. The chronic carrier state may be driven by immune evasion strategies or incomplete clearance, sustaining pathogen reservoirs in endemic zones.

33.3.1 Cutaneous and Granulomatous Immune Responses

Dermatophilosis, caused by *Dermatophilus congolensis*, engages both humoral and innate skin immune defenses. Infected donkeys develop antibodies against multiple antigenic determinants, yet recurrent infections suggest insufficient protective immunity. Local immune responses include keratinocyte activation, neutrophil infiltration, and cytokine production. The cutaneous immune microenvironment, possibly shaped by genetic or environmental factors, remains poorly defined.

Mycobacterial infections such as those caused by Mycobacterium bovis activate a classical type IV hypersensitivity reaction. Donkeys form granulomas rich in epithelioid macrophages and multinucleated giant cells, with surrounding lymphocytes and fibrotic encapsulation. The dominant Th1 response, marked by IFN-γ secretion, contains but does not eliminate infection, allowing for latent disease. Reactivation in older donkeys may be linked to immunosenescence, and diagnostic difficulties arise from the nonspecific nature of clinical signs and poor performance of traditional tests in equids (Collins et al. 2011).

34 Skin Tumors in Donkeys: Sarcoids and Melanocytic Neoplasms

Donkeys develop cutaneous neoplasms similar to those seen in horses, particularly sarcoids associated with bovine papillomavirus (BPV) and melanocytic tumors in aging gray animals. Despite the focus of equine oncology research on horses, donkeys exhibit comparable tumor biology and immune evasion mechanisms, making them valuable comparative models for studying skin cancer (Jones 2021; Davis et al. 2016).

34.1 Immunopathogenesis of Sarcoids in Donkeys

Equine sarcoids are locally invasive fibroblastic tumors caused by BPV-1 and BPV-2, which infect dermal fibroblasts. Viral oncoprotein E5 downregulates MHC class I, impairing $CD8^+$ T-cell recognition and promoting viral persistence (Ashrafi et al. 2008). Histologically, sarcoids show hyperkeratosis, acanthosis, and spindle-cell proliferation. Infected tissues recruit macrophages and $CD4^{+}FoxP3^{+}$ regulatory T cells, which secrete IL-10 and TGF-β, creating an immunosuppressive microenvironment (Jones 2021).

PCR detection of BPV DNA in donkey sarcoids confirms the viral etiology, paralleling findings in horses (Ashrafi et al. 2008). Treatments such as surgical excision, cryotherapy, and cisplatin injections often result in high recurrence rates (Semieka et al. 2012). Topical immunomodulators like imiquimod aim to stimulate local Th1 responses but show variable success. Preventive strategies, including fly control and minimizing skin trauma, remain the most effective (Jones 2021).

34.2 Immunology of Melanocytic Tumors in Donkeys

Melanocytic tumors originate from dermal melanocytes and are prevalent in aging gray donkeys. These tumors typically appear as firm nodules in perineal, parotid, or supraorbital regions. Benign melanocytomas may progress to malignant melanomas, which can metastasize to lymph nodes and internal organs (Mosbah and Awadin 2016).

Immunologically, melanoma cells in donkeys downregulate MHC class I and upregulate PD-L1, facilitating escape from cytotoxic T-cell responses. Tumor-associated macrophages adopt an M2 phenotype, secreting IL-10 and VEGF, which promote angiogenesis and suppress immune responses. Regulatory T cells accumulate in the peritumoral stroma, further dampening antitumor immunity (Mosbah and Awadin 2016). Diagnosis involves fine-needle aspiration, histopathology, and Fontana–Masson staining. Surgical excision is preferred for isolated lesions, while advanced cases may benefit from intralesional chemotherapy or checkpoint inhibitor immunotherapy (Cuitiño 2011) (Fig. 10).

34.3 Donkeys as Comparative Oncology Models

Donkeys' susceptibility to sarcoids and melanocytic tumors reflects key features of human cutaneous oncology, including immune evasion, regulatory T-cell recruitment, and cytokine-driven microenvironmental modulation. Their outbred genetics, large size, and spontaneous tumor development make them ideal models for testing novel immunotherapies, such as checkpoint inhibitors and BPV-targeted vaccines, under conditions more representative than rodent models (Alkayyal 2025; LeBlanc et al. 2025).

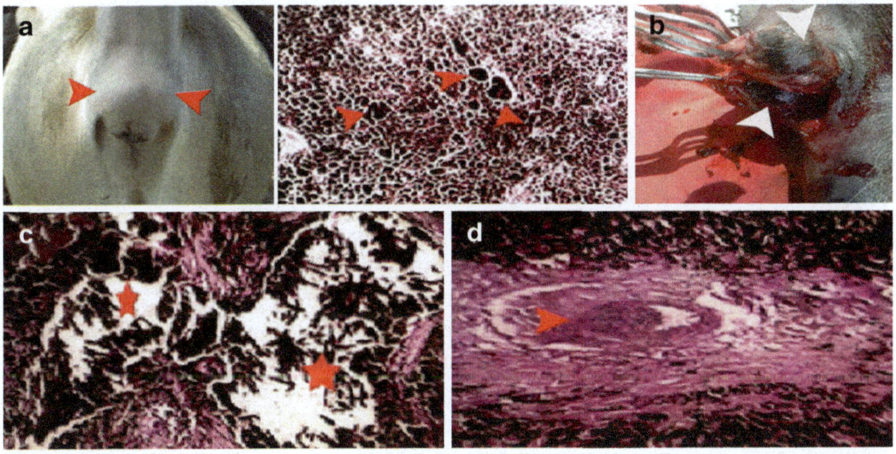

Fig. 10 A 2-yr-old intact male donkey (*Equus asinus*) shows two 5 × 10 and 5 × 8 cm diameters, round and firm subcutaneous masses under the tail (red arrowheads) (Upper left), the masses consist of a highly pleomorphic population of heavily pigmented cells. These cells had marked anisocytosis and anisokaryosis with single to multiple nuclei (arrowheads) (upper middle), Shows two primary firm and pigmented masses dissected from the surrounding subcutaneous and muscular tissues (white arrowheads) (upper right); Neoplastic cells are present within blood vessel (arrowhead) and multifocal areas of necrosis and dystrophic calcification are present (red asterisks) (lower left) and shows thrombosed blood vessel (red arrowhead) (H&E, 100×) (lower right)

35 Donkey Immunology in Socioeconomic and Production Environments

35.1 Socioeconomic Stressors, Welfare, and Immune Health in Working Donkeys

In many low-income regions, donkeys are essential to household economies—transporting goods, tilling fields, and hauling water. However, their critical role often comes at the expense of their own health. In brick-kiln communities of India, donkeys endure heavy loads, long hours, extreme temperatures, and poor nutrition. These chronic stressors activate the hypothalamic–pituitary–adrenal (HPA) axis, elevate cortisol, and suppress cell-mediated immunity (Pritchard et al. 2005; Watson et al. 2020).

Undernutrition compounds these effects. Deficiencies in protein, energy, and micronutrients like zinc, selenium, and vitamin A impair lymphocyte proliferation, antibody production, and neutrophil function, weakening both innate and adaptive immunity (Pritchard et al. 2005). Simultaneously, heavy workloads increase gastrointestinal nematode burdens and ectoparasitism, leading to eosinophilia and a skewed Th2 cytokine bias, which undermines resistance to bacterial and viral pathogens (Geiger et al. 2021).

The Brooke's brick-kiln project found that owners with greater financial security provided better donkey care—regular deworming, supplemental feeding, and rest periods—resulting in improved body condition scores, packed-cell volumes, and neutrophil oxidative bursts (Pritchard et al. 2018). Long-term owner–animal bonds correlated with lower basal cortisol and reduced parasitic loads, suggesting that stable human–animal relationships enhance immunocompetence (Watson et al. 2020).

Conversely, households under financial strain often lack access to veterinary services. Delayed treatment allows subclinical infections—such as *Salmonella abortus equi* or *Brucella* spp.—to persist, exhausting lymphoid reserves and blunting vaccine responsiveness (Kithuka et al. 2025). Poor donkey health reduces work capacity and deepens poverty, creating a feedback loop of disease and economic hardship.

Integrated development programs that subsidize veterinary outreach, provide fortified feeds, and train owners in basic husbandry have broken this cycle in pilot regions. Recognizing donkeys not merely as draft power but as sentinels of rural livelihoods allows policymakers to leverage equid welfare improvements to advance Sustainable Development Goals, including poverty reduction and food security (Rodrigues 2024).

35.2 Immunological Surveillance in Large-Scale Donkey Farming

The rise of commercial donkey farming—driven by demand for meat and Ejiao (donkey-hide gelatin)—has reshaped disease risks in regions like Liaocheng City, Shandong Province, China. High-density operations (>15 animals/hectare) with poor ventilation and constant contact favor the spread of respiratory viruses, especially equine influenza virus (EIV) H3N8 (Yongfeng et al. 2020) (Fig. 11).

Infected donkeys show fever, nasal discharge, and cough, but often with subtler signs and delayed viral shedding compared to horses, complicating early detection (Yongfeng et al. 2020). Initial defenses—mucosal barrier integrity, surfactant protein D, and type I interferons—can temporarily contain the virus, but high pathogen loads and stress-induced immunosuppression may lead to systemic spread and secondary pneumonia in up to 10% of unvaccinated herds (Ji et al. 2023).

A 2020 survey in Shandong found 32.5% seropositivity by hemagglutination-inhibition assay and 21.5% RT-PCR positivity for H3 and M gene fragments, confirming active infection. Sequencing revealed 99.77% homology with equine isolates from Northeast China, indicating cross-regional transmission via animal movement (Yongfeng et al. 2020).

Antigenic drift in the hemagglutinin (HA) glycoprotein alters viral epitopes, enabling immune escape and undermining herd immunity. No EIV vaccines are licensed specifically for donkeys in China, and off-label equine vaccines show inconsistent protection (Ruan et al. 2025).

To safeguard herd health, farms should implement rigorous immunological surveillance:

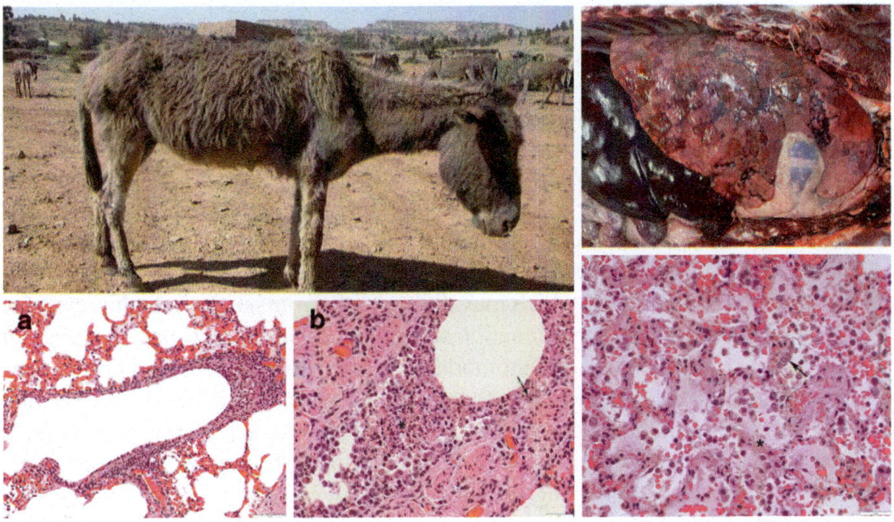

Fig. 11 Equine influenza virus (upper left), lung, donkey foal. Multifocally, the left lung lobe parenchyma is red-dark red and collapsed, and interlobular septa are often expanded by edema. There are ill-defined rib impressions along the craniodorsal pulmonary surfaces; Equine influenza virus, lung, donkey foal (upper right), Photomicrograph of affected alveolar spaces and septa. Alveolar spaces contain variable amounts of eosinophilic proteinaceous fluid, neutrophils, and fewer macrophages. Adjacent alveolar septa are necrotic, expanded by karyorrhectic debris, frequently lined by eosinophilic hyaline membranes (*), and rare areas of type II pneumocyte hyperplasia (black arrow). (Hematoxylin and eosin stains, bar = 20 μm) (lower left); Equine influenza virus, lung, donkey foal. (**a**) Photomicrograph of an affected terminal airway in which the bronchiole mucosa is denuded. (**b**) Photomicrograph of a more severely affected bronchiole that includes karyorrhectic epithelial cells (arrow). Exudate in bronchiole lumen is composed of necrotic debris and degenerate neutrophils (asterisk). (Hematoxylin and eosin stains, bar = 20 μm) (lower right)

Quarterly HI and microneutralization assays to monitor antibody titers and define seroprotective thresholds

Routine RT-PCR screening of nasal swabs during high-risk seasons (autumn, spring)

Whole-genome sequencing of emergent isolates to track antigenic evolution and guide vaccine updates

Biosecurity measures: all-in/all-out stocking, improved ventilation, quarantine protocols, and disinfection routines

Strengthening farm-to-authority disease reporting will enable real-time outbreak mapping and coordinated vaccine development, transitioning donkey production from reactive to proactive herd health management under One Health and One Welfare frameworks (Khan et al. 2025).

36 Conclusions

From a clinical perspective, advancing our understanding of donkey and mule immunology has direct implications for improving veterinary care, disease prevention, and welfare in these often-overlooked equids. Recognizing species-specific differences in neonatal passive transfer, hematological parameters, and immunological thresholds is critical for accurate diagnosis and timely intervention in young animals. Practitioners should be aware of the distinct presentation of immune-mediated conditions such as neonatal isoerythrolysis, hypersensitivity reactions, and autoimmune syndromes, which may differ subtly from horses in both prevalence and clinical course. Tailored vaccination schedules, careful monitoring of colostrum intake, and judicious use of immunomodulatory therapies are essential components of effective health management. Furthermore, incorporating donkey-specific diagnostic reference values and considering their unique immune responses to infectious agents, parasites, and vaccines will enhance both preventive and therapeutic strategies. Ultimately, improving immunological awareness in clinical practice not only supports better outcomes for donkeys and mules but also strengthens their role in agriculture, animal-assisted interventions, and global health systems where these equids remain indispensable.

Self-Evaluation Questionnaire

Q1. **In neonatal donkeys, what physiological process most determines the success of passive transfer of maternal immunoglobulins?**
   A. The onset of endogenous IgG synthesis at 2 weeks postpartum
   B. The duration of intestinal epithelial permeability to macromolecules
   C. The concentration of neonatal complement proteins at birth
   D. The expansion of neonatal $CD4^+$ T-cell populations
   Answer: B

Q2. **Which immunological feature best illustrates "inflammaging" in geriatric donkeys?**
   A. Decreased serum IgG and increased mucosal IgA
   B. Expansion of $FoxP3^+$ regulatory T cells and altered naïve-to-memory T-cell ratio
   C. Increased thymic output of naïve lymphocytes
   D. Elevated acute-phase proteins without regulatory T-cell involvement
   Answer: B

Q3. **Why are mule foals at particularly high risk for neonatal isoerythrolysis (NI)?**
   A. They inherit the "donkey factor" antigen absent in horse mares
   B. They lack maternal colostrum intake during the first 12 hours
   C. They fail to mount endogenous complement activity after birth
   D. They are unable to synthesize IgM until after weaning
   Answer: A

Q4. **Which property of donkey milk most directly contributes to its use as a hypoallergenic alternative in children with cow's milk protein allergy (CMPA)?**
   A. Enrichment in casein fractions compared to bovine milk
   B. Low levels of αs1- and κ-caseins with high lysozyme activity
   C. Absence of lactoferrin and β-lactoglobulin
   D. Reduced lactose concentration relative to goat milk
   Answer: B

Q5. **Compared with horses infected by Equine Infectious Anemia Virus (EIAV), donkeys typically show:**
   A. Higher viremia, severe clinical anemia, and rapid disease progression
   B. Viral loads approximately 1000-fold lower with subclinical presentation
   C. Complete viral resistance due to absence of $CD8^+$ T-cell responses
   D. Equivalent viral loads but greater reliance on innate immunity
   Answer: B

Q6. **From a One Health perspective, why are donkeys significant in zoonotic disease epidemiology?**
   A. Their immunological naivety prevents them from harboring pathogens
   B. They serve as silent reservoirs for pathogens such as Brucella spp. and tick-borne protozoa
   C. They exhibit universal resistance to bacterial zoonoses
   D. Their immune responses eliminate zoonotic carriage more rapidly than in horses
   Answer: B

References

Ahearne MM, Pentzke-Lemus LL, Romano AM, Larsen ED, Watson AM, O'Fallon EA, Landolt GA (2022) Disease progression, pathologic, and virologic findings of an equine influenza outbreak in rescue donkeys. J Vet Intern Med 36(6):2230–2237. https://doi.org/10.1111/jvim.16563. Epub 2022 Oct 7. PMID: 36205917; PMCID: PMC9708454

Aiello, S. E., Moses, M. A. (Eds.). (2016). Merck Veterinary Manual (11th ed.). Merck & Co., Inc.

Alkayyal AA (2025) Insights from veterinary models for advancing oncolytic virotherapy through comparative oncology. Front Mol Biosci 12:1615393

Allen WR (1982) Immunological aspects of the endometrial cup reaction and the effect of xenogeneic pregnancy in horses and donkeys. J Reprod Fertil Suppl 31:57–94. PMID: 6962845

Allen PZ, Dalton EJ (1975) Studies on equine immunoglobulins. IV. Immunoglobulins of the donkey. Immunology 28(1):187–197. PMID: 1116869; PMCID: PMC1445747

Allen WR, Letoble A, Ricketts SW (1985) Immunological factors influencing interspecies embryo transfer in equids. Equine Reprod Sci 12(4):345–360

Antczak DF, Davies CJ, Kydd J, Allen WR (1985) Immunological aspects of pregnancy in mules. Equine Vet J Suppl 3:68–71. https://doi.org/10.1111/j.2042-3306.1985.tb04596.x

Ashrafi GH, Piuko K, Burden F, Yuan Z, Gault EA, Müller M et al (2008) Vaccination of sarcoid-bearing donkeys with chimeric virus-like particles of bovine papillomavirus type 1. J Gen Virol 89(1):148–157

Attia MM et al (2022) Evaluation of immune responses and oxidative stress in donkeys: Immunological studies provoked by Parascaris equorum infection. Saudi J Biol Sci 29(4):2173–2179.

https://doi.org/10.1016/j.sjbs.2021.11.044. https://www.madbarn.compubmed.ncbi.nlm.nih.govresearchgate.neteuropepmc.org

Ayo JO, Ake AS, Abimbola AA (2023) Hematological and biochemical responses of Donkeys (Equus asinus) to packing and trekking during the hot-dry season. J Equine Vet Sci 122:104203. https://doi.org/10.1016/j.jevs.2022.104203

Banyard AC, Tordo N (2018) Rabies pathogenesis and immunology. Revue Scientifique et Technique—Office International des Epizooties 37(2):323–330

Bradley C (2022) Recognising and treating uveitis in donkeys. UK Vet Equine 6(1):14–23

Burden FA, Du Toit N, Trawford AF et al (2016) Reference intervals for biochemical and haematological parameters in mature domestic donkeys (Equus asinus) in the UK. Equine Vet Educ 28(3):134–139. https://doi.org/10.1111/eve.12512

Cocco R, Rizzo M, Carta C, Arfuso F, Piccione G, Luridiana S, Crovace A, Passino ES, Sechi S (2023) Immune and inflammatory response of donkeys (Equus asinus) tested positive to tick-borne pathogens. J Equine Vet Sci 120:104190. https://doi.org/10.1016/j.jevs.2022.104190. Epub 2022 Dec 9. PMID: 36509242

Cequier A, Sanz C, Rodellar C, Barrachina L (2021) The Usefulness of Mesenchymal Stem Cells beyond the Musculoskeletal System in Horses. Animals (Basel). Mar 25;11(4):931. https://doi.org/10.3390/ani11040931. PMID: 33805967; PMCID: PMC8064371

Collins SN, Dyson SJ, Murray RC et al (2011) Radiological anatomy of the donkey's foot: objective characterisation of the normal and laminitic donkey foot. Equine Vet J 43(4):478–486

Corpa JM, Pérez V, García Marín JF (1999) Lesions in the placenta and foetus in equine abortion due to Salmonella abortus equi. Journal of Comparative Pathology 121(1):1–10. https://doi.org/10.1053/jcpa.1999.0293

Costa EA, da Silva MC, Guilherme JE, Mota RA (2019) Serum amyloid A and other acute phase proteins in donkeys with experimentally induced endotoxemia. Vet Res Commun 43(3): 189–197. https://doi.org/10.1007/s11259-019-09747-y

Cray C, Zaias J, Altman NH (2009) Acute phase response in animals: a review. Comp Med 59(6): 517–526

Cuitiño MC (2011) Neoplasias melanocíticas en caninos: evaluación de marcadores inmunohistoquímicos que contribuyen a la formulación del diagnóstico y del pronóstico. Doctoral dissertation, Universidad Nacional de La Plata

Davis CR, Valentine BA, Gordon E, McDonough SP, Schaffer PA, Allen AL, Pesavento P (2016) Neoplasia in 125 donkeys (Equus asinus): literature review and a survey of five veterinary schools in the United States and Canada. J Vet Diagn Invest 28(6):662–670

Di Francesco CE, Smoglica C, De Amicis I, Cafini F, Carluccio A, Contri A (2020) Evaluation of colostral immunity against Equine Herpesvirus Type 1 (EHV-1) in Martina Franca's Foals [Original Research]. Front Vet Sci 7. https://doi.org/10.3389/fvets.2020.579371

Podliachouk L (1957) Groupes sanguins des équidés. Annales de l'Institut Pasteur, 93, 1–12

Eckersall PD, Bell R (2010) Acute phase proteins: biomarkers of infection and inflammation in veterinary medicine. Vet J 185(1):23–27

Eckersall PD (2008) Proteins, proteomics, and the dysproteinemias. Clinical biochemistry of domestic animals 6:117–155

El-Ashker MR et al (2017) The influence of experimentally-induced endotoxaemia on clinical variables and markers of systemic inflammation in donkeys (Equus asinus). Vet Med 62(03): 117–124. https://doi.org/10.17221/133/2016-VETMED

Felippe JB (2018) Equine neonatal isoerythrolysis. In: Pusterla N, Higgins J (eds) Interpretation of equine laboratory diagnostics. Wiley, pp 251–255. https://doi.org/10.1002/9781119279338.ch42

French H, Peterson E, Schulman M, Roth R, Crampton M, Conan A, Marchi S, Knobel D, Bertschinger H (2020) Efficacy and safety of native and recombinant zona pellucida immunocontraceptive vaccines in donkeys. Theriogenology 153:27–33. https://doi.org/10.1016/j.theriogenology.2020.04.041

Gabay C, Kushner I (1999) Acute-phase proteins and other systemic responses to inflammation. N Engl J Med 340(6):448–454

Garg SK, Delaney C, Toubai T, Ghosh A, Reddy P, Banerjee R, Yung R (2014) Aging is associated with increased regulatory T-cell function. Aging Cell 13(3):441–448. https://doi.org/10.1111/acel.12191

Geiger M, Hockenhull J, Buller H et al (2021) Comparison of the socioeconomic value and welfare of working donkeys in rural and urban Ethiopia. Anim Welf 30(3):269–277. https://doi.org/10.7120/09627286.30.3.004

Giguère S, Polkes AC (2005) Immunologic disorders in neonatal foals. Vet Clin N Am Equine Pract 21(2):241–272. https://doi.org/10.1016/j.cveq.2005.04.004

Giorgis V, Rolla G, Raie A et al (2018) A case of work-related donkey milk allergy. J Invest Allergol Clin Immunol 28(3):197–199. https://doi.org/10.18176/jiaci.0237

Girardi A, Toledo C, Silva P, Marques L (2017) The effect of age and sex on serum proteins in the Pega donkey (Equus asinus). Vet Med 62:10–15. https://doi.org/10.17221/397/2014-VETMED

Gömer A, Puff C, Reinecke B, Bracht S, Conze M, Baumgartner W, Steinmann J, Feige K, Cavalleri JMV, Steinmann E, Todt D (2022) Experimental cross-species infection of donkeys with equine hepacivirus and analysis of host immune signatures. One Health Outlook 4(9). https://doi.org/10.1186/s42522-022-00065-y

Grove VS (2008) Conditions of the eye. In: Svendsen ED, Duncan J, Hadrill D (eds) The professional handbook of the donkey, 4th edn. Whittet Books, pp 202–221

Gutiérrez JP, Marmi J, Goyache F, Jordana J (2005) Pedigree information reveals moderate to high levels of inbreeding and a weak population structure in the endangered Catalonian donkey breed. J Anim Breed Genet 122(5):378–386. https://doi.org/10.1111/j.1439-0388.2005.00590.x

Horohov DW, Adams AA, Chambers TM (2010) Immunosenescence of the equine immune system. J Comp Pathol 142(Suppl 1):S78–S84. https://www.pubmed.ncbi.nlm.nih.gov.

Huang J, Zhao Y, Bai D, Shiraigol W, Li B, Yang L et al (2015) Donkey genome and insight into the imprinting of fast karyotype evolution. Sci Rep 5:14106. https://doi.org/10.1038/srep14106

Jamieson CA, Baillie SL, Johnson JP (2022) Blood transfusion in equids—a practical approach and review. Animals (Basel) 12(17):2162. https://doi.org/10.3390/ani12172162. https://www.mdpi.commdpi.com

Ji Y, Xu D, Zhao X, Liu W (2023) Epidemiological investigation of equine herpesvirus in large-scale donkey farms. Society Preprint. https://sciety.org/articles/activity/10.21203/rs.3.rs-3420287/v1

Jirillo F, Magrone T (2014) Anti-inflammatory and anti-allergic properties of donkey's and goat's milk. Endocr Metab Immune Disord Drug Targets 14(1):27–37. https://doi.org/10.2174/1871530314666140122104703

Jones SE (2021) An epidemiological and molecular study of equine sarcoids at The Donkey Sanctuary UK. Doctoral dissertation, University of Glasgow

Kahn CM (2010) The Merck veterinary manual, 10th edn. Merck and Co., Inc.

Karaj I (2020) Common variable immunodeficiency: epidemiology, pathogenesis, clinical manifestations, diagnosis, classification and management. J Investig Allergol Clin Immunol 30(1)

Katz ISS, Guedes F, Fernandes ER, Silva SR (2017) Immunological aspects of rabies: a literature review. Arch Virol 162(12):3251–3268

Khan SA, Fröysa B, Diczfalusy E (1984) Biological and immunological characterization of purified follicle-stimulating hormone preparations from monkey pituitaries. J Med Primatol 13(5):269–282. PMID: 6439872

Khan MZ, Li Y, Zhu M, Li M, Wang T, Zhang Z, Liu W, Ma Q, Wang C (2025) Advances in donkey disease surveillance and microbiome characterization in China. Microorganisms 13(4):749. https://doi.org/10.3390/microorganisms13040749

Kim HJ, Shin JH, Ha J et al (2014) Pharmacokinetics and bioavailability of cimicoxib in horses and donkeys. J Vet Pharmacol Ther 37(6):624–628. https://doi.org/10.1111/jvp.12118

Kirui G, Mulei CM, Abuom TO (2011) Rabies in donkeys—A review. Presented at the Kenya Veterinary Association 45th Scientific Conference, Kisumu, Kenya. Retrieved from https://erepository.uonbi.ac.ke/handle/11295/43836

Kithuka JM, Wachira TM, Onono JO, Ngetich W (2025) The burden of brucellosis in donkeys and its implications for public health and animal welfare: a systematic review and meta-analysis. Vet World 18(2):367–378. https://doi.org/10.14202/vetworld.2025.367-378

Kydd J, Miller J, Antczak DF, Allen WR (1982) Maternal anti-foetal cytotoxic antibody responses of equids during pregnancy. J Reprod Fertil Suppl 32:361–369

Lavoie JP, Maghni K, Desnoyers M, Taha R, Martin JG, Hamid QA (2001) Neutrophilic airway inflammation in horses with heaves is characterized by a Th2-type cytokine profile. Am J Respir Crit Care Med 164(8 Pt 1):1410–1413. https://doi.org/10.1164/ajrccm.164.8.2012091. PMID: 11704587

Lazić S, Savić S, Petrović T, Lazić G, Žekić M, Drobnjak D, Lupulović D (2023) Serological examinations of significant viral infections in domestic donkeys at the Special Nature Reserve "Zasavica", Serbia. Animals 13(2056). https://doi.org/10.3390/ani13132056

LeBlanc A, Mazcko CN, Mason NJ, Chambers MR, Brockington DM, Pluhar GE, Naik S (2025) Comparative oncology in action: vignettes on immunotherapy development. Vet Oncol 2:5

LeCuyer TE, Rink A, Bradway DS, Evermann JF, Nicola AV, Baszler T, Haldorson GJ (2015) Abortion in a Mediterranean miniature donkey (Equus asinus) associated with a gammaherpesvirus similar to Equid herpesvirus 7. J Vet Diagn Invest 27(6):749–753. https://doi.org/10.1177/1040638715611444

Li L, Li S, Ma H, Akhtar MF, Tan Y, Wang T et al (2024) An overview of infectious and non-infectious causes of pregnancy losses in equine. Animals 14(13):1961

Liu B, Gong S, Tulafu H, Zhang R, Tao W, Adili A et al (2024) Analysis of the genetic relationship and inbreeding coefficient of the Hetian Qing donkey through a simplified genome sequencing technology. Genes 15(5):570. https://doi.org/10.3390/genes15050570

Liu J, Wang X, Yang Z, Chen Y, Li H (2025) Proteomic profiling of donkey milk exosomes: insights into neonatal immunity. J Proteome Res 24(1):45–59

Magdesian KG (2024) Immunodeficiencies in Foals. Veterinary Clinics: Equine Practice 40(2), e1–e19

Malalana F, Stylianides A, McGowan C (2015) Equine recurrent uveitis: human and equine perspectives. Vet J 206(1):22–29. https://doi.org/10.1016/j.tvjl.2015.06.017. Epub 2015 Jul 3. PMID: 26188862

McClure JJ, Koch C, Traub-Dargatz J (1994) Characterization of a red blood cell antigen in donkeys and mules associated with neonatal isoerythrolysis. Anim Genet 25(2):119–120

McEwen BS, Wingfield JC (2003) The concept of allostasis in biology and biomedicine. Horm Behav 43(1):2–15. https://doi.org/10.1016/S0018-506X(02)00024-7

McKenzie E (2012) Immune-mediated and metabolic myopathies

McLean AK, Falt T, Abdelfattah EM, Middlebrooks B, Gretler S, Spier S, Turoff D, Navas Gonzalez FJ, Knych HK (2023) Transdermal Flunixin Meglumine as a pain relief in donkeys: a pharmacokinetics pilot study. Metabolites 13(7):776. https://doi.org/10.3390/metabo13070776. PMID: 37512483; PMCID: PMC10383245

Mealey RH (2013) Equine infectious anemia. Equine Infectious Diseases E-Book: Equine Infectious Diseases E-Book 40,228

Mendoza Garcia FJ, Gonzalez-De Cara C, Aguilera-Aguilera R, Buzon-Cuevas A, Perez-Ecija A (2020) Meloxicam ameliorates the systemic inflammatory response syndrome associated with experimentally induced endotoxemia in adult donkeys. J Vet Intern Med 34(4):1631–1641

Mendoza FJ, Toribio RE (2025) An overview of donkey neonatology. Animals 15(13):1986. https://doi.org/10.3390/ani15131986

Mendoza FJ, Toribio RE, Pérez-Écija A (2018a) Donkey internal medicine—Part I: metabolic, endocrine, and alimentary tract disturbances. J Equine Vet Sci 65:66–74. https://pmc.ncbi.nlm.nih.gov

Mendoza FJ, Toribio RE, Pérez-Écija A (2018b) Donkey internal medicine—Part II: cardiovascular, respiratory, neurologic, urinary, ophthalmic, dermatology, and musculoskeletal disorders. J Equine Vet Sci 65:75–83. https://pmc.ncbi.nlm.nih.gov

Mendoza FJ, Toribio RE, Perez-Ecija A (2019) Metabolic and endocrine disorders in donkeys. Vet Clin North Am Equine Pract 35(3):399–417. https://doi.org/10.1016/j.cveq.2019.07.001. Epub 2019 Oct 3. PMID: 31587976

Mendoza FJ, Toribio RE, Perez-Ecija A (2024) Metabolic and endocrine insights in donkeys. Animals 14(4):590

Mosbah E, Awadin W (2016) Diagnosis and management of a malignant dermal melanoma in a donkey (Equus asinus). Int Vet Surg Assoc J

Mosca M, Randleff-Rasmussen P, Milhau N, Panzuti P, Meylan M, Desjardin I, Pin D (2020) Long-term successful treatment of a donkey with cutaneous lupus erythematous with methotrexate. Vet Dermatol 31(4):313–e78. https://doi.org/10.1111/vde.12847. Epub 2020 Apr 15. PMID: 32293082

Mozaffari AA, Derakhshanfar A, Mohammadpour O, Hadadipour Zarandi M (2025) First report of idiopathic advanced follicular atrophy of the thyroid gland in three miniature donkeys. Vet Comparative Biomed Res:30–34

Navas González FJ, Jordana Vidal J, Camacho Vallejo ME, León Jurado JM, de la Haba Giraldo MR, Barba Capote C, Delgado Bermejo JV (2018) Risk factor meta-analysis and Bayesian estimation of genetic parameters and breeding values for hypersensibility to cutaneous habronematidosis in donkeys. Vet Parasitol 252:9–16

Nervo T, Bertero A, Donato GG, Suffat EP, Vincenti L (2021) Analysis of factors influencing the transfer of passive immunity in the donkey foal. Ital J Anim Sci 20(1):1947–1956. https://doi.org/10.1080/1828051X.2021.1963863

Ozkaraca M, Ozdemir S, Comakli S, Timurkan MO (2022) Roles of apoptosis and autophagy in natural rabies infections. Vet Med 67(1):1–12

Paraschou G, Xue C, Egan R, Bolfa P (2024) Multicentric lymphoma in a donkey with intestinal and bone marrow involvement. BMC Vet Res 20(57). https://doi.org/10.1186/s12917-024-03903-5

Peart S (2022) The birth of a silver-gray donkey colt comes as a complete surprise! Marion duPont Scott Equine Medical Center, Virginia Tech. Retrieved from https://vetmed.vt.edu

Peeters C, Herman A, Baeck M (2017) Donkey's milk allergy. Br J Dermatol 177(6):1760–1761. https://doi.org/10.1111/bjd.15367

Perzyna M, Gręzdzicka J, Milczek-Haduch D, Dąbrowska I, Trela M, Pawliński B, Witkowska-Piłaszewicz O (2025) Immunological responses to tetanus and influenza vaccination in donkeys. J Vet Intern Med 39(2):e70137. https://doi.org/10.1111/jvim.70137

Peterson EW, Segabinazzi LGTM, Gilbert RO, Bergfelt DR, French HM (2022) Evaluation of stress accompanying immunocontraceptive vaccination in donkeys. Animals 12(457). https://doi.org/10.3390/ani12040457

Podliachouk L (1958) Les groupes sanguins des équidés (cheval, mulet, ane) [Blood groups of Equidae (horses, mules, asses)]. Ann Inst Pasteur (Paris) 95(1):7–22. French. PMID: 13545630

Podliachouk L (1965) The blood groups of equidae. In: Blood groups of animals: proceedings of the 9th European Animal Blood Group Conference (First Conference Arranged by ESABR) held in Prague, August 18–22, 1964. Springer Netherlands, Dordrecht, pp 229–235

Podliachouk L, Eyquem A (1953) Les groupes sanguins des equides. III. Les groupes sanguins des anes [Blood groups in Equidae. III. Blood groups in donkey]. Ann Inst Pasteur (Paris) 84(5): 966–968. Undetermined Language. PMID: 13125006

Pritchard JC, Lindberg AC, Main DCJ, Whay HR (2005) Assessment of the welfare of working horses, mules and donkeys. Prev Vet Med 69(3):265–283. https://doi.org/10.1016/j.prevetmed.2005.02.002

Pritchard J, Upjohn M, Hirson T (2018) Improving working equine welfare in 'hard-win' situations. PLoS One 13(2):e0191950. https://doi.org/10.1371/journal.pone.0191950

Purdy SR (2010) Herd health for donkeys. In: Proceedings of the North American Veterinary Conference. University of Massachusetts, Amherst. Retrieved from https://www.ivis.org
Qiu Y, Jiang J, Yi X, Wang S, Sun X (2025) Exploration of the differential expression patterns of immunoglobulin heavy chain genes in horses and donkeys. Dev Comp Immunol 166:105360. https://doi.org/10.1016/j.dci.2025.105360
Ramirez S, Gaunt SD, McClure JJ, Oliver J (1999) Detection and effects on platelet function of anti-platelet antibody in mule foals with experimentally induced neonatal alloimmune thrombocytopenia. J Vet Intern Med 13(6):534–539. https://doi.org/10.1892/0891-6640(1999)013<0534:daeopf>2.3.co;2. PMID: 10587252
Rickards KJ, Thiemann AK (2019) Respiratory disorders of the donkey. Vet Clin N Am Equine Pract 35(3):561–573. https://doi.org/10.1016/j.cveq.2019.08.009
Rocha JM, Ferreira-Silva JC, Veloso Neto HF, Moura MT, Nunes Ferreira H, da Amaro Silva V Jr, Manso Filho HC, de Lemos Oliveira MA (2018) Immunocastration in donkeys: clinical and physiological aspects. Pferdeheilkunde–Equine Med 34(1):xxx
Rodrigues JB (2024) Innovative approaches for better health and welfare of working donkeys. WEVA International Congress. https://www.ivis.org/library/weva/weva-international-congress-chile-2023/innovative-approaches-for-better-health-and-welfare-of-working-donkeys
Rodriguez-Garcia M, Patel MV, Wira CR (2013) Innate and adaptive anti-HIV immune responses in the female reproductive tract. J Reprod Immunol 97(1):74–84. https://doi.org/10.1016/j.jri.2012.10.010. PMID: 23432874; PMCID: PMC3581821
Ruan L, Li L, Yang R et al (2025) Equine herpesvirus-1 induced respiratory disease in Dezhou donkey foals. Vet Sci 12(1):56. https://doi.org/10.3390/vetsci12010056
Semieka MA, Ali MM, Al-Lethie AA (2012) Sarcoids in donkeys: common types and available treatment. J Adv Vet Res 2(4):276–283
Sobrero L, Dalla Costa E, Minero M (2024) Management of donkeys in assisted interventions: a snapshot. Animals 14(670). https://doi.org/10.3390/ani14050670
Stewart MC (2011) Cell-based therapies: current issues and future directions. Vet Clin North Am Equine Pract 27(2):393–399. https://doi.org/10.1016/j.cveq.2011.07.001. PMID: 21872767
Strid J, Hourihane J, Kimber I, Callard R, Strobel S (2005) Epicutaneous exposure to peanut protein prevents oral tolerance and enhances allergic sensitization. Clin Exp Allergy 35(6):757–766. https://doi.org/10.1111/j.1365-2222.2005.02239.x
Sumano López H, Lizárraga Madrigal I, Ocampo Camberos L, Obregón Jurgens K (2020) Reacciones adversas de los fármacos en los equinos. Veterinaria México OA 7(3):e925. https://doi.org/10.22201/fmvz.24486760e.2020.3.925
Tallmadge RL, Such KA, Miller KC, Matychak MB, Felippe MJ (2012) Expression of essential B cell development genes in horses with common variable immunodeficiency. Mol Immunol 51(2):169–176. https://doi.org/10.1016/j.molimm.2012.03.018
Taylor SE, Smith RK, Clegg PD (2007) Mesenchymal stem cell therapy in equine musculoskeletal disease: scientific fact or clinical fiction? Equine Vet J 39(2):172–180. https://doi.org/10.2746/042516407x180868. PMID: 17378447
Thiemann AK, Bell NJ (2001) The peculiarities of donkey respiratory disease. In: Equine respiratory diseases. Retrieved from https://ivis.org/library/equine-respiratory-diseases/peculiarities-of-donkey-respiratory-disease
Thiemann AK, Sullivan RJE (2019) Gastrointestinal disorders of donkeys and mules. Vet Clin North Am Equine Pract 35:419–432
Tizard IR (2018) Veterinary immunology: an introduction, 10th edn. Elsevier
Traub-Dargatz JL, McClure JJ, Koch C, Schlipf JW (1995) Neonatal isoerythrolysis in mule foals. J Am Vet Med Assoc 206(1):67–70
Turini L, Bonelli F, Nocera I, Meucci V, Conte G, Sgorbini M (2021) Evaluation of different methods to estimate the transfer of immunity in donkey foals fed with colostrum of good IgG quality: a preliminary study. Animals 11(507). https://doi.org/10.3390/ani11020507

University of California, Davis Veterinary Medical Teaching Hospital (2018) Neonatal isoerythrolysis in horse and mule foals [Pamphlet]. UC Davis School of Veterinary Medicine. Revised 10/2018. https://www.vetmed.ucdavis.edu

Valberg S, McKenzie E (Eds.) (2025) Muscle Disorders of Horses, An Issue of Veterinary Clinics of North America: Equine Practice (Vol. 41, No. 1). Elsevier Health Sciences. ISBN 9780443293177

Veronesi F, Morganti G, Ravagnan S et al (2014) Molecular and serological detection of tick-borne pathogens in donkeys (Equus asinus) in Italy. Vet Microbiol 173(3–4):348–354. https://doi.org/10.1016/j.vetmic.2014.08.017

Villa L, Cafiso A, Cialini C, Olivieri E, Allievi C, Pintore E, Garippa G, Manfredi MT, Bazzocchi C (2025) Serological and molecular insights into tick-borne pathogens in wild donkeys from an unexplored Mediterranean nature reserve. Curr Res Parasitol Vector Borne Dis. May 7;7: 100267. https://doi.org/10.1016/j.crpvbd.2025.100267. PMID: 40503032; PMCID: PMC12152590

Vincenzetti S, Foghini L, Pucciarelli S et al (2014) Hypoallergenic properties of donkey's milk: a preliminary study. Vet Ital 50(2):99–107. https://doi.org/10.12834/VetIt.219.125.5

Vitale V, Bonelli F, Briganti A, Sgorbini M (2021) Bronchoalveolar lavage fluid cytology in healthy Amiata donkeys and those with respiratory disease. Open Vet J 11(1):160–164. https://doi.org/10.4314/ovj.v11i1.23

Wafy MN, Hassan EA, Ahmed KA, Aboelmaaty AM, Abu Seida AM (2023) Ultrasonographic and histopathologic features associated with common ocular diseases in donkeys (Equus asinus). Vet Res Commun 47(3):1479–1491. https://doi.org/10.1007/s11259-023-10102-4

Wang Z, Zeng S, Wang Y, Wang T, Qu H, Zhu Y, Li J (2024) Reproductive disorders in donkeys: current evidence and update. Animals 14(2494). https://doi.org/10.3390/ani14172494

Watson TL, Kubasiewicz LM, Chamberlain N et al (2020) Cultural 'blind spots', social influence and the welfare of working donkeys in brick kilns in Northern India. Front Vet Sci 7:214. https://doi.org/10.3389/fvets.2020.00214

Watt KA, Nussey DH, Maclellan R, Pilkington JG, McNeilly TN (2015) Fecal antibody levels as a noninvasive method for measuring immunity to gastrointestinal nematodes in ecological studies. Ecol Evol 6:56–67. https://doi.org/10.1002/ece3.1858

Williams R, Du Plessis DH, Van Wyngaardt W (1993) Group-reactive ELISAs for detecting antibodies to African horsesickness and equine encephalosis viruses in horse, donkey, and zebra sera. J Vet Diagn Invest 5(1):3–7. https://doi.org/10.1177/104063879300500102. PMID: 8385502

Worku A, Molla W, Kenubih A et al (2024) Seroprevalence and associated risk factors of equine herpesvirus type-1/−4 in selected districts of Northwest Amhara, Ethiopia. Comp Immunol Microbiol Infect Dis 107:102155. https://doi.org/10.1016/j.cimid.2024.102155

World Health Organization (2017) The immunological basis for immunization series: module 17 – rabies vaccines. WHO, Geneva

Wynn TA, Chawla A, Pollard JW (2013) Macrophage biology in development, homeostasis, and disease. Nature 496(7446):445–455

Yin X, Guo M, Gu Q (2014) Antiviral potency and functional analysis of tetherin orthologues encoded by horse and donkey. Virology Journal 11,151. https://doi.org/10.1186/1743-422X-11-151

Yongfeng Y, Xiaobo S, Nan X, Jingwen Z, Wenqiang L (2020) Detection of the epidemic of the H3N8 subtype of the equine influenza virus in large-scale donkey farms. Int J Vet Sci Med 8(1): 26–30. https://doi.org/10.1080/23144599.2020.1739844

Hepatic Disorders in Donkeys and Mules

F. J. Mendoza, R. E. Toribio, and R. A. Perez-Ecija

Abstract

Although anatomic and physiologic differences for some body systems have been described between donkeys, mules, and horses, these species share similar liver disorders. Hyperlipemia is the most common metabolic disorder in donkeys, with high morbidity and mortality in hospitalized animals. Viral and parasitic hepatitis occurs in donkeys and mules. Treatment of hepatic disorders is similar for donkeys, mules, and horses, although specific pharmacologic studies have not been performed in donkeys and mules for drugs commonly used in liver diseases.

Keywords

Cholangiohepatitis · Fluke · Hyperlipidemia · Liver biopsy · Liver enzymes · Viral hepatitis

1 Introduction

Liver diseases are frequent in donkeys, representing a major cause of morbidity and mortality, mainly hepatic lipidosis secondary to hyperlipemia. Although epidemiological information about liver conditions is lacking, viral and parasite-related hepatitis seems to be common causes of primary liver diseases in donkeys. Disorders

F. J. Mendoza (✉) · R. A. Perez-Ecija
Department of Animal Medicine and Surgery, College of Veterinary Medicine, University of Cordoba, Cordoba, Spain
e-mail: fjmendoza@uco.es; alejandro.perez.ecija@uco.es

R. E. Toribio
Department of Veterinary Clinical Sciences, College of Veterinary Medicine, The Ohio State University, Columbus, OH, USA
e-mail: toribio.1@osu.edu

A. K. McLean et al. (eds.), *Donkey and Mule Medicine: Clinical Medicine and Reproduction*, https://doi.org/10.1007/978-3-032-18540-2_7

such as cholangiohepatitis, hepatic neoplasia, and hemochromatosis are also reported. Mule information is even more scarce than in donkeys, and epidemiological information is not available. However, they are affected by similar disturbances as horses and donkeys.

2 Hyperlipemia

2.1 Definition

Most dyslipidemias (abnormal distribution of blood lipids) are secondary to a primary disorder, often associated with a negative energy balance (primary disease or stressful situation). The underlying condition leads to a rapid and high mobilization of fat deposits and increased blood triglyceride concentrations (hypertriglyceridemia). This type of dyslipidemia is called hyperlipidemia, whereas hyperlipemia refers to the clinical condition associated with hypertriglyceridemia. Likely, hyperlipemia is the most common systemic metabolic disorder in donkeys, with a high mortality rate.

2.2 Epidemiology

Studies on the prevalence of hyperlipemia in donkeys are scarce, but it is estimated to be around 10–20% in sick animals. Data are not available for mules, but the prevalence is likely lower than in donkeys and ponies, falling somewhere between ponies and horses.

Predisposing risk factors for hyperlipemia include obesity (large fat deposits favor rapid lipid mobilization and severe hyperlipemia), age (older donkeys have higher risk, in part due to higher serum glucagon concentrations and lower insulin sensitivity), gender (jennies are more prone to be obese, can be pregnant and lactating), and the presence of a primary disease or stressful situation (Mendoza et al. 2024). Incidence appears to be higher in miniature donkeys than regular-size donkeys, although a breed effect has not been demonstrated.

Primary conditions causing hyperlipemia include disorders of the gastrointestinal tract (colic, enteritis, endoparasitism), liver (hepatitis), respiratory tract (pleuropneumonia), musculoskeletal system (laminitis, pedal abscesses, hip luxation, fractures), stress (transportation, separation from bonded mates), and anorexia due to husbandry practices (dental floating, farriery). Abnormalities in other body systems can also lead to hyperlipemia. Physiologic conditions that result in a negative energy balance (pregnancy, lactation) promote lipid mobilization and hyperlipemia (Mendoza et al. 2024).

Donkeys in developed countries or in regions where they are fed hypercaloric diets or have access to good quality pasture are often obese, which makes them prone to hyperlipemia.

2.3 Pathophysiology

Similar pathophysiologic mechanisms lead to hyperlipemia in donkeys, horses, and ponies. Briefly, in adipocytes, hormone-sensitive lipase is activated by glucagon, adrenocorticotropic hormone (ACTH), growth hormone, pro-inflammatory mediators, catecholamines, and glucocorticoids, releasing free fatty acids (FFA) into the bloodstream. FFA are re-esterified in the liver to triglycerides (TGL) that are released into circulation as very-low-density lipoproteins (VLDL) (Mendoza et al. 2019). It seems that donkeys are very efficient in this metabolic transformation compared to other equids, resulting in hypertriglyceridemia (hyperlipidemia) that in severe cases causes fat infiltration of the liver (hepatic lipidosis, Fig. 1b) as well as other organs, impairing their function. Donkeys, in contrast to ruminants, do not produce ketone bodies in these situations.

It has been proposed that glucagon can contribute to hyperlipemia in donkeys (Perez-Ecija et al. 2021) by increasing the activity of hormone-sensitive lipase, independently of plasma insulin. Furthermore, obese donkeys have higher serum glucagon concentrations than normal sized donkeys and horses (Mendoza et al. 2015, 2018).

Lipoprotein lipase removes TGL from circulation. In other species, this enzyme is activated by insulin and heparin; however, information in donkeys and mules is lacking, and at least in ponies, the activity of this enzyme is maximal, making the use of these drugs questionable. In addition, the use of heparin can increase the risk of bleeding, in particular in animals with liver dysfunction. In donkeys with severe hypertriglyceridemia, lipids can be deposited in other tissues (liver, heart, muscle, pancreas, intestine, kidney), impairing their function (hepatic lipidosis, renal failure, ileus, pancreatitis, myocarditis, etc.).

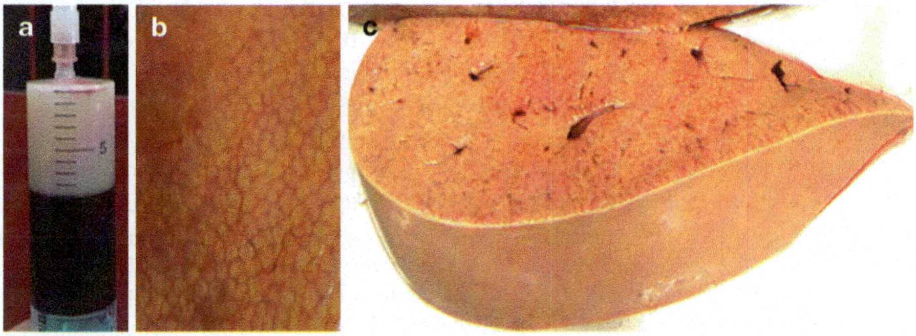

Fig. 1 (**a**) Gross macroscopic lipemia in an anorectic jenny 4 days postpartum. (**b**, **c**) Fat infiltration of the liver parenchyma (hepatic lipidosis) secondary to severe hyperlipemia

2.4 Clinical Signs

Signs of hyperlipemia are not specific and usually manifest according to the primary problem (colic, pneumonia, etc.). In hyperlipemia secondary to stressful situations, signs are subtle and vague, including lethargy, anorexia, and behavioral changes (lower physical activity or reluctance to move, low head posture, etc.). Fatty infiltration of the pancreas or intestine can result in diarrhea or ileus, dysrhythmias if the myocardium is affected, and renal dysfunction with subsequent azotemia if the kidneys are involved.

2.5 Diagnosis

Hyperlipemia should be suspected in any ill, lethargic, or anorectic donkey. In these animals, it is recommended to measure blood TGL concentrations. Blood TGL concentration can be measured using point-of-care devices or more dedicated analyzers.

Healthy donkeys have higher serum TGL concentrations than horses and mules (upper limit of 150 versus 50 and 80 mg/dL, respectively); thus species-specific reference ranges should be used to avoid misinterpretation (Table 1). A small seasonal effect has been observed in mules (Gehlen et al. 2020).

Table 1 Reference ranges of blood parameters used in the diagnosis of liver disease in donkeys and mules

Parameter	Units	Donkeys	Mules
Albumin	g/dL	2–3.6	2–4
ALP	IU/L	98–252	<350[a]
Ammonia	µg/dL	13–108[a]	
AST	IU/L	223–555	200–400
Bile acids	µmol/L	<19	<20
Direct bilirubin	mg/dL	<0.1	<0.1[a]
GGT	IU/L	7–70	5–30
GLDH	IU/L	0–9	<13[a]
Indirect bilirubin	mg/dL	<0.2	<0.2[a]
SDH	IU/L	2–10	<5.1[a]
Total bilirubin	mg/dL	<0.3	0.8–1
Transferrin saturation	%	27–56[a]	
TIBC	µg/dL	289–535[a]	
Triglycerides	mg/dL	25–150	25–80
Serum iron	µg/dL	100–300[a]	

ALP alkaline phosphatase, *AST* aspartate aminotransferase, *GGT* gamma-glutamyl transferase, *GLDH* glutamate dehydrogenase, *SDH* sorbitol dehydrogenase, *TIBC* total iron binding capacity

[a]Horse reference ranges since donkey and mule-specific ranges are not currently available

Donkey and mule data from Burden et al. (2016), McLean et al. (2016), Gehlen et al. (2020), and Goodrich and Webb (2024)

Blood lipoprotein concentrations are not routinely measured in equids, but hyperlipemic donkeys will have higher VLDL, LDL, and cholesterol concentrations, and lower HDL concentrations (Meng et al. 2024). Lipoprotein reference ranges differ between donkeys and horses, with donkeys having lower low-density lipid (LDL), higher high-density lipid (HDL), but similar concentrations of VLDL (Goodrich and Behling-Kelly 2022). Donkeys with higher body condition scores have higher TGL and VLDL concentrations.

Although hormones such as glucagon, leptin, and adiponectin play crucial roles in energy regulation, they are not routinely measured in the clinical setting.

In cases where organ dysfunction secondary to fat infiltration is present, related laboratory findings such as azotemia and increased liver enzymes activities can be found. In mild to severe hepatic fatty infiltration, activities of alkaline phosphatase (ALP), gamma-glutamyl transferase (GGT), and sorbitol dehydrogenase (SDH) are often elevated, although there is not a specific liver enzyme profile for hyperlipemia. The magnitude of these increases has no prognostic value in donkeys and mules. Increased lipase and amylase activities or high serum cardiac troponin I concentrations can be present due to pancreatic or myocardial fatty infiltration.,

High serum TGL concentrations can result in gross lipemia (opalescent sample, Fig. 1a), which can affect other analytes (mainly those measured by spectrophotometry), such as creatinine, sodium, chloride, potassium, phosphorus, and calcium that can be falsely decreased. Ultracentrifugation or other techniques to separate lipids may be required for accurate measurements.

Similar to horses and ponies, fatty liver infiltration increases hepatic echogenicity, but this is not pathognomonic of hyperlipemia. In severe cases, an enlarged liver with rounded borders can be observed. Ultrasound-guided liver biopsy would be needed to differentiate a primary hepatopathy (e.g., viral hepatitis) inducing secondary hyperlipemia or secondary liver failure from hepatic lipidosis. However, liver biopsy in animals with hepatic lipidosis has the risk of liver rupture and excessive bleeding (see details in liver biopsy section below).

2.6 Treatment

The treatment of hyperlipemia is based on halting lipolysis, restoring a positive energy balance, promoting blood lipid clearing, and addressing the primary problem (Durham and Thiemann 2015).

1. *Halt lipolysis by solving/treating the primary disorder and controlling the stressful situation:*
    - Treatment of primary problems (colic, pneumonia, etc.).
    - Perform dental work in a phased manner if extensive care is needed.
    - Analgesic administration if pain is a contributing factor.
    - In donkeys that cannot reach the food due to musculoskeletal issues (e.g., pain associated with extensive farriery work), facilitate feeding.
    - Provide additional nutritional support to pregnant or lactating jennies.

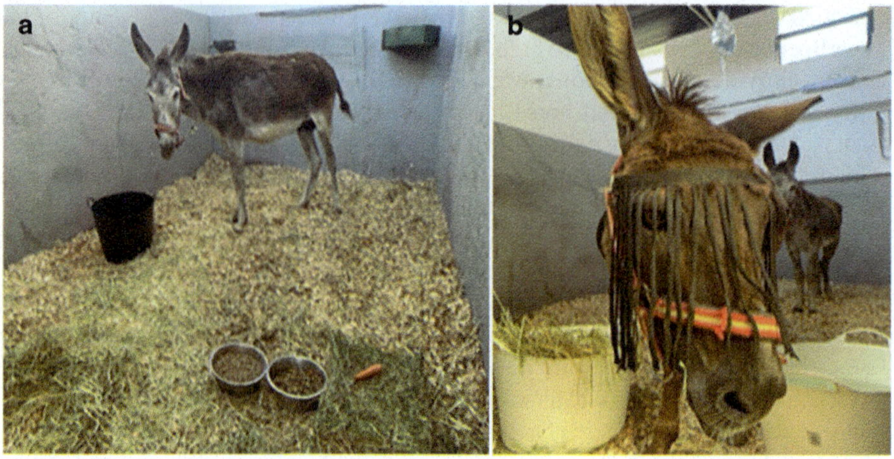

Fig. 2 (**a**) Donkey with hyperlipemia secondary to pneumonia. Note the multiple feedstuffs (alfalfa and oak hay, straw, pellet beep pulp, mixed grain, carrot, etc.) to stimulate appetite. (**b**) Companion-bound donkey transported to the hospital to accompany the sick mate. Note the fluid therapy supplemented with glucose (partial parenteral nutrition) in the ill anorectic donkey

- Induction of abortion should be considered in some animals.
- Consider weaning if lactation is advanced and hyperlipemia is severe.
- In donkeys requiring transportation (e.g., hospitalization, sales, etc.), the presence of a mate can reduce stress and facilitate handling (Fig. 2b).
- Avoid any further husbandry modifications since that could impact the donkey's behavior (e.g., it is better to use buckets in only one color, etc.).
- Administration of antipyretic drugs (metamizole or acetaminophen [paracetamol]) since anorexia is common in febrile animals.

2. *Restoring a positive energy balance.* This will depend on gastrointestinal function and the primary condition. Provide at least half (40–50 kJ DE/kg/day [9.5–12 kcal DE/kg/day]) the daily resting energy requirements (80–100 kJ DE/kg/day [19–24 kcal DE/kg/day]), which are lower than horses. Consider the following recommendations:

- *Donkeys with appetite and a functional gastrointestinal tract.*
  - Offer palatable feedstuffs. Fasting must be avoided (Fig. 2a).
    Tip: It may be enough to switch off the lipolysis.
  - Provide treats such as apples, carrots, mints, bananas (skin), molasses, corn syrup, etc.
  - Use food flavor enhancers such as apple or carrot juices and mints.
  - Provide oral honey (10–20 mL) every 6–8 h.

- *Anorectic donkeys with a functional gastrointestinal tract: enteral nutrition*
  - Through the mouth:
    - Provide oral honey (20–30 mL) or glucose every 4–6 h. Caution in donkeys with endocrinopathic laminitis since the effect of insulin peak can potentially aggravate signs.
    - Daily vitamin and mineral supplementation are recommended.
  - Through a small-bore nasogastric tube:
    - Administer commercial enteral formulations.
    - *Tip*: Preferably use those marketed for ponies.
    - Administer a homemade blended gruel: dissolve 250 g of pelleted alfalfa in 5 L of plain warm water. Administer 3 L every 2 h for a 250 kg donkey.
    - *Tip*: tube obstructions are common if the mixture is not correctly blended. Diarrhea or soft feces can be observed secondary to excessive water in the gruel or intestinal dysbiosis.
    - Administer a solution of dextrose/glucose powder (100–150 g [5–10% oral solution]). Sodium bicarbonate, potassium chloride, magnesium sulfate, and sodium phosphate can also be added to the mixture.
    - *Tip:* sodium phosphate enema solutions for humans can be administered orally to donkeys to correct hypophosphatemia.
    - *Tip*: in donkeys younger than 1 year of age, milk replacer or starter grain can be added.
- *Donkeys with a non-functional gastrointestinal tract or severe hyperlipemia: partial parenteral nutrition (PPN, without lipids)*
  Tip: Follow indications for PPN in horses, considering that donkeys have lower energy maintenance requirements.
  - At home approach (performed by the veterinarian on the farm):
    - Intravenous bolus administration of balanced isotonic polyionic fluids (Hartmann's or lactate Ringer's solution) supplemented with dextrose/glucose at a concentration of 2–5%. Solutions containing amino acids, vitamins, and minerals can be added (e.g., Duphalyte/Metabolik®, 100 ml/50 kg) if anorexia is longer than 48 h or in animals with a high catabolic rate.
  - At hospital approach:
    - Close monitoring of blood glucose concentrations (every 2–4 h).
    - Continuous rate infusion (CRI) of 2–5% dextrose/glucose at a 2–3 ml/kg/h rate. If hyperglycemia (>160 mg/dL) occurs during the first 4 h, decrease the CRI rate. If hyperglycemia persists (endotoxemia or insulin dysregulation), insulin administration should be considered (Table 2), with caution to avoid hypoglycemia.
    - Intravenous solutions of essential amino acids should be considered if PPN (carbohydrates) is used longer than 48 h or in animals with a high catabolic rate.
    - Magnesium sulfate and potassium chloride supplementation can be required since hypomagnesemia and hypokalemia is frequent in these animals. Some may also require phosphate administration (orally, parenterally).

Table 2 Common drugs used for treatment of liver diseases in donkeys and mules

Drugs	Dose	Route	Interval	Observations
Anti-inflammatory drugs				
Dimethyl sulfoxide (DMSO)[a]	1 mg/kg, >10% solution	IV	8–12	Able to dissolve some choleliths. Hemolysis at parenteral concentrations higher than 10%
Pentoxifylline[a]	10–15 mg/kg	IV-PO	8–12	Antifibrotic effect
***Anti-parasitic drugs** (against parasites causing liver disease)*				
Closantel	20 mg/kg	PO	Once, repeat at 8–10 weeks	Fluke treatment Overdosing: blindness, anorexia, and ataxia
Fenbendazole	5–10 mg/kg	PO	5 consecutive days	Roundworms (Cyathostomins resistance has been observed) Caution in early pregnant jennies due to potential teratogenicity
Ivermectin	0.2 mg/kg	PO	Once	Roundworms Overdosing: blindness, staggering, and seizures. No use in younger than 6 months
Moxidectin	0.2–0.4 mg/kg	PO	Once	Roundworms Overdosing: staggering and seizures
Praziquantel	1–2.5	PO	Once	Roundworms
Pyrantel pamoate	19 mg/kg	PO	Once	Roundworms
Triclabendazole	18 mg/kg	PO	Once	Fluke treatment
Antimicrobials				
Chloramphenicol[a]	50–60 mg/kg	PO	6 h	Where legally allowed
Doxycycline[a]	10 mg/kg	PO	12 h	
Enrofloxacin	5–7.5 mg/kg	IV-PO	24	Overdosing: lethargy or seizures
Penicillin	20,000 IU/kg	IV	6–8–12	
Sulfonamide/ Trimethoprim	30 mg/kg	PO	12	Likely this dose is not adjusted for trimethoprim
Corticosteroids				
Dexamethasone[a]	0.05–0.2 mg/kg	IV-IM-PO	24	Laminitis, immunosuppression
Prednisolone[a]	0.5–1 mg/kg	PO	12–24	Laminitis, immunosuppression
Chemotherapeutic drugs[b]				
Chlorambucil	20–30 mg/m^2	PO	14	BMS
Cyclophosphamide	150–800 mg/m^2	IV	7–30	

(continued)

Table 2 (continued)

Drugs	Dose	Route	Interval	Observations
				BMS, hemorrhagic cystitis
Cytosine arabinoside	200–300 mg/m^2	IV-IM-SC	7–14	
Doxorubicin	35–70 mg/m^2	IV	7–30	Tissue necrosis with extravasation, nephrotoxicity, hypersensitivity
L-Asparaginase	10,000–40,000 IU/m^2	SC	7–30	Hypersensitivity
Lomustine	65 mg/m^2	IG	7–30	BMS
Vincristine	0.5–0.7 mg/m^2	IV	7–30	BMS
Heparins *(increases LPL activity)*				
Dalteparin[a]	50–100 IU/kg	SC	24 h	
Enoxaparin[a]	40–80 IU/kg	SC	24 h	
Hemochromatosis *(iron chelator)*				
Deferoxamine mesylate[a]	10 mg/kg	IM	24–48 h	
Insulins *(increases LPL activity and inhibits HSL)*				
Regular insulin	0.05–01 IU/kg	IV-IM	6 h or CRI	
Protamine zinc insulin	20–60 IU	IM	24 h	
Others				
Silymarin (milk thistle)[a]	5–10 g/500 kg	PO	24 h	Hepatoprotective, antioxidant, and anti-inflammatory effects. Hyperlipemia treatment.
Ursodiol[a]	15 mg/kg	PO	24 h	Cholestatic disease

BMS bone marrow suppression, *CRI* constant rate infusion, *HSL* hormone-sensitive lipase, *IG* intragastric, *IM* intramuscularly, *IV* intravenously, *LPL* lipoprotein lipase, *PO* per os, *SC* subcutaneous

[a]Dosage extrapolated from horses

[b]Dose regimen and potential toxicity obtained from Luethy et al. (2019) and Samuels and Burns (2024)

3. *Clear the circulating TGL in the bloodstream.* Although the effect of insulin suppressing hormone-sensitive lipase activity has not been evaluated in hyperlipemic donkeys, insulin and low molecular weight heparin (enoxaparin or dalteparin) at similar doses as described in horses (Table 2) can be considered in these animals. It is important to note that donkeys with liver dysfunction are prone to bleed and the risk is increased with heparin administration. Silymarin (milk thistle) is used by some clinicians due to its apparently hepatoprotective, antioxidant, and anti-inflammatory effects (Table 2).

Tip: Insulin administration should be avoided in donkeys and mules with insulin dysregulation due to risk of laminitis.

4. *Treatment of organ dysfunction secondary to fatty infiltration:*
   - Intravenous fluid therapy to treat acute renal failure and decrease azotemia.
   - Treatment of cardiac dysrhythmias.
   - Avoid high-protein diets (<8% of crude protein) if there is liver dysfunction.

2.7 Prognosis

The presence of hyperlipemia aggravates the prognosis of any primary condition. In general, the mortality rate is around 40–50%, but it can be as high as 80% in severe cases. Rapid response to treatment, with decreasing TGL concentrations in the first 24 h, is a good indicator. No information is available on mules.

3 Common Clinical Signs for all Liver Diseases

Clinical signs of hepatic disease are nonspecific and can go unnoticed for a long time. Poor and unpredictable appetite and dullness are the most common signs in acute cases, whereas poor body condition, weight loss, and recurrent colic are common in chronic cases. These clinical signs can be seen with any liver disease, independently of the cause. Sometimes hyperlipemia is the first indicator of a liver problem in donkeys and mules. Icterus can be noticed, and other differential diagnoses should be considered (e.g., piroplasmosis in endemic areas or other causes of hemolytic anemia). Other signs, such as diarrhea and photodermatitis, can be observed less frequently in donkeys. Neurological signs (behavior changes, obtundation, head pressing, circling, ataxia, etc.) from hepatic encephalopathy could also be observed.

4 Diagnosis of Liver Diseases

- *Blood analysis: hematology and biochemistry*

Hematology is of little help in the diagnosis, and changes are usually not specific. Polycythemia has been documented in horses with hepatic carcinomas but has not been reported in donkeys or mules at this moment. Leukocyte count is variable and can be above, below, or within the reference range. Leukocytosis with neutrophilia, lymphocytosis, and/or monocytosis can be observed with liver abscesses.
Serum biochemistry can be useful in the diagnosis of hepatic diseases but cannot differentiate causes of liver disease and lacks prognostic value.

Liver specific enzymes include glutamate dehydrogenase (GLDH), sorbitol dehydrogenase (SDH), and gamma-glutamyl transferase (GGT). Nonspecific markers of liver damage, such as alkaline phosphatase (ALP) and aspartate aminotransferase (AST), can be measured and are often part of biochemistry profiles. Liver function can be evaluated by measuring serum bile acids, bilirubin (total, indirect, and direct) and ammonia concentrations.

Similar to horses, concentrations of these analytes are not correlated with illness severity, and they cannot be used as prognostic tools. Nonetheless, values can be useful to evaluate disease progression and response to treatment. No biochemical variable is disease-specific, and two donkeys with the same condition and severity can show different results (Divers 2015). Increased serum ALP and GGT activities can be suggestive of biliary obstructive disease (cholangitis, cholestasis, cholelithiasis, or flukes), but additional diagnostic tests are required for further confirmation.

Healthy donkeys have higher plasma GGT activity than horses and mules (70 versus 25 and 30 IU/L, respectively) (Goodrich and Webb 2024); thus, species-specific reference ranges should be used to avoid misdiagnosis and unnecessary treatments. Mules have higher serum GGT activities in August (up to 68 IU/L); thus, caution should be taken when results are interpreted in this species in this month (Gehlen et al. 2020). Donkeys have lower serum bilirubin concentrations than horses. No differences have been described for the rest of the hepatic parameters (Table 1).

Changes in others biochemical parameters are not indicative of liver disease because hypoalbuminemia or prolonged clotting times are not frequently observed even in severe liver insufficiency. Serum amyloid A and fibrinogen concentrations are difficult to interpret in donkeys with hepatic diseases (since they are synthesized in the liver but also act as inflammatory markers). Hyperglobulinemia can be observed in animals with liver abscesses or neoplasia, although this is not specific, and other differential diagnoses must be taken into consideration.

If hemochromatosis (chronic liver disease due to iron toxicity) is suspected, transferrin saturation (>80%), total iron binding capacity (TIBC), ferritin, and serum iron should be measured (Table 1).

Several serum markers of hepatic fibrosis are used in human medicine; however, serum hyaluronic acid is the only one that has shown a correlation with the degree of liver fibrosis in donkeys, even higher than the equine-adapted enhancer liver fibrosis score (eELF) (Potier et al. 2023). Thus, this serum marker could potentially be used in donkeys where a liver biopsy cannot be performed, for example, in those with a concomitant hemostatic disorder.

In donkeys with severe hypertriglyceridemia (gross lipemia), analyte results obtained by spectrophotometry should be interpreted with caution because lipemia can interfere with their results. Ultracentrifugation or techniques that separate lipids from plasma should be used. Other problems related to sample processing affecting interpretation should also be considered, such as the effect of fasting, immediate measurement of ammonia due to its volatility, and use of chilled samples for GLDH and SDH measurement.

- *PCR and serology*

Polymerase chain reaction (PCR) of serum (or liver tissue, preferably), is the confirmatory technique for viral hepatitis. Similar primers and protocols described for horses are used in donkeys. It is advisable to contact the reference laboratory to check availability and preferred type of samples for testing hepatic viruses.
Serology (ELISA tests validated for horses) can be used in the diagnosis of viral hepatitis in donkeys and mules, although this technique has a lower specificity than PCR, only confirming previous viral exposure but not a causative relationship. In fact, more donkeys are positive by serology than by PCR. Serology for *Fasciola hepatica* (fluke) is also available for horses and could be used in donkeys.

- *Liver ultrasound*

Principles of liver and abdominal ultrasound are similar between donkeys, mules, and horses. In healthy animals, the liver has a homogeneous echogenicity along the entire parenchyma, with lower echogenicity than the spleen. There should be no abnormal structures, heterogeneous echogenicity, or bile duct dilation (Divers 2015). Hepatic lipidosis is associated with increased echogenicity and rounded liver edges. Cholangitis may be suspected if there is dilation of the bile ducts (obstruction sign) with hyperechogenic biliary canaliculi, with choleliths displaying acoustic shadows. Hyperechogenicity, dilated vasculature, and heterogenicity (starry sky appearance) can be nonspecific findings of liver fibrosis. Color doppler can be used to distinguish between dilated hepatic sinuses and dilated bile ducts. Enlarged liver edges and hyperechogenic parenchyma have been reported with hemochromatosis. Liver abscesses or neoplasia will appear as masses with heterogeneous echogenic patterns disrupting the normal parenchyma structure. Diffuse, anechoic-filled rounded cysts with thick walls and disrupting the parenchyma can be found in donkeys with *Echinococcus* spp. infection (hydatid cysts) (Piotrowski et al. 2024). Other parasitic infections cannot be diagnosed with ultrasound.

- *Liver biopsy*

Liver biopsy is a highly reliable tool to reach a final diagnosis. The procedure (preferably ultrasound-guided) and tools (Tru-cut needle or automatic gun biopsy) are similar to horses. Histology scores for liver damage previously described in horses can be used in donkeys, with a score higher than 2 points (out of 4) reflecting a poor prognosis (Potier et al. 2023). This score system evaluates the degree of fibrosis, hemosiderin accumulation, irreversible or reversible cytopathology, inflammatory infiltration, and biliary hyperplasia (Table 3) (Durham et al. 2003). This score is better to assess the fibrosis degree than eELF alone (Potier et al. 2023).
Although liver biopsy is generally a safe procedure, complications such as bleeding resulting in hemoabdomen, bowel puncture, peritonitis, lung puncture, pneumothorax, endotoxemia, and even death can occur.

Table 3 Histology score for grading severity and prognosis of liver diseases in donkeys and mules

Parameter	*Severity degree*			
	Absent	Mild	Moderate	Severe
Biliary hyperplasia	0	0	2	4
Fibrosis	0	1	2	2
Hemosiderin accumulation	0	0	1	2
Inflammatory infiltrate	0	0	0	2
Irreversible cytopathology	0	0	2	4

Adapted from Durham et al. (2003)
Score from 0 (good prognosis) to 14 (poor prognosis due to extensive damage)

Clotting times should be measured prior to biopsy, and this procedure should be avoided or delayed in animals with hemostasia disorder, but the risk must be weighed, and this should not prevent sampling if platelet count is normal or if the biopsy results might prove to be important to case management. Biopsy is not recommended if there is suspicion of liver abscess or hydatid cysts to avoid peritonitis or spread of hydatid protoscoleces (Sullivan 2024).

- *Complementary imaging techniques: CT and MRI*

In small-breed donkeys or donkey foals, computed tomography (CT) or magnetic resonance imaging (MRI) can be performed to assess liver masses (e.g., abscess, neoplasia), parenchymal abnormalities (cholelithiasis), extension of the lesion (e.g., choleliths in duodenal duct), portosystemic shunts, involvement of other abdominal organs, and to assist with surgical decisions.

- *Fecal egg counts.*

Since parasites are a common cause of hepatic disease in donkeys, identifying parasites and their burden can be useful. *Fasciola hepatica* may lay eggs intermittently, and some may not reach patency. Thus, repeated sedimentation analysis is recommended (Bell et al. 2021). The McMaster technique is recommended to find *Parascaris equorum* in young animals where liver damage due to migrating larvae is suspected.

- *Abdominocentesis*

Evaluation of the peritoneal fluid is not directly useful to diagnose hepatic disease, but changes in the cell population (and cytomorphology) and total protein concentrations can be suggestive of abscessation or neoplasia (commonly in exfoliative neoplasia). In addition, the sample can be submitted for bacterial culture.

5 Liver Diseases Other than Hyperlipemia

- Viral hepatitis

It has been proposed that the prevalence of viral hepatitis is higher in donkeys, although controlled comparative studies are required.

Viral hepatitis has a worldwide distribution, and several hepatotropic viruses have been found: equine hepacivirus (EqHV), equine parvovirus-hepatitis virus (EqPV-H), hepatitis B virus (EqHBV), hepatitis C virus (EqHCV), and hepatitis E virus (Rasche et al. 2021).

Some of these hepatotropic viruses may induce chronic hepatitis characterized by liver bridging fibrosis, a similar lesion as described in humans with hepatitis B or C.

Treatment: No specific treatment is available. Antiviral drugs have not been evaluated in this species and are not recommended. Based on horse studies, pentoxifylline administration could have some beneficial anti-fibrotic effect (Table 2). See section on nutritional support.

Prognosis: Depends on the degree of fibrosis and parenchymal involvement, with a histology score higher than 2 (out of 4) having a poor prognosis. Concurrent severe hyperlipemia worsens the prognosis.

- Liver parasites

Fasciola hepatica (fluke) and migrating nematodes (roundworms) such as *Parascaris equorum* and *Strongylus* spp. are the most common parasites causing cholangiohepatitis and hepatitis, respectively. Donkeys and mules co-grazing with ruminants have a higher risk of fluke and other parasitosis (Howell et al. 2020).

Treatment: No flukicides are marketed for donkeys or mules, but triclabendazole and closantel (Table 2) are the recommended drugs. Anorexia, ataxia, and blindness can be observed with closantel toxicity. Fenbendazole or a combination of moxidectin and praziquantel are effective options against roundworms and *Echinococcus* spp. in donkeys. Ideally, a fecal egg count should be performed before deworming to assess effectiveness but also potential anti-helminthic resistance, which is becoming an emerging problem in developed countries (e.g., cyathostomins). Proper husbandry practices, a clean environment, manure removal, pasture rotation, animal grouping (age, activity, etc.), avoiding co-grazing with ruminants, and regular fecal egg counts, are essential to decrease parasite prevalence. Excessive or unjustified use of anthelmintics should be avoided.

Prognosis: Prognosis is generally good, although secondary disturbances such as hyperlipemia or diarrhea can worsen it. Weight gain, brighter demeanor, and return of liver-specific biochemical parameters to the normal range are good markers of clinical improvement.

- Cholangitis/Cholelithiasis

Information on risk factors for cholangitis or cholelithiasis for donkeys or mules is scarce. It seems that the prevalence is lower than other primary hepatic disorders such as viral hepatitis or liver parasites. It is assumed that calcium is the main component of choleliths, as described in horses in contrast to humans (bile, cholesterol).
Descriptions about cholangitis due to ascendent infections through the choledochal duct in donkeys are lacking. Other causes, such as ingesta-associated choledocholithiasis, could also be involved (Linn-Peirano et al. 2023).

Treatment: Long-term administration (4–8 weeks) of antimicrobials excreted in active form through the bile (such as sulfonamides, enrofloxacin, third-generation cephalosporins, or chloramphenicol) is recommended to treat infectious cholangitis (Table 2). Dimethyl sulfoxide (DMSO) has been proposed to dissolve choleliths in horses (Table 2). Long-term oral pentoxifylline (Table 2) could potentially halt hepatic fibrosis and improve the outcome. See section on nutritional support. Although bile salts chenodeoxycholic and ursodeoxycholic (ursodiol) have no effect on calcium bilirubinate calculi, their administration could make bile more liquid and easier to excrete (Divers 2015) (Table 2). The bioavailability of this drug has not been studied in donkeys or mules.
In cases of extrahepatic choleliths, endoscopic-guided retrieval can be tried, although the balloon dilation technique (described in humans) is not performed in equids. A surgical attempt can be an option in small-breed donkeys.

Prognosis: Depends on the size, number, and location of choleliths, as well as the severity of liver dysfunction. Extrahepatic choleliths have a worse prognosis than intrahepatic ones.

- Hepatic neoplasia

Secondary (metastatic) neoplasia is more common than primary one. Epidemiological data of liver neoplasms in donkeys and mules is lacking, but lymphoma (multicentric) seems to be the most common one. Other neoplasms such as hepatic carcinoma and hemangiosarcoma have been reported.

Treatment: In cases of primary focal neoplasia without extensive surrounding damage, surgical excision with adjuvant chemotherapy is recommended. Cyclophosphamide and vincristine have been used to treat intestinal lymphoma in donkeys (Luethy et al. 2019). However, additional chemotherapeutic drugs could also be used, such as prednisolone (Table 2).

Prognosis: Regardless of the type of hepatic neoplasia, primary or secondary, the prognosis is poor.

- Hepatic abscess

Abscesses involving the hepatic parenchyma are rare in donkeys or mules. They are more common in donkey neonates secondary to umbilical vein infections (omphalophlebitis).

Treatment: Oral long-term (1–3 months depending on sensitivity and size) antimicrobials with good penetration is the recommended therapeutic management in these cases. Chloramphenicol and doxycycline are the first choices, although enrofloxacin and sulfonamide could also be used depending on culture and sensitivity results (Table 2).

Prognosis: Depends on the abscess size and involvement of surrounding abdominal organs (adhesions). Abscess leakage and secondary septic peritonitis are serious complications with poor prognosis.

- Toxicosis: iron, mycotoxins, and pyrrolizidine alkaloids

Chronic iron overload causing hemochromatosis can occur in donkeys consuming water with high iron content, as it has been observed in farms closely located to iron mines (soil filtration) or in some regions of the Netherlands or the British Islands, diets with high iron content, and/or due to excessive supplementation (Theelen et al. 2019). Old donkeys can accumulate iron in the liver without associated damage (fibrosis or inflammation) (Brown et al. 2011).

Liver failure secondary to pyrrolizidine alkaloid (PA) toxicity has been described in donkeys. Some plants containing PA include rattlebox (*Crotalaria sagittalis*), houndstongue (*Cynoglossum officinale*), ragwort (Senecio jacobaea), Salvation Jane (*Echium plantagineum*), and heliotrope (*Heliotropium* spp.).

Chronic liver disease secondary to long-term mycotoxin consumption is rarely described in donkeys. It is unknown if the low prevalence is the result of some type of species-specific tolerance or simply a lack of information.

Treatment: No specific treatment is available for these toxicoses other than the elimination of the source (plants, water, etc.). Pentoxifylline could be administered to halt liver fibrosis (Table 2). In animals with PA-related photosensitization (sunburn sores), direct sunlight exposure should be avoided, and antimicrobial creams applied to prevent secondary pyoderma and to facilitate skin repair.

Intravenous deferoxamine administration has not been evaluated in donkeys, but it could be useful to chelate blood iron and increase its clearance (Table 2). See below information for nutritional management in these chronic liver diseases.

Prognosis: Liver damage can be extensive, with high fibrosis scores having a poor prognosis. Hyperlipemia can further worsen the prognosis. Cirrhosis and hepatic encephalopathy are indicators of a bad prognosis.

6 Nutritional Management in Liver Diseases

Nutritional management of donkeys and mules with liver diseases follows similar principles as those for horses (Divers 2015); however, some considerations must be taken into account (The Donkey Sanctuary 2020). Donkeys have lower daily energy (80–90 kJ digestible energy/kg body weight/day) and protein requirements than horses and ponies. However, the risk of hyperlipemia should be reduced since it can increase the mortality risk in animals with hepatic disease.

The main goal with nutritional management is to supply enough energy to cover the animal needs, and to prevent hyperlipemia without overwhelming the liver capacity and function. Daily needs should be supplied with high-quality fiber, a low-protein content diet (<8%), and adequate levels of vitamins and minerals. Do not restrict protein content excessively, as it may promote muscle mass mobilization. These principles apply to any donkey or mule with liver disease.

Free access to cereal straw is a good source of long fiber, but access to water and dental care must be provided to avoid secondary complications such as choke, gastric, or large bowel impactions. Legume forages such as alfalfa or soy hay must be avoided due to their high protein content. If a diet is based exclusively on cereal straw, supplementation with a commercial balancer containing essential amino acids, minerals (selenium) and vitamins (A, D and E) is required. Avoid supplements with iron.

Grazing on grass pasture can be allowed (12–18 h a day), which has additional mental and exercise benefits. If sunburn sores develop, overnight turnout is recommended.

If daily needs are not satisfied with cereal straw, or the donkey cannot be fed with this type of forage (e.g., dental or mandibular problems, etc.), alternative high quality fiber such as unmolassed beep pulp, pelleted fiber, haylage or short-chopped chaff can also be provided, but protein concentrations must be evaluated to not exceed the daily protein requirements (approx. 40 g crude protein/100 kg of body weight). Feeding several times a day (3–4 times/d) is recommended, with care not to exceed the total daily caloric needs.

If concurrent problems are present, such as insulin dysregulation, pituitary pars intermedia disfunction, or endocrinopathic laminitis, high non-structural carbohydrate (starch) content diets and those based on grains must also be avoided.

If a high-quality source of oil is supplemented (e.g., linseed oil), serum triglycerides concentrations must be monitored closely, in order to avoid hyperlipemia.

Regular body weight control (loss of weight must be rapidly addressed) and serial blood analysis to check serum urea, triglyceride, and ammonia concentrations, can be used as guidance to correct/modify the ration or diet.

7 Conclusions

Hyperlipidemia is the most common metabolic disorder in donkeys, with the prevalence in mules being almost similar to horses. Hyperlipemia has high mortality in donkeys; thus, halting the negative energy balance, and treating or removing the primary cause is an emergency in this species. Viral hepatitis can go unnoticed in donkeys due to their more stoic behaviour, and blood or liver tissue PCR is the diagnostic method of choice. The incidence of liver flukes is increasing in some countries, mainly in donkeys and mules close to ruminants or sharing natural water sources. Species-specific studies have not been performed for most drugs commonly used for liver disorders, thus doses reported for horses should be used meanwhile.

Questions

1. Which of the following drugs is the recommended one for fluke treatment in donkeys and mules?
   (a) Fenbendazole
   (b) **Closantel**
   (c) Moxidectin
   (d) Pyrantel
2. Which of the following diagnostic tools should be performed in a donkey with high liver enzyme activities and fever?
   (a) Liver ultrasound
   (b) PCR panel for viral hepatitis
   (c) Liver biopsy
   (d) **All of them**
3. What should be avoided in a donkey with hyperlipemia and prone to laminitis?
   (a) To treat the primary disorder
   (b) To administer intravenous fluid with glucose (dextrose), closely controlling serum glucose concentration to avoid hyperglycemia.
   (c) **To administer insulin to facilitate the clearance of triglycerides from bloodstream, independently of blood insulin concentrations.**
   (d) To administer feedstuff with a high sugar content (i.e., honey) in non-laminitic donkeys.
4. The recommended diet for a mule with liver disease could be:
   (a) **Cereal straw supplemented with unmolassed beep pulp.**
   (b) A high protein concentration ration based on legumes.
   (c) Grasses pastures and balanced supplements based on selenium, iron, and vitamin D.
   (d) Cereal straw without protein content supplemented with olive oil.

References

Bell N, Thiemann A, Sullivan R (2021) Endoparasite control for donkeys in the UK. UK-Vet Equine 5:84. https://doi.org/10.12968/ukve.2021.5.2.84

Brown PJ, Whitbread TJ, Burden FA, Bailey M, Carley K, Daniell L, Jones C, Bell NJ (2011) Haemosiderin deposition in Donkey (Equus asinus) livers: comparison of quantitative histochemistry for iron and liver iron content. Res Vet Sci 90:284–287. https://doi.org/10.1016/j.rvsc.2010.06.016

Burden FA, Hazell-Smith E, Mulugeta G, Patrick V, Trawford R, Brownlie HWB (2016) Reference intervals for biochemical and haematological parameters in mature domestic donkeys (Equus asinus) in the UK. Equine Vet Educ 28:134–139. https://doi.org/10.1111/eve.12512

Divers TJ (2015) The equine liver in health and disease. In: 61st Annual Convention of American Association of Equine Practitioners, Las Vegas, Nevada, USA, 5–9 December

Durham AE, Thiemann AK (2015) Nutritional management of hyperlipaemia. Equine Vet Educ 27: 482–488. https://doi.org/10.1111/eve.12366

Durham AE, Smitht KC, Newton JR, Hillyer MH, Hillyer LL, Smith MR, Marr CM (2003) Development and application of a scoring system for prognostic evaluation of equine liver biopsies. Equine Vet J 35:534–540. https://doi.org/10.2746/042516403775467171

Gehlen H, Twickel S, Stockle S, Weber C, Bartmann CP (2020) Diagnostic orientation values for ACTH and other parameters for clinically healthy donkeys and mules (insulin, triglycerides, glucose, fructosamines, and -GT). J Anim Physiol Anim Nutr 104:679–689. https://doi.org/10.1111/jpn.13279

Goodrich EL, Behling-Kelly E (2022) Particle size distribution of plasma lipoproteins in donkeys from Death Valley compared to a sampling of horses. Animals 12:2746. https://doi.org/10.3390/ani12202746

Goodrich EL, Webb JL (2024) Complete blood count and biochemistry reference intervals for healthy adult donkeys in the United States. Animals 14:2018. https://doi.org/10.3390/ani14142018

Howell AK, Malalana F, Beesley NJ, Hodgkinson JE, Rhodes H, Sekiya M, Archer D, Clough HE, Gilmore P, Williams DJL (2020) Fasciola hepatica in UK horses. Equine Vet J 52:194–199. https://doi.org/10.1111/evj.13149

Linn-Peirano SC, Hepworth-Warren K, Kinsella H, Diaz-Campos D, Brenseke BM, Cianciolo RE, Schroeder E, Schreeg ME (2023) Ingesta-associated choledocholithiasis in horses: 2 cases and literature review. J Vet Diagn Invest 35:417–424. https://doi.org/10.1177/10406387231177251

Luethy D, Frimberger AE, Bedenice D, Byrne BS, Groover ES, Gardner RB, Lewis T, Macdonald VS, Proctor-Brown L, Tomlinson JE, Rassnick KM, Johnson AL (2019) Retrospective evaluation of clinical outcome after chemotherapy for lymphoma in 15 equids (1991–2017). J Vet Intern Med 33:953–960. https://doi.org/10.1111/jvim.15411

Mclean AK, Wang W, Navas-Gonzalez FJ, Rodrigues JB (2016) Reference intervals for hematological and blood biochemistry reference values in healthy mules and hinnies. Comp Clin Pathol 25:871–878. https://doi.org/10.1007/s00580-016-2276-3

Mendoza FJ, Estepa JC, Gonzalez-De Cara CA, Aguilera-Aguilera R, Toribio RE, Perez-Ecija A (2015) Energy-related parameters and their association with age, gender, and morphometric measurements in healthy donkeys. Vet J 204:201–207. https://doi.org/10.1016/j.tvjl.2015.03.004

Mendoza FJ, Gonzalez-Cara CA, Aguilera-Aguilera R, Toribio RE, Perez-Ecija A (2018) Effect of intravenous glucose and combined glucose-insulin challenges on energy-regulating hormones concentrations in donkeys. Vet J 240:40–46. https://doi.org/10.1016/j.tvjl.2018.09.002

Mendoza FJ, Toribio RE, Perez-Ecija A (2019) Metabolic and endocrine disorders in donkeys. Vet Clin North Am Equine Pract 35:399–417. https://doi.org/10.1016/j.cveq.2019.07.001

Mendoza FJ, Toribio RE, Perez-Ecija A (2024) Metabolic and endocrine insights in donkeys. Animals 14:590. https://doi.org/10.3390/ani14040590

Meng Q, Shao Y, Li W, Lu J, Wang X, Deng L (2024) Evaluation of serum lipids, biochemical parameters, selected antioxidant elements and oxidative stress profiles in late pregnant jennies with hyperlipemia. Vet Sci 11:664. https://doi.org/10.3390/vetsci11120664

Perez-Ecija A, Gonzalez-Cara C, Aguilera-Aguilera R, Toribio RE, Mendoza FJ (2021) Energy hormone response to fasting-induced dyslipidemia in obese and non-obese donkeys. Vet J 271: 105652. https://doi.org/10.1016/j.tvjl.2021.105652

Piotrowski IL, Fabian R, Ohlerth SM, Grimm F, Wehrlieser ME (2024) Cystic echinococcosis due to Echinococcus equinus in a Swiss donkey. Equine Vet Educ 37(4):e71–e76. https://doi.org/10.1111/eve.14046

Potier JFN, Durham AE, Modi R, Rosenberg W, Dash SA (2023) Investigation of derum markers of hepatic fibrosis in equids. J Equine Vet Sci 131:104937. https://doi.org/10.1016/j.jevs.2023.104937

Rasche A, Lehmann F, Goldmann N, Nagel M, Moreira-Soto A, Nobach D, De Oliveira Carneiro I, Osterrieder N, Greenwood AD, Steinmann E, Lukashev AN, Schuler G, Glebe D, Drexler JF (2021) A hepatitis B virus causes chronic infections in equids worldwide. Proc Natl Acad Sci U S A 118:e2013982118. https://doi.org/10.1073/pnas.2013982118

Samuels A, Burns TA (2024) Lymphoma and myeloproliferative disease. Vet Clin North Am Equine Pract 40:441–454. https://doi.org/10.1016/j.cveq.2024.07.009

Sullivan R (2024) Hepatic disease in the donkey. Equine Vet Educ 37(4):174–176. https://doi.org/10.1111/eve.14075

The Donkey Sanctuary (2020) Feeding the donkey with liver disease. https://www.thedonkeysanctuary.org.uk/for-owners/owners-resources/feeding-donkeys-with-liver-disease

Theelen MJP, Beukers M, Grinwis GCM, Sloet Van Oldruitenborgh-Oosterbaan MM (2019) Chronic iron overload causing haemochromatosis and hepatopathy in 21 horses and one donkey. Equine Vet J 51:304–309. https://doi.org/10.1111/evj.13029

Neurologic Examination of Donkeys and Mules

Cynthia Xue

Abstract

Neurologic examination of the donkey and their hybrids (mules and hinnies) requires solid knowledge in normal neuroanatomy and neurophysiology. A systematic approach is key to preventing premature closure on differential diagnoses. A good history and collation of all clinical signs to determine one neuroanatomic area as the "common denominator" for all clinical signs should be the goal, with a multifocal or diffuse neurolocalization considered only if neuro-anatomic localization to one area is impossible. This ensures that the eventual differentials list truly aligns with the patient's clinical signs and disease process.

Keywords

Neurologic · CNS · Nervous system

1 General

Neurologic assessment of the donkey, mule, and hinny is as with a horse. In fact, much of the information here is translated over from horses in the absence of donkey, mule, and hinny specific studies related to neurology. However, performance of the examination can be complicated in donkeys, mules, and hinnies that are not halter trained. In donkeys, gait assessment for ataxia can be hampered by a donkey's natural inclination to freeze in place when alarmed. Thus, while a systematic approach to neurologic testing is necessary to avoid missing lesions, performance of the neurologic exam should prioritize handler/examiner safety above all. Further,

C. Xue (✉)
Department of Clinical Sciences, Ross University School of Veterinary Medicine, Basseterre, Saint Kitts and Nevis, West Indies
e-mail: cxue@rossvet.edu.kn

A. K. McLean et al. (eds.), *Donkey and Mule Medicine: Clinical Medicine and Reproduction*, https://doi.org/10.1007/978-3-032-18540-2_8

Table 1 Neurologic examination

Static	
	General mentation and head/body posture
	Cranial nerve evaluation
	Muscle mass, tone, and symmetry
	Cutaneous reflexes
Dynamic	
	Gait/Proprioceptive analysis

the neurologic testing should be executed in a manner that does not worsen an animal's anxiety towards being handled, as increased anxiety often leads to increased protective behaviors (freezing, aggression) that make progressing through the neurologic exam more difficult.

After obtaining a history, the author generally likes to break down the neurologic exam into static and dynamic segments. The static portion of the neurologic exam is performed with the animal at a standstill whilst the dynamic portion is performed with the animal in motion (Table 1).

The primary goal of the neurologic exam is to arrive at a reasonable neuroanatomic diagnosis derived from the neurologic abnormalities found on exam (Fig. 1). The neuroanatomic diagnosis then directly informs the differential diagnoses list, subsequent diagnostic testing, and treatment plan as necessary. In the author's opinion, proceeding to a diagnostic/treatment plan without having systematically gone through the process of first arriving at a neuroanatomic diagnosis is an inefficient strategy. Although this may be a realistic outcome in donkeys[1], particularly with animals that are not compliant, an observant clinician should still be able to make reasonable conjecture using clues from owner history and whatever neurologic information can be obtained from a limited exam.

In the same vein, it is important not to skip either static or dynamic examination as it may not be possible to accurately neurolocalize a lesion from abnormalities detected in only one portion of the examination alone. For example, ataxia detected on the dynamic examination can suggest a lesion in the spinal cord, but the neuroanatomic location of the lesion would change to the forebrain if there was also concurrent information about cortical blindness, seizures, or mentation deficits from the static examination. Thus, the most important aspect of a neurologic exam, in the author's opinion, is avoiding premature closure of their differentials list by collecting as much information as possible prior to making a neuroanatomic diagnosis.

[1]For brevity, the author will use "donkeys" to encompass donkeys, mules, and hinnies in the remainder of the chapter if the information is applicable to all three species and only utilize specific terminology to elaborate on major differences.

GENERAL ASSESSMENT

Mentation/Behavior	
Head/Neck Posture	
Stance	

CRANIAL NERVE EXAMINATION

	Left	Right	Comments
Vision			
Menace			
Pupils (room light)			
PLR (direct/consensual)			
Strabismus			
Facial Sensation			
Jaw Tone			
Palpebral Reflex			
Facial Symmetry			
Nystagmus			
Head Tilt			
Tongue Tone			
Slap Test			
Swallow			

MUSCLE MASS TONE AND SYMMETRY

	Left	Right
Head/Neck		
Thoracic Limbs		
Epaxial Muscles		
Pelvic Limbs		
Tail Tone		

CUTANEOUS SENSATION AND REFLEXES

	Left	Right
Cutaneous Trunci		
Cervical		
Cervicofacial/auricular		
Perineal		
Spinal Flexion		

GAIT AND PROPRIOCEPTIVE ASSESSMENT

	LF	RF	LH	RH
Placement Tests				
Walk (straight, head neutral)				
Walk (straight, head elevated)				
Backing				
Circling (left)				
Circling (right)				
Hill				
Curb				
Tail Pull				
Other/Comments				
Ataxia Grade				

NEUROANATOMIC DIAGNOSIS AND COMMENTS	

Fig. 1 Neurologic evaluation record, adopted from the University of Georgia

2 History

As with any species or workup, collecting a thorough history from the owner can facilitate neurolocalization of lesions. A comprehensive anamnesis can eliminate confounding factors or facilitate prioritization of differentials. For example, neurologic abnormalities in an animal with an incomplete vaccine history can increase the index of suspicion of a viral encephalitis. In the absence of donkey specific vaccine recommendations, donkeys are often anecdotally vaccinated against viral etiologies of encephalitis following recommendations put forth for horses (American Association of Equine Practitioners 2025). However, it must be noted that no vaccines are licensed specifically for non-horse equids yet despite this practice being performed out of an abundance of caution for donkeys and donkey hybrids. Neurologic abnormalities in a donkey with a past history of *Streptococcus equi* subs *equi* respiratory disease should prompt consideration for metastatic abscessation (i.e. bastard strangles). Donkeys with a history of trauma and acute onset seizures should have the cerebrum prioritized as a neurolocalization and traumatic brain injury prioritized as the differential with subsequent diagnostic imaging focusing on finding lesions in the cerebrum (Junge et al. 2017). Seropositivity to vector-born neurologic etiologies (e.g. Japanese Encephalitis Virus, West Nile Virus) may prompt consideration for differentials of infectious encephalitis in an animal with concomitant neurologic signs (García-Bocanegra et al. 2012; Mendoza et al. 2018; Idoko et al. 2021; Rusenova et al. 2024). Animals with gastrointestinal disease (Fielding et al. 2015; Giannitti et al. 2015) may develop hyperammonemia, secondary hepatic encephalopathy, and seizures which indicate neurolocalization to the cerebrum. Realistically, incomplete anamnesis is common, especially in animals that are not frequently handled, or in individual donkeys belonging to a herd under production-style management. Nonetheless, the index of suspicion for neurologic disease can increase in an observant clinician and can inform selection of specific diagnostic tests even in animals with vague clinical signs.

3 Static Exam

3.1 Mentation

Donkeys are normally considered more stoic compared to horses and may be misinterpreted as being QAR or lethargic even when they are not. It is often easier to see a more honest manifestation of a donkey's true mentation from a distance observing its interaction with herd mates and its environment.

Any acute mentation change should narrow down the neuroanatomic location to the cerebral cortex or the brainstem. Mentation changes with seizures, ataxia, cortical blindness, and absent menace suggest cerebral cortical disease, with differentials such as the encephalidities and herpesvirus (Vengust et al. 2008). Mentation changes with multiple cranial nerve deficits suggest a brainstem lesion (Jula et al. 2024) given that the brainstem is the origin of all cranial nerves.

Difficulties with temperature regulation should prompt consideration for a central lesion as the hypothalmus acts as the thermoregulatory center of the brain (Hampton 2022). However, this is clinically difficult to interpret in a field setting and is likely more practically useful in hospitalized donkeys where there is much more consistent monitoring of vital parameters.

3.2 Head/Body Posture

Animals standing at rest with their feet in an abnormal posture (base-wide, crossed over) or with a head tilt can be considered to have deficits to sensory input that inform. In the former, this would be known as spinal, or proprioceptive, ataxia. In the latter, this would be considered vestibular ataxia. Spinal ataxia does not distinguish between lesions in the cerebrum, brainstem, or spinal cord. Thus, the ability to neurolocalize further would require the presence of other clinical signs specific to those locations. An animal with seizures, blindness, and spinal ataxia likely has a lesion in the cerebrum. An animal with lethargy, and clinical signs of central vestibular disease or multiple cranial nerve deficits likely has a lesion within the brainstem. For purely spinal cord lesions, no mentation deficits are expected. Animals that have spinal ataxia and intention tremors have cerebellar ataxia arising from a lesion in the cerebellum. If the lesion extends beyond the cerebellopontine angle or flocculonodular lobe to involve a larger area of the cerebellum, other clinical signs such as delayed to absent menace may also be present (Raphel 1982).

A head tilt and spontaneous nystagmus, often with ataxia and a base-wide stance, is indicative of vestibular disease. Subtle vestibular disease can manifest with ipsilateral ventrolateral strabismus as the only clinical sign (Mayhew 2010). Vestibular disease can be further broken down into peripheral vestibular disease, central vestibular disease, and central vestibular disease with a *paradoxical* head tilt (often shortened to paradoxical vestibular disease) neurolocalized to the vestibulocochlear nerve, the brainstem, and the flocculonodular node in the cerebellum, respectively (Table 2).

Table 2 Key clinical signs useful for distinguishing between vestibular diseases

Clinical Signs	Peripheral	Central	
	CN VIII	*Brainstem*	*Cerebellum*
Head tilt	Ipsilateral	Ipsilateral	Contralateral*
Falling/leaning/circling	Ipsilateral	Ipsilateral	Either
Nystagmus	Horizontal/rotary	Positional	Positional
Proprioceptive deficits	–	Ipsilateral	Ipsilateral
Mentation changes	–	Present	+/–

The (*) is to highlight the major differentiating factor between central vestibular disease and paradoxical vestibular disease that results in a paradoxical head tilt contralateral to the lesion

3.3 Cranial Nerves

Assessment of the cranial nerves can be grouped together by their functional regions, or by the individual nerves. The author finds the former more practical, whereas the latter prevents cranial nerves from getting overlooked. Regardless of which method the clinician adapts, being systematic and consistent with the evaluation is key. The author will list what they observe for when evaluating the functional groupings below (Table 3) but discuss the individual nerves in the text with a focus on clinically pertinent details.

3.3.1 CN I: Olfactory Nerve

Objective evaluation of this nerve is difficult. The presumption is made that the olfactory nerve is functional if the patient shows interest in food, or if they react negatively to noxious smells (e.g., alcohol-soaked gauze). Perfunctory evaluation of the olfactory nerve is often performed by collating behavioral data observed while performing the rest of the physical exam.

3.3.2 CN II: Optic Nerve

The optic nerve carries afferent information that governs the dazzle reflex, pupillary light reflex (PLR), menace response, and vision. Animals should blink in response to light shone into the eyes (indicative of an intact dazzle reflex), have direct and consensual pupillary contraction in responses to light (indicative of an intact PLR), and blink or attempt evasive maneuvers when a menacing gesture is made at the eyes (indicative of an intact menace response). Vision loss related to optic nerve pathology can similarly result in an absent menace response. However, donkeys that have had insidious vision loss over time may not act overtly blind due to having memorized their environment. When evaluating for the menace response in donkeys, some donkeys may normally respond in a delayed manner compared to horses (Mendoza et al. 2018), although the author has not found it in their case when evaluating for the menace response. This must be taken into account not to be

Table 3 Cranial nerve evaluation by functional groupings

Smell	Subjective interest in and/or ability to track food
Ocular functions	Eye and pupil orientation Pupil size and response to light/darkness Presence of physiologic nystagmus Eyelid position and blinking in response to stimuli
Facial sensation, musculature, and symmetry	Ear, nose, and muzzle symmetry Masseter and temporalis symmetry
Head position and ear movement	Head perpendicular to the ground; absence of head tilt
Mouth, tongue, swallowing	Ability to prehend, masticate, and swallow food and water Symmetrical tongue tone and musculature Tongue retraction

overinterpreted as an abnormality in an animal that is systemically stable and otherwise neurologically normal.

Note that a lack of any of these responses to the associated stimuli does not immediately indicate optic nerve pathology. Rather it should inform the clinician to consider the normal afferent and efferent signaling pathways, the anatomic locations that govern those pathways, and any other clinical signs that would further narrow down which part of the signaling pathway is dysfunctional. For example, although a malfunctioning optic nerve is a reasonable differential for blindness, loss of vision can result from any pathology in the afferent arm of the menace pathway (from the retina to the visual cortex in the cerebrum). In a patient that is blind in the right eye, circling to the left, and demonstrating seizures, the optic nerve no longer becomes a reasonable differential diagnosis because the other clinical signs (seizures, circling to the left) have greater specificity for the cerebrum as the neuroanatomic diagnosis. To put it in other words, a lesion in the cerebrum is the "common denominator" that can explain the presence of three clinical signs, whereas a lesion in the optic nerve can only explain one clinical sign. Therefore, always prioritize an aim to make one neuroanatomic location explain all the clinical signs. If it seems impossible for one neuroanatomic location to explain all the clinical signs, then consider that there is a multifocal or diffuse lesion.

3.3.3 CN III: Oculomotor Nerve

Pathology in the oculomotor nerve can result in mydriasis, loss of direct PLR (with an intact consensual PLR), ptosis of the upper eyelid, and fixed lateral and slightly ventral strabismus of the affected side (Mayhew 2010). This is because the oculomotor nerve normally innervates the pupillary constrictors; efferent arm of the PLR; the *levator palpebrae superioris*; and the dorsal rectus, medial rectus, ventral rectus, and ventral oblique muscles. In equids, an ipsilateral ventrolateral strabismus is indicative of vestibular disease until proven otherwise (Mayhew 2010). However ventrolateral strabismus from vestibular disease is not fixed compared to the type of strabismus manifested in oculomotor nerve disease. In vestibular strabismus, the eye can be returned to its normal position with head movement. The ability of the eye to move back to its normal position with induced head movement should be used to inform a clinician of the appropriate neuroanatomic diagnosis.

3.3.4 CN IV: Trochlear Nerve

There is only one muscle that the trochlear nerve innervates—the dorsal oblique. Lesions affecting the trochlear nerve result in a dorsomedial strabismus, named after the manner in which the globe rotates so that the medial aspect of the horizontal equid pupil sits dorsomedially.

3.3.5 CN V: Trigeminal Nerve

The trigeminal nerve governs afferent information about facial sensation and efferent input to the muscles of mastication (temporalis and masseters). Therefore, trigeminal nerve issues typically manifest as absent palpebral reflex and atrophy of the temporalis and masseter muscles. Atrophy to those muscles may also affect normal mastication due to their weakness. In cases where the normal eating function in an equid is affected, it is prudent to evaluate which aspect of eating is abnormal in order to differentiate between lesions affecting the trigeminal nerve, the glossopharyngeal and vagus nerves, or the hypoglossal nerve.

3.3.6 CN VI: Abducens Nerve

Much like the trochlear nerve, the abducens nerve serves only one function—to innervate the lateral rectus muscle. Therefore, lesions will manifest in medial strabismus.

3.3.7 CN VII: Facial Nerve

Facial nerve trauma results in ipsilateral ear drop, ipsilateral ptosis, and contralateral muzzle deviation. Injury to the rostral most buccal branches of the nerve can result in muzzle deviation alone without other accompanying signs of facial nerve paralysis whereas proximal injury results in complete facial paresis or paralysis (Prange 2019). Other clinical signs include absent dazzle reflex, menace response, and palpebral reflex. A history of compression injury during anesthesia, a tight-fitting halter or halter-related trauma, otitis media/interna, temporohyoid osteoarthropathy, and guttural pouch mycosis have all been reported as causes of facial nerve paralysis in horses (Borges and Watanabe 2011; Prange 2019).

3.3.8 CN VIII: Vestibulocochlear Nerve

As discussed earlier in the chapter, lesions in the vestibulocochlear nerve can manifest peripheral vestibular disease. In horses, causes of peripheral vestibular disease include head trauma, temporohyoid osteoarthropathy, and otitis media/ interna (Aleman 2024; Wilson et al. 2023). The close proximity of the vestibulocochlear nerve to the facial nerve also makes it possible for facial nerve paralysis to be present in horses with vestibulocochlear nerve injured (Aleman 2022). Although donkeys and horses have very similar anatomy of that region as it relates to structures implicated in the previously described differentials (Alsafy et al. 2008), peripheral vestibular disease is rarely reported in donkeys, mules, and hinnies. It is unknown whether this is related to logistical difficulties in performing detailed diagnostic workup required to obtain definitive diagnosis in resource limited areas where these animals usually reside, species-specific differences in disease presentation and/or etiology, or inability to obtain follow up with further diagnostic testing after initial presentation for non-specific neurologic signs. Nonetheless, knowing the clinical signs associated with peripheral vestibular disease (ipsilateral ventrolateral strabismus or ipsilateral head tilt, nystagmus with a contralateral fast phase, and ataxia) will be key at facilitating neurolocalization and investigation of differentials. Hearing is also governed by this nerve, but it is often difficult to detect hearing loss in horses without performing brainstem auditory evoked responses (BAER) for definitive diagnostic testing in horses (Aleman 2022). No studies have yet been performed in donkeys and their hybrids on BAER testing.

3.3.9 CN IX: Glossopharyngeal Nerve and CN X: Vagus Nerve

The glossopharyngeal and vagus nerves are discussed together as dysfunction leads to pharyngeal issues such as dysphagia, dysphonia (roaring), and regurgitation (Large animal internal medicine 6E). Dysphagia and regurgitation can be more easily detected than roaring in donkeys as they are not often used for high speed, intensive exercise work (unlike horses). An abnormal slap test can suggest dysfunction of the vagus nerve, which can then be corroborated by upper airway endoscopy findings. Historical information regarding quidding, coughing, feed colored discharge expelled from the nares, or choke-like episodes should prompt further evaluation of normal pharyngeal function if other differentials have been ruled out.

3.3.10 CN XI: Accessory Nerve

In horses, accessory nerve dysfunction results in atrophy of the trapezius, sternocephalicus, and brachiocephalicus muscles (Van Metre and MacKay 2020), although it is rarely reported. Because of the muscles it innervates, accessory nerve issues may also result in delays to the cervical spinal reflexes, or possibly forelimb gait deficits. These deficits are presumed to be the same in donkeys and their hybrids.

3.3.11 CN XII: Hypoglossal Nerve

Because the hypoglossal nerve provides motor/efferent input to the muscle of the tongue, hypoglossal nerve pathology results in muscle atrophy and weakness of the tongue. Lesions can be unilateral or bilateral. Evaluating tongue tone during the static portion of the neurologic examination, by pulling the tongue out laterally through the diastema on both sides of the mouth, will be most helpful in visualizing muscle atrophy and detecting deficiencies of tongue tone. Note that pathology of the hypoglossal nerve may clinically mimic difficulties in mastication, prehension, or deglutition by an owner. It is important for a clinician to carefully distinguish the subtleties between the different stages of eating that normally work in cohort animals to permit accurate neurolocalization.

3.4 Muscle Mass Tone and Symmetry

Abnormalities in muscle mass, tone, or symmetry can be suggestive of lower motor neuron (LMN) disease such as peripheral neuropathies (i.e. radial nerve paralysis, sweeney, equine motor neuron disease) or disease with a multifocal distribution (i.e. polyneuritis equi, equine protozoal myeloencephalitis). However, chronic upper motor neuron (UMN) lesions may also eventually manifest in muscle atrophy. Differentiating between LMN and UMM associated muscle atrophy will necessitate pairing information about the animal's gait as LMN weakness and UMN weakness have very different clinical presentations (discussed further under Sect. 4). Further, some diseases (i.e. polyneuritis equi) may have a combination of LMN and UMN clinical signs depending on the extent of the lesion.

Diseases affecting the neuromuscular junction can also manifest in dysfunctional muscle activity. Tetanus (*Clostridium tetani*) should be prioritized in any unvaccinated donkey and mule presenting with muscle rigidity affecting the gait and/or facial features (e.g. retracted ears, tense muzzle), and prolapsed nictitans. Trauma or surgical wounds with concomitant clinical signs are highly correlated with disease. However, absence of trauma does not rule out disease (Kay and Knottenbelt 2010) as wounds many not always be found on physical examination particularly if small or healed over (The Donkey Sanctuary 2017). This is a differential that is particularly important to evaluate for in donkeys and their hybrids that, as working animals, are at increased risk of being unvaccinated and sustaining wounds from their daily activities. In mild to moderate disease, animals can still ambulate and eat. As disease becomes more severe, animals become progressively lose voluntary function of walking, eating, and drinking (Kay and Knottenbelt 2010). Terminal cases present recumbent with complete trismus or "lockjaw" (Kay and Knottenbelt 2010). In donkeys, it may be important to re-vaccinate with a tetanus toxoid even after resolution of clinical disease as recurrent tetanus has been reported in this species (Kay et al. 2022). Although likely a rare situation and only ever reported in a donkey, it suggests that natural anti-tetanus immunity should not be presumed in donkeys, mules, and hinnies following exposure (Kay et al. 2022). Protection is likely best conferred through annual vaccination much like in horses (van Galen 2024). No vaccination recommendations specifically for donkeys, mules, and hinnies currently exist—however, the high correlation of clinical tetanus in unvaccinated animals, the aggressive nursing care required to manage it, and the often fatal nature of this disease should be strongly considered when tailoring preventative care plans to optimize these patients' welfare.

Botulism (*Clostridium botulinum*) results in muscle flaccidity that is often characterized by weakness and dysphagia. Reduced reflex responses are also possible cranial nerve examination (Lanci et al. 2019). Because donkeys, mules, and hinnies can be stoic, and the clinical manifestation of subtle weakness deficits (laying down) and dysphagia (changes in appetite) can overlap with the non-specific signs of colic, choke, or guttural pouch disease, the author recommends evaluating for these differentials as well in the workup of a patient with suspected botulism. Three forms of botulism exist in horses—forage poisoning, toxicoinfectious, and wound botulism (Whitlock and Buckley 1997; Johnson et al. 2010; Wilkins and Palmer 2003; Schoenbaum et al. 2000). Clinical cases have been reported in donkeys and mules (Schoenbaum et al. 2000; Lanci et al. 2019). The reciprocal cross, hinnies, should not be presumed to be immune from disease even though no reports of clinical botulism have been described in them yet. Signalment is helpful in determining which form of botulism to prioritize. Adult equids are at greatest risk of botulism from forage poisoning (Whitlock and Buckley 1997), and foals are at greatest risk of toxicoinfectious botulism (Whitlock and Buckley 1997; Wilkins and Palmer 2003). Forage poisoning can be prioritized in a patient with an appropriate concomitant history: multiple affected animals with the same clinical signs that eat from the same feed source; access to decaying, rotten vegetable matter, haylage, or silage; and carrion in the environment. Pasture proximity to birds (e.g. hens and geese), or the environment required by waterfowl, may also represent a risk factor in donkeys (Lanci et al. 2019; Xue personal communication). Although this has been reported in cattle (Smart et al. 1987; Haagsma 1991), more studies are needed to determine the association between proximity to birds and development of botulism in donkeys and other equids. Wound botulism is the rarest form of botulism but should be considered if the other forms have been ruled out and wounds are found on physical examination or diagnostic testing (Whitlock and McAdams 2006; Liguori et al. 2008).

3.5 Cutaneous Reflexes

The cervicofacial reflex, cervical reflex, and cutaneous trunci reflexes are all readily testable on a donkey. The cervicofacial and cervical reflexes are performed by using a sharp object to poke along the cervical vertebral bodies to elicit a grimace or

4 Dynamic Exam

In donkeys, mules, and hinnies, dynamic neurologic evaluation of an animal may be complicated by their anxiety and behavior. If portions of this examination procedure described below cannot be safely performed by an examiner, it is recommended to watch an animal move freely on its own to obtain some information rather than skipping the dynamic exam completely. In this scenario, adjustments to the terrain can also be made to elicit deficiencies in postural reactions and proprioception without physically maneuvering an animal that is fractious to being handled (e.g. putting logs in the animal's path; placing them in a small pen or paddock with divots in the ground, inclines, or declines; etc). The author prefers to collect information about proprioception and postural reactions by observing donkeys from a distance before attempting foot placement in donkeys that are amenable to being handled. This is because observation of an equid in motion is helpful for detecting subtle proprioceptive deficits that may only be elicited when they are asked to acutely stop. It also eliminates safety issues associated with physically maneuvering legs into abducted or adducted positions in an animal not amenable to having their limbs touched.

4.1 Gait Assessment

For dynamic assessment of gait, animals should be:

1. Walked in a straight line on a flat surface with the head in neutral carriage
2. Walked in a straight line on a flat surface with the head elevated
3. Backed in a straight line on a flat surface
4. Circled to the left and right on a flat surface, taking care to start with large circles and progressively tightening the circle
5. Walked in a straight line on an incline/decline with the head in neutral carriage

6. Walked in a straight line on an incline/decline with the head elevated
7. Challenged to walk on and off a curb, ideally in an infinity loop or circling in both directions for a greater challenge

If there continues to be persistent questions about the presence of neurologic abnormalities, greater challenge can be induced by asking the animal to perform the above at a trot. In some situations, this can be helpful in distinguishing between neurologic gaits and musculoskeletal lameness.

While in motion, animals should be assessed for normal foot placement, evidence of ataxia, and dysmetria (hyper or hypometria). Neurologic gaits often have an irregularly irregular gait, whereas gait deficits arising from musculoskeletal issues tend to be more regularly irregular (Aleman et al. 2018). Gait deficits associated with spinal ataxia can include a combination of sensory ataxia (dysfunction in afferent input) and upper motor neuron (UMN) or lower motor neuron (LMN) weakness (both of which arise from dysfunction in efferent output). All of these can occur in one or all legs depending on whether a lesion is unilateral or bilateral, and where the lesion is located in the spinal cord. Sensory ataxia results in general proprioceptive (GP) deficits that can clinically look like floating (particularly when stepping off curbs), delayed initiation of protraction, irregular foot placement, circumduction in the hind limbs, pivoting on circles, and truncal sway. UMN weakness classically appears as spasticity, increased muscle tone, exaggerated reflex responses, and a hypermetric gait whereas LMN weakness appears as the opposite—increased flaccidity, decreased muscle tone, diminished reflex responses, and profound muscle atrophy with secondary muscle fasciculations (Furr and Reed 2015).

There can be a combination of GP deficits and UMN weakness or GP deficits and LMN weakness depending on the lesion location. The clinical differentiation between GP deficits and UMN weakness is often difficult and rarely matters for neurolocalization given that spinal compression will often simultaneously disrupt both afferent and efferent signaling pathways. General proprioceptive deficits in motion (such as abnormal foot placement) can be less obvious, too, when there is LMN weakness (hypometria, toe dragging, knuckling, stumbling, buckling). The author has found using algorithm, modified from Furr and Reed (2015) and based on the *clinical appearance* of various neurologic deficits, to be the most helpful for neurolocalization (Fig. 2):

In a straight line moving forward, GP and UMN deficits may manifest as a floating hypermetric gait where the hind feet may almost step on the heels of the front feet, uncoordinated steps and/or inconsistent foot placement, and truncal sway. Typically, with cervical spinal cord compression (C1–C5), the hind limb deficits are more obviously prominent than forelimb deficits (if forelimb deficits are even noticeable) and can be worsened with head elevation. This is because the general proprioceptive tracts governing the hind limbs are more superficial along the spinal cord, and a greater degree of compression is necessary to elicit forelimb deficits. If an animal allows itself to be backed, ataxia often appears as uncoordinated, inconsistent foot placement and a lack of parallel movement of the contralateral forelimbs and hindlimbs (e.g. left forelimb moving back with right hindlimb). When circled,

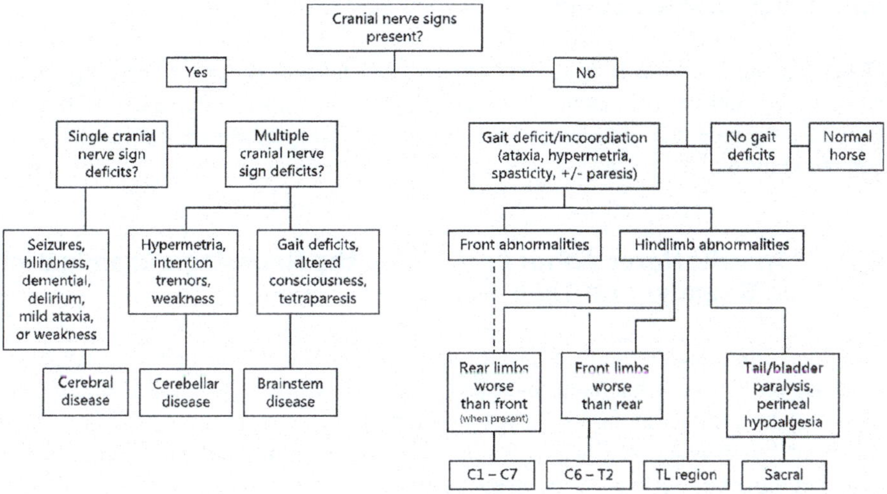

Fig. 2 Modified graph of algorithm used to aid neurolocalization based on clinical manifestation of neurologic deficits. Note that true cranial nerve abnormalities can only occur from disease in a cranial nerve or from the brainstem (where all the cranial nerve originate). However, abnormalities such as cortical blindness (from a cerebral lesion) or vestibular ataxia (from a cerebellar lesion) may be imperfectly interpreted as stemming from the cranial nerves if other neurologic abnormalities are subtle or missed during the initial workup. The modifications to Furr and Reed's original Neurolocalization algorithm are intended to take this into account

circumduction of the outside hind legs, pivoting of the inside leg, and interference between inside and outside leg may occur. When asked to stop acutely from a circle, the animal may stop without standing square.

For lesions at the cervical intumescence or brachial plexus (C6–T2), cranial lesions in this region will manifest with all four leg deficits that look like the deficits previously described in the cervical spine. However, caudal lesions in this region will cause GP and LMN deficits to the forelimbs that clinically manifest as the forelimbs appearing worse than the hind legs. The hind legs will continue to manifest GP and UMN deficits. Clinically, this can look like short, choppy gait with toe dragging in the forelimb, and overreaching from the hind limbs when walked in a straight line.

For lesions at the thoracolumbar spine, there will only be GP and UMN deficits in the hind limb. For lesions at the sacral plexus, there will only be GP and LMN deficits in the hind limbs. All lesions caudal to the sacral plexus do not manifest in gait abnormalities. Rather, they present as LMN bladder and tail issues such as urinary or fecal incontinence and tail flaccidity, along with sensory issues to the perineal region.

4.2 Postural Reactions

General proprioceptive deficits may appear like delayed return to standing square when a leg is deliberately abducted or adducted. While this is helpful in isolating out general proprioceptive deficits, it may not be necessary based on the results of the gait evaluation.

4.3 Specific Notes About Performing the Neurologic Examination in Donkeys and Their Hybrids

In animals that do not want to move forward or backwards to halter pressure—or have not yet learned what halter pressure means—a whip can be used as a visual cue to stimulate animals to move on a loose lead line. However, some animals may freeze or hyper-react to a whip, leading to the appearance of an altered gait. A clinician must be cautious about interpreting results in such scenarios and weigh the cost-benefit analysis of creating undue stress on an animal that may subsequently become less easy to handle for the rest of the examination.

5 After Neurolocalizing the Lesion, What Next?

Differential diagnoses can be made based on the neuroanatomic location of the lesion, which then informs diagnostic tests. In many situations, a cerebrospinal fluid (CSF) tap is helpful in evaluating for central nervous system inflammation, and radiographs and ultrasonography can be centered around the neuroanatomic area of interest. CSF fluid reference ranges for donkeys are available for centesis from the atlantooccipital (AO) and lumbosacral (LS) spaces (Table 4), but not yet available from samples obtained via atlantoaxial centesis (i.e. C1–C2 tap).

Unfortunately, no studies specifically evaluating reference ranges for CSF analytes exist for mules or hinnies yet. In one case report describing a mule with equine protozoal myeloencephalitis and equine motor neuron disease (Finno et al. 2010), the reference ranges for leukocytes, erythrocytes, and total protein used to interpret that mule's CSF sample were <8 cells/uL, <1 cell/uL, and <80 mg/dL, respectively. In the absence of specific reference ranges established for donkey hybrids, it can be common for clinicians to utilize horse references instead.

Table 4 Reference ranges for various parameters in cerebrospinal fluid collected from atlantooccipital and lumbosacral centeses

Parameter	Atlantooccipital	Lumbosacral
Specific gravity	1.004–1.005	1.004–1.006
Leukocytes (cells/ul)	1.5–2.7	1.3–2.7
Erythrocytes (cells/ul)	1.5–2.7	1.3–2.7
Total protein (mg/dl)	15–30	10–26

Modified from Abdelhakiem and Hussein (2021)

CSF collection from the AO space necessitates general anesthesia, while centesis from the C1–C2 and LS spaces can be performed under standing sedation. C1–C2 centesis in particular can be performed easily and accurately in the field using portable ultrasound systems in a standing, sedated animal. As in other species, the addition of 4 drops of autologous serum preserves cell morphology of horse CSF (Quattrini et al. 2023). This has not yet been evaluated on donkey CSF but has the potential to improve diagnostic accuracy in situations where there may be a delay between sample collection and sample interpretation in a lab. Additional advanced diagnostics such as MRI or CT are recommended in situations where owners can afford referral and are motivated to find a definitive diagnosis.

6 Conclusion

Neurologic examination in a donkey and their hybrids is heavily facilitated by strong foundational knowledge of neuroanatomy, neurophysiology, and deductive reasoning. A definitive diagnosis is reached more effectively after accurate neurolocalization as the specificity of differentials is highly dependent on the neuroanatomic area suspected to be diseased.

References

Abdelhakiem MAH, Hussein HA (2021) Collection of cerebrospinal fluid in 50 adult healthy donkeys (Equus asinus): clinical complications, and cytological and biochemical constituents. BMC Vet Res 17:302

Aleman M (2022) Vestibular disease. Vet Clin N Am Equine Pract 38:397–407

Aleman M, Nout-Lomas YS, Reed SM (2018) Disorders of the neurologic system. In: Reed SM, Bayly WM, Sellon DC (eds) Equine internal medicine, 4th edn. Elsevier, St. Louis, pp 580–687

Aleman M (2024) How to perform a neurologic exam in the horse. WEVA International Congress Proceedings, IVIS

Alsafy MAM, El-Kammar MH, El-Gendy SAA (2008) Topographical anatomy, computed tomography, and surgical approach of the guttural pouches of the donkey. J Equine Vet Sci 28:215–222

American Association of Equine Practitioners (2025) Vaccination guidelines. https://aaep.org/guidelines-resources/vaccination-guidelines/. Accessed 27 June 2025

Borges AS, Watanabe MJ (2011) Guttural pouch diseases causing neurologic dysfunction in the horse. Vet Clin N Am Equine Pract 27:545–572

The Donkey Sanctuary (2017) Neurological disease and disorders in donkeys. The Donkey Sanctuary Research Programme, Sidmouth

Fielding CL, Higgins JK, Higgins JC, McIntosh S, Scott E, Giannitti F, Mete A, Pusterla N (2015) Disease associated with equine coronavirus infection and high case fatality rate. J Vet Intern Med 29:307–310

Finno CJ, Eaton JS, Aleman M, Hollingsworth SR (2010) Equine protozoal myeloencephalitis due to *Neospora hughesi* and equine motor neuron disease in a mule. Vet Ophthalmol 13:259–265

Furr M, Reed SM (2015) Examination of the nervous system. In: Furr M, Reed SM (eds) Equine neurology, 2nd edn. Wiley, Ames, IA, pp 65–78

García-Bocanegra I, Arenas-Montes A, Jaén-Téllez JA, Napp S, Fernández-Morente M, Arenas A (2012) Use of sentinel serosurveillance of mules and donkeys in the monitoring of West Nile virus infection. Vet J 194:262–264

Giannitti F, Diab S, Mete A, Stanton JB, Fielding L, Crossley B, Sverlow K, Fish S, Mapes S, Scott L, Pusterla N (2015) Necrotizing enteritis and hyperammonemic encephalopathy associated with equine coronavirus infection in equids. Vet Pathol 52:1148–1156
Haagsma J (1991) Botulism in cattle, a review. Tijdschr Diergeneesk 116:663–669
Hampton CE (2022) Thermoregulation. In: Doherty T, Valverde A, Reed R (eds) Manual of equine anesthesia and analgesia, 2nd edn. Wiley, Hoboken, pp 173–183
Idoko IS, Schvartz G, Tirosh-Levy S, Erster O, Jibril JY, Adamu AM, Enem SI, Omeje JN, Nafarnda WD, Steinman A (2021) West Nile virus neutralizing antibody prevalence in donkeys from northern Nigeria. Trans R Soc Trop Med Hyg 115:566–568
Johnson AL, McAdams SC, Whitlock RH (2010) Type a botulism in horses in the United States: a review of the past (1998–2008). J Vet Diagn Invest 22:165–173
Jula C, Buechner-Maxwell V, Southard T, LeCuyer T (2024) Listeria monocytogenes encephalitis in a donkey foal. Equine Vet Educ 36(3):e79–e84
Junge HK, Mitchell KJ, Hoey S (2017) Diagnosis of severe traumatic brain injury using multimodal imaging in a donkey. Equine Vet Educ 29:539–543
Kay G, Knottenbelt DC (2010) Neurological disorders of donkeys and mules. In: Svendsen ED (ed) The Professional Handbook of the Donkey, 4th edn. The Donkey Sanctuary, Sidmouth
Kay G, Souhail ML, Mazan M (2022) Recurrent tetanus in a donkey. Equine Vet Educ 34:e418–e421
Lanci A, Rinnovati R, Anniballi F, Auricchio B, Scalfaro C, Menchetti M, Spadari A, Mariella J (2019) The first case of botulism in a donkey. Vet Sci 6:2–6
Liguori V, De Luliis P, Fenicia L, Anniballi F, Aureli P (2008) A case of wound botulism in a foal affected by gastric ulcers in Italy. J Equine Vet Sci 28:476–478
Mayhew IGJ (2010) Neuro-ophthalmology: a review. Equine Vet J 42:80–88
Mendoza FJ, Toribio RE, Perez-Ecija A (2018) Donkey internal medicine—part II: cardiovascular, respiratory, neurologic, urinary, ophthalmic, dermatology, and musculoskeletal disorders. J Equine Vet Sci 65:86–97
Prange T (2019) Peripheral nerve injuries. In: Auer JA, Stick JA, Kümmerle JM, Prange T (eds) Equine surgery, 5th edn. Elsevier, St. Louis, pp 900–909
Quattrini C, Scalco R, Vernau W, Dini P, Aleman M (2023) Effect of time and autologous serum addition on the analysis of cerebrospinal fluid in horses. J Vet Intern Med 37:713–717
Raphel CF (1982) Brain abscess in three horses. J Am Vet Med Assoc 180:874–877
Rusenova N, Rusenov A, Monaco F (2024) A retrospective study on the Seroprevalence of West Nile virus among donkeys and mules in Bulgaria. Vector Borne Zoonotic Dis 24:274–277
Schoenbaum MA, Mark Hall S, Glock RD, Grant K, Jenny AL, Schiefer TJ, Sciglibaglio P, Whitlock RH (2000) An outbreak of type C botulism in horses and a mule. J Am Vet Med Assoc 217:365–368
Smart JL, Jones TO, Clegg FG, McMurtry MJ (1987) Poultry waste associated type C botulism in cattle. Epidemiol Infect 98:73–79
van Galen G (2024) AAEP tetanus (*Clostridium tetani*) guidelines. AAEP infectious disease guidelines: tetanus. pp 1–5. https://aaep.org/wp-content/uploads/2025/02/Tetanus-Guidelines-Final.pdf
Van Metre DC, MacKay RJ (2020) Localization and differentiation of neurologic diseases. In: Smith BP, Van Metre DC, Pusterla N (eds) Large animal internal medicine, 6th edn. Elsevier, St Louis, pp 131–139
Vengust M, Wen X, Bienzle D (2008) Herpesvirus-associated neurological disease in a donkey. J Vet Diagn Invest 20:820–823
Whitlock RH, Buckley C (1997) Botulism. Vet Clin N Am Equine Pract 13:107–128
Whitlock RH, McAdams BS (2006) Equine Botulism. Clin Tech Equine Pract 5:37–42
Wilkins PA, Palmer JE (2003) Botulism in foals less than 6 months of age: 30 cases (1989–2002). J Vet Intern Med 17:702–707
Wilson A, Gonçalves R, Carslake H, Malalana F (2023) Suspected idiopathic vestibular syndrome in the horse: eight cases (2014–2021). Equine Vet Educ 36:e67–e72

Clinical Ophthalmology of Donkeys and Mules

Andrea Marchegiani, Marilena Bazzano, and Fulvio Laus

Abstract

Donkey ophthalmology is a new branch of veterinary ophthalmology, born in very recent years. Up until a few years ago, donkeys were believed to suffer from the same diseases of horses and, for this reason, were treated similarly. However, in very recent years, donkey ophthalmology emerged as a proper field focused on the diagnosis and treatment of eye conditions in donkeys. This chapter reports the most recent scientific information regarding the diseases and treatments of the donkey and mule's eye, enriched by the personal experiences of the authors.

Keywords

Donkey · Mule · Ophthalmology · Corneal disease · Anterior uveitis · Ocular procedure

1 Introduction

Donkeys and mules can suffer from similar ocular diseases than horses but with a different prevalence. The incidence of ocular disease in donkeys has been estimated to be 9% at the Veterinary Teaching Hospital of Camerino University (Italy) in a three-year period (personal author data), much lower than that reported in horses, in which the incidence can be up to 76% (Hurn and Turner 2006; Rushton et al. 2013; Wada et al. 2010). Donkeys and mules have a quieter attitude than horses, making ocular traumatic injuries less frequent. Because of their stoic response to pain,

A. Marchegiani · M. Bazzano · F. Laus (✉)
School of Biosciences and Veterinary Medicine, University of Camerino, Camerino, Italy
e-mail: andrea.marchegiani@unicam.it; marilena.bazzano@unicam.it; fulvio.laus@unicam.it

A. K. McLean et al. (eds.), *Donkey and Mule Medicine: Clinical Medicine and Reproduction*, https://doi.org/10.1007/978-3-032-18540-2_9

donkeys and mules do not show early clinical signs of ocular disease. Corneal ulcers, infectious kerato-conjunctivitis, uveitis, and eyelid trauma are reported to be the most common ocular conditions reported in donkeys and mules (Wafy et al. 2023).

2 Key Concepts

Donkeys and mules are unique equids with special considerations for many disciplines, including ophthalmology. Horses, donkeys, and mules belong to the genus Equus, but important differences exist between the different species considering the eye and its annexa, many of which affect their management and welfare.

3 Physiology

From an anatomical point of view, the **orbit** of donkeys and mules is slightly different from that of horses (Merkies et al. 2020), making it more difficult to open the upper eyelid in the former, also due to stronger palpebral muscles (Maśko et al. 2022). The lateral aspect of the frontal bone (the supraorbital ridge, forming the dorsal part of the orbit) is more laterally pronounced in donkeys and mules than in horses, making them more prone to flystrike at the medial canthus of the eye. In donkeys and mules, the nasal aperture of the **nasolacrimal duct**, called the nasolacrimal orifice, is located laterally on the flare of the nostril, and is slightly dorsal (Adams et al. 2013). In the horse, the opening is usually located at the mucocutaneous junction on the floor of the nostril, slightly medially. This anatomical difference has to be kept in mind in when lavage of the nasolacrimal duct is deemed necessary, for example in the case of fly strike. Anatomical differences in conformation between horses and donkeys also influence **ultrasound measurement** of the globe and adnexa, as better detailed forward in this chapter.

4 Clinical Evaluation

4.1 Predisposition

Donkeys and mules have a quieter attitude than horses, making ocular traumatic injuries less common. Because of their stoic response to pain, donkeys and mules do not show early clinical signs of ocular disease.

Corneal ulcers, infectious keratoconjunctivitis, uveitis, and eyelid trauma are reported to be the most common ocular conditions in donkeys and mules.

Being more stoic, with a muted pain response, donkeys and mules tend to present with ocular disease only in the late stages and initial examination may reveal a more severe or chronic problem.

Donkeys and mules may be more prone to organic foreign bodies in the eye, due to their tendency to feed with their head buried into hay or straw. Additionally, this is more common in winter due to their tendency to develop a thin coat and but a hairy face.

At the time of writing this chapter, neither predisposing nor genetic factors toward ophthalmic diseases in donkeys and mules have been discovered.

4.2 Patient History

Acquiring a thorough history is essential for both routine and urgent ophthalmological examination in horses, but unfortunately it does not always provide the same information in the case of donkeys and mules. Their stoic nature may impede the owner's ability to notice ophthalmic problems, further contributing to their tendency to present with ocular disease only later in the course. Regardless, the owner should be carefully questioned to try to obtain a reliable history. The history should include information about vaccination, deworming, travel, and exposure to other animals (both wild and captive) as well as any current or previous medical problems or administration of medication. Disorders involving the head and neck region should be recorded, including past or present respiratory signs and nasal discharge. In addition, the duration and severity of the presenting signs should be noted.

4.3 Clinical Examination of the Eye

Clinical examination of the eye can be more difficult in donkeys and mules than in horses. In fact, the lateral aspect of the frontal bone (the supraorbital ridge, forming the dorsal part of the orbit) is more laterally pronounced in donkeys and mules than in horses and consequently, the direct evaluation of the eye is uncomfortable for the examiner, since the procedure to lift the upper eyelid is more difficult and sometimes impossible to perform, especially in the donkey (Fig. 1). A head support (Fig. 2) may facilitate the veterinary clinician and make the patient more comfortable during ophthalmological examination.

When approaching the ophthalmological clinical examination of donkeys and mules, the veterinary surgeon should keep in mind the differences described above and otherwise follow the same approach to the examination of the horse's eye. The examination should be detailed and methodical, with a detailed recording of the history, to formulate an appropriate treatment strategy and monitor response. Sedation is less frequently needed than horses, but their strong palpebral muscles make periocular nerve blocking (Figs. 3, 4, 5, and 6) a frequent and essential part of examining a painful eye.

The supraorbital nerve represents the frontal nerve branch of the trigeminal nerve (sensory). To block it, it is necessary to locate the supraorbital foramen (that can be quite easily detected as a boney depression) and inject 1 ml of lidocaine over the foramen (Fig. 4).

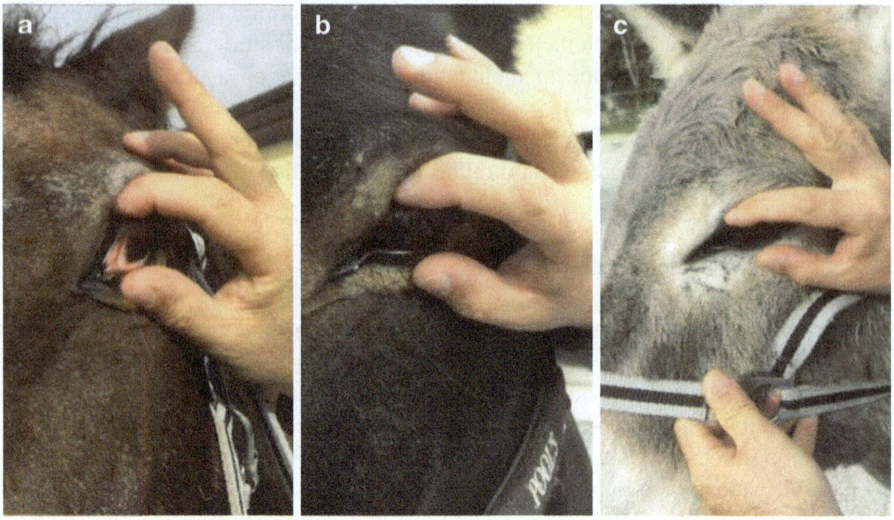

Fig. 1 Difference in conjunctival evaluation in horses (**a**), mule (**b**), and donkey (**c**). Due to the more pronounced supraorbital rim, opening of the upper eyelid in a donkey (**c**) is more difficult than in a mule and horse. To facilitate conjunctival inspection, a head support (as the one in Fig. 2) can be used to flip the head horizontally to evert the upper eyelid and uncover the underlying conjunctiva. The head support also helps the veterinary clinician during the ultrasound examination of the eye, since it may facilitate the positioning of ultrasound probe to visualize the retrobulbar region, which is not easy to display due to the obstacle caused by the frontal bone

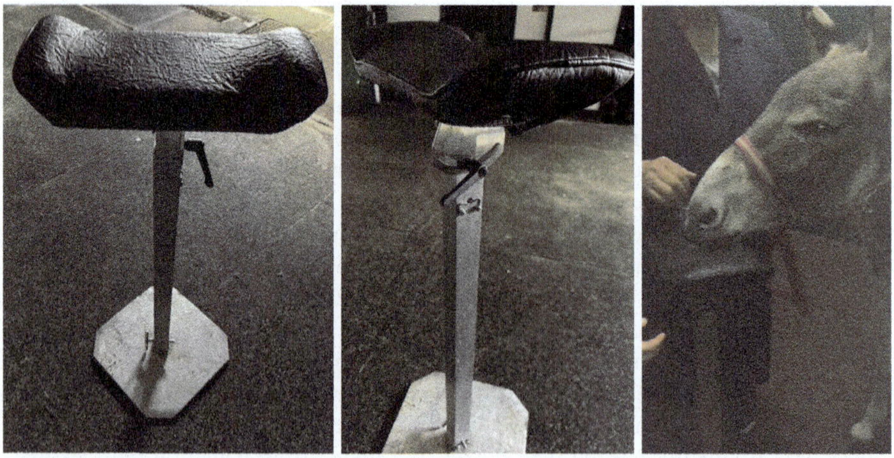

Fig. 2 Head support may help during examination and therapeutic procedures, useful for having donkeys and mules more comfortable and make some procedures (i.e. eyelid opening and ultrasound examination) straightforward for the examiner. The head support used in this case is in aluminum (to be lightweight but robust and durable). The stand can be elevated to fit any mule and donkey size, and padded support can be flipped horizontally to ensure comfort for the patient

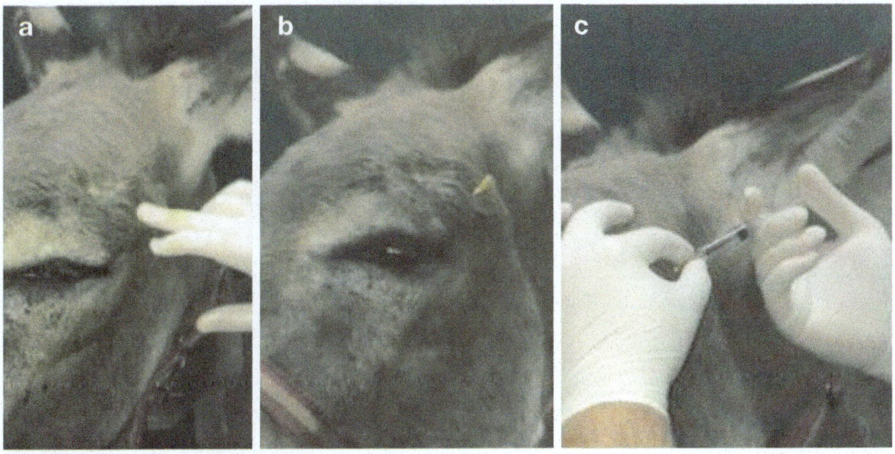

Fig. 3 Auricolopalpebral nerve blocking. The auricolopalpebral nerve, a branch of the facial nerve that has only motor function, can be easily identified as a small cord that runs halfway between the ear base and the lateral canthus of the eye (**a**). Once the nerve has been identified, a 25 g needle can be inserted subcutaneously (**b**) and a volume of 0.5 to 2 ml of lidocaine is injected perineurally to perform the nerve block (**c**). There are no significant differences in techniques between horses, donkeys, and mules.

The lacrimal nerve (sensory) can be blocked either by placing a 25 g needle in the lacrimal notch (Fig. 5), which is palpated on the dorsolateral rim of the orbit, or by using a linear block on the lateral third of the dorsal orbital rim. In each case, a total volume of 1–1.5 ml of lidocaine is enough to block the nerve.

The infratrochlear nerve is blocked in an opposite position from the lacrimal nerve, as it runs through the trochlear notch on the medial aspect of the dorsal orbital rim (Fig. 6). The infratrochlear nerve block is performed to anesthetize the medial canthus including the third eyelid. The same procedure can be used in horses and mules.

The zygomatic nerve (sensory) block is performed using a linear block along the ventrolateral orbital rim with 1 to 1.5 ml of lidocaine, enough to anesthetize the lower eyelid (Fig. 7). The same procedure can be applied in horses and mules without any notable differences.

Once periocular anesthesia has been performed as necessary, the examination technique is the same for all equids. After an exhaustive history acquisition, the animals should be observed from a distance in their environment to try to detect any possible changes in facial symmetry, abnormal supraorbital fossa movements, periocular swelling (unilateral or bilateral), or ocular discharge. Then, the examiner should observe the donkey or mule closely to evoke neuro-ophthalmic reflexes (menace response, direct and indirect pupillary light reflexes, swinging flashlight reflex, dazzle reflex, corneal and palpebral reflexes). After, the examination should

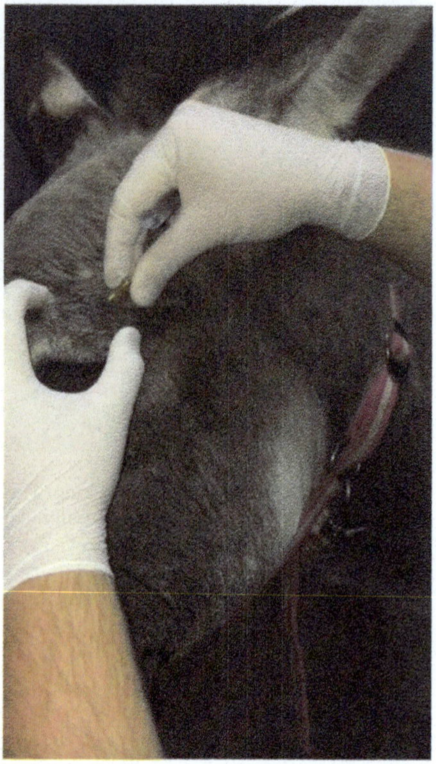

Fig. 4 Supraorbital nerve blocking

continue with inspection of the periorbital region, orbit, and adnexa, performing direct illumination and transillumination of the anterior segment to carefully evaluate and examine the eyelids, third lid, conjunctiva (including conjunctival fornix), cornea, irido-corneal angle, anterior chamber, and lens. Ocular surface staining or tear production tests can be performed next, when indicated, before examination of the vitreous body and retina. Ophthalmological textbooks claim that the basic set of dedicated tools should include a Finoff transilluminator with magnifying loupe to visualize any possible changes in the periocular area and anterior chamber. In the authors' experience, such tools can be replaced by a portable slit-lamp that emits a more intense and focused light beam (than smartphone flashlight, for example) which can be used to examine the anterior segment, using an indirect ophthalmoscope with related hand-held lens to complete the clinical examination. To record findings and to take pictures to share with colleagues or to be used to monitor the condition over time, a mobile phone or tablet should be accessible. If available, a portable tonometer completes the instruments needed for a comprehensive ocular examination (Fig. 8).

Fig. 5 Lacrimal nerve block

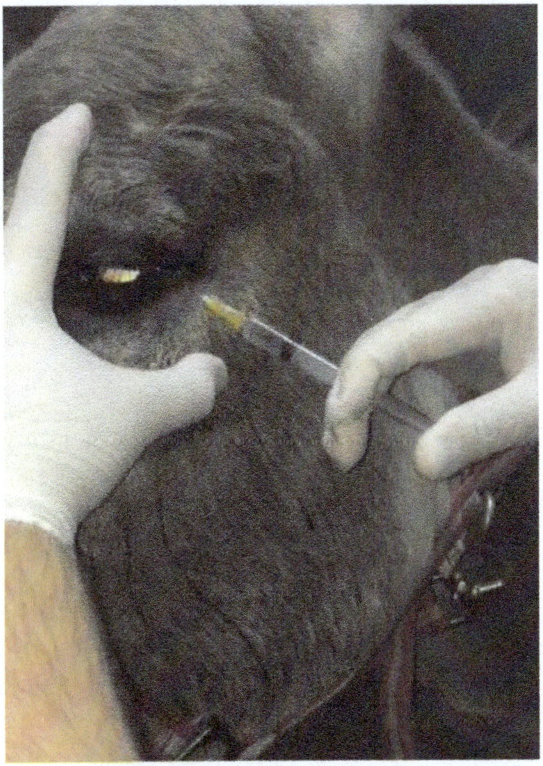

Fig. 6 Infratrochlear nerve blocking

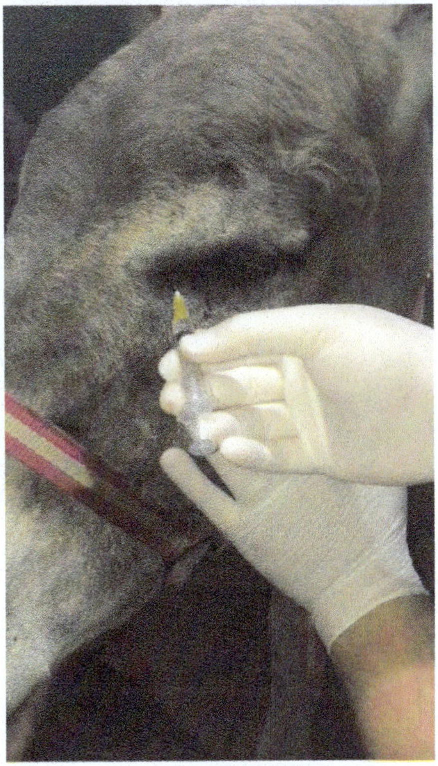

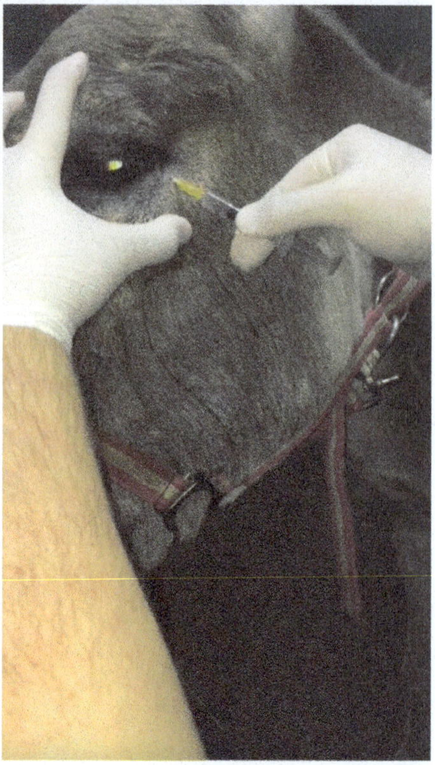

Fig. 7 Zygomatic nerve blocking

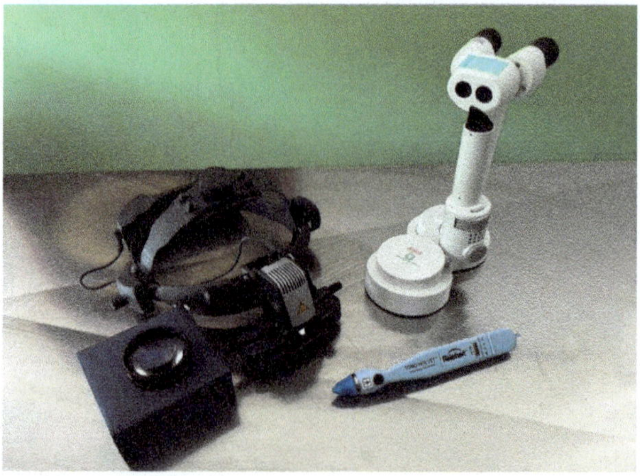

Fig. 8 Tools needed for donkey and mule ocular examinations. From left to right: indirect ophthalmoscope with hand-held lens, portable slit lamp and tonometer

The following flowchart describes all the steps needed to carefully and comprehensively examine the globe and adnexa in donkeys and mules.

Acquire a thorough **history** as complete as possible →

Start with **examination from a distance** in good light

↓

In case of blepharospasm or partially closed eyelid, avoid the use of force to open and perform nerve blocks to easily proceed with the examination →

Get close to the animal in a dark environment for examination of the **neurophthalmic reflexes**

↓

In the absence of corneal ulceration, evaluate intraocular pressure with a tonometer. Then, if appropriate, evaluate the lacrimal secretion using a Schirmer tear test →

Assess the anterior chamber using a light source (i.e. slit lamp or Finoff transilluminator) and magnification

↓

Assess the posterior chamber using an ophthalmoscope and then examine the vitreous body and fundus →

in the case of loss of transparency, perform ultrasonography to visualize the above described ocular structures

↓

if a corneal abrasion or ulceration is suspected, apply fluorescein staining or other staining (i.e. rose bengal or lissamine green for fungi and viruses, respectively) to highlight the affected area →

considering all the findings from previous steps, formulate a **presumptive clinical diagnosis** and run ancillary tests, if needed, to confirm

4.4 Main Problems & Clinical Findings

Corneal ulcers, infectious keratoconjunctivitis, uveitis, and eyelid trauma are reported to be the most common ocular conditions in donkeys and mules (Teferi and Adem 2019). According to the UK Donkey Sanctuary assessment, donkeys also suffer from other ocular diseases such as medial canthal wounds and lacerations due to habronemiasis and flystrike, conjunctivitis and corneal ulcers, scars and opacities generally attributed to trauma (Evans and Crane 2018).

Comprehensive examination with direct visualization of the intraocular structures is a key-point for accurate diagnosis, prognosis and therapy (Wafy et al. 2023).

One of the most frequently diagnosed ophthalmological diseases in working equids is **eyelid laceration**. Despite the anatomy of the orbital rim that tends to protect the globe of the donkey and mule from the environment in which they live and work, the prominent lateral positioning of their eyes makes them particularly vulnerable to periocular trauma. Eyelid lacerations are frequently encountered in working equids. Generally, when lacerations occur acutely, they are easily noted by the caretaker so veterinary services can be promptly required. Seldomly, if donkeys and mules are not closely watched, the veterinarian may only be alerted several days after the injury occurred and secondary infection/inflammation is already established, resulting in severe mucopurulent ocular discharge and possibly wound contraction. In many cases, the clinical severity of the appearance of the lesion does not correlate well with overall prognosis. Thanks to the superb blood supply, most eyelid lacerations heal, achieving a relatively well-functioning eyelid, even in the presence of significant secondary clinical signs. As a rule of thumb, when eyelid trauma or laceration is suspected, veterinary intervention should occur promptly to repair the defect and allow the healing process to begin as soon as possible. Whenever a donkey or mule presents with an eyelid laceration, a thorough ocular examination should be performed to identify and properly handle any possible associated ocular lesions (i.e. corneal ulcer or uveitis). Due to their feeding habits, donkeys and mules can be prone to **corneal foreign bodies**, mainly due to plants or seeds. The primary clinical manifestation of corneal foreign bodies includes epiphora, blepharospasm, and corneal opacity and vascularization, the extent of each may vary upon the size, depth, and duration of the foreign body. Sometimes foreign bodies are lodged in the conjunctival fornices or under the third eyelid, and, less commonly, superficial foreign bodies retained on the corneal surface can rapidly scratch the corneal epithelium as they are moved across the corneal surface by the action of the lid. This may result in local erosion and ulceration corresponding to the area of contact of the foreign body or more extensive erosion, usually manifested, at least in the very initial days, as conjunctivitis. The clinical appearance of **conjunctivitis** is characterized by the "red eye" which is one of the most common presenting complaints associated with many ocular problems in donkey and mule medicine. It is essential to determine if inflammation of the cornea and conjunctiva is a primary problem (i.e. due to bacterial or fungal contamination and proliferation) or a secondary problem (i.e. due to blunt trauma or eyelid laceration), as conjunctivitis is a non-specific response to many types of ocular disease.

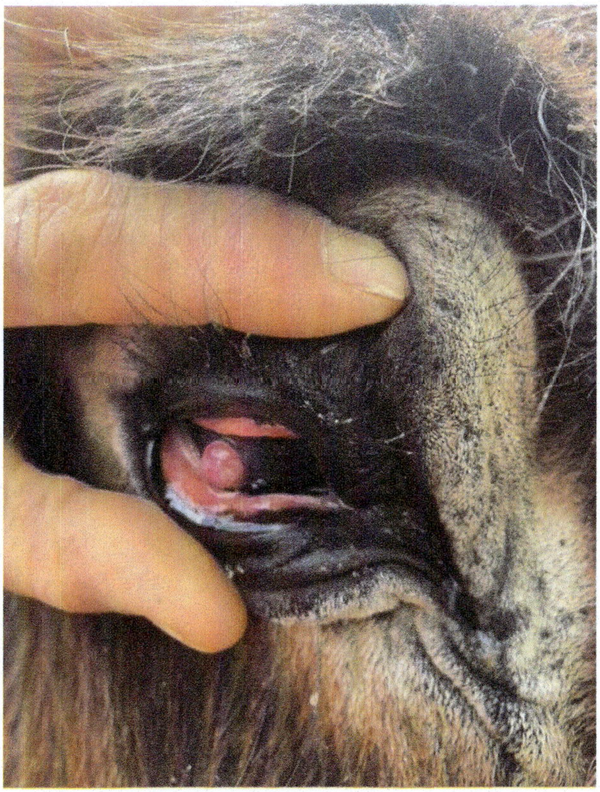

Fig. 9 Anterior uveitis secondary to corneal perforation. Note the conjunctival involvement which may lead to a misdiagnosis of conjunctivitis

Conjunctivitis may be associated with other severe diseases such as uveitis, glaucoma, or corneal disease which may cause the loss of vision if not promptly treated. Depending on the underlying cause, the clinical appearance of conjunctivitis ranges from very mild conjunctival changes to significant chemosis and hyperemia. A combination of both significant hyperemia and abundant ocular discharge can be indicative of secondary conjunctivitis due to uveal involvement (uveitis or glaucoma, Fig. 9).

Conjunctivitis is a clinical sign and not an etiologic diagnosis, therefore treatment should be directed at the primary or secondary cause for this ocular finding. Medical therapy for conjunctivitis requires topical ophthalmic ointments administered three to four times per day, possibly using a subpalpebral lavage system (SPL). **Corneal erosions and lacerations** are usually initially associated with various degrees of pain, blepharospasm, and epiphora. As previously recommended in this chapter, at least an auriculopalpebral nerve block is suggested to facilitate a safe examination in most cases and forced manual opening of the eyelids must be avoided to prevent iatrogenic rupture of the globe.

Anterior uveitis may initially present as a severe conjunctivitis, clinically characterized by the presence of ocular pain and very closed eyelids (blepharospasm) and discharge (epiphora), as well as other findings consistent with anterior uveitis such as chemosis, mitotic pupils, atrophy of the corpora nigra, aqueous flare, and the presence of proteinaceous material in the anterior chamber called keratic precipitates. In some cases, anterior uveitis may be accompanied by vitreal opacity in association with cataract. Vitreal opacity can be easily recognized as small mobile flakes that floating within the vitreous when directly visualized or during ultrasound examination. Hypopyon describes the presence of pus in the anterior chamber and can be found with both uveitis and corneal lacerations, causing impairment of vision if the whitish material and fibrin threads cause extended opacification of the anterior chamber and adhere to corneal endothelium. As a consequence of uveitis or corneal laceration, cataract occurs with the presence of varying degrees of lens opacity, depending on the density and severity of the disorder. When the lens luxates, for example after a trauma or because of uveitis, it can be seen in the anterior chamber or lying on the vitreous body, which tends to liquefy due to the pressure of the lens itself.

4.5 Diagnostic Tests

- One of the most frequently used diagnostic tests in veterinary ophthalmology are **vital dyes** (Brooks et al. 2000; Srinivas and Rao 2023). The term "vital stain" refers to the ability of a dye to penetrate the living cells of the cornea without inducing any immediate evident degenerative staining. Fluorescein sodium is probably the most used vital dyes in clinical practice since the nineteenth century. Fluorescein is not able to penetrate the intact layer of the corneal epithelium (which is lipidic in nature), and therefore, it does not stain normal, non-ulcerated cornea. Fluorescein stains the cornea whenever there is disruption of the cell-to-cell junctions, as in the case of abrasion and ulceration. Rose Bengal and lyssamine green are stains that can be used alone, or after fluorescein staining, to highlight the presence of fungi or viruses within the corneal stroma.
- **Tear testing** is only seldom required for donkeys and mules. When indicated, it can be carried out using disposable and sterile tear test strips (Ibrahim 2018). The hands of the examiner should be clean and dry on manipulation of the strip from its package to avoid any contamination or bias in the tear test. The strip is inserted in the middle of the lower conjunctival sac with (Schirmer tear test II) or without (Schirmer tear test I) using any topical anesthetics in the eye. It is left in place for 1 minute and then removed and read immediately. Currently, the normal average values for healthy eyes are 22 ± 1.35 mm/min for STT I and 18 ± 0.70 mm/min for STT II. Female donkeys tend to respond differently from males regarding tear testing. In fact, jennies usually have greater values (+2 mm/min) than donkey stallions (Selk Ghaffari et al. 2017).

- As for equines, the **ultrasonographic evaluation** of the donkey and mule eye is a manageable and easy-to-perform procedure which can provide information not always obtainable with direct ocular examination, especially when ocular structures are not clear enough to allow a direct visualization of anterior and posterior chambers. Ultrasound allows for the imaging of intraocular and retrobulbar structures and the diagnosis of some important disorders involving these areas and it is indicated when it is not possible to directly visualize posterior structures of the globe in cases of corneal edema or ulceration, cataract or ocular masses. Ultrasound can be used to investigate enophthalmos, buphthalmia or exophthalmia in cases of ocular protrusion, and in the suspicion of disparity in globe size. It can be performed with a normal ultrasound device (i.e. the same used for reproduction purposes). When performing ultrasound examination of the retrobulbar space, it should be kept in mind that the anatomical conformation of the supraorbital arch may impede a complete visualization of the retrobulbar structures, which is different from horses. The most common diseases that can be detected or confirmed with ultrasound are corneal diseases, cataract, lens luxation, intraocular cysts or masses, glaucoma and retinal detachment. Ocular globe structures are usually displayed at a depth of 4–6 cm, depending on the animal size. The transducer is usually placed on the closed upper eyelid, positioning the lateral canthus to the left of the screen and the medial canthus to the right. A long-axis scan in a horizontal plane is performed using the probe approximately parallel to the eyelid margin (Fig. 1). Minimal pressure needs to be applied to the eyelid to avoid deforming the globe. Minimal physical or mechanical restraint should be used, and sedation is only seldom required. Globe measurements should be taken in a horizontal plane corresponding to the maximum diameter of the eye, at the point where the anterior chamber, the anterior and posterior lens capsule, and the layers of the fundus are all included and aligned in the image. Average values for ocular measurements can be different among different breeds of donkeys. In 2014, Laus and collaborators (Laus et al. 2014) performed ultrasonography in healthy Italian adult donkeys of different breeds and sizes (from 71 to 306 kg). They determined that ocular measurements vary among different classes of body weights. They evaluated structure classes or body weight as follows:

Weight category (kg)	GAL	ACD	VD	LDi	LDe
Less than 100 kg	31.09 ± 0.45	2.01 ± 0.35	18.72 ± 0.70	16.33 ± 1.03	10.35 ± 0.56
Between 100 and 200 kg	34.17 ± 1.87	3.07 ± 0.53	20.10 ± 1.52	18.26 ± 1.46	11.16 ± 0.72
More than 200 kg	36.35 ± 1.20	3.30 ± 0.23	22.00 ± 1.40	17.01 ± 2.39	10.91 ± 0.25

GAL Globe axial length; *ACD* anterior chamber depth; *VD* vitreous body depth; *LDi* lens diameter; *LDe* lens depth. Data are expressed in mm as mean value ± standard deviation

More recently, Abdelbaset-Ismail and collaborators performed a similar study using Egyptian donkeys (Abdelbaset-Ismail et al. 2022). The ultrasound biometric values for Egyptian donkeys' eyes are reported in the following table:

GAL	GD	ACD	VD	LDe
33.7 ± 1.7	39.8 ± 2.1	3.2 ± 0.7	19.3 ± 1.6	10.8 ± 0.7

GAL Globe axial length; *GD* globe diameter; *ACD* anterior chamber depth; *VD* vitreous body depth; *LDe* lens depth. Data are expressed in mm as mean value ± standard deviation

The mean nasolacrimal duct length estimated by Abdelbaset-Ismail and collaborators was 193.0 ± 9.8 mm by radiography and 206.0 ± 20.4 mm by gross assessment. One nasolacrimal orifice was noticed on each side, with a diameter of 3.0 ± 0.1 mm and located 12.1 ± 2.1 mm from the dorsal commissure of the nostril.

4.6 Treatment

- Due to anatomical and physiological differences with the horse, careful clinical inspection of the eye in donkeys and mules often requires sedation, nerve blocks and local anesthesia that are not necessarily required in horses in similar conditions. Prolonged treatment protocols involving ocular topical medications is also challenging in donkeys and mules. The difficulty associated with opening the eye, the small size, the inexperience with being handled (in some cases), all make administration of ocular topical treatment challenging in donkeys and mules (Mendoza et al. 2018). For this reason, the application of a subpalpebral lavage system (SPL) in donkeys and mules is more often indicated than in horses. Furthermore, since sedation, nerve blocks (injecting 0.5 to 2 ml of lidocaine perineurally) to block the strong palpebral muscles) and topical anesthesia are usually necessary for examination, SPL can be directly applied after diagnosis, if indicated. The technique for the application of the SPL in donkeys and mules is the same as in horses; the zygomatic process is a bit larger in donkeys and mules and should be considered when this part of the bone is palpated to located the supraorbital foramen for anesthetic injection.
- In cases of **corneal foreign body**, the goal of the treatment is to remove the foreign body as soon and as atraumatically as possible: after proper corneal desensitization (oxybuprocaine eye drops or similar) to make the animal amenable without any discomfort, the foreign body can be removed either using ophthalmic forceps or by irrigation of the corneal surface with sterile saline solution, especially in the case of plant seeds. If the removal of foreign body results in corneal abrasion or erosion, the proper management of corneal ulcers should be implemented (Evans and Crane 2018).
- In cases of **corneal ulcer**, after proper identification and addressing of the primary underlying disease, the therapy should treat or prevent corneal infection, using antiseptics or antibiotics depending on the extent of infection, to be administered three to four times per day. In cases of suspected Gram-positive bacteria (i.e. *Staphylococcus* spp. or *Streptococcus* spp.) it is possible to use moxifloxacin, chloramphenicol, neomycin, and bacitracin, while in cases of

Gram-negative bacteria (i.e. *Enterobacteriaceae*) ciprofloxacin, ofloxacin, aminoglycosides, chloramphenicol, neomycin, and polymyxin B can be useful. The therapy should also prevent the liquefaction of corneal stroma by administering topical antiprotease agents containing N-acetylcysteine or autologous serum three to four times per day. In many cases it is necessary prevent or treat secondary uveitis (please refer to uveitis section for proper medical management) and provide structural support to cornea, applying temporary tarsorrhaphy (which means join, by surgical suture, part or all the upper and lower eyelids so as to partially or completely close the eye and then help the cornea heal or to protect the cornea during a short period of exposure or disease) or conjunctival flaps (Briefly, after complete removal of the corneal epithelium and debridement of necrotic tissue from the cornea, a mobile, pedunculated or not, thin conjunctival flap that contains minimal Tenon capsule and without any tears, is sutured just above the affected corneal tissue preventing tears, proteolytic enzymes, and proinflammatory mediators from reaching the corneal ulcer and causing stromal lysis). In all cases in mandatory to prevent self-trauma using dedicated facial masks (i.e. those for protection from flies) and provide pain- and inflammation relief administering topical mydriatic and cycloplegic agents such as tropicamide 1% or atropine or, in case of systemic pain flunixin meglumine (Teferi and Adem 2019). During corneal ulceration, avoid administering a topical ointment if the ulcer extends beyond half of the corneal thickness.

- **Corneal lacerations** require surgical management. Superficial lacerations can be managed with debridement of any detached epithelium, with the exposed stroma subsequently treated as a corneal ulcer. In uncomplicated cases (those not exceeding the first third of stromal thickness in depth), quick re-epithelialization of the ulcer is expected within 7 days, irrespective of the extent. In cases where the laceration extends beyond the first third of the stromal thickness, and for perforating lacerations, in addition to treating the corneal ulceration, it is recommended to provide structural support to the cornea by applying temporary tarsorrhaphy or conjunctival flaps, in addition to providing pain and inflammation relief (Brooks et al. 2017).
- As with horses, donkeys and mules with any degree of corneal disease (foreign bodies, ulcerations, lacerations, etc.) can also develop **anterior uveitis**. The goal of the treatment is to provide relief from pain and prevent the dangerous cascade of inflammation by applying topical cycloplegics (e.g. 1% atropine) to cause pupillary dilatation and stabilization of the blood–aqueous barrier to reduce aqueous flare. Topical cycloplegics also cause relaxation of the ciliary muscles and eliminate ciliary spasm, which is a major cause of ocular pain. Obtaining a proper pupillary dilation is essential in cases of uveitis, to protect the visual axis from occlusion and to minimize development of synechiae, which can occur rapidly in equids, and have the potential to compromise vision in the healed eye by preventing dilation of the pupil. In the presence of synechiae, topical atropine (1%) and phenylephrine (2.5%) in combination can dilate some pupils and breakdown the synechiae already present. Topical atropine is usually administered once or twice daily and animals receiving more than four times per day should be watched closely for signs of reduced gut motility. In addition to

topical cycloplegics, systemic anti-inflammatory drugs (mainly NSAIDs) should be administered to reduce iridocyclitis and control inflammation (Brooks et al. 2017).

4.7 Prevention

- Preventing eye diseases in donkeys and mules involves proactive care and regular monitoring. When a veterinarian performs a health check, she/he should remember that a thorough physical exam includes the eyes. This is the only way to detect ocular issues that need to be addressed and provide recommendations for preserving eye health.
- Especially with working equids, it is essential to use protective eyewear during activities where they risk eye injury, such as hiking through thick vegetation.

4.8 New Insights and Future Directions

- The work done by different research groups on donkey and mule ophthalmology revealed a general lack of knowledge of the potential consequences of eye disease or a low perception of its importance among donkey owners and keepers.
- Increased knowledge is required to understand the social and economic reasons for owner's recognition and perceptions of ocular health and disease, and the motivating and deterring factors behind ophthalmology research. This would accompany the creation of educational strategies for improving ocular health and general welfare in donkeys and mules.
- Further investigation is needed to better understand the etiology and risk factors for the major sight-threatening pathologies among working equine populations, further elucidating the impacts and epidemiology of ocular pathology in the equids involved in industries worldwide.

5 Conclusion

- Donkey and mule ophthalmology is a relatively new branch of veterinary medicine, since owners are frequently unaware of the presence of ocular abnormalities in their equids and consequently treatment is seldom sought.
- Regular ocular examinations are recommended as routine management in working equids, to achieve early diagnoses in these animals, whose stoicism and ability to thrive with chronic and dangerous pathologies make ophthalmic disease generally a late diagnosis.

Annex

- **Doses Chart:** the following chart Includes dosage recommendations for common medications used in the treatment of ocular disorders

Family of medical agents	Active principles	Route of administration	Dose	Frequency	References
Topical diagnostic and therapeutic desensitizing	Oxybuprocaine 0.4%	Topical eyedrops	2–4 eyedrops	Use only for diagnostic or therapeutic procedures	Little et al. (2016)
	Lidocaine	Perineural injection	0.5-2 ml	Use only for diagnostic or therapeutic procedures	Hoerdemann et al. (2017)
Topical diagnostic and therapeutic mydriatic cycloplegic	Atropine 1%	Topical eyedrops	1–2 eyedrops	Depends on degree of mydriasis achieved between doses	Lizarraga et al. (2014)
	Tropicamide 0.5% or 1%	Topical eyedrops	1–2 eyedrops	Use only for diagnostic or therapeutic procedures	Budde and McCluskey (2023)
Antibiotics	Gram-positive: Moxifloxacin, chloramphenicol, neomycin, bacitracin	Topical eyedrops or ointments	0.1–0.2 ml or ¼ in strip	q2–4h	Brooks et al. (2017)
	Gram-negative: Ciprofloxacin, ofloxacin, aminoglycosides, neomycin, polymyxin B	Topical eyedrops or ointments	0.1–0.2 ml or ¼ in strip	q2–4h	Brooks et al. (2017)
Antiprotease	N-acetylcysteine 5–10%	Topical eyedrops	2–4 eyedrops	q2–4h	Brooks et al. (2017)
Topical antibiotic and corticosteroid	Neomycin/polymyxin/ dexamethasone 0.1%	Topical ointment	¼ in strip	q6–8h	Brooks et al. (2017)
Systemic NSAID	Flunixin meglumine	Systemic (paste or solution)	1.1 mg/kg PO or IV	q12h	Brooks et al. (2017)
Systemic corticosteroid	Prednisone	Systemic (tablet)	0.5–1.0 mg/kg PO	q12–24 h, and then taper	Brooks et al. (2017)

Review Questions

1. Which are the main anatomical differences between horses and donkey/mules to keep in mind when performing an ophthalmological examination?
   A: orbit, nasolacrimal duct aperture, globe measurements by ultrasound examination
2. List the main pathological conditions affecting the donkey and mule's eye?
   A: Corneal ulcer, infectious keratoconjunctivitis, uveitis, eyelid trauma
3. Are there any tips or tricks to put in place and remember when performing an ophthalmological examination?
   A: Difference in conjunctival evaluation technique, use of head support, nerve blocking if clinically necessary
4. When an eyelid laceration is diagnosed, which are the main steps of diagnostic and therapeutic approach?
   A: Comprehensive ophthalmological evaluation, including cornea and posterior segment to highlight any possible associated laceration; promptly medical or surgical treatment, depending on the severity and extent of lesions; prevention of corneal ulcer (if not already present) and uveitis
5. Why corneal diseases (such as foreign bodies or ulcers) can speedily proceed into uveitis?
   A: due to the innervation of cornea and proximity of ocular structure. Prompt treatment is mandatory to prevent such instances.
6. What is the ocular condition to be feared when dealing with donkey and mule ophthalmology and why?
   A: uveitis, because (i) it can be challenging to diagnose, (ii) it can be mistaken with other pathological conditions such as corneal disease, and (iii) can be vision-threatening due to the development of cataract or retinal involvement

References

Abdelbaset-Ismail A, Aref M, Ezzeldein S, Eisa E, Gugjoo MB, Abdelaal A, Emam H, Al Syaad K, Ahmed AE, Alshati A, El Raouf MA (2022) Ultrasound, dacryocystorhinography and morphological examination of Normal eye and lacrimal apparatus of the donkey (Equus Asinus). Animals 12(2). https://doi.org/10.3390/ani12020132

Adams MF, Castro JR, Morandi F, Reese RE, Reed RB (2013) The nasolacrimal duct of the mule: anatomy and clinical considerations. Equine Vet Educ 25(12):636–642. https://doi.org/10.1111/EVE.12077

Brooks DE, Andrew SE, Denis HM, Strubbe DT, Biros DJ, Cutler TJ, Samuelson DA, Gelatt KN (2000) Rose bengal positive epithelial microerosions as a manifestation of equine keratomycosis. Vet Ophthalmol 3(2–3):83–86. https://doi.org/10.1046/J.1463-5224.2000.00128.X

Brooks DE, Matthews A, Clode AB (2017) Diseases of the cornea. In: Gilger BC (ed) Equine ophthalmology, 3rd edn. John Wiley & Sons, Ltd, pp 252–368

Budde JA, McCluskey DM (2023) Plumb's veterinary drug handbook, 10th edn. 1568. https://www.wiley.com/en-sg/Plumb's+Veterinary+Drug+Handbook%2C+10th+Edition-p-9781394172207

Evans L, Crane M (2018) The clinical companion of the donkey (L Evans, M Crane, eds). The Donkey Sanctuary

Hoerdemann M, Smith RL, Hosgood G (2017) Duration of action of mepivacaine and lidocaine in equine palmar digital perineural blocks in an experimental lameness model. Vet Surg 46(7): 986–993. https://doi.org/10.1111/VSU.12689

Hurn SD, Turner AG (2006) Ophthalmic examination findings of thoroughbred racehorses in Australia. Vet Ophthalmol 9(2):95–100. https://doi.org/10.1111/J.1463-5224.2006.00444.X

Ibrahim A (2018) Reference values of Schirmer tear test I and II in clinically healthy adult donkeys (Equus asinus): effect of sex. J Equine Vet Sci 67:23–26. https://doi.org/10.1016/J.JEVS.2018.02.019

Laus F, Paggi E, Marchegiani A, Cerquetella M, Spaziante D, Faillace V, Tesei B (2014) Ultrasonographic biometry of the eyes of healthy adult donkeys. Vet Rec 174(13):326. https://doi.org/10.1136/VR.101436

Little E, Yvorchuk-St. Jean K, Little W, Sithole F, St. Jean G (2016) Degree of corneal anesthesia after topical application of 0.4% oxybuprocaine ophthalmic solution in normal equids. Can J Vet Res 80(4):329. /pmc/articles/PMC5052886/

Lizarraga I, Castillo-alcala F, Stucke D, Hall S, Morrone B, Lebelt D, Loughran CM, Raisis AL, Secombe CJ, Lester GD, Parker RA, Hollis RA, Mariella J, Castagnetti C, Prosperi A, Scagliarini A, Adriani G, Peli A et al (2014) Topical ophthalmic atropine in horses—an experimental study. Equine Vet J 46(S47):23–23. https://doi.org/10.1111/EVJ.12323_53

Maśko M, Wierzbicka M, Zdrojkowski Ł, Jasiński T, Sikorska U, Pawliński B, Domino M (2022) Comparison of donkey, pony, and horse dorsal profiles and head shapes using geometric morphometrics. Animals 12(7). https://doi.org/10.3390/ANI12070931

Mendoza FJ, Toribio RE, Perez-Ecija A (2018) Donkey internal medicine—part II: cardiovascular, respiratory, neurologic, urinary, ophthalmic, dermatology, and musculoskeletal disorders. J Equine Vet Sci. W.B. Saunders 65:86–97. https://doi.org/10.1016/j.jevs.2018.02.025

Merkies K, Paraschou G, McGreevy PD (2020) Morphometric characteristics of the skull in horses and donkeys—a pilot study. Animals 10(6):1–11. https://doi.org/10.3390/ani10061002

Rushton J, Tichy A, Brem G, Druml T, Nell B (2013) Ophthalmological findings in a closed herd of Lipizzaners. Equine Vet J 45(2):209–213. https://doi.org/10.1111/J.2042-3306.2012.00630.X

Selk Ghaffari M, Sabzevari A, Ghamsari S, Shad H (2017) Determination of reference values for intraocular pressure and Schirmer tear test results in clinically normal domestic donkeys (Equus asinus). Vet Rec 181(21):565. https://doi.org/10.1136/VR.104226

Srinivas SP, Rao SK (2023) Ocular surface staining: current concepts and techniques. Indian J Ophthalmol 71(4):1080. https://doi.org/10.4103/IJO.IJO_2137_22

Teferi K, Adem A (2019) Common ophthalmic problems in equines. Glob Vet 21(5):246–254. https://doi.org/10.5829/idosi.gv.2019.246.254

Wada S, Hobo S, Niwa H (2010) Ulcerative keratitis in thoroughbred racehorses in Japan from 1997 to 2008. Vet Ophthalmol 13(2):99–105. https://doi.org/10.1111/J.1463-5224.2010.00767.X

Wafy MN, Hassan EA, Ahmed KA, Aboelmaaty AM, Abu-Seida AM (2023) Ultrasonographic and histopathologic features associated with common ocular diseases in donkeys (Equus asinus). Vet Res Commun 47(3):1479–1491. https://doi.org/10.1007/s11259-023-10102-4

Respiratory Disease in Donkeys and Mules

Fulvio Laus, Marilena Bazzano, and Andrea Marchegiani

Abstract

In this chapter the most relevant respiratory conditions that may affect donkeys and mules have been addressed. Anatomical and physiological differences in comparison with the horse have been considered from a clinical standpoint. Clinical evaluation focused on donkey and mule features may assist clinicians dealing with these animals to promptly identify the respiratory condition. Descriptions of the most frequent respiratory diseases are provided, together with principles of therapy and prevention.

Keywords

Respiratory diseases · Infectious disease · Upper airway · Lower airway · Lung

1 Introduction

- Due to the stoic nature of donkeys and mules, most respiratory diseases may go undiagnosed and untreated at early stages, therefore most respiratory conditions may be advanced at the time of clinical investigation. Donkeys have a less sensitive cough reflex and are not always required to perform exercise that can exacerbate respiratory signs. For these reasons, donkey and mule owners should be advised to seek veterinary service in the presence of even minor respiratory signs.
- Clinicians dealing with donkeys and mules should always keep in mind the specific anatomical and physiological differences of the respiratory tract

F. Laus (✉) · M. Bazzano · A. Marchegiani
School of Biosciences and Veterinary Medicine, University of Camerino, Camerino, Italy
e-mail: fulvio.laus@unicam.it; marilena.bazzano@unicam.it; andrea.marchegiani@unicam.it

A. K. McLean et al. (eds.), *Donkey and Mule Medicine: Clinical Medicine and Reproduction*, https://doi.org/10.1007/978-3-032-18540-2_10

compared to horses, as well as the use of these animals (work, pet-assisted therapy, companion, etc), to promptly diagnose and treat the respiratory condition.
- Infectious respiratory diseases of donkeys and mules can be transmitted to horses and *vice versa*. Cohabitation of equines can facilitate the spread of diseases. The gregarious habits of donkeys make them more prone to the transmission of diseases from one subject to another. Additionally, routine vaccination may not occur in donkeys as in horses, which can further facilitate the spread of disease.
- Acute respiratory diseases can lead to hyperlipidemia and a rapid worsening of the clinical condition. Some chronic forms result in fibrosis of the lung with a significant decrease in respiratory function.

2 Glossary

BAL: BronchoAlveolar Lavage
TW: Tracheal Wash
Chondroids: "stones" of inspissated pus found in guttural pouches
NSAIDS: Non-Steroidal Anti-Inflammatory Drugs

- Additional resources:
  - The Donkey Sanctuary https://www.thedonkeysanctuary.org.uk
  - The Peculiarities of Donkey Respiratory Disease. Thiemann A.K. and Bell N.J. (Authors). https://www.ivis.org

3 Anatomy and Physiology of Respiratory System

The respiratory system delivers oxygen to tissues and removes carbon dioxide from the blood. Through the upper (nares, nasal cavity, pharynx, larynx, and trachea) and lower (bronchi, and bronchioles) airways, air is delivered to the lungs where the alveoli lined by an extensive pulmonary capillary network allow oxygen and carbon dioxide exchanges (McGorum et al. 2007).

Many features of the respiratory system of donkeys and mules are similar to that of horses (Thiemann and Bell 2001). However, there are specific anatomical and physiological differences which influence the presentation, incidence, and treatment of respiratory disease in donkeys and mules (Thiemann and Bell 2001; Burnham 2002).

Anatomical differences should be considered during clinical examinations. Starting from nares inspection, the location of the nasolacrimal orifice in the mule can be easily found within the internal cutaneous tissue of the lateral wall of the external nares (Adams et al. 2013). In the donkey, the nasolacrimal orifice is positioned dorsally in the false nostril approximately 1 cm from the mucocutaneous junction, and the nasal cavity is narrower compared to the horse, requiring particular attention when performing endoscopy or passing a nasogastric tube, to avoid bleeding (Thiemann and Bell 2001; Mendoza et al. 2018). Another feature of the donkey upper airway anatomy that should be considered during endoscopy include the guttural pouch openings, which are horizontally oriented, affecting the technique required to gain access to the pouches. (Mendoza et al. 2018). The donkey larynx differs from horses due to the presence of the pharyngeal recess between the openings of the guttural pouches and the angle of the airway and aditus laryngis (Fig. 5). The donkey's pharyngeal recess or diverticulum, located caudo-medially to the guttural pouches (Fig. 6), is in a direct line to entrap the nasogastric tube, preventing passage through to the esophagus. Another unique feature of the donkey is the angle of the opening of the airway from the pharynx to the larynx, which tilts on average 5.5° caudally, possibly causing difficulties in visualizing the trachea with an endoscope (Burnham 2002). Finally, donkeys have a narrow pharynx that can predispose to collapse, and epiglottis and aryepiglottic folds that are usually smaller than horses (Mendoza et al. 2018).

The respiratory system anatomy of the mule is similar to horses with the exception of a muscular sphincter that surrounds the entrance into the dorsal pharyngeal recess (McGorum et al. 2007).

Considering the mechanical ventilation parameters of donkeys, they differ from horses being more similar to those of ruminants with higher resistance, lower compliance and higher maximum pleural pressure variation. The "respiratory strategy" adopted by certain donkeys is different from that of horses: there are two to three phases of inspiration with a brief pause between each phase. Expiration is the same as for horses (Delvaux et al. 2001).

4 Clinical Evaluation

4.1 Predisposition and Patient History

Signalment and history provide essential information to reach a diagnosis.

Age, management (environment, diet, type of activity), vaccination and anthelminthic history, previous and current diseases and medications, and transportation history need to be considered when dealing with respiratory diseases (McGorum et al. 2007).

In geriatric donkeys, the onset of acute dyspnea is frequently associated with idiopathic pulmonary fibrosis or tracheal collapse. In donkeys and mules, mild respiratory signs may pass unnoticed until advanced stages develop, resulting in respiratory emergencies.

Vaccination history is also important when dealing with respiratory disease. Many donkeys are not vaccinated against equine influenza and equine herpesvirus, but they are susceptible, and disease outbreaks can occur in unvaccinated populations (Archer 2013).

4.2 Clinical Signs

Donkeys rarely undergo intensive physical exercise, and respiratory function testing is not routinely performed in this species. Furthermore, these animals are stoic, and clinical signs are often noticed in the late stages, so that acute respiratory diseases are rarely observed in donkeys. The majority of respiratory disturbances in donkeys involve more chronic and severe clinical conditions than horses, and generally carry a poorer prognosis as a consequence (Mendoza et al. 2018).

Donkeys presenting with respiratory signs should always have a general physical examination first, as multiple pathologic conditions can coexist (Thiemann and Bell 2001).

The clinical examination of the respiratory system in the donkey is similar to horses. However, clinicians should always consider some important differences such as the anatomical peculiarities mentioned in the previous section, and the relatively stoic nature and insensitive cough reflex typical of donkeys.

Furthermore, the smaller thoracic compartment in the donkey compared to the horse determines more audible lung sounds through the thoracic wall during auscultation. However, in cases of obese donkeys, respiratory sounds are not easily audible, and the use of a re-breathing bag can help accentuate subtle sounds. (Thiemann and Bell 2001).

The following charts provide a systematic guide for clinical examination of upper and lower respiratory tracts in donkeys and mules.

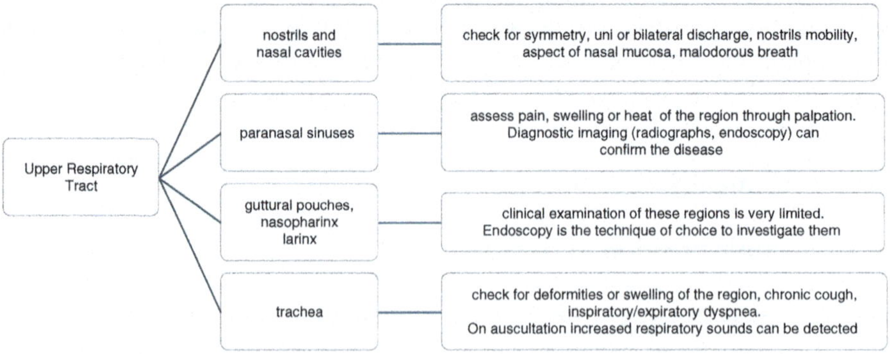

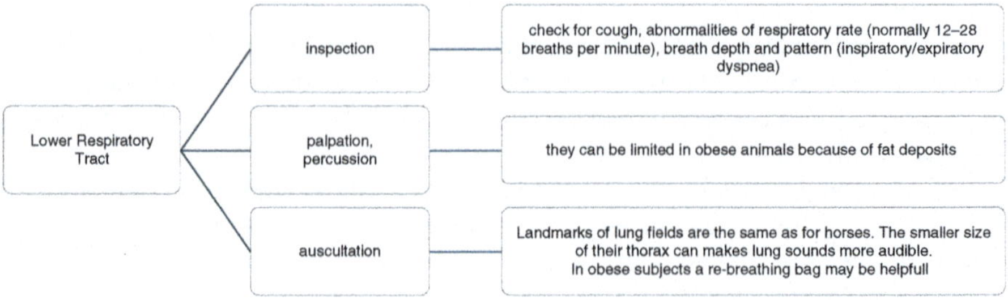

4.3 Respiratory Problems

4.3.1 Upper Airways

Rinoestrosis is a parasitic disease caused by larvae of *Rhinoestrus* spp., and has been detected in donkeys (Laus et al. 2022). L2 and L3 larvae located in the nasal cavity, pharynx, and larynx of donkeys and can cause mucoid nasal discharge and sneezing, and rarely pulmonary complications or neurological disorders because of intracranial migration of larvae through the ethmoid bone (Laus et al. 2022). The disease has been detected in several countries but it is thought that in some parts of the world, where equines are not used for human consumption, it is underestimated as it is a frequent accidental finding during slaughter (Laus et al. 2022; Otranto et al. 2004). Diagnosis can be reached by endoscopy of the upper airway and oral treatment with ivermectin or moxidectin is effective.

Mycotic rhinitis or sinusitis due to *Aspergillus* spp. or *Cryptococcus* spp., can cause mono or bilateral mucopurulent and/or hemorrhagic nasal discharge. Endoscopically, yellow/gray plaques and/or haemorrhagic erosions can be seen (Fig. 1). Radiology is necessary to diagnose a sinus infection. Samples can be collected for fungal culture by gently scraping the lesions with a swab. Daily irrigation with antimycotic drugs for several days (e.g. enilconazole, miconazole, itraconazole, and others), based on susceptibility testing is usually effective.

Strangles is a disease caused by the gram-positive, beta hemolytic bacteria *Steptococcus equi* subsp. *equi (S. equi)*. The infection is restricted to equids (horse, donkey, and mule) and has a worldwide distribution (Barrandeguy and Carossino 2018). It can affect animals of all ages but tends to be more severe in young animals. Morbidity rates can reach 100% (Barrandeguy and Carossino 2018). After infection and replication in the lymphoid tissue of the pharynx, *S. equi* reaches submandibular and retropharyngeal lymph nodes where it causes swelling and abscessation, easily detected on clinical examination. Fever and nasal discharge are other clinical signs and complication can include systemic dissemination (with

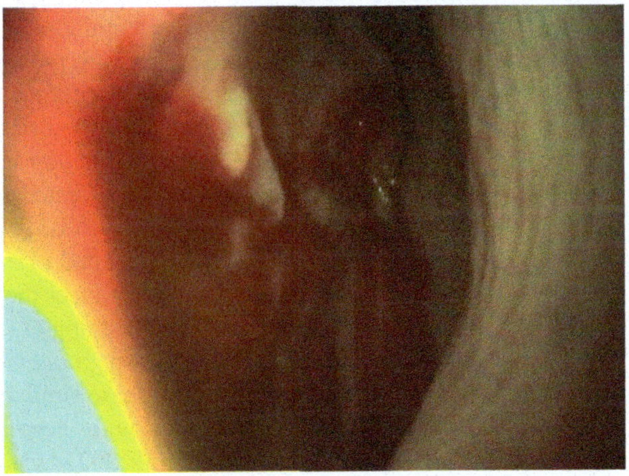

Fig. 1 Endoscopic imagine of a mycotic plaque (Aspergillus fumigatus) in the rostral part of the left nasal cavity in a donkey

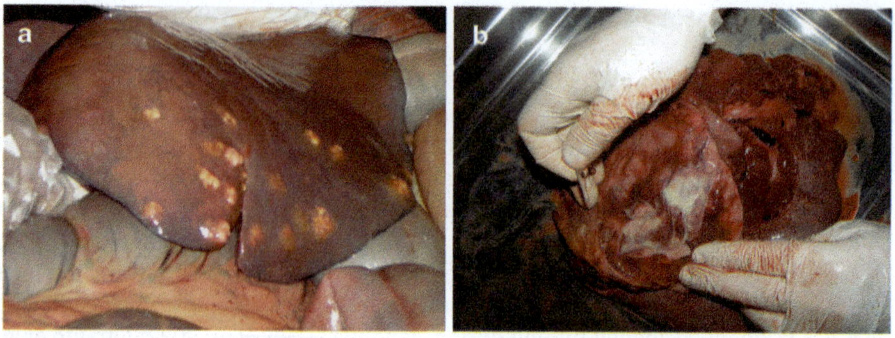

Fig. 2 Bastard strangles: abscessation in the liver (**a**) and Kidney (**b**) caused by systemic dissemination of *Streptococcu equi equi*

Fig. 3 Chondroids in the right guttural pouch

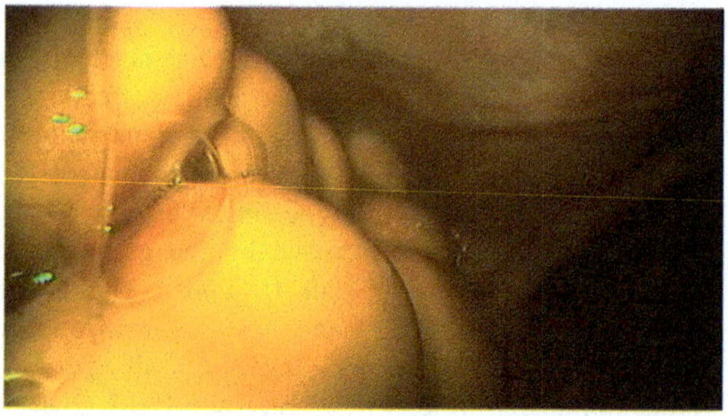

abscess formation in other lymph nodes and organs) (Fig. 2), hemorrhagic purpura (mediated by immunocomplex formation) and, in some cases, severe respiratory distress caused by the enlarged lymph nodes.

Diagnosis can be made by bacterial isolation or polymerase chain reaction (PCR) detection of *S. equi* from nasal secretions or pus from draining abscesses. Antipyretics and supportive therapy are recommended. Antibiotics should only be administered in the early stage of the disease or in more severe cases. *S. equi* can also cause empyema of guttural pouches and subsequent formation of **chondroids** (Fig. 3).

These animals are an important source of infection acting as inapparent, long term carriers. Donkeys and mules are frequently affected by this disease (Boyle et al. 2018; Firew and Pal 2015) and it is important to isolate infected animals. For other preventive measures in an outbreak it is suggested to refer to the detailed guidelines reported by Boyle et al. (2018). Vaccination reduces the severity of clinical signs and the spread of infection but does not prevent strangles. As with horses, donkeys are also susceptible to *Streptococcus equi* subsp. *zooepidemicus* infection. This is an opportunistic pathogen which can cause disease similar to strangles, and also severe respiratory disease often in association with herpesvirus infection, with lung abscesses and possible death (Evans et al. 2022; Cantelmi et al. 2023; Rickards

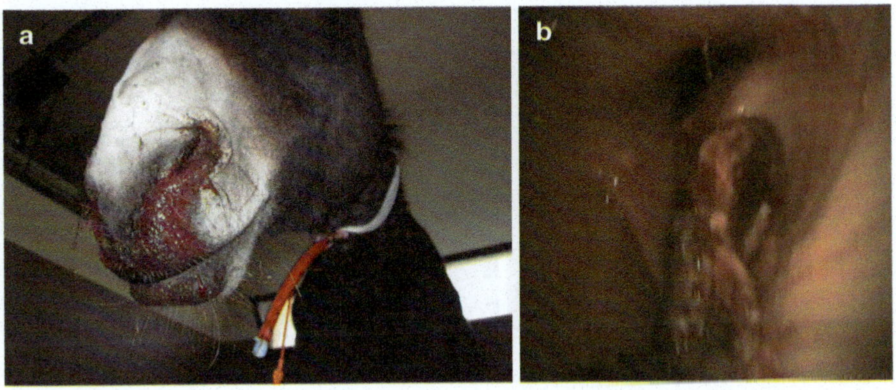

Fig. 4 Guttural pouch mycosis. Notice the epistaxis and the emergency tracheostomy (**a**) and the large blood clots from the left pouch occupying space in upper airways (**b**). (Laus et al. 2010)

and Thiemann 2019). One should always test respiratory secretions for *S. zooepidemicus* in animals showing signs of bronchopneumonia since it is a zoonotic agent and prompt treatment and precautionary measures are advised.

Guttural pouch mycosis in donkeys has been anecdotally reported (Evans et al. 2022) but also described in detail in a case report (Laus et al. 2010). In the latter, a profuse unilateral nasal hemorrhage was described. The coagulated blood almost occluded the larynx causing respiratory distress and requiring an emergency tracheostomy (Fig. 4).

The previously described peculiarities of the donkey upper airway including the angle of the airway from the pharynx to the larynx, the narrow nasopharynx, the tendency of the pharynx to collapse (detected in 43% of normal donkey in a previous study (Delvaux et al. 2001)), and the natural position of epiglottis (higher than horses) all contribute to the severe dyspnea reported in donkeys,. Endoscopically, mycotic plaques can be seen on large arteries in guttural pouches and should not be removed (Fig. 5).

Topical administration of antifungal agents through the endoscope (15 ml of a 2% miconazole gel solution applied for 6 timesevery 48 h for the first two times and every 72 h for the next 4 applications) was effective in this case (Laus et al. 2010). Medical treatment approach usually does not work in horse's mycotic plaques (Aisworth and Hackett 2004; Hensel et al. 2020, Thiemann and Bell (Authors). https://www.ivis.org).

Among equids, donkeys are the most susceptible to **Glanders** (Khan et al. 2013), an infectious disease caused by the gram negative bacterium *Burkholderia mallei* (Barrandeguy and Carossino 2018). National control programs adopted in several countries have drastically reduced its prevalence, but due to its zoonotic potential, early recognition of the disease is crucial. Although to a slightly lesser extent, mules are also susceptible, while horses appear to show some level of resistance (Khan et al. 2013; World Organization for Animal Health (OIE) 2024). Respiratory secretions or exudate from skin lesions of infected animals are rich in *B. mallei*.

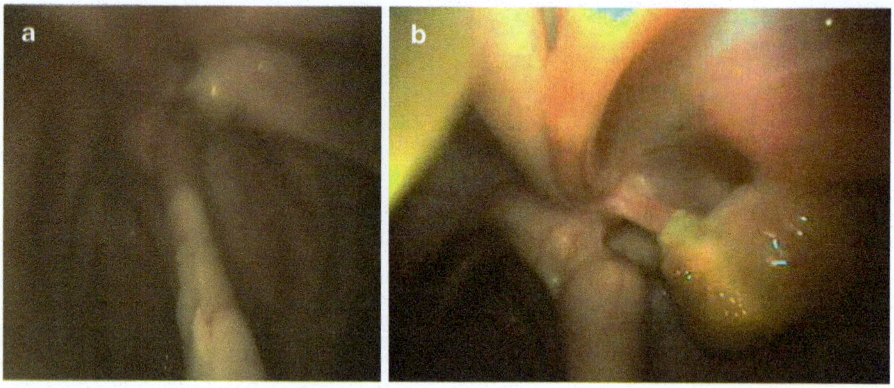

Fig. 5 Fungal plaque on maxillary artery and on stylohyoid bone (**a**) and detail of the plaque on the artery (**b**). (Laus et al. 2010)

Aerosolization through the respiratory route or direct contact with abrasions on the skin or mucous membranes are the main routes of transmission. However, venereal, vertical transmission and ingestion are alternative routes of infection (Barrandeguy and Carossino 2018). Nasal, pulmonary, or cutaneous forms of the disease exist. The acute form (nasal and pulmonary) of the disease is the most common in donkeys, while the chronic form (cutaneous) is more frequently seen in horses (in endemic areas) (Torres 2025). Mules are susceptible at both chronic and acute forms but the susceptibility of mules to glanders has been suggested to be somewhat less compared to horses and donkeys (Khan et al. 2013). In acute disease, fever, anorexia, depression, ulcerated nodules in the nasal cavity, cough, and dyspnea are the main signs. Subsequently, dyspnea worsens, and the disease often progresses to severe bronchopneumonia with lung abscessation. These cases are fatal in days to weeks. The chronic form can last months to years and cutaneous ulcerated nodules, intermittent fever, progressive weakness, wasting, and intermittent cough are the main signs. Swelling of the joints in the hindlegs with lameness, hematuria, polyuria, diarrhea, epistaxis, and orchitis can also develop (Barrandeguy and Carossino 2018). Subclinical cases may occur and are potential sources of infection (World Organization for Animal Health (OIE) 2024). Diagnosis can be reached by bacterial culture, the mallein test, or serological assays (World Organization for Animal Health (OIE) 2024). Glanders persists in several African, Asian, and South American countries but has been eradicated from most parts of the world thanks to surveillance testing, culling of infected animals, and restriction of movement (World Organization for Animal Health (OIE) 2024). Because it is considered a dangerous emerging disease, treatment is prohibited and vaccines are not currently available.

In elderly donkeys **tracheal collapse** is a recognized problem (Rickards and Thiemann 2019) especially at the level of thoracic inlet where the tracheal rings are naturally flattened dorsoventrally (Fig. 6).

Integrity of tracheal cartilage gradually declines with age and can be exacerbated by increased respiratory effort due to underlying respiratory pathology. Discomfort

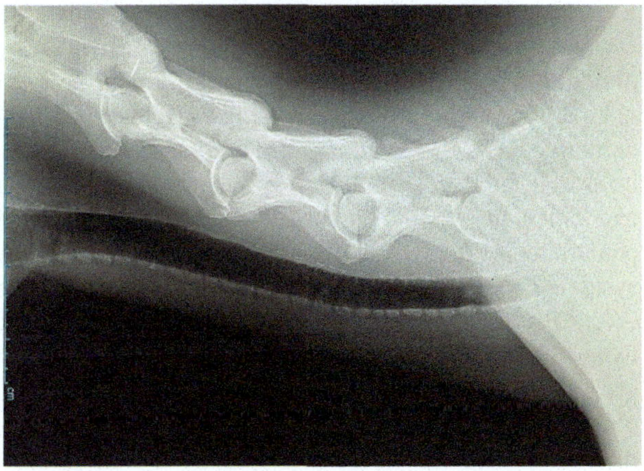

Fig. 6 latero-lateral radiography of the trachea in a donkey affected by tracheal collapse. A slight narrowing of the trachea before the thoracic inlet can be seen

can be severe and clinical signs become more evident with progression. These donkeys do not tend to tolerate endoscopic examination and may show signs of respiratory distress during the examination. Chronic cough, dyspnea, loud tracheal sounds, and nostril flare are the typical clinical signs. Flattening of the trachea can also be seen radiographically. Treatment is supportive and is mainly aimed at alleviating dyspnea due to concomitant causes (e.g. with bronchodilators or corticosteroids, see asthma treatment).

4.3.2 Lower Airways

Donkeys and mules are the natural host for the parasite ***Dictyocalus arnfieldi***, commonly referred to as lungworms. Infection rarely produces clinical signs even in cases with high parasite burdens, possibly because of a down regulatory mechanism that is absent in horses (Reed et al. 2018; Laus et al. 2022). The latter often have a marked inflammatory response to *D. arnfieldi* and they are more susceptible to infection if stabled together with donkeys (Reed et al. 2018; Laus et al. 2022). However, some authors suggest that *D. arnfieldi* infection may make donkeys more susceptible to viral infections. The disease is more frequently seen in tropical countries but has also been reported in Europe, Australia and North America. (Laus et al. 2022). Infection occurs mainly on pasture but is also possible in paddocks or stalls due to contamination of forage or bedding with L3 larvae. Clinical signs in donkeys and mules are absent or limited to slight dyspnea and some increased sounds on auscultation of caudal pulmonary lobes. Diagnosis in donkeys and mules can be achieved using the Baermann test on feces to detect L1 stage larvae, but this technique does not work in horses as they don't tend to develop patent infections (Laus et al. 2022). Larvae can also sometimes be detected in tracheal washes (TW) or bronchoalveolar lavage (BAL) fluid (rarely in horses) (Laus et al. 2022). Oral treatment with ivermectin (0.2 mg/Kg) or moxidectin (0.4 mg/Kg) is typically effective in donkeys (Laus et al. 2022). No data for treatment of mules are available but based on efficacy of ivermectin in mules against other parasites, the same protocols could be suggested (Bazzano et al. 2020).

Echinococcus granulosus is a tapeworm with an indirect, two host life cycle causing **Pulmonary Hydatidosis**. Genetic analysis can provide further classification of other species of the parasite including *Echinococcus equinus* but the adult form is not distinguishable from *E. granulosus* (Laus et al. 2022). Dogs and foxes represent definitive hosts of *E. equinus* while horses, donkeys, and mules are the intermediate hosts that can become infected grazing on pasture contaminated with eggs (Rickards and Thiemann 2019). Once in the small intestine the eggs hatch and larvae, after penetration of the intestinal mucosa, reach the liver and lung through the circulatory system (Laus et al. 2022). In these tissues hydatid cysts develop, growing slowly (1 cm per year) and reaching a maximum diameter of 10 cm, causing pulmonary hydatidosis. Definitive hosts can become infected by ingestion of the protoscolices present in infected offal (Rickards and Thiemann 2019). Echinococcus infection is a zoonosis but *E. equinuus* is rarely diagnosed in humans (Laus et al. 2022). The disease has a worldwide distribution, and cases have been reported from Europe, Asia, Africa, and North America (Laus et al. 2022). In the UK and some countries of Europe, echinococcosis is endemic (Boufana et al. 2015). Echinococcosis in donkeys is similar to horses (Lahmar et al. 2014) and tends to be asymptomatic unless the number and size of cysts result in a significant compromise to organ function (Rickards and Thiemann 2019). Diagnosis can be reached by ultrasonographic/radiographic visualization of cysts in the lung and liver. Long term albendazole administration (4 mg/kg for 90 days) was effective in the treatment of sheep but is the focus should be on prevention: control and deworming of infected dogs and not permitting feeding of or exposure of infected offal to canines (Laus et al. 2022; Rickards and Thiemann 2019).

The most common respiratory infection in donkeys is **influenza** (Mendoza et al. 2018) with clinical signs similar to horses: fever, depression, anorexia, ocular/nasal discharge, and coughing (Fig. 7).

Donkeys are more susceptible to influenza infection than horses, frequently showing more severe clinical signs, a higher mortality rate, and a propensity to develop secondary bacterial infections. For this reason, antibiotic administration is recommended in donkeys with influenza. The real incidence of influenza in mules is not known and no data are available regarding susceptibility. Hyperlipemia is a serious complication in sick donkeys especially when bacterial infections and anorexia occur. As with other diseases, monitoring and prevention of hyperlipemia is essential to prevent the situation from worsening dramatically.

Herpesvirus infection in donkeys can be caused by equine herpesviruses (EHV-1, EHV-4) and asinine herpesvirus (AHV-1, 2, 3). The role of other AHVs (AHV-4, 5 and 6) in the pathogenesis of disease is still unclear (Mendoza et al. 2018; Rickards and Thiemann 2019; Duncan and Hadrill 2008). Both EHV-1 and EHV-4 can cause respiratory disease in donkeys, and EHV-1 can also cause abortion (Barrandeguy and Carossino 2018). A case of neurological disease has been associated with a virus most closely resembling AHV-5 in donkeys (Vengust et al. 2008). The same virus has also been implicated as a cause of interstitial pneumonia in donkeys (Kleiboeker et al. 2002). EHV-7 (AHV-2) can be isolated from healthy donkeys and mules (Bell et al. 2008), but it has also been detected in the placenta

Fig. 7 Ocular and nasal discharge in a donkey affected by Influenza

from an abortion in a miniature donkey (LeCuyer et al. 2015). The nature of respiratory disease caused by herpesviruses is not too different from that of horses. The disease usually begins with clear bilateral nasal discharge and fever and severity of clinical signs can range from mild to severe. Cough is usually less evident in donkeys than in horses. Mules can become infected with the same virus as donkeys, but rarely do they exhibit clinical signs (Bell et al. 2008). Neurological signs, abortion and coital exanthema are also possible in donkeys as in horses. To diagnose, nasal swabs and EDTA whole blood samples should both be collected for PCR (when available) or virus isolation. No vaccines for asinine herpesviruses exist.,and the current literature lacks any studies regarding cross protection from the commercially available equine herpesvirus vaccines. That said, the vaccine products for EHV-1 and EHV-4, approved for horses, can and should be used to help prevent disease in donkeys (Mendoza et al. 2018).

In sub-Saharan Africa an arthropod-borne virus (orbivirus) causes **African horse sickness** in equids (Rickards and Thiemann 2019). The disease can be transmitted by *Culicoides* spp., mosquitos or ticks. Horses are more susceptible to developing severe disease than donkeys and mules. The pulmonary form is characterized by dyspnea, spasmodic coughing, and abundant foamy nasal discharge, while the cardiac form causes swelling of the supraorbital fossa, eyelids, neck, thorax, and shoulders (Rickards and Thiemann 2019). In endemic regions of Africa, infected donkeys are often asymptomatic and considered potential reservoirs (Barrandeguy and Carossino 2018). On the other hand, donkeys living in non-endemic regions are susceptible to both forms of the disease. In some endemic regions of Africa, vaccines have been utilized to aid in prevention, despite being complicated by the presence of several serotypes of the virus.

Equine Viral Arteritis can also cause respiratory signs in donkeys (Barrandeguy and Carossino 2018). For reproductive consequences of arteritis, management of outbreak and prevention please refer to the specific chapter.

4.3.3 Equine Asthma

Equine asthma is an environmental asthma-like disease of adult equids characterized by a recurring bronchoconstriction, mucus secretion, and airway inflammation, with alternating periods of remission and exacerbation (Leclere et al. 2011; Bazzano et al. 2018). Clinical signs are not evident in low-grade airway obstruction but in severe cases, cough, nasal discharge, increased respiratory effort, and weight loss can be observed. Equids affected by equine asthma show disease exacerbation when exposed to organic dust particulates in hay.

Auscultation of the thorax can be improved by the use of a rebreathing bag to accentuate respiratory sounds. In severe cases the inspiratory effort can dramatically exacerbate the potential for tracheal collapse in geriatric patients, increasing the severity of dyspnea. Right sided heart hypertrophy and failure can occur. Clinical examination is often highly suggestive of this diagnosis but respiratory endoscopy and cytology of BAL fluid (see sectione 4.4) are useful to confirm and to monitor progression.

Treatment includes antigen avoidance, administration of systemic or inhaled corticosteroids, bronchodilators and mucolytic drugs (Fig. 8).

A well-recognized cause of both inspiratory and expiratory dyspnea in donkeys is **pulmonary fibrosis**. The etiology of this condition has not yet been defined but shows similarities with pleuroparenchymal fibroelastosis in humans (Miele et al. 2014). Furthermore, AHV-4 and AHV-5 have been isolated from the lungs of affected donkeys (Kleiboeker et al. 2002). Fibrosis can be localized to small lung areas or can be more diffuse, affecting large parts of the lung parenchyma. Increased resting respiratory effort, tachypnea, and cough are features of this condition. (Mendoza et al. 2018). Decreased or absent lung sounds may be detected on auscultation. Diagnosis can be confirmed with radiographs or ultrasound where a diffuse interstitial pattern may be identified. If the lesion is subpleural, comet tail artifacts may be observed (Fig. 9).

There is no specific treatment for this condition and the aim of therapy is to minimize further airway compromise by a combination of clean air management, regular vacci- nations, and medications, including bronchodilators and steroids administered by inhalation or systemically. Donkeys with the condition need regular monitoring with the aim of defining a suitable time for euthanasia, avoiding a respiratory crisis and possible sudden death. There is no evidence that antiviral drugs provide any benefit.

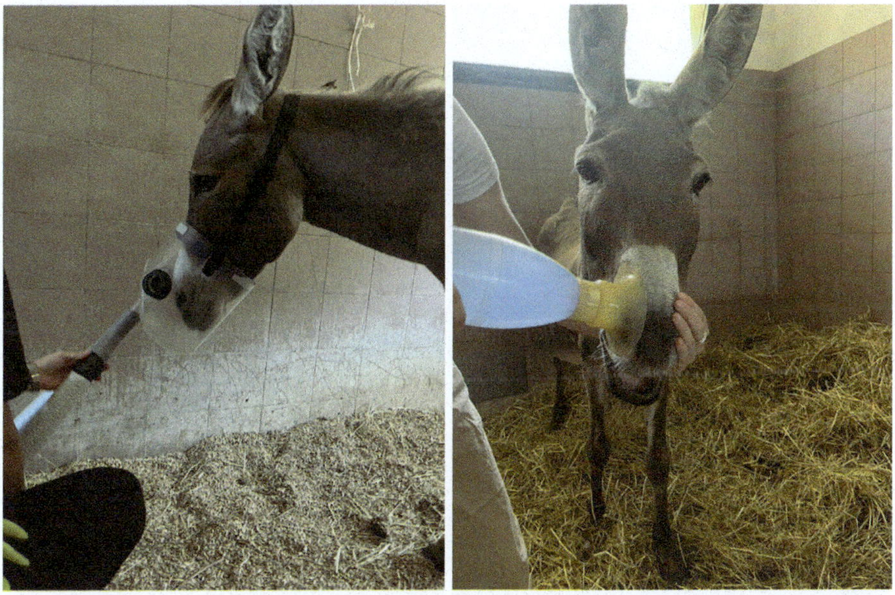

Fig. 8 different use of inhaled medication in donkey. Temporary occlusion of the contralateral nares (right side of the picture) during the inspiratory phase ensures the best result

4.4 Diagnostic Tests

Radiography, ultrasonography, and endoscopy may support the clinician in the localization of respiratory diseases in donkeys.

The thoracic radiographic appearance is similar in donkeys compared to ponies and foals (Thiemann and Bell 2001). In small donkeys, a single radiograph may be sufficient. In larger donkeys and in mules, lateral and dorsoventral radiographs may be necessary as in horses.

In the author's experience, an approximate distance from the donkey of about 60 cm and exposure parameters of 60–70 KV, 100–150 mA, 150 mSec can be used to obtain diagnostic images in most donkeys.

Ultrasonography can be performed as in horses, following dorsoventrally each intercostal space bilaterally. Clipping can be necessary in donkeys and mules with long hair but cleaning with alcohol and application of ultrasonographic gel, allows for good visualization in most of the cases.

Endoscopy can be challenging in donkeys affected by respiratory disease, especially those with asthma and/or tracheal collapse. Unlike horses and mules, these animals do not tolerate the presence of the endoscope in their airways due to the anatomical narrowed larynx and, if affected by obstructive diseases, they can have respiratory distress, preventing further examination.

Once the affected area of the respiratory tract has been defined, swabs, aspirates, or biopsies from different sites in the upper respiratory tract, as well as tracheal wash

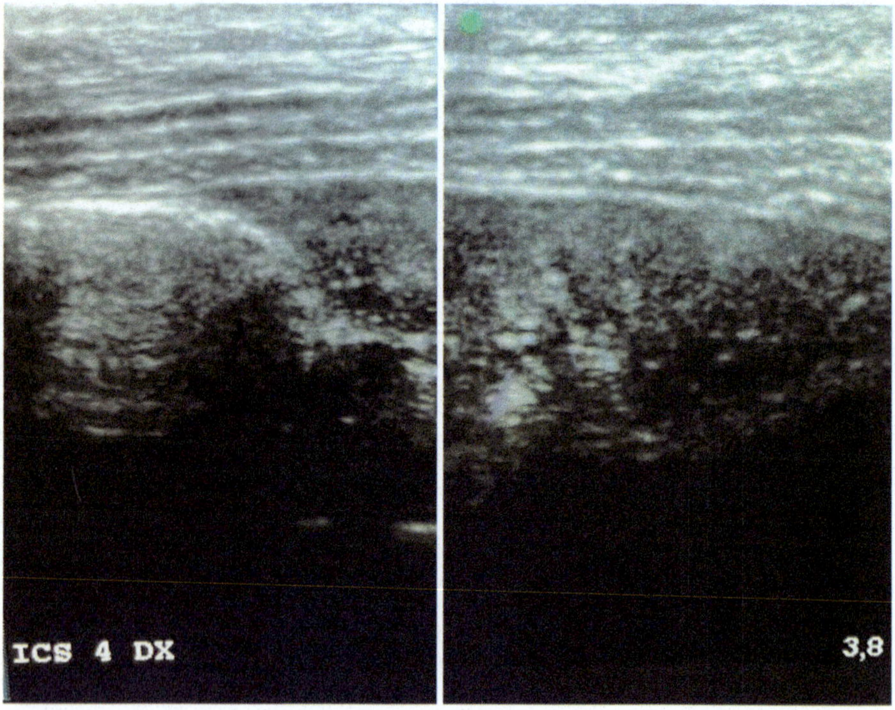

Fig. 9 Ultrasonographic image of pulmonary fibrosis. Notice the "comet tail" artifact (right side of the picture) and fibrotic parenchyma (left side of the picture) with air entrapped in it

and bronchoalveolar lavage in the lower respiratory tract can be performed to obtain appropriate samples for cytology, bacterial culture, viral or bacterial PCR and histopathology, based on the history and differential diagnoses. Additionally, basic bloodwork may be performed to aid in the diagnosis and treatment approach.

The normal findings of cytologic analysis of TW and BAL fluids in donkeys are different compared to horses. In donkeys, these fluids show different leukocyte counts, particularly BAL which tends to have slightly higher macrophage percentages, whereas TW shows higher neutrophil and eosinophil counts (Mendoza et al. 2018).

Dynamic pulmonary tests have also been described in donkeys, showing higher total pulmonary resistance and respiratory rates, whereas pulmonary compliance was lower than in horses (Delvaux et al. 2001).

4.5 Treatment

From a pharmacological standpoint, donkeys are different from horses, whereas mules share some common characteristics to both (Archer 2013). However, pharmacokinetic drug studies in donkeys and mules are still limited compared to the literature for horses.

In general, most medications can be used in donkeys and mules at the same dosage for horses, but some sedatives, non-steroidal anti-inflammatory drugs (NSAIDs), and antimicrobials should be administered at specific doses (see annex).

For example, the dose of alpha-2-agonists is usually higher compared to horses to obtain the same level of sedation when performing diagnostics procedures like endoscopy or radiology. Mules in particular may need up to 50% higher dosage than horses (Archer 2013).

The most common systemic medications used in respiratory disease include three main classes: (i) corticosteroids (dexamethasone), (ii) bronchodilators (clenbuterol) and, (iii) antibiotics (for bacterial infections; administered based on antimicrobial susceptibility testing whenever possible). In cases of respiratory distress, corticosteroids reduce pulmonary inflammation and bronchodilators rapidly relieve clinical signs of airway obstruction.

Corticosteroids and bronchodilators can also be used for nebulization. Flucatisone and beclomethasone are aerosolized corticosteroids commonly used in horses but no specific dosage is available for donkeys and mules. Unlike aerosolized corticosteroids, aerosolized bronchodilators typically remain effective regardless of the severity of disease. Aerosolized bronchodilator therapy is more rapid and powerful than oral administration of bronchodilators, and is the preferred route of administration during crises (McGorum et al. 2007).

4.6 Prevention

Animals susceptible to respiratory conditions should be kept outdoor or housed in a well-ventilated stable with low levels of dust.

Ideally, owners should provide bedding with minimal dust such as wood chips and large wood shavings, or rubber instead of straw. Regular removal of fecal material and urine-soaked bedding is essential.

Donkeys and mules suffering from or susceptible to respiratory disease should be fed with low-dust forage sources, including soaked or pelleted hay instead of dry hay.

Clinicians should inform owners about the importance of regularly vaccinating donkeys and mules against respiratory pathogens, when available, especially when they have regular contact with other equids.

4.7 New Insights and Future

Some of the previously described diseases have zoonotic potential (e.g. glanders and *S. zooepidemicus*). From a one health perspective, interdisciplinary communication is essential. Cases of diseases in humans may be missed if the history of contact with equines is not appropriately reported. Furthermore, in many countries donkeys are involved in "animal assisted therapy" (onotherapy) because of their peaceful and collaborative nature. They are often in close contact with people and children, which makes routine healthcare and disease prevention even more important.

5 Conclusion

- The stoic nature of donkeys and mules may allow for respiratory conditions at early stages to go undiagnosed. Donkeys have a less sensitive cough reflex and are not often involved in exercise that can exacerbate respiratory signs. Owners should be advised to seek veterinary advice even in the presence of minor respiratory signs.
- Clinicians dealing with donkeys and mules should always keep in mind the specific anatomical and physiological differences of the respiratory tract compared to horses.
- Cohabitation of donkeys and mules with horses can facilitate the spread of some infectious diseases. Some chronic forms of disease may lead to lung fibrosis in donkeys, resulting in a significant decrease in respiratory function.
- Acute respiratory diseases can lead to secondary hyperlipidemia. Prevention, early recognition and prompt intervention is recommended.
- Clinicians should make owners aware that vaccination of donkeys and mules is as important as in horses, to limit the spread of diseases.
- Advances in diagnostic imaging (ultrasonography, radiology, endoscopy) in mule and donkey medicine are essential, to detect differences with horses.

Questions

1. Which disease do you consider as differential diagnosis in a geriatric donkey with dyspnea?
   - Differential diagnosis includes asthma, pulmonary fibrosis and tracheal collapse.
2. Which are useful tools to diagnose inflammatory airways disease in donkeys and mules?
   - BAL and/or TW cytology can be used for the diagnosis of inflammatory airways disease.
3. How do you prevent the onset/worsening of clinical signs in animals affected by chronic respiratory conditions?
   - Maintaining low levels of dust and improving stable ventilation are essential in the management of animals with chronic respiratory conditions.

Annex

Doses Charts

Family of medical agents	Active principles	Route of administration	dose	frequency	references
Antibiotics	Ampicillin	IV	10 mg/ kg	q 4–6 h	Archer (2013)
	Penicillin G (sodium)	IV	20,000 UI/kg	q 4–6 h	Archer (2013)
NSAID	Flunixin meglumine	IV	1.1 mg/kg PO or IV	q 12–24 h	Archer (2013)
Corticosteroid	Dexamethasone	IV	0.1–0.2 mg/kg	–	Thiemann and Bell (2001)
Bronchodilators	Clenbuterol	IV or OS	0.8–3.2 μg/kg	q 12 h	Thiemann and Bell (2001)
Sedatives	Xylazine	IV	1-1.5 mg/kg	–	Archer (2013)
	Detomidine	IV	0.02–0.03 mg/ kg	–	Archer (2013)
	Butorfanol	IV	0.02–0.04 mg/ kg	–	Archer (2013)

References

Adams MF, Castro JR, Morandi F, Reese RE, Reed RB (2013) The nasolacrimal duct of the mule: anatomy and clinical considerations. Equine Vet Educ 25(12):636–642. https://doi.org/10.1111/eve.12077

Aisworth DM, Hackett RP (2004) Disorders of the respiratory system. In: Reed SM, Bayly WM, Sellon DC (eds) Equine internal medicine, 2nd edn. Saunders, St. Louis, pp 289–353

Archer DC (2013) Handbook of equine emergencies. Elsevier Health Sciences. SBN: 9780702045455

Barrandeguy ME, Carossino M (2018) Infectious diseases in donkeys and mules: an overview and update. J Equine Vet Sci 65:98–105

Bazzano M, Laghi L, Zhu C, Magi GE, Serri E, Spaterna A, Tesei B, Laus F. Metabolomics of tracheal wash samples and exhaled breath condensates in healthy horses and horses affected by equine asthma. J Breath Res 2018 27;12(4):046015

Bazzano M, Di Salvo A, Diaferia M, Veronesi F, Galarini R, Paoletti F, Tesei B, McLean A, Veneziano V, Laus F (2020) Anthelmintic efficacy and pharmacokinetics of Ivermectin paste after oral administration in mules infected by Cyathostomins. Animals 10:934

Bell SA, Pusterla N, Balasuriya UBR, Mapes SM, Nyberg NL, MacLachlan NJ (2008) Isolation of a gammaherpesvirus similar to asinine herpesvirus-2 (AHV-2) from a mule and a survey of mules and donkeys for AHV-2 infection by real- time PCR. Vet Microbiol 130:176e83

Boufana B, Lett WS, Lahmar S, Buishi I, Bodell AJ, Varcasia A, Casulli A, Beeching NJ, Campbell F, Terlizzo M, McManus DP, Craig PS (2015) Echinococcus equinus and Echinococcus granulosus sensu stricto from the United Kingdom: genetic diversity and haplotypic variation. Int J Parasitol 45(2–3):161–166

Boyle AG, Timoney JF, Newton JR, Hines MT, Waller AS, Buchanan BR (2018) Streptococcus equi infections in horses: guidelines for treatment, control, and prevention of strangles-revised consensus statement. J Vet Intern Med 32(2):633–647. https://doi.org/10.1111/jvim.15043. Epub 2018 Feb 9. PMID: 29424487; PMCID: PMC5867011

Burnham SL (2002) Anatomical differences of the donkey and mule. In: Proceedings of the annual convention of the AAEP 2002

Cantelmi MC, Merola C, Averaimo D, Chiaverini A, Cito F, Cocco A, Di Teodoro G, De Angelis ME, Di Bernardo D, Auzino D et al (2023) Identification of the novel Streptococcus equi subsp. zooepidemicus sequence type 525 in donkeys of Abruzzo region, Italy. Pathogens 12:750

Delvaux V, Kirschvink N, Amory H, Busoni V, Art T, Lekeux P (2001) Characteristics of cardio-respiratory function of the donkey. Pratique Vétérinaire Équine 33(129):21–28. ref. 15

Duncan J, Hadrill D (2008) The professional handbook of the donkey, 4th edn. The Donkey Sanctuary

Evans L, Crame M, Preston E (Eds): The clinical companion of the donkey, 2, The Donkey Sanctuary, 2022

Firew S, Pal M (2015) Clinical and microbiological observations on strangles in donkeys. Haryana Vet 54:64e6

Hensel M, Meason-Smith C, Plumlee QD, Myers AN, Coleman MC, Lawhon S, Rodrigues Hoffmann A, Rech RR (2020) Retrospective analysis of aetiological agents associated with pulmonary mycosis secondary to enteric salmonellosis in six horses by panfungal polymerase chain reaction. J Comp Pathol 174:1–7

Khan I, Wieler LH, Melzer F, Elschner MC, Muhammad G, Ali S, Sprague LD, Neubauer H, Saqib M (2013) Glanders in animals: a review on epidemiology, clinical presentation, diagnosis and countermeasures. Transbound Emerg Dis 60(3):204–221. https://doi.org/10.1111/j.1865-1682.2012.01342.x. Epub 2012 May 27. PMID: 22630609

Kleiboeker SB, Schommer SK, Johnson PJ, Ehlers B, Turnquist SE, Boucher M et al (2002) Association of two newly recognized herpesviruses with interstitial pneumonia in donkeys (Equus asinus). J Vet Diagn Invest 14:273–280

Lahmar S, Boufana B, Jebabli L, Craig PS, Ayari H, Basti T, Dhibi M, Torgerson PR (2014) Modelling the transmission dynamics of cystic echinococcosis in donkeys of different ages from Tunisia. Vet Parasitol 205(1–2):119–124

Laus F, Paggi E, Cerquetella M, Spaziante D, Spaterna A, Tesei B (2010) Guttural pouch mycosis in a donkey (Equus asinus): a case report. Veterinarni Med 55:561–565

Laus F, Veneziano V, Veronesi F (2022) Malattie parassitarie del cavallo e dell'asino, guida alla diagnosi e alla gestione. Point Veterinaire Italie, Milan

Leclere M, Lavoie-Lamoureux A, Lavoie JP (2011) Heaves, an asthma-like disease of horses. Respirology 16:1027–1046

LeCuyer TE, Rink A, Bradway DS, Evermann JF, Nicola AV, Baszler T et al (2015) Abortion in a Mediterranean miniature donkey (Equus asinus) associated with a gammaherpesvirus similar to Equid herpesvirus 7. J Vet Diagn Invest 27:749e53

Mendoza FJ, Toribio RE, Perez-Ejica A (2018) Donkey internal medicine-part II: cardio- vascular, respiratory, neurologic, urinary, ophthalmic, dermatology, and musculo- skeletal disorders. J Equine Vet Sci 65:86–97

Miele A, Dhaliwal K, Du Toit K et al (2014) Chronic pleuropulmonary fibrosis and elastosis of aged donkeys: similarities to human pleuroparenchymal fibroelastosis. Chest 145(6):1325–1332

McGorum BC, Dixon PM, Robinson NE, Shumacher J (2007) Equine Respiratory Medicine and Surgery. Saunders-Elesevier, Philadelphia

Otranto D, Colwell DD, Milillo P, Di Marco V, Paradies P, Napoli C, Giannetto S (2004) Report in Europe of nasal myiasis by Rhinoestrus spp. in horses and donkeys: seasonal patterns and taxonomical considerations. Vet Parasitol 122:79–88

Reed SM, Bayly WM, Sellon DC (2018) Equine internal medicine. Elsevier, St. Louis

Rickards KJ, Thiemann AK (2019) Respiratory disorders of the donkey. Vet Clin North Am Equine Pract 35(3):561–573

Thiemann A.K. and Bell N.J. 2001. The peculiarities of donkey respiratory disease. In: Equine respiratory diseases by Lekeux P. IVIS

Torres AG (2025) Glanders: an ancient and emergent disease with no vaccine or treatment on site. PLoS Negl Trop Dis 19(6):e0013160. https://doi.org/10.1371/journal.pntd.0013160. PMID: 40498787; PMCID: PMC12157308

Vengust M, Wen X, Bienzle D (2008) Herpesvirus-associated neurological disease in a donkey. J Vet Diagn Invest 20:820e3

World Organization for Animal Health (OIE) (2024) Manual of diagnostic tests and vaccines for terrestrial animals, 13th edn. World Organization for Animal Health (OIE), Paris

Urogenital Disorders

Lorenzo Segabinazzi

Abstract

This chapter explores congenital and inherited urogenital abnormalities in equids, with a particular emphasis on donkeys—a species often underrepresented in veterinary literature. Conditions discussed include intersex and sex-reversal syndromes, cryptorchidism, inguinal hernias, ectopic ureters, and various penile, testicular, and epididymal anomalies. While many of these disorders are well documented in horses, reports in donkeys remain scarce, raising questions about true prevalence versus underdiagnosis. Intersex conditions—particularly male pseudohermaphroditism—are among the more commonly reported anomalies, while true hermaphroditism and sex reversal are exceedingly rare. Cryptorchidism appears to be the most prevalent testicular abnormality in donkeys, primarily presenting as unilateral inguinal retention. Mosaicism, ectopic testes, and polyorchidism are described through isolated case studies. Female reproductive tract tumors, such as granulosa cell tumors and leiomyomas, have also been identified in a limited number of donkeys. Urolithiasis is rarely reported but likely underdiagnosed. Diagnosis of these conditions typically involves physical examination, ultrasonography, histopathology, and genetic testing. Despite limited data, the fundamental mechanisms underlying these disorders appear consistent across equid species. Increased awareness, research, and case documentation are needed to improve diagnostic accuracy and treatment outcomes in donkeys.

Keywords

Urogenital abnormalities · Donkeys · Intersex conditions · Cryptorchidism · Sex-reversal syndrome

L. Segabinazzi (✉)
Department of Clinical Sciences, Ross University School of Veterinary Medicine, Basseterre, Saint Kitts and Nevis
e-mail: lsegabinazzi@rossvet.edu.kn

A. K. McLean et al. (eds.), *Donkey and Mule Medicine: Clinical Medicine and Reproduction*, https://doi.org/10.1007/978-3-032-18540-2_11

1 Introduction

The urogenital system consists of the urinary and reproductive organs. Urogenital disorders can be inherited or congenital anomalies, and infectious or neoplastic in origin. These abnormalities are caused by abnormal genes or produced by injury, disease, or exposure to toxic substances in the womb and they may or may not cause health problems. Although chromosomal aberrations are a well described cause of sex developmental anomalies in horses, reports in donkeys are lacking. It is unclear if there is a true low prevalence in donkey species, or a lack of recognition due to inapparent clinical signs or lack of research in this species. This chapter describes congenital and inherited urogenital abnormalities observed in equids and practical aspects for the diagnosis and treatment of these conditions.

2 Urogenital System Abnormalities

The most common congenital and inherited abnormalities of the urogenital system observed in equids include inguinal hernias, cryptorchidism, intersex, sex-reverse, and ectopic ureter. Although some of them have not been described in donkeys yet, all the cited abnormalities and some more uncommon aberrations will be discussed in this chapter. It is worth noting that the term "disorders of sexual development" has been proposed to replace the terms "intersex," and "true and pseudo hermaphrodite" in the scientific world. However, this chapter will use informal nomenclature to differentiate these conditions more easily.

3 Intersex

Intersex is a broad term describing a range of conditions in which reproductive or sexual anatomy does not align with typical definitions of male or female. Among equids, intersex conditions have been documented, with male pseudohermaphroditism being the most frequently reported, while female pseudohermaphroditism and true hermaphroditism are exceedingly rare.

Pseudohermaphroditism refers to a condition where an animal's external genitalia differ from their genetic sex, but their internal genitalia align with their genotype. In male pseudohermaphrodites, individuals are genetically male (XY) but exhibit female or ambiguous external genitalia. These animals typically possess intra-abdominal testicles, and their external genitalia may include an enlarged clitoris resembling a penile structure. Vaginal structures can appear normal or abnormal. Behaviorally, male pseudohermaphrodites may exhibit stallion-like traits or display nymphomanic behaviors similar to mares in estrus.

In contrast, female pseudohermaphrodites are genetically female (XX) with ovaries but present with external genitalia resembling male structures. This condition is even rarer than true hermaphroditism in equids. Pseudohermaphroditism in equids often occurs in mosaic individuals, where chromosomal anomalies result in mixed cellular genotypes—a topic discussed in subsequent sections.

True hermaphroditism is a condition in which an individual possesses both ovarian and testicular tissue, either as separate structures or combined into a single gonad (ovotestis). This condition is exceptionally rare but has been described in horses (Meyers-Wallen et al. 1997). Clinical signs include poorly developed external genitalia, such as an underdeveloped penis or absent scrotum. Definitive diagnosis relies on histological examination to confirm the presence of ovotestes.

Reports of pseudohermaphrodites and true hermaphrodites in donkeys are exceedingly rare and less documented than in horses. Isolated case studies have been described in donkeys with ambiguous genitalia or abnormal reproductive anatomy, but comprehensive data is scarce. Despite this, the fundamental developmental biology underlying intersex conditions is shared across equids, making it plausible that similar cases can occur in donkeys.

Affected animals may exhibit masculinized phenotypic traits and require detailed diagnostic evaluations, including clinical examinations, genetic testing, and imaging to assess gonadal structures. Castration is typically recommended as the primary treatment to address undesirable behaviors and prevent medical complications associated with intersex conditions (Fig. 1).

4 Sex-Reverse Syndrome

Unlike the broader term "intersex," sex-reversal syndrome refers specifically to a mismatch between an individual's genetic sex and their gonadal development. For instance, an individual with an XX karyotype but possessing testicular tissue instead of ovarian tissue would represent a case of XX sex reversal. Typically, XX sex-reversed individuals are classified as either XX males, having bilateral testes, or XX true hermaphrodites, possessing ovotestes. Conversely, in XY sex reversal, the development of oocytes and follicles during fetal life often leads to degeneration, resulting in gonadal dysgenesis. In such cases, ovarian architecture may be unrecognizable by the time of clinical presentation.

In equids, both XX and XY sex-reversal syndromes have been reported, though these conditions are exceedingly rare or undocumented in donkeys. The diagnosis of sex-reversal syndrome requires a karyotype analysis that reveals a normal XX or XY karyotype, combined with histopathological confirmation that the gonadal tissue does not align with the genetic sex. In cases where gonadectomy is not performed, ultrasonographic examination of the gonads and internal reproductive structures can offer valuable insights, enabling a presumptive diagnosis. While not definitive, this approach can be highly beneficial for case management and decision-making regarding treatment and care.

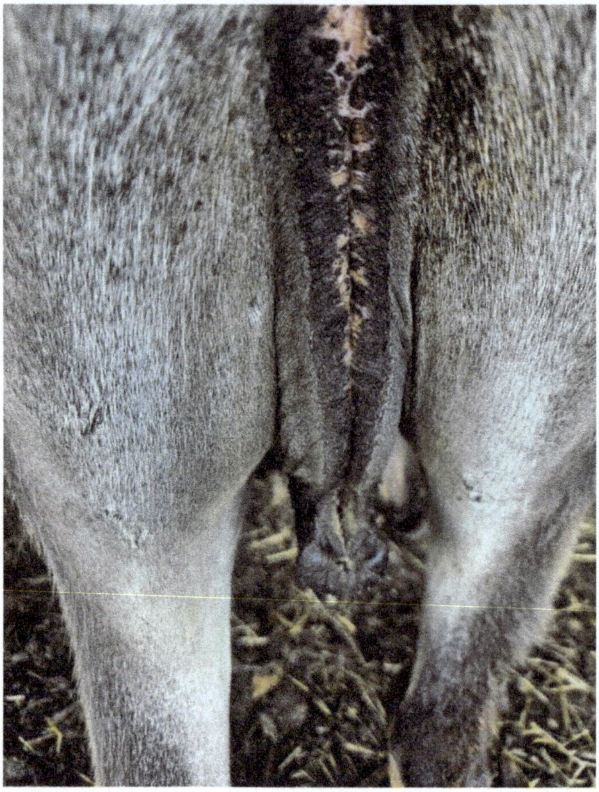

Fig. 1 Donkey presenting elongated and fused vulva and an enlarged clitoris reminiscent of a glans penis. (Picture courtesy of Marianne Lemelin DeCondi)

5 Mosaicism

Mosaicism is characterized by the presence of two or more genetically different sets of cells in the body. In horses, mosaicism is commonly associated with reproductive tract anomalies, and gonadal dysgenesis is one of the most often reported conditions (Paget et al. 2001; Sato et al. 2012). In donkeys, there is one reported case of a confirmed mosaic jenny (62,XX/61,X) gonadal dysgenesis (Dorado et al. 2016). The jenny presented with a normal external morphology and a regularly sized and shaped vulva but absent sexual behavior. Further examination revealed a short blind vaginal vestibule with no cervical *os* and an abnormal ball-shaped structure inside (Fig. 2), and hypoplastic ovaries.

Cases of mosaicism have also been associated with other anomalies, such as angular limb deformities, growth retardation, and neurological deficits (Bowling and Millon 1990; Buoen et al. 1997), and in some cases no phenotypic abnormalities were observed (Kubień and Tischner 2002).

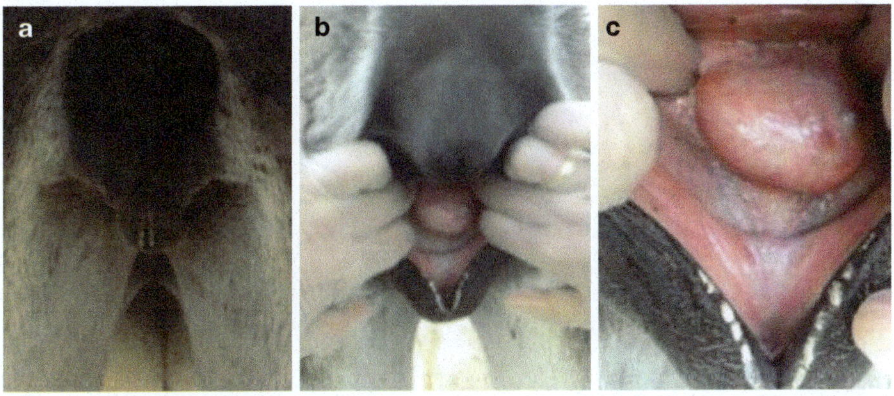

Fig. 2 External reproductive tract of a mosaic jenny (62, XX/61,X). (**a**) Normal-sized vulva; (**b**) Abnormal ball-shaped structure protruding from the vaginal vestibule; (**c**) Blind vagina with absence of cervical os. (Figure from Dorado et al. (2016))

6 Ectopic Ureter

An ectopic ureter is a congenital condition where a ureter, the tube that carries urine from the kidney to the bladder, doesn't connect properly to the bladder. The most common clinical sign is urinary incontinence with subsequent skin scalding and a persistently wet perineum and hind legs. In horses, this condition is more commonly reported in fillies. Surgical repair is often the best approach to correct this abnormality. Anecdotal reports of ectopic ureter in donkeys have been described.

7 Donkey Jack

Very little information is available about reproductive disorders of the jack. In male donkeys specifically, similarly to horses, inguinal hernias, cryptorchidism, and testicular abnormalities appear to be the most frequently observed conditions. It is worth noting that castration in donkeys tends to have a higher risk of hemorrhage complications due to the enlarged blood supply in this species compared to horse. More information about that can be found in chapter " Anesthesia and Pain management in Donkeys and Mules ". Other abnormal conditions, such as ectopic testicles, sperm granulomas, and diphallia have also been reported in donkeys.

8 Inguinal Hernia

An inguinal hernia occurs when tissue, such as part of the intestine, protrudes through a weak spot in the abdominal muscles or the inguinal canal. While this condition can affect both males and females, most reports in the veterinary literature

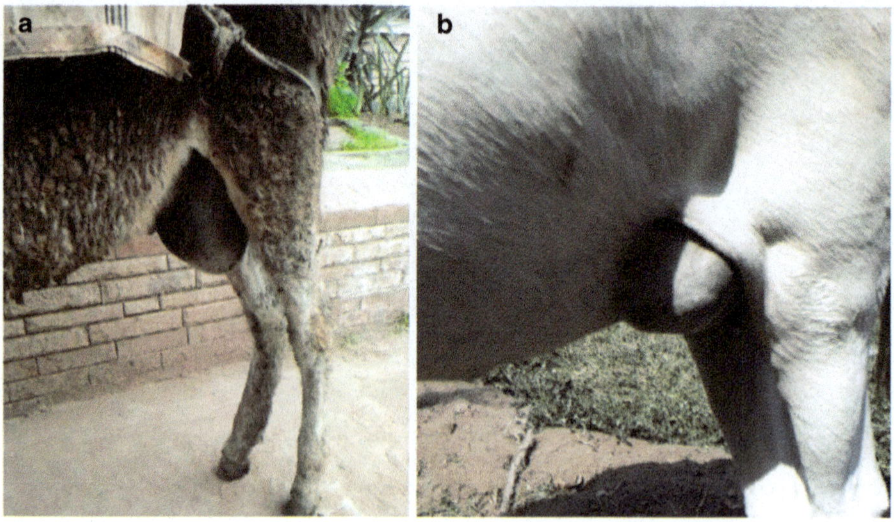

Fig. 3 Jack (**a**) and Jenny (**b**) with an acquired left-sided scrotal/inguinal hernia (Marzok and Moustafa 2023)

describe cases in males, often involving herniated abdominal contents extending into the scrotum. The anatomical structure of the inguinal canal plays a significant role in the development and potential incarceration of inguinal hernias, particularly in equids.

Although reports of inguinal hernias in donkeys are scarce, one documented study identified 11 cases in donkeys, including an unusual presentation in a jenny (Marzok and Moustafa 2023). This jenny exhibited a large swelling near the groin, highlighting that this condition, while rare in females, is not exclusive to males (Fig. 3). Further research is needed to understand the prevalence and characteristics of inguinal hernias in donkeys and their clinical management.

Inguinal and scrotal herniation in equids can be classified as either acquired or congenital. Acquired hernias are more common in adult males and are often accompanied by signs of colic, necessitating prompt medical attention. In contrast, congenital herniation typically occurs in newborn colt foals, is relatively uncommon, and is usually bilateral and non-incarcerated, with or without associated signs of colic. In donkeys, most reported cases involve acquired hernias of traumatic origin (Marzok and Moustafa 2023).

In horses, inguinal herniation frequently manifests as scrotal enlargement and is considered a veterinary emergency. Early detection of herniated abdominal contents into the scrotum is critical to reduce the risk of colic or the need for emergency surgical intervention. Ultrasonography, in conjunction with a thorough physical examination, is the primary diagnostic approach for confirming herniation. Although there are no documented cases of incarcerated inguinal hernias in donkeys, adopting a similar diagnostic and treatment strategy is advisable to prevent complications in these animals.

Congenital inguinal hernias in foals are most commonly seen in newborn males, presenting as an enlarged scrotum caused by the presence of fluid and/or intestines. These hernias typically do not require surgical repair unless they cannot be easily reduced or are associated with intestinal strangulation. Severe cases may present with scrotal pain, swelling, and, in advanced stages, skin necrosis. Immediate identification and surgical intervention are necessary to prevent further complications. In both congenital and acquired cases involving intestinal incarceration or strangulation, bilateral castration is often recommended to prevent recurrence.

Mild congenital inguinal hernias, where the peritoneum remains intact and the hernia is reducible, can often be managed conservatively with daily reduction. Most uncomplicated cases resolve spontaneously as the foal matures. However, due to limited research interest in donkeys, there are currently no reports of congenital inguinal hernias in this species. Further studies are needed to better understand the presentation and management of this condition in donkeys.

9 Cryptorchidism

Cryptorchidism is characterized by the abnormal location of one or both testes in a location other than the scrotum (Fig. 4). In horses, the undescended testis may be located in the inguinal region or in the abdomen. Usually, the testis located in the inguinal region can be palpated externally during a physical exam. In horses, males with inguinal retained testis are anecdotally referred as "high flankers". However, true inguinal cryptorchid males should be differentiated from males with a testis that can be palpated in the inguinal area but manipulated into the scrotum.

Cryptorchidism is one of the most prevalent urogenital conditions reported in donkeys. In one study evaluating 274 donkeys, 30% (74/274) were cryptorchid (Kumi-Diaka et al. 1981). The majority (71/74) of the cases consisted of unilateral cryptorchidism, whereas 6 donkeys were bilateral. Interestingly, all cryptorchid donkeys had the retained testis in the inguinal region.

Bilateral cryptorchid males are infertile, whereas unilateral cryptorchid males are often fertile. However, although the heritability of cryptorchidism remains an enigma, it is usually undesirable to maintain cryptorchid males for reproduction; therefore, males with this condition should be castrated and not used for reproduction. It is important to note bilateral cryptorchid males show normal libido and behavior since the androgen-producing cells of Leydig of an undescended testis remain functional and produce testosterone.

10 Ectopic Testes

Ectopic tests are those that are not present in the scrotum, inguinal region, or abdomen. The ectopic-located testis may lie alongside the penis, cranial to the scrotum, in the ventral abdomen. Interestingly, although ectopic testis has not been well documented in horses, and has been considered much more common in cattle,

Fig. 4 Caribbean donkey with only one scrotal testicle, a possible unilateral cryptorchid

one report evaluating 274 donkeys described this condition in three donkeys (Kumi-Diaka et al. 1981). All animals exhibited complete absence of a scrotum and were uniformly aggressive, and all exhibited excessive libido. In one donkey, the testes were located subcutaneously on the ventral abdomen, about 19 cm apart, equidistant from the penis. In the other two donkeys, the testes were located in the femoral canal and were embedded in fatty tissue. In all three donkeys the penis was normally developed, and rectal palpation revealed vestigial seminal vesicles and a prostate gland.

11 Abnormal Number of Testicles

Monorchidism describes the complete absence of one testis, whereas polyorchidism is the presence of more than two testicular structures. Both conditions are rare in equids. The differentiation between agenesis and a cryptorchid with total testicular degeneration is difficult. The diagnostic approach to monorchid and cryptorchid horses should be similar. A definitive diagnosis of monorchidism is based on history, behavior, and physical examination and may require hormonal testing and surgical exploration.

12 Testicular Hypoplasia

Testicular hypoplasia is a condition characterized by underdeveloped testes, resulting in reduced fertility or complete infertility. This condition can occur unilaterally or bilaterally. In cases of bilateral involvement, affected animals are typically sub-fertile to sterile due to azoospermia, severe oligospermia, or the production of dysfunctional spermatozoa. Testicular hypoplasia is generally considered congenital and may have a hereditary component. The reduced size of the testes is attributed to a diminished amount of seminiferous epithelium.

While specific cases of testicular hypoplasia in donkeys are not well-documented, similar reproductive abnormalities have been reported in other equids. In horses, it is advised not to finalize a diagnosis of testicular hypoplasia before the stallion reaches two years of age, as this condition could be misdiagnosed in individuals with delayed testicular development. The diagnostic criteria and age considerations for donkeys remain unexplored and warrant further investigation.

13 Abnormalities of the Epididymis

Epididymal abnormalities are uncommon in equids. In stallions, congenital hypoplasia of the epididymis and segmental aplasia of the mesonephric duct, which can result in the absence of the epididymal tail, have been reported. In donkeys, epididymal sperm granulomas have been documented (Podico et al. 2021). These abnormalities often lead to azoospermia despite continued sperm production. The accumulation of sperm due to obstruction (spermatostasis) can trigger the formation of antisperm antibodies, tubular dilation, and eventual testicular atrophy, all of which compromise fertility.

14 Penile/Preputial Abnormalities

Congenital abnormalities of the penis in equids are rare. In donkeys, an exceptionally rare case (Abu-Seida and Torad 2014) of diphallia (double penis and double prepuce) with double scrota has been reported in a two-year-old donkey (Fig. 5). In this case, the posterior penis was functional and structurally normal but displayed ventral deviation and was positioned caudally to the posterior scrotum and testes. The testes were atrophied, asymmetrical, and located anterior to the posterior penis in a transverse plane. The anterior penis was rudimentary, non-functional, replaced by a fibrous band, fixed to the prepuce, and oriented cranially. Unfortunately, the karyotype of the animal was not determined, leaving the underlying cause of these phenotypical abnormalities unclear.

No other penile abnormalities have been documented in donkeys. However, several congenital conditions affecting the penis and prepuce have been described in horses. For example, constriction or stenosis of the preputial ring, which prevents penile extrusion, is a congenital disorder clinically consistent with phimosis. Clinical

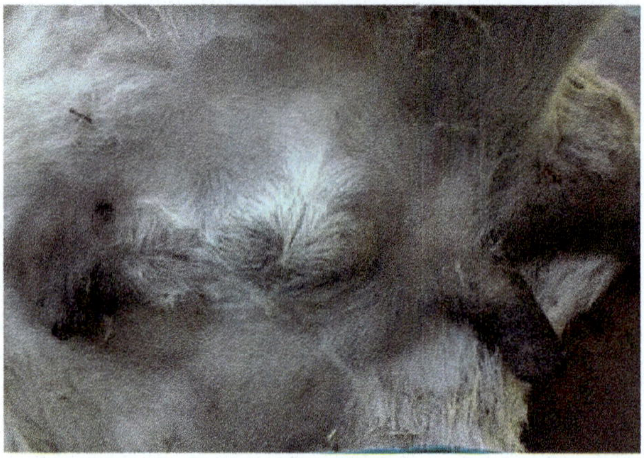

Fig. 5 A two-year-old donkey showing double peni, double scrota, double prepuces and two asymmetrical testes (Abu-Seida and Torad 2014).

signs may be evident at birth or emerge later in life. Additionally, congenital aplasia or hypoplasia of the prepuce has been reported in horses. In these cases, animals may present with paraphimosis, which is due to insufficient preputial tissue to retain the penis within the prepuce. Most congenital anomalies of the penis and prepuce in horses are thought to occur in male pseudohermaphrodites.

15 Testicular Tumors

To the best of the authors' knowledge, there are no documented cases of testicular tumors in donkeys. A study examining post-mortem findings in 1444 aged donkeys found no lesions in the male reproductive tract (Morrow et al. 2011). Nevertheless, it is hypothesized that testicular tumors in donkeys would resemble those reported in horses. In equines, seminomas and interstitial cell tumors are the most common testicular neoplasms in mature stallions, while teratomas are more frequently associated with young and/or cryptorchid stallions. Regardless of the type, surgical removal is the recommended treatment for such conditions.

16 Tumors of the Female Reproductive Tract

Not many reports have described tumors of the jenny's reproductive tract. In one report from 2001 (Sokkar et al. 2001), the genital tracts of 165 female donkeys were collected and examined with special concern to neoplasms. Ovarian Neoplasms, including a granulosa cell tumor, were found in 12 cases (7.72%). Uterine neoplasms were found in 14 cases (8.84%), which were cavernous haemangioma (10 cases, 6.06%), endometrial polyps and fibroleiomyoma (one case each, 0.61%), and leiomyoma (2 cases, 1.21%). In addition, 5 cases (3.03%) of cervical polyps were also observed. In another report, a tubular adenocarcinoma in a 6-year-old jenny was diagnosed but no primary genital tract tumors were observed in a survey of 125 cases of neoplasia in donkeys (Davis et al. 2016).

17 Urolithiasis

Urolithiasis (bladder stones) is a condition that can affect donkeys and mules, although it is infrequently reported in the literature. This may reflect a genuinely low prevalence or result from subtle clinical signs and under recognition by veterinarians or owners. Affected animals may show signs of dysuria, hematuria, pollakiuria, and, in severe cases, complete urethral obstruction.

Few clinical reports of urolithiasis in donkeys and mules are currently available (Abdel-Hady 2014; Robert et al. 2016; Mendoza et al. 2018; Campebell et al. 2021; Lazim et al. 2017). Diagnosis typically involves rectal palpation, ultrasonographic evaluation of the bladder and urethra, and, in some cases, endoscopic examination of the lower urinary tract. In the authors' experience, discomfort in jennies and mares may be misinterpreted as estrus behavior, as these animals often assume a urination stance repeatedly.

Timely diagnosis is crucial for a favorable outcome. Prognosis is generally better in females due to their shorter and wider urethra, which facilitates treatment (Abdel-Hady 2014). Management most often requires surgical intervention, including urethrostomy, cystotomy, or, in select cases, laser lithotripsy. In female animals, depending on the size and location of the stone, transurethral removal via bladder lavage and endoscopy may be possible (Fig. 6).

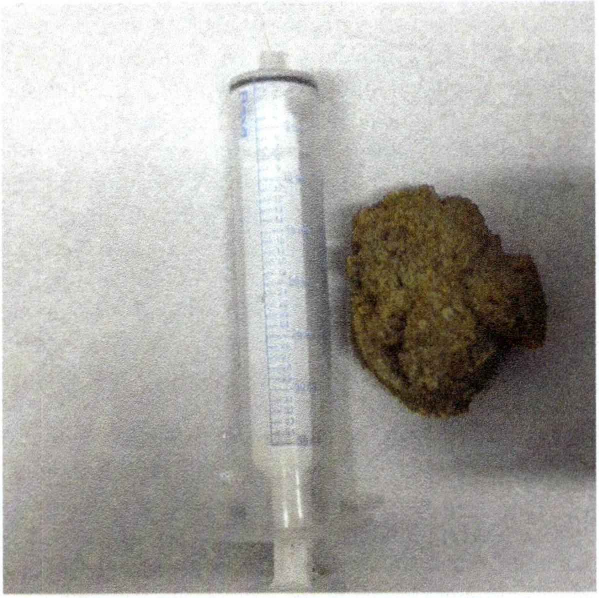

Fig. 6 Urolith removed from the urethra of a Pêga jenny. The owner reported persistent signs of estrus

18 Conclusion

Congenital and inherited disorders of the urogenital system in equids represent a diverse and complex group of conditions with significant implications for reproductive function, animal welfare, and veterinary care. While many of these conditions have been relatively well described in horses, there remains a marked lack of data regarding their occurrence, diagnosis, and management in donkeys. This knowledge gap likely stems from a combination of underreporting, subtle or inapparent clinical signs, and limited research attention given to this species, despite donkeys' global importance in agriculture, transport, and subsistence livelihoods.

Among the most commonly observed abnormalities are cryptorchidism and inguinal hernias. Cryptorchidism, particularly unilateral inguinal retention, appears to be the most prevalent urogenital anomaly reported in donkeys, affecting approximately 30% in some populations. Although fertility may be retained in unilateral cases, these individuals are generally not recommended for breeding due to potential heritability and behavioral concerns. Inguinal hernias, though less frequently documented in donkeys, have been reported in both male and female animals and can present as veterinary emergencies if associated with intestinal incarceration.

Intersex conditions, including male and female pseudohermaphroditism and true hermaphroditism, while rare, highlight the complexity of sex development disorders in equids. These conditions often involve mosaic chromosomal patterns and require a combination of genetic testing, histopathology, and clinical evaluation for diagnosis. Behavioral anomalies, such as increased libido or ambiguous external genitalia, may be the first clinical signs, and surgical intervention—especially castration—is often necessary for management and welfare.

Sex-reversal syndromes, although uncommon, are critical to recognize, particularly in animals presenting with atypical gonadal tissue not consistent with chromosomal sex. Diagnostic tools such as karyotyping and gonadal imaging provide essential information to guide case management. Mosaicism further complicates diagnosis and may be associated with both reproductive and non-reproductive anomalies, including growth and limb deformities.

In male donkeys, additional rare conditions such as ectopic testes, polyorchidism, and testicular hypoplasia can affect fertility and behavior. Ectopic testicles may be located outside the normal descent pathway and are often associated with high libido and aggressive behavior. Testicular hypoplasia and abnormalities of the epididymis, though uncommon, can lead to infertility and often require detailed reproductive evaluations.

Female donkeys are not exempt from urogenital pathology. Neoplasms of the reproductive tract, while infrequently reported, include ovarian tumors such as granulosa cell tumors and uterine conditions like leiomyomas and haemangiomas. Additionally, urolithiasis, though rare, poses significant clinical challenges due to potential urinary obstruction, especially in males. Early recognition and surgical intervention are essential for a favorable outcome.

In summary, while donkeys share many urogenital pathologies with horses, species-specific considerations—such as anatomical differences, behavior, and vascular supply—necessitate tailored diagnostic and treatment approaches. Increased awareness, research focus, and clinical reporting are critical to bridging the knowledge gap in donkey reproductive and urinary health. By enhancing our understanding of these conditions, veterinary practitioners will be better equipped to support the health, welfare, and reproductive management of donkeys globally.

Questions

1. **What is the most commonly reported congenital urogenital abnormality in donkeys?**
   Answer:
   Cryptorchidism is the most frequently reported congenital urogenital abnormality in donkeys. In a study of 274 donkeys, 30% were cryptorchid, with most cases involving unilateral inguinal retention of the testis.
2. **How does male pseudohermaphroditism present in equids?**
   Answer:
   Male pseudohermaphrodites are genetically male (XY) but exhibit female or ambiguous external genitalia. They often have intra-abdominal testicles and may display an enlarged clitoris resembling a penis. Behaviorally, they may show stallion-like traits or nymphomanic behavior.
3. **What diagnostic methods are recommended for evaluating intersex or sex-reversed animals in equids?**
   Answer:
   Recommended diagnostics include clinical examination, ultrasonography of gonads and internal reproductive structures, genetic testing (karyotyping), and histopathological analysis to confirm discrepancies between chromosomal sex and gonadal tissue.
4. **Why is surgical castration often advised in donkeys with intersex conditions or ectopic testes?**
   Answer:
   Castration is advised to manage undesirable behaviors, prevent potential complications, and because many of these animals are infertile or should not be used for breeding due to genetic anomalies.

References

Abdel-Hady AAA (2014) Spontaneous repelling of a large urocystolith in a Working She-donkey. Sch J Agric Vet Sci 1(3):105–106

Abu-Seida A, Torad F (2014) Diphallia and double scrota in a donkey: a case report. Res Opin Anim Vet Sci 4:117–119

Bowling AT, Millon LV (1990) Two autosomal trisomies in the horse: 64,XX,-26,+t(26q26q) and 65,XX,+30. Genome 33(5):679–682. https://doi.org/10.1139/G90-101

Buoen LC, Zhang TQ, Weber AF, Turner T, Bellamy J, Ruth GR (1997) Arthrogryposis in the foal and its possible relation to autosomal trisomy. Equine Vet J 29(1):60–62. https://doi.org/10.1111/j.2042-3306.1997.tb01638.x

Campebell RC, Silva FB, Silva JFB et al (2021) Obstructive urethrolithiasis in a mule. Acta Sci Vet 49(Suppl 1):674

Davis CR, Valentine BA, Gordon E, McDonough SP, Schaffer PA, Allen AL, Pesavento P (2016) Neoplasia in 125 donkeys (Equus asinus): literature review and a survey of five veterinary schools in the United States and Canada. J Vet Diagn Invest 28(6):662–670. https://doi.org/10.1177/1040638716665659

Dorado J, Anaya G, Bugno-Poniewierska M, Molina A, Mendez-Sanchez A, Ortiz I, Moreno-Millán M, Hidalgo M, Peral García P, Demyda-Peyrás S (2016) First case of sterility associated with sex chromosomal abnormalities in a jenny. Reprod Domest Anim 52(2):227–234. https://doi.org/10.1111/rda.12884

Kubień EM, Tischner M (2002) Reproductive success of a mare with a mosaic karyotype: 64,XX/65,XX,+30. Equine Vet J 34(1):99–100. https://doi.org/10.2746/042516402776181240

Kumi-Diaka J, Kwanashie GG, Ogwu D (1981) Cryptorchidism and malformations of the scrotum and testes of donkeys. Theriogenology 15(3):241–243

Lazim et al (2017) J Tradit Med Clin Natur 6:1. https://doi.org/10.4172/2573-4555.1000211

Marzok MA, Moustafa AM (2023) Acquired non-incarcerated inguinal/scrotal herniation in working donkeys (Equus Asinus). Bulg J Vet Med 26(2):280–288. https://doi.org/10.15547/bjvm.2396

Mendoza FJ et al (2018) J Equine Vet Sci 65:86e97

Meyers-Wallen VN, Hurtgen J, Schlafer D, Tulleners E, Cleland WR, Ruth GR, Acland GM (1997) Sry-negative XX true hermaphroditism in a Pasa Fino horse. Equine Vet J 29(5):404–408. https://doi.org/10.1111/j.2042-3306.1997.tb03148.x

Morrow LD, Smith KC, Piercy RJ, du Toit N, Burden FA, Olmos G, Gregory NG, Verheyen KLP (2011) Retrospective analysis of post-mortem findings in 1,444 aged donkeys. J Comp Pathol 144(2–3):145–156. https://doi.org/10.1016/j.jcpa.2010.08.005

Paget S, Ducos A, Mignotte F, Raymond I, Pinton A, Séguéla A, Berland HM, Brun-Baronnat C, Darré A, Darré R, Tamzali Y, Bergonier D, Berthelot X (2001) 63,XO/65,XYY mosaicism in a case of equine male pseudohermaphroditism. Vet Rec 148(1):24–25. https://doi.org/10.1136/vr.148.1.24

Podico G, Lago-Alvarez Y, Carossino M, Ferrer MS, Arnold CE, Canisso IF (2021) Epididymal sperm granuloma and antisperm antibodies in donkeys. J Equine Vet Sci 101:103450. https://doi.org/10.1016/j.jevs.2021.103450

Robert et al (2016) Rectal prolapse secondary to long standing urolithiasis in a donkey. Equine Vet Educ 28(11):625–630

Sato F, Hirota K, Tozaki T, Ito K, Dhakal P, Taya K, Endo Y, Murase H, Nambo Y (2012) A case of ambiguous external genitalia in a thoroughbred male horse with the 63,XO/64,XY Mosaic Karyotype. J Vet Med Sci 74(10):1327–1331. https://doi.org/10.1292/jvms.11-0473

Sokkar SM, Hamouda MA, Abdel-Rahman SM (2001) Histopathological studies on neoplasms in the genital tract of female donkeys in Egypt. Berl Munch Tierarztl Wochenschr 114:166–168

Part II

Diagnostics, Treatment, and Surgery

Veterinary Diagnostic Sample Collection, Laboratory Submission and Testing Strategies (Antemortem and Postmortem Considerations) for Donkeys and Mules

Erin L. Goodrich and Mason Jager

Abstract

The stoic nature of donkeys and mules can prove to be challenging for the practitioner who is tasked with diagnosing an illness or disease condition in these animals. Testing to obtain a specific diagnosis serves as a crucial tool for veterinarians who are examining sick donkeys and mules. Testing of both live and dead equids allows for the identification of pathogens, which can aid in the development of targeted prevention measures to reduce transmission in a group or herd. It can also assist in the detection of zoonotic disease agents, thereby allowing for appropriate post exposure or preventive measures to be implemented. Although a gross necropsy examination may provide all the answers that are needed surrounding the death of a donkey or mule, further diagnostic testing utilizing specimens collected from the necropsy may be warranted in many situations. These tests can only be properly performed and interpreted when the diagnostic specimens are collected appropriately. This chapter will describe the proper diagnostic specimens to collect from both live and dead equids in a variety of scenarios, for submission to a laboratory for testing. It will also describe the equipment and techniques required to perform a proper field necropsy.

Keywords

Diagnostic testing · Specimens · Necropsy · Laboratory

E. L. Goodrich (✉) · M. Jager
Department of Population Medicine and Diagnostic Sciences, Cornell College of Veterinary Medicine, Ithaca, NY, USA
e-mail: elg25@cornell.edu

A. K. McLean et al. (eds.), *Donkey and Mule Medicine: Clinical Medicine and Reproduction*, https://doi.org/10.1007/978-3-032-18540-2_12

1 Introduction

Veterinary diagnostic services are crucial to animal health, and the associated technology is constantly evolving. The specific services offered, and the areas of expertise vary across laboratories in relation to the needs of the veterinary industry in the region and the resources available. In addition, there are various accreditation programs available to veterinary laboratories, depending on their services and location. In the United States (U.S.), many state and university-affiliated laboratories benefit from receiving accreditation through the American Association of Veterinary Laboratory Diagnosticians (AAVLD) and/or the American Association of Laboratory Accreditation (A2LA), both of which ensure a high standard of quality across the laboratory is maintained and conveyed in the diagnostic tests that are performed and the results that are generated. Full-service laboratories may have specialized units offering tests in a variety of disciplines, including anatomic pathology, clinical pathology, bacteriology/mycology, virology, serology/immunology, molecular diagnostics, parasitology, toxicology, endocrinology, coagulation, and more. When choosing a diagnostic laboratory to test samples obtained from donkeys and mules, it is important to understand the testing capabilities of that lab. While donkeys and mules do get some of the same diseases as horses, they can also suffer from unique diseases and infections with pathogens that are not typically seen in horses, such as asinine herpesviruses. Additionally, interpretation of some test results relies on comparing those results to population-based reference intervals that have been established for that specific species and laboratory. Caution must be taken when reference intervals appear on result reports for donkey and mule tests as some laboratories will utilize intervals established for horses, which can result in erroneous interpretation of results for donkeys and mules. Communicating directly with your veterinary diagnostic laboratory or checking for information on their website is necessary to ensure that proper samples are collected, appropriate tests are requested, and results are interpreted correctly.

2 Sample Handling

In veterinary medicine, diagnostic specimens are often collected from stressed and/or sick equids in remote, rural areas. Controlling the temperature of the specimens and ensuring prompt delivery to a laboratory is challenging in these conditions. The environmental conditions, sample handling techniques, and the equid-associated variables that may impact the integrity of the diagnostic specimens prior to testing are described as preanalytical factors. It is imperative that we minimize the impacts of these whenever possible.

2.1 Collection

First, check the expiration date of the tube or media that you will be using for your sample collection. Expired blood collection tubes may have lost their vacuum, and the additives may no longer be effective. Likewise, expired blood culture media or other transport media may not be effective and can impact the results of the tests.

Follow instructions from tube and media manufacturers to ensure proper sample collection techniques are followed. This may include filling tubes to a minimum volume, utilizing aseptic techniques to obtain specimens, and/or gently inverting tubes or media immediately after collection.

Label all diagnostic specimens clearly and legibly. Each specimen should be identified with the name of the patient, the date it was collected, and the site or specimen type. For example, plasma transferred from an EDTA whole blood sample and placed in a plain tube may be mistaken for any other type of plasma, or even serum, if not properly labeled. Likewise, swabs collected and placed in tubes or transport media must be labeled with the site that was swabbed.

2.2 Temperature

Many diagnostic specimens must be kept chilled (refrigeration temperature, roughly 2–4 °C, about 35–39 °F) but some necessitate freezing (−18 °C, 0 °F), while others must be kept close to room temperature (roughly 20 °C, 68 °F) prior to arrival in a diagnostic laboratory for testing. Consultation with experts at the diagnostic laboratory prior to sample collection is the best way to ensure proper handling techniques. Blood culture bottles and blood smears or other slides are a few examples of diagnostic specimens that should be maintained at room temperature after collection, and during transport to the lab.

2.3 Packaging and Shipping

In the U.S., the Department of Transportation and International Air Transportation Association regulates the packaging and shipping requirements for clinical samples. Specimens must be packaged in leakproof containers with enough absorbent material to absorb all the enclosed liquids. Triple layer packaging is recommended or may be required in some instances. Specimens should be packaged appropriately to withstand the anticipated jostling as well as the pressure and temperature changes that may occur during transport.

Avoid the shipment of medical sharps such as needles and scalpel blades as these pose a risk to carriers and laboratory personnel. To ship specimens that must be kept cool, include ice packs in the package, but be sure to provide padding between the hard frozen ice packs and any breakable specimen containers. To ship specimens at room temperature during hot or cold months, including room temperature ice packs can increase the ambient mass and help to insulate the specimens from the external temperatures that the package is exposed to. It is always advised that diagnostic specimens be shipped for overnight delivery to the laboratory whenever possible.

3 Common Diagnostic Specimens

3.1 Live Animal (Antemortem)

3.1.1 Blood Samples

One of the most common sample types obtained from live equids for diagnostic testing is blood. Blood samples are characterized as whole blood (collected into a tube with an anticoagulant such as EDTA, heparin or citrate), serum, or plasma. Depending on the tube or media used for blood collection, the sample may be used for clinical pathology testing, micronutrient assays, toxicology tests, endocrinology tests, hemostasis and thrombosis assays, antibody titers, or for pathogen detection in the form of polymerase chain reaction (PCR) assays, or even culture. See Table 1 for a summary of some of the common types of collection tubes, blood samples or blood components associated with each, and some examples (not an exhaustive list) of diagnostic tests that can be performed on each sample.

Blood collected into a tube containing anticoagulant (EDTA, heparin, or citrate) is described as whole blood. To avoid inappropriate proportions of blood to anticoagulant and erroneous results, blood tubes must be filled to the minimum fill lines. Tubes should be gently agitated immediately after filling to ensure proper mixing of anticoagulant with blood, and to avoid clotting. When whole blood in a tube containing anticoagulant is centrifuged, the supernatant is plasma. Plasma is further characterized by the anticoagulant additive contained in the tube. When submitting EDTA whole blood to a laboratory for a hemogram (complete blood count or CBC), it is also advisable to send along 2–3 unstained, air-dried blood smears prepared from the blood sample, at the time of collection. These slides can be examined by a clinical pathologist and will aid in interpretation of the hemogram, and in the diagnosis of hemoparasites. See Table 2 for the hematology reference intervals established from 109 donkeys at the Animal Health Diagnostic Center (AHDC) (Goodrich and Webb 2024).

When blood is collected into a non-additive tube, such as a red top tube or a white top tube, and allowed to clot, the supernatant remaining after that clotted blood is centrifuged is called serum. Donkeys are especially prone to hyperlipidemia and while the incidence in mules is unknown, it is thought to occur more frequently than in horses, so triglyceride concentration is a key component of the biochemistry panel that MUST be included in the diagnostic work up of every sick donkey or mule. (Mendoza et al. 2019) Take note of the appearance of the serum, as opalescent serum or plasma (both shown in Fig. 1 below) help to establish a preliminary diagnosis of hyperlipidemia. Triglyceride concentration should still be quantified though as it can be utilized as a prognostic indicator and monitored throughout the course of treatment. Likewise, if the serum or plasma appears normal, hyperlipidemia cannot be ruled out without measuring the triglyceride concentration.

Table 1 Some of the common types of collection tubes, blood samples or blood components, and examples of diagnostic tests that can be performed on each

Blood collection container or media	Blood or blood component	Diagnostic test or testing modality
Lavender (purple) top tube 	EDTA whole blood	Hemogram Polymerase chain reaction (PCR) tests Selenium testing Lead & other heavy metal testing
Blood smear made from EDTA whole blood 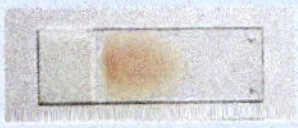	Blood smear	Blood smear examination (should accompany whole blood submission for hemogram)
Supernatant removed from lavender top tube and placed in new plain tube 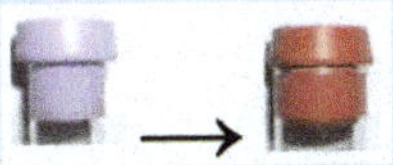	EDTA plasma	Endogenous ACTH
Supernatant removed from green top tube and placed in new plain tube 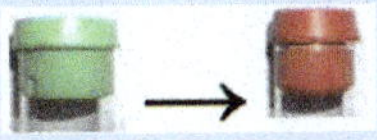	Heparinized plasma	Pre-purchase drug screening Biochemistry panel
Supernatant removed from blue top tube and placed in new plain tube 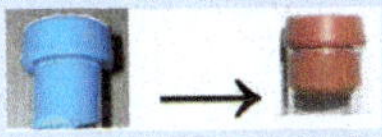	Citrated plasma	Comparative coagulation tests
Supernatant removed from red top or tiger top tube and placed in new plain tube 	Serum	Biochemistry panel Vitamin quantification Antibody detection tests Insulin concentration Thyroid testing Leptin
Inoculated aerobic or anaerobic blood culture bottle with up to 5 mls of blood, collected <u>aseptically</u> 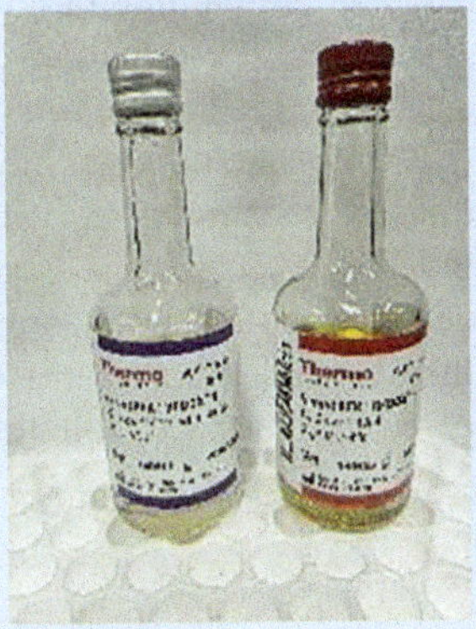	Whole blood	Aerobic or anaerobic blood culture

Table 2 Hematology reference intervals established from 109 apparently healthy adult donkeys and at least 120 apparently healthy adult horses at the AHDC (Goodrich and Behling-Kelly 2019; Goodrich and Webb 2024)

Measurand	Donkey Lower reference limit	Donkey Upper reference limit	Horse Lower reference limit	Horse Upper reference limit
Hematocrit (%)	28	53	34	46
RBC (10^{12}/L)	4.4	8.8	6.6	9.7
Hemoglobin (g/L)	93	175	118	159
MCV (fL)	53	68	43	55
MCHC (g/L)	320	350	330	360
MCH (pg)	18	23	15	20
RDW (%)	16.1	19.4	16.3	19.3
WBC (10^9/L)	5.8	16.0	5.2	10.1
Neutrophil (10^9/L)	2.2	8.5	2.7	6.6
Lymphocyte (10^9/L)	1.5	9.9	1.2	4.9
Monocyte (10^9/L)	0	1.0	0	0.6
Eosinophil (10^9/L)	0	1.7	0	1.2
Basophil (10^9/L)	0	0.1	0	0.2
Platelet (10^9/L)	84	380	94	232
MPV (fL)	4.8	8.1	5.3	8.4
TP (g/L)	60	81	52	78

It is also important to realize that hypertriglyceridemia can interfere with some hematology and serum biochemistry tests so results must be interpreted with caution. Depending on the techniques and analyzers used, hypertriglyceridemia may falsely increase mean cell hemoglobin (MCH), mean cell hemoglobin concentration (MCHC), total protein, and platelet count. Additionally, hypertriglyceridemia may falsely decrease some serum chemistry analytes such as electrolytes (sodium, chloride and potassium). See Table 3 for the serum biochemistry reference intervals established at the AHDC from 117 donkeys (Goodrich and Webb 2024).

When blood is collected into a serum separator or tiger top tube, the supernatant remaining after centrifugation is called serum, however the gel component of these tubes can be problematic for some tests and laboratory equipment. Consult the laboratory before using these tubes for serum collection.

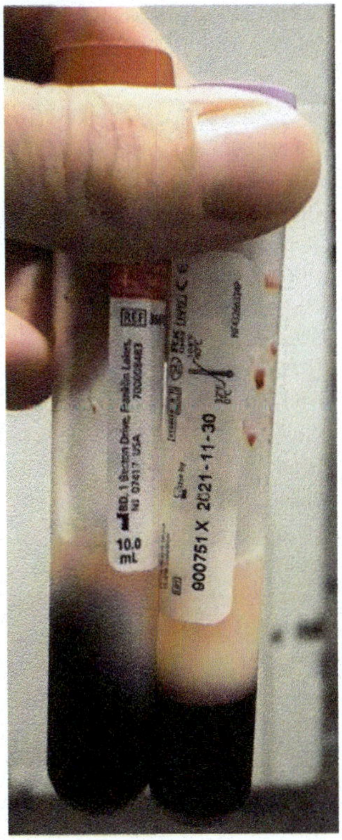

Fig. 1 Opalescent serum (left, in a red top tube) and opalescent plasma (right, in a purple top tube) taken from a donkey with hyperlipidemia

Blood can also be a useful sample for diagnosing bacteremia in the live, sick equid. To do so, it is best to collect the sample during the acute phase of the illness, prior to administration of any antimicrobials. The sample must be collected aseptically and inoculated into aerobic and/or anaerobic blood culture bottles (Dawson 2014). It is often recommended to collect samples for blood culture three times within 24 h to maximize the chances of pathogen detection (Young 1990). For animals who are currently being treated with antibiotics, it is best to collect the sample immediately prior to the timing of the next dose of antibiotic or 3 days after the cessation of treatment.

3.1.2 Enteric/Fecal Samples

Fecal samples and/or swabs of feces placed in transport media may be collected from donkeys and mules for a variety of testing purposes, including endoparasite assessments, enteric cultures, PCRs for specific pathogen detection, and even to check for the presence of sand.

A fecal sample (1–2 normal fecal balls, or about 10 g of feces) collected into a leak proof container and refrigerated may be submitted to a lab within 7 days for a fecal egg count (FEC). Samples should not be frozen as that can result in damage to the eggs and a falsely decreased egg count. FECs are used to screen for endoparasites

Table 3 Serum biochemistry reference intervals established for 117 apparently healthy adult donkeys and at least 120 apparently healthy adult horses at the AHDC (Goodrich and Behling-Kelly 2019; Goodrich and Webb 2024)

Measurand	Donkey Lower reference limit	Donkey Upper reference limit	Horse Lower reference limit	Horse Upper reference limit
Sodium (mmol/L)	127	140	134	142
Potassium (mmol/L)	3.4	5.7	2.4	4.8
Chloride (mmol/L)	94	106	95	104
Bicarbonate (mmol/L)	16	27	24	31
Anion gap (mmol/L)	12	25	12	19
Na/K ratio	23	40	–	–
Urea nitrogen (mmol/L)	1.1	7.5	1.7	3.7
Creatinine (μmol/L)	53	133	71	133
Calcium (mmol/L)	2.7	3.2	2.7	3.2
Phosphorus (mmol/L)	0.7	1.6	2.1	4.7
Magnesium (mmol/L)	0.5	1.0	1.2	1.9
Total protein (g/L)	57	78	54	70
Albumin (g/L)	20	36	29	36
Globulin (g/L)	29	48	23	38
A/G ratio	0.5	1.1	0.8	1.5
Glucose (mmol/L)	3.2	6.5	3.9	6.8
AST (U/L)	223	555	222	489
SDH (U/L)	2	10	1	6
GLDH (U/L)	0	9	2	10
GGT (U/L)	10	55	8	33
Bilirubin (μmol/L)	0	3.4	0.5	2.1
Direct bilirubin (μmol/L)	0	1.7	0.1	0.5
Indirect bilirubin (μmol/L)	0	1.7	0.3	2.0
CK (U/L)	95	500	171	567
Cholesterol (mmol/L)	1.4	2.4	1.8	3.4
Triglycerides (mmol/L)	0.3	2.2	0.2	0.7

and to determine the shedding status (high, moderate or low). It is generally recommended to perform FECs one to two times yearly. When anthelmintics are indicated and administered, fecal egg counts should be repeated 14 days later to assess the percent reduction in the endoparasite population of interest. This is called a fecal egg count reduction test (FECRT), and it can be used to assess for resistance of endoparasites to various anthelmintic drug classes in herds or groups of equids ("Internal-Parasite-Guidelines_Updated.Pdf" n.d.). See Volume 1 for more detailed information regarding parasites of donkeys and mules.

Donkeys and mules are capable of developing patent, subclinical infections with lungworms (*Dictyocaulus arnfieldi*) (Saadi et al. 2018). Horses sharing a pasture or paddock with donkeys or mules are at risk of exposure to lungworm, which often does present with coughing in horses but rarely develops into a patent infection. Therefore, screening of donkeys for lungworm is important and can be achieved using sugar flotation or the Baermann technique on a fecal sample. The sugar floatation method can distort the lungworm larvae but is often more sensitive for detecting them. The Baermann technique, however, allows for a better estimate of the actual number of larvae present in the feces.

Fecal samples can also be collected for the purposes of pathogen detection via PCR. In that case, placing the fecal sample in a leak proof, clean container is appropriate. Alternatively, swabs of the fecal sample may be submitted in plain red or white top tubes with 0.5 mls of sterile saline added. Swabs placed in gel-based transport media are typically not appropriate for PCR tests.

Fecal samples may also be collected for the purposes of bacterial culture. In this instance, it is necessary to communicate to the laboratory the specific organisms you are considering as differential diagnoses. Some labs require specific culture techniques or other tests for various enteric pathogens and can make recommendations based on the differentials on your list.

3.1.3 Nasal Swabs and Swabs of Other Sites

Nasal swabs, swabs from draining abscesses, and swabs of other lesions are also common diagnostic specimens collected from live equids.

Swabs collected for the purposes of PCR testing must be placed either in plain (non-additive) tubes with a few drops of sterile saline or in viral transport media, or other media intended for the purposes of PCR. Gel-based transport media is generally not acceptable for PCR testing.

Swabs collected for the purposes of culture must be placed in appropriate transport media that maintains the viability of the organisms. Some transport medias are appropriate only for aerobic and fungal cultures while others are intended to maintain oxygen-free conditions to maintain the viability of obligate anaerobes. Consult the transport media guidance from the manufacturer for specific instructions related to inoculation, storage temperature, and maximum time allowed from inoculation to plating for culture, to ensure viability.

3.1.4 Diagnostic Testing Plans

The most appropriate diagnostic laboratory testing strategy depends on the history and signalment of the donkey or mule, the clinical presentation and physical examination findings, financial constraints of the owner, the geographical region where the equid is located, and the resources available to the veterinarian. Some common diagnostic plans to consider in donkeys and mules are described in Appendix. Please note that whenever possible laboratories with established donkey or mule-specific reference intervals should be used since interpreting results compared to values established for horses, or values established by other laboratories may result in misdiagnoses. These diagnostic plans do not include other diagnostic tests or procedures that may be performed in the clinic or in the field such as radiographs, ultrasound examinations, rectal palpations, etc.

3.2 Dead Animal (Postmortem)

3.2.1 General Considerations

Postmortem examinations are a vital diagnostic tool for determining the cause of illness or death in animals. For complex cases or those involving potential zoonotic diseases, postmortem examinations should ideally be conducted at a diagnostic laboratory. However, field-based examinations can be valuable when access to laboratories is limited. A clinician or trained technician can perform a targeted postmortem examination of the thoracic and abdominal cavities or collect tissue samples for further analysis. Samples collected for histopathology can be submitted to veterinary diagnostic laboratories for evaluation by a pathologist, while other samples can be sent for microbiology, toxicology, parasitology or other testing.

Challenges specific to field postmortem examinations include environmental factors (such as ambient temperature), limited access to tools, personnel, and hydraulic tables, public onlookers, and the need to maintain appropriate biosecurity measures. Discussions with clients ahead of time can help ensure they are informed and willing to allow a postmortem examination when the need arises. Selecting an appropriate location is also crucial; it should offer privacy, accessibility for heavy machinery to move remains, and protection from scavenging animals.

3.2.2 Equipment

Maintaining a dedicated postmortem kit in the veterinary clinic or vehicle ensures that all necessary tools are readily available. Many of these supplies can be found in hardware stores, medical supply companies, and meat processing suppliers. A basic field postmortem kit should include:

- Sharp knives (at least two) and sharpeningspiepr146 steel
- Steel loppers for cutting ribs
- A short handsaw
- Scissors and forceps
- Skull key

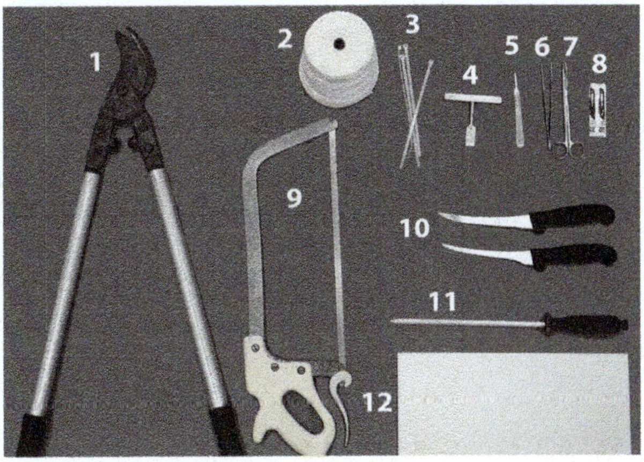

Fig. 2 An overview of tools that can be used to perform a postmortem examination in the field. 1. Steel loppers 2. String 3. Zip ties 4. Skull key 5. Scalpel 6. Forceps 7. Scissors 8. Scalpel blades 9. Handsaw 10. Knives 11. Sharpening steel and 12. Small cutting board

- String or zip ties
- Scalpel blade
- Expired suture material or umbilical tape (for cosmetic closures, when indicated)

See Fig. 2 for examples of these tools.

Additional supplies:

- Digital camera or cell phone for photo documentation (placed in a sealed plastic bag to prevent contamination)
- Pen, paper and clipboard to take notes on findings
- Tissue collection bags or other non-breakable, leak proof containers for individual fresh (non-formalin fixed) tissues
- Plain (non-additive) tubes and syringes to collect liquid samples (when appropriate)
- Small cutting board
- Non-breakable, leak proof jar(s) of formalin for sample preservation
- Permanent marker for labeling diagnostic specimens

Personal protective equipment (PPE):

- Disposable face shield or protective eyewear and face mask (to cover mucous membranes)
- Plastic sleeves
- Gloves
- Shoe covers or boots that can be easily cleaned and disinfected
- Cut-resistant mesh glove (optional)

See Fig. 3 for examples of PPE.

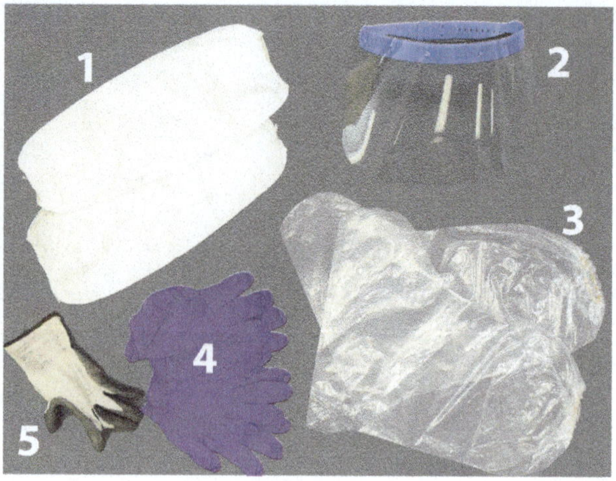

Fig. 3 Examples of personal protective equipment (PPE) that can be used during a postmortem examination. 1. Plastic sleeves 2. Face shield 3. Shoe/boot covers 4. Gloves and 5. Cut-resistant glove

3.2.3 Procedure

A field postmortem examination can range in thoroughness depending on time, weather, and client preferences. Begin by documenting a thorough history as obtained by the owner or handler, noting any recent administration of vaccines or medications, the clinical signs observed prior to death, and any other important herd information. Perform a visual external inspection of the carcass, paying close attention to the hooves, mucous membranes, body condition and noting any abnormalities. Positioning the animal in left lateral recumbency is ideal when feasible, as this approach allows access to core organs and facilitates cosmetic closure, when desired.

1. **Initial Incision**
   - Make a longitudinal incision from the right paralumbar fossa to a point on the right lateral thorax just caudal to the elbow. Extend the incision dorsally toward the shoulder blade. See Fig. 4 for a diagram illustrating the locations for this incision.
   - To avoid dulling the knife on hair, make cuts with the blade facing outward and under the skin. Take care to avoid perforating the intestines by using slow, shallow cuts parallel to the abdominal wall. Once the peritoneum is accessed, the non-cutting hand can press down on the viscera while extending the incision with the knife.
2. **Reflect the Skin and Subcutaneous Tissues**
   - Reflect the skin and subcutaneous tissues dorsally. An assistant can provide tension by pulling the skin edge upward to facilitate cutting.
3. **Open the Thoracic Cavity**
   - Use the steel loppers to cut the ribs. Position the lopper blades perpendicular to the short, transverse axis of the rib for efficient cutting.
   - Remove the ribs to expose both the abdominal and thoracic cavities. This can be facilitated by cutting a slit in the intercostal muscles to create a handle. This

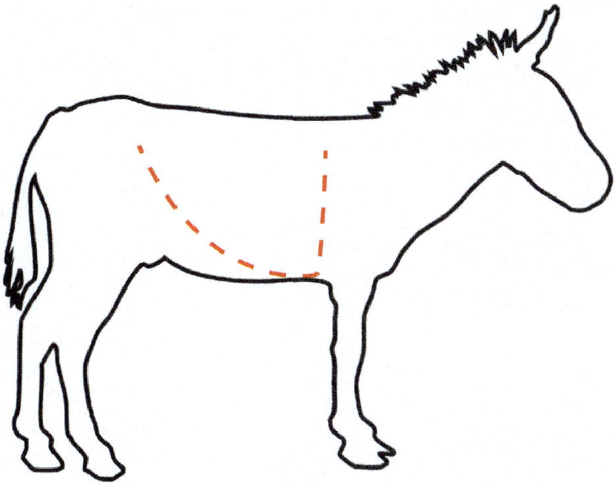

Fig. 4 An initial incision for a postmortem examination is made from the right paralumbar fossa to a point on the right lateral thorax just caudal to the elbow. Extend the incision dorsally toward the shoulder blade

can be used to grasp the rib plate while you cut remaining soft tissue attachments. The removed rib plate can even serve as a cutting board and work surface for processing samples when no other clean surface is available in the field. At this point, take an overview photo of the exposed viscera for documentation.

4. **Collect Two Sets of Tissue Samples: Formalin fixed & fresh (chilled, or frozen)**
   - **Formalin fixed tissues**
     - It is important to collect a full set of formalin fixed tissues from each necropsy, for the purpose of histologic examination by a pathologist. Collect representative samples from major organs, including the liver, kidneys, heart, lungs, spleen, and intestines. Additional samples from adrenal glands, lymph nodes, reproductive tract, stomach, and brain should be collected when possible. See Table 4 for an example of tissues that may be important to include in the formalin fixed tissue set.
     - Prioritize tissues with obvious lesions. Remove larger tissue sections from the body cavity and section them into smaller pieces (1 cm or thinner) for adequate formalin fixation. For larger organs (e.g., kidneys), ensure sections include representative regions such as the cortex and pelvis.
     - The collection jar used for formalin fixed tissues must be leakproof and non-breakable and the volume of formalin should be adequate to ensure a 1: 10 tissue to formalin ratio.
   - **Fresh (not formalin fixed) chilled or frozen tissues**
     - It is also important to collect a complete set of fresh tissues, meaning tissues that will NOT be fixed in formalin. These tissues will be used for ancillary diagnostic tests such as cultures, PCRs, fluorescent antibody tests, etc. Table 4 highlights some of the fresh tissues that may be appropriate to collect during a necropsy, depending on the history, and gross examination

Table 4 Examples of formalin fixed and fresh (chilled or frozen) diagnostic specimens that may be appropriate to collect during a necropsy

Formalin fixed tissues	Fresh (chilled, or frozen) specimens
Lung (section from each lobe)	Lung
Liver (section from each lobe)	Liver
Spleen	Spleen
Kidney	Kidney
Lymph node	Lymph node
Stomach	Stomach contents
Jejunum, ileum, cecum, colon	Intestine (ligated)
Heart	Heart
Brain	Brain
Portions of reproductive tract	Heart blood (do not freeze)
Skeletal muscle	Feces
Adrenal gland	Urine
	Portions of reproductive tract
	Skeletal muscle

findings. Freezing fresh tissues in individual containers, preferably at −80 °C (−112 °F), will preserve the integrity of the samples until a testing strategy is determined. This allows for testing to be ordered in a stepwise fashion, based on the results of previous tests or histologic examination of tissues.

- Fresh tissues should be placed in individual sterile containers, and labeled with the tissue type, the patient's identification and date of collection.
- Fresh tissue samples may be used for culture. For non-enteric samples that are to be submitted for anaerobic culture, the tissue should be at least 2–3 cm^3 in size to ensure anaerobic conditions are maintained in the center of the tissue.
- Enteric (bowel) samples collected for the purposes of anaerobic culture should be ligated at each end, maintaining anaerobic conditions in the lumen.
- It is generally assumed that the surface of most organs will be contaminated during a necropsy, so it is not recommended to swab the outer surface of organs for culture submission. Instead, fresh tissues should be collected and sent to a laboratory for overnight delivery (with ice packs). If it is not possible to immediately ship samples, fresh tissues can be frozen as described above, until shipment to a lab is possible. When tissues are received by a laboratory, efforts are made by the technical staff to decontaminate the outer surface prior to collecting a swab of the interior of the tissue for culture.

5. **Cardiopulmonary System**
   - For a thorough examination of the cardiopulmonary system, extend the incision cranially through the axilla to reflect the right forelimb dorsally. Continue the rib cuts cranially to the thoracic inlet.
   - When indicated, collect samples directly from the trachea, lungs, esophagus, and heart, or remove the entire pluck (trachea, lungs, heart) for closer examination and sampling outside the body.
6. **Brain Removal**
   - Brain removal is indicated for necropsies performed on equids who had neurological signs prior to death. In these instances, the transmission of zoonotic diseases such as rabies virus is a risk. To minimize risk, proper PPE should be worn, such that all mucous membranes are protected. Additionally, the examiner should refrain from using any tools that require electricity as they will increase the risk of aerosolization of virus particles. Only tools such as knives, handsaws and bone rongeurs should be used for the brain removal.
   - Remove the skin from the dorsal aspect of the skull and disarticulate the head at the atlantooccipital joint.
   - Using the handsaw, make a transverse cut across the calvaria just caudal to the orbits. Next, make two sagittal cuts that extend from the lateral edges of the transverse cut to the lateral aspects of the foramen magnum at the base of the skull. See Fig. 5 which illustrates the location for these cuts. If the head is attached to the body for a cosmetic examination, more angled cuts may be required. Alternatively, a sharp hatchet can be used, but extreme care must be taken to maintain biosecurity and avoid exposure to infectious materials.

     When rabies virus is the top differential, the entire brain can be submitted to a laboratory for that purpose, adhering to local health department protocols as applicable, however, in some instances, the laboratory who performs the rabies test may not be a full-service veterinary diagnostic laboratory capable of running tests to rule out other common differentials. In that scenario, it is necessary to submit 1/3 of the cerebellum that includes samples from all three lobes and a complete cross section of the brainstem for rabies FA testing. The remainder of the brain can be sampled for additional testing with portions also placed in formalin for histopathology, as described in the Neurologic Plan in Appendix.
7. **Closure and Carcass Disposal**
   - To facilitate transportation and provide a cosmetic appearance, stab incisions along the skin edges can be used to anchor suture material, string, or twine to close the body cavities.
   - Dispose of the carcass in compliance with local regulations.

3.2.4 Additional Considerations

- **Photographic Documentation**: Photographs are a crucial part of every postmortem examination. Even autolyzed tissues may retain diagnostic value when evaluated alongside images.

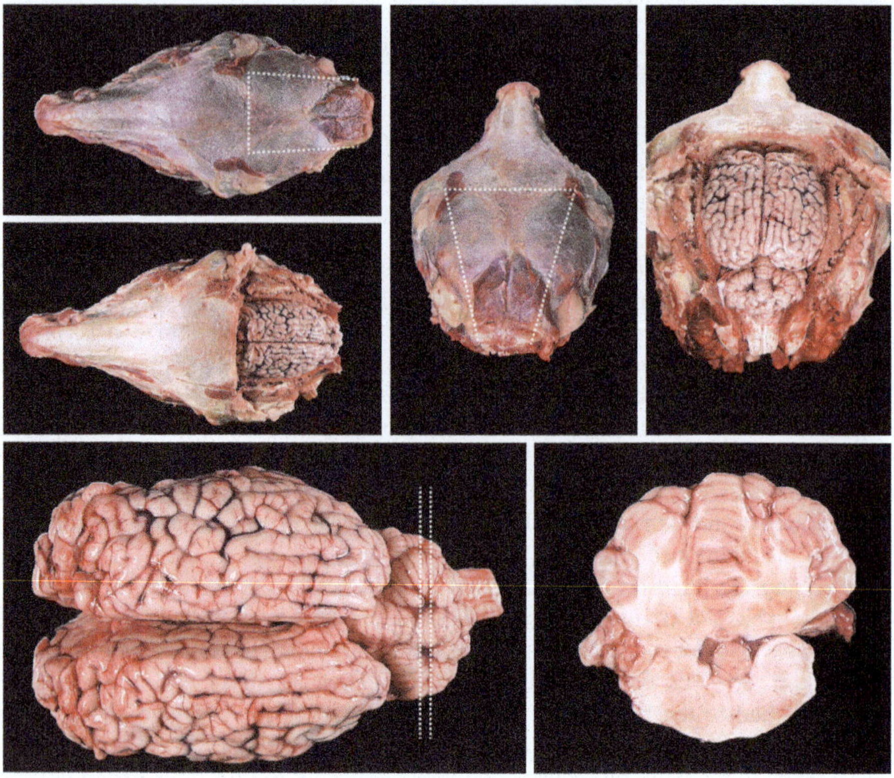

Fig. 5 An overview of removing the brain during a postmortem examination with dotted lines to highlight cut trajectories. The skin has been removed from the dorsal aspect of the skull for illustrative purposes. Using the hacksaw, make a transverse cut across the calvaria just caudal to the orbits. Next, make two sagittal cuts that extend from the lateral edges of the transverse cut to the lateral aspects of the foramen magnum at the base of the skull. When collecting samples for rabies testing, collect a section of the cerebellum that includes samples from all three lobes and a complete cross section of the brainstem for rabies FA testing (bottom row)

- **Collection Artifacts in Formalin Fixed Tissues**: Common artifacts in field samples include acid hematin formation (from unbuffered formalin), freeze artifact, and inadequate fixation due to overly large tissues or insufficient formalin.
- **Biosecurity**: Follow strict biosecurity protocols, particularly when dealing with potential zoonotic pathogens.

By following these guidelines, field postmortem examinations can provide valuable diagnostic information even in challenging conditions.

4 Conclusion

Diagnostic testing is an important and challenging part of veterinary medicine. It allows clinicians to arrive at accurate conclusions when faced with scenarios involving sick or even dead equids. Donkeys and mules are notoriously stoic, requiring the veterinarian to rely even more heavily on objective test results when history and physical examination findings alone are inconclusive. Adhering to the guidance in this chapter regarding sample collection and handling as well as necropsy technique will greatly improve the likelihood of obtaining accurate results on the diagnostic tests performed.

Appendix

Donkey and Mule Diagnostic Testing Plans

Wellness Plan	
Some or all of these tests may be appropriate as part of routine health monitoring, depending on the geographic location, age and diet of the equid.	
Tests	**Samples**
Hemogram (complete blood count, CBC)	EDTA whole blood and 2–3 air-dried, unstained blood smears
Serum biochemistry panel[a]	Serum (separated) or Heparinized plasma (separated)
Fecal egg count	Feces
Selenium concentration	EDTA whole blood or Heparin whole blood
Vitamin E concentration	Serum (separated, cooled, & protected from light)

[a]*Request a separate triglyceride concentration if not included in the biochemistry panel*

Metabolic/Endocrine Plan	
Some or all of these tests may be indicated for equids suspected of having equine (donkey) metabolic syndrome or pituitary pars intermedia dysfunction (PPID), however consideration of clinical signs, age, geographic location, diet and season must all be considered when selecting tests and interpreting results.	
Tests	**Samples**
Triglyceride concentration	Serum (separated) or Heparinized or EDTA plasma (separated)
Leptin concentration	Serum (separated) or EDTA plasma (separated)
Insulin concentration[a]	Serum (separated) or EDTA plasma (separated)
Glucose concentration	Serum (separated)
Baseline ACTH[b]	EDTA plasma (separated and frozen within 4 h of collection)

[a]*Consider dynamic insulin tests when baseline results are inconclusive*

[b]*Consider the TRH-stimulation test if baseline ACTH concentration is inconclusive, or if equid may be in the early stage of disease*

Chronic Ill Thrift Plan	
Some or all of these tests may be useful for equids who have non-specific signs of ill thrift such as weight loss, cachexia, anemia, edema, etc. Note that this is not an exhaustive list. Differential diagnoses and therefore relevant diagnostic testing strategies will vary by geographic region, signalment and clinical presentation	
Tests	**Samples**
Insulin concentration[a]	Serum (separated) or EDTA plasma (separated)
Baseline ACTH[b]	EDTA plasma (separated and frozen within 4 h of collection)
Fecal egg count	Feces
Sand recovery analysis	Feces (8–10 fecal balls)
Hemogram (complete blood count, CBC)	EDTA whole blood and 2–3 air-dried, unstained blood smears
Serum biochemistry panel[c]	Serum (separated) or Heparinized plasma (separated)
Selenium concentration	EDTA whole blood or Heparin whole blood
Vitamin E concentration	Serum (separated, kept cool, & protected from light)
Peritoneal fluid cytology	Peritoneal fluid in EDTA tube and 2–3 unstained, air-dried smears of fluid
Thyroid panel (thyroxine or T4, triiodothyronine or T3, and free T4) (Brown et al. 2024)	Serum (separated)
Besnoitia species PCR[d] *Besnoitia* species immunofluorescent antibody test (IFA)	Biopsy of skin with lesions (pinpoint cysts) for PCR Serum (separated) for IFA
Equine infectious anemia ELISA[e]	Serum (separated)
Equine Piroplasmosis serology (CFT, IFA, WB or ELISA)	Serum (separated)

[a]*Consider dynamic insulin tests when baseline results are inconclusive*

[b]*Consider the TRH-stimulation test if baseline ACTH concentration is inconclusive, or if equid may be in the early stage of disease*

[c] *Request a separate triglyceride concentration if not included in the biochemistry panel*

[d]*Besnoitiosis may rarely result in severe generalized disease and weight loss/cachexia in donkeys, however dermatologic disease alone is more common. To the authors' knowledge, besnoitiosis has not been described in mules*

[e]*Unlikely to cause clinical disease in donkeys, but clinical disease has been described in mules*

Neurologic Plan	
The tests and diagnostic samples listed here are appropriate to rule out some of the infectious agents that may be responsible for neurologic disease in donkeys and mules. Note that this is not an exhaustive list. Differential diagnoses and therefore relevant diagnostic testing strategies will vary by geographic region, signalment and clinical signs	
Tests	**Samples**
Eastern equine encephalitis IgM ELISA Western equine encephalitis PRNT Venezuelan equine encephalitis IgM ELISA West Nile virus IgG/IgM ELISA Japanese encephalitis ELISA or PRNT (Chakravarti et al. 1981; Zohaib et al. 2015) Saint Louis encephalitis virus HI or PRNT (Carrera et al. 2018; Costa et al. 2019; Pauvolid-Corrêa et al. 2014)	Serum (separated)
Eastern equine encephalitis PCR	EDTA whole blood (may be negative due to transient viremia) Brain (preferred sample)
Venezuelan equine encephalitis PCR West Nile virus PCR	Brain
Aerobic culture and/or fungal culture	Swab inoculated with CSF and placed in media suitable for both aerobic and fungal cultures Brain
PCRs for relevant equine and asinine herpesviruses: EHV-1 PCR (Pusterla et al. 2016) EHV-8 / AHV-3 PCR (van Maanen et al. 2017) EHV-4 PCR (Mureşan et al. 2022) Pan herpesvirus PCR with sequencing or virus isolation (when specific EHV or AHV PCRs are not available) (Barrandeguy and Carossino 2018; Vengust et al. 2008)	Antemortem: EDTA whole blood AND nasal swab in plain (non-additive) tubes with a few drops of sterile saline Postmortem: CNS tissue AND nasal swab in plain (non-additive tubes with a few drops of sterile saline)
Borna virus IFA, WB, or ELISA (Binz et al. 1994; Dauphin et al. 2002) Borna virus PCR	Serum (separated) for IFA, WB or ELISA Brain for PCR
Equine enteric coronavirus PCR	Feces Colon tissue
Serum:CSF ratio using *Sarcocystis neurona* SAG 2/4/3 ELISA or IFAT (Llano et al. 2021) *Neospora hughesi* IFAT or ELISA (Javanmardi et al. 2020)	Serum (separated) AND CSF in a plain red top tube
Botulism (*Clostridium botulinum* Types A, B, C) toxin testing (mouse bioassay)	30 grams of stomach contents, feces or feed
Rabies direct FA	Brain (1/3 of the cerebellum that includes samples from all three lobes and a complete cross section of the brainstem) See Fig. 4
Cytology	CSF in EDTA tube
Histopathology	Complete set of formalin fixed tissues

Note that the diagnosis of tetanus (Clostridium tetani) is typically presumptive, based on clinical signs in conjunction with poor, absent or unknown vaccination history for C. tetani

Respiratory Plan	
The tests and diagnostic samples listed here are appropriate to rule out some of the infectious agents that may be responsible for respiratory disease in donkeys and mules. Note that this is not an exhaustive list. Differential diagnoses and therefore relevant diagnostic testing strategies will vary by geographic region, age of equid, and specific clinical signs and history	
Tests	**Samples**
Aerobic culture and/or fungal culture	Swab inoculated with fluid from TTW or BAL and placed in media suitable for both aerobic and fungal cultures Lung
PCRs for relevant equine and asinine herpesviruses: EHV-1 (Barrandeguy and Carossino 2018) EHV-4 (Barrandeguy and Carossino 2018) EHV-2 (Rickards and Thiemann 2019) EHV-5 (Rickards and Thiemann 2019) EHV-7/AHV-2 (Barrandeguy and Carossino 2018) AHV-3 (Rickards and Thiemann 2019) AHV-4 (Kleiboeker et al. 2002) AHV-5 (Kleiboeker et al. 2002) Pan herpesvirus PCR, exploratory sequencing or virus isolation (when specific EHV or AHV PCRs are not available)	Nasal swab or swab of draining abscess in plain (non-additive) tube with a few drops of sterile saline TTW or BAL fluid in plain (non-additive) tube Lung
Influenza A virus PCR (equine influenza, H3N8, most common) (Barrandeguy and Carossino 2018) (note that Avian influenza, H5N1, has been reported in Egyptian donkeys) (Abdel-Moneim et al. 2010)	Nasal swab in plain (non-additive) tube with a few drops of sterile saline TTW or BAL fluid in plain (non-additive) tube Lung
Equine viral arteritis (EVA) PCR (Carossino and Timoney 2018)	Nasal swab in plain (non-additive) tube with a few drops of sterile saline TTW or BAL fluid in plain (non-additive) tube Lung
Streptococcus equi subspecies *equi* PCR or culture (Rickards and Thiemann 2019)	Nasal swab or swab of draining abscess in plain (non-additive) tube with a few drops of sterile saline (for PCR) Nasal swab or swab of draining abscess in bacterial transport media (for culture) TTW, BAL, Guttural pouch lavage (GPL) or nasal wash fluid in plain (non-additive) tube (for PCR) Swab inoculated with fluid from TTW, BAL, or GPL and placed in bacterial transport media (for culture) Lung (for culture or PCR)
Streptococcus equi subspecies *zooepidemicus* PCR or culture (Rickards and Thiemann 2019; Cantelmi et al. 2023) (Considered a commensal but also pathogenic in donkeys in some instances)	Guttural pouch lavage sample (for culture or PCR) Lung (for culture or PCR)

(continued)

Respiratory Plan	
Fecal flotation using sugar solution or the Baermann fecal technique (for detection of *Dictyocaulus arnfieldi*) (Patent infections more commonly seen in donkeys and mules, but clinical signs are less common compared to horses)[b]	Feces
Cytology	TTW or BAL fluid in EDTA tube
Histopathology	Complete set of formalin fixed tissues

Diarrhea Plan	
The tests and diagnostic samples listed here are appropriate to rule out *some* of the common infectious agents that may be responsible for diarrheal disease in donkeys and mules. Note that this is not an exhaustive list. Differential diagnoses and therefore relevant diagnostic testing strategies will vary by geographic region, age of equid, and specific clinical signs and history	
Tests	**Samples**
Equine enteric coronavirus PCR	Feces Colon tissue
Clostridium perfringens PCR (or anaerobic culture to isolate) & PCRs for toxins: Alpha toxin (CPA), beta toxin (CPB), beta 2 toxin (CPB2), cytotoxin (netF), and enterotoxin (CPE)	Feces for PCRs Colon tissue for PCRs Ligated loop of intestine or swab of feces or intestine inoculated into anaerobic bacterial transport media (for anaerobic culture)
Clostridium difficile toxins A and B PCRs Anaerobic culture to isolate *Clostridium difficile*	Feces for PCRs Colon tissue for PCRs Ligated loop of intestine or swab of feces or intestine inoculated into anaerobic bacterial transport media (for anaerobic culture)
Fecal flotation (for detection of internal parasites)	Feces
Lawsonia intracellularis PCR	Feces Intestine
Salmonella spp. culture	Feces Intestine
Rotavirus (PCR, FA or ELISA)	Feces for PCR or ELISA Intestine for PCR or FA
Potomac Horse Fever (*Neorickettsia risticii* or *Neorickettsia findlayensis*) PCR	Feces EDTA whole blood Spleen
Aerobic culture for detection of septicemia	Antemortem: Blood culture bottles inoculated using aseptic technique Postmortem: Lung, spleen, kidney, lymph node
Histopathology	Complete set of formalin fixed tissues

Abortion Plan	
The tests and diagnostic samples listed here are appropriate to rule out *some* of the common infectious agents that may be responsible for abortion in donkeys and mules. Note that this is not an exhaustive list. Differential diagnoses and therefore relevant diagnostic testing strategies will vary by geographic region, age of equid, and specific clinical signs and history	
Tests	**Samples**
Aerobic and/or fungal culture	Submit placenta, fetal lung and fetal stomach contents for individual aerobic and/or fungal culture on each (Ensure that the aerobic culture performed at the laboratory is capable of detecting *Salmonella abortus equi*)
Equine arteritis virus (EAV) FA or PCR	Fresh fetal tissue: Placenta, liver, lung, kidney
PCRs for relevant equine and asinine herpesviruses: EHV-1, EHV-4, EHV-8 (Wang et al. 2022; Balasuriya et al. 2020) Pan herpesvirus PCR, exploratory sequencing or virus isolation (when specific EHV or AHV PCRs are not available)	Fresh fetal tissue: Lung (preferred), placenta acceptable
Leptospira PCR	Submit both placenta and fetal kidney (fetal stomach contents acceptable) for PCR on each
Histopathology	Complete set of formalin fixed fetal tissues (placenta, liver, lung, brain, adrenal, heart, thymus, intestine, kidney, and fetal skin)

Questions

1. If you were to collect blood into a tube with EDTA anticoagulant (purple top tube), and then you centrifuged it; the supernatant should be appropriately labeled as:
   (a) Serum
   (b) Whole blood
   (c) EDTA plasma
   (d) Citrated plasma
2. True or False: Swabs placed in gel-based media are always appropriate for PCR testing
3. When removing the brain during a postmortem examination on a donkey who had displayed neurologic disease prior to death, what tool below should never be utilized?
   (a) Handsaw
   (b) Bone rongeurs
   (c) Electric saw
   (d) Knife

4. When performing a necropsy on a fetus, which tissues should be collected for potential ancillary diagnostic testing
   (a) Fetal kidney
   (b) Placenta
   (c) Fetal lung
   (d) None of the above
   (e) All of the above

References

Abdel-Moneim AS, Abdel-Ghany AE, Shany SAS (2010) Isolation and characterization of highly pathogenic avian influenza virus subtype H5N1 from donkeys. J Biomed Sci 17(1):25–25. https://doi.org/10.1186/1423-0127-17-25

Balasuriya UBR, Câmara RJF, Bueno BL, Resende CF, Balasuriya UBR, Sakamoto SM, dos Reis JKP (2020) Viral diseases that affect donkeys and mules. Animals 10(12):2203. https://doi.org/10.3390/ani10122203

Barrandeguy ME, Carossino M (2018) Infectious diseases in donkeys and mules: an overview and update. J Equine Vet Sci SI:Donkey Sci 65(June):98–105. https://doi.org/10.1016/j.jevs.2018.02.026

Binz T, Lebelt J, Niemann H, Hagenau K (1994) Sequence analyses of the P24 gene of Borna disease virus in naturally infected horse, donkey and sheep. Virus Res 34(3):281–289. https://doi.org/10.1016/0168-1702(94)90128-7

Brown K, Louie EW, Pinn-Woodcock T, Pearson E, Pearson GB, Marr J, Hackett ES, Brown LR, Mitchell KJ (2024) Cardiac disease related to primary hyperthyroidism in a 20-year-old mule. Animals 14(11):1660. https://doi.org/10.3390/ani14111660

Cantelmi MC, Merola C, Averaimo D, Chiaverini A, Cito F, Cocco A, Di Teodoro G et al (2023) Identification of the novel streptococcus Equi Subsp. Zooepidemicus sequence type 525 in donkeys of Abruzzo region, Italy. Pathogens 12(6):750. https://doi.org/10.3390/pathogens12060750

Carossino M, Timoney PJ (2018) Equine viral arteritis: a respiratory and reproductive disease of significant economic importance to the equine industry. Equine Vet Educ 30(9):497–512. https://doi.org/10.1111/eve.12672

Carrera J-P, Bagamian KH, Travassos da Rosa AP, Wang E, Beltran D, Gundaker ND, Armien B et al (2018) Human and equine infection with alphaviruses and Flaviviruses in Panamá during 2010: a cross-sectional study of household contacts during an encephalitis outbreak. Am J Trop Med Hyg 98(6):1798–1804. https://doi.org/10.4269/ajtmh.17-0679

Chakravarti SK, Sardana DN, Mukherjee KK, Mitra AC, Chakrabarty MS (1981) Serological evidence of infection with Japanese encephalitis virus in mules of Eastern Himalayan region. Indian J Med Res 73:4–7

Costa LRR, Aleman M, Davis E (2019) Comprehensive report of the caseload of donkeys and mules presented to a veterinary medical teaching hospital over a ten-year period. Animals 9(7).: undefined-undefined. https://doi.org/10.3390/ani9070413

Dauphin G, Legay V, Pitel P-H, Zientara S (2002) Borna disease: current knowledge and virus detection in France. Vet Res 33(2):127–138. https://doi.org/10.1051/vetres:2002002

Dawson S (2014) Blood culture contaminants. J Hosp Infect 87(1):1–10. https://doi.org/10.1016/j.jhin.2014.02.009

Goodrich EL, Behling-Kelly E (2019) Clinical pathology of donkeys and mules. Vet Clin North Am Equine Pract 35(3):433–455. https://doi.org/10.1016/j.cveq.2019.08.002

Goodrich EL, Webb JL (2024) Complete blood count and biochemistry reference intervals for healthy adult donkeys in the United States. Animals 14(14):2018. https://doi.org/10.3390/ani14142018

"Internal-Parasite-Guidelines_Updated.Pdf" (n.d.) https://aaep.org/wp-content/uploads/2024/05/Internal-Parasite-Guidelines_Updated.pdf. Accessed 6 Jan 2025

Javanmardi E, Majidiani H, Shariatzadeh SA, Anvari D, Shamsinia S, Ghasemi E, Kordi B, Shams M, Asghari A (2020) Global Seroprevalence of *Neospora* Spp. in horses and donkeys: a systematic review and meta-analysis. Vet Parasitol 288(December):109299. https://doi.org/10.1016/j.vetpar.2020.109299

Kleiboeker SB, Schommer SK, Johnson PJ, Ehlers B, Turnquist SE, Boucher M, Kreeger JM (2002) Association of two newly recognized herpesviruses with interstitial pneumonia in donkeys (Equus Asinus). J Vet Diagn Invest 14(4):273–280. https://doi.org/10.1177/104063870201400401

Llano HAB, Soares RM, Acevedo-Gutierrez LY, Rodas JD, Polo G, Borges-Silva W, Jesus RF, Gondim LFP (2021) Seroepidemiology of *Sarcocystis Neurona* and *Neospora* Spp. in horses, donkeys, and mules from Colombia. Acta Trop 220(August):105970. https://doi.org/10.1016/j.actatropica.2021.105970

Mendoza FJ, Toribio RE, Perez-Ecija A (2019) Metabolic and endocrine disorders in donkeys. Vet Clin North Am Equine Pract Dis Donkeys Mules 35(3):399–417. https://doi.org/10.1016/j.cveq.2019.07.001

Mureşan A, Mureşan C, Siteavu M, Avram E, Bochynska D, Taulescu M (2022) An outbreak of equine Herpesvirus-4 in an ecological donkey milk farm in Romania. Vaccine 10(3):468. https://doi.org/10.3390/vaccines10030468

Pauvolid-Corrêa A, Campos Z, Juliano R, Velez J, Nogueira RMR, Komar N (2014) Serological evidence of widespread circulation of West Nile virus and other Flaviviruses in equines of the Pantanal, Brazil. PLoS Negl Trop Dis 8(2):e2706. https://doi.org/10.1371/journal.pntd.0002706

Pusterla N, Mapes S, Akana N, Barnett C, MacKenzie C, Gaughan E, Craig B, Chappell D, Vaala W (2016) Prevalence factors associated with equine herpesvirus type 1 infection in equids with upper respiratory tract infection and/or acute onset of neurological signs from 2008 to 2014. Vet Rec 178(3):70–70. https://doi.org/10.1136/vr.103424

Rickards KJ, Thiemann AK (2019) Respiratory disorders of the donkey. Vet Clin North Am Equine Pract 35(3):561–573. https://doi.org/10.1016/j.cveq.2019.08.009

Saadi A, Tavassoli M, Dalir-Naghadeh B, Samiei A (2018) A survey of Dictyocaulus Arnfieldi (Nematoda) infections in equids in Urmia region, Iran. Ann Parasitol 64(3):235–240. https://doi.org/10.17420/ap6403.158

van Maanen K, van der Zaag E, Buter R, van den Wollenberg L, Sloet M, van Oldruitenborgh-Oosterbaan. (2017) Asinine Herpesvirus-3 (equine Herpesvirus-8)-associated neurological disease in a donkey. Vet Rec Case Rep 5(4):e000498. https://doi.org/10.1136/vetreccr-2017-000498

Vengust M, Wen X, Bienzle D (2008) Herpesvirus-associated neurological disease in a donkey. J Vet Diagn Invest 20(6):820–823. https://doi.org/10.1177/104063870802000620

Wang T, Hu L, Wang Y, Liu W, Liu G, Zhu M, Zhang W, Wang C, Ren H, Li L (2022) Identification of equine herpesvirus 8 in donkey abortion: a case report. Virol J 19(1):10. https://doi.org/10.1186/s12985-021-01738-2

Young A (1990) Blood culture: indications and technique. Equine Vet Educ 2(1):38–40. https://doi.org/10.1111/j.2042-3292.1990.tb01377.x

Zohaib A, Saqib M, Beck C, Hussain MH, Lowenski S, Lecollinet S, Sial A et al (2015) High prevalence of West Nile virus in equines from the two provinces of Pakistan. Epidemiol Infect 143(9):1931–1935. https://doi.org/10.1017/S0950268814002878

Pharmacological and Therapeutical Considerations in Donkeys and Mules

Francisco J. Mendoza, Alejandro Perez-Ecija, and Ramiro E. Toribio

Abstract

Donkeys and mules have been gaining popularity worldwide, not only as work animals, but also as pets, for therapy of adults and children with motor or psychologic disorders, sport activities, and for consumption. Pharmacologic information for drugs used in equine practice is scarce for these animals. Thus, clinicians often extrapolate dosing protocols from horses, even when differences have been described. This can result in the wrong diagnosis and treatment, but also on unnecessary expenses. Therefore, species-specific pharmacologic studies for donkeys and mules are necessary. The metabolism for many drugs is faster in donkeys, needing shorter dosing intervals and higher doses, but they can also be more susceptible to some drugs.

Keywords

Anti-inflammatories · Antimicrobials · Antiparasitic · Pharmacokinetics · Pharmacodynamic

F. J. Mendoza (✉) · A. Perez-Ecija
Department of Animal Medicine and Surgery, College of Veterinary Medicine, University of Cordoba, Cordoba, Spain
e-mail: fjmendoza@uco.es; alejandro.perez.ecija@uco.es

R. E. Toribio
Department of Veterinary Clinical Sciences, College of Veterinary Medicine, The Ohio State University, Columbus, OH, USA
e-mail: toribio.1@osu.edu

A. K. McLean et al. (eds.), *Donkey and Mule Medicine: Clinical Medicine and Reproduction*, https://doi.org/10.1007/978-3-032-18540-2_13

1 Introduction

As described in other chapters of this book, donkeys and mules have a myriad of physical, metabolic, and physiological idiosyncrasies. Most current information on equid pharmacology is based on studies performed in horses, with specific studies in donkeys and mules being scarce. Moreover, information on drug distribution, excretion, and accumulation is lacking and could impact human health from the consumption of byproducts.

In general terms, donkeys and mules have higher metabolic rates and cellular water content than horses, which may influence drug distribution, metabolism, and excretion. Since these factors can influence the pharmacokinetic and pharmacodynamic properties of drugs, there are notable differences in doses, dosing intervals, efficacy and side effects among donkeys, mules, and horses (Lizarraga et al. 2004; Grosenbaugh et al. 2011). Therefore, it is important to consider potential differences in drug dynamics when extrapolating from horses to donkeys or mules to avoid inadequate dosing and toxicity (Toutain et al. 2010). Moreover, having an accurate body weight is crucial (refer to nutrition chapter).

In this chapter, we compiled available information from donkey and mule pharmacology. We will discuss some considerations about commonly used drugs (antimicrobial, anti-inflammatory, and antiparasitic drugs) and we will group them based on body system.

2 General Considerations

2.1 Antimicrobials

Few pharmacokinetic (PK) and pharmacodynamic (PD) studies on antimicrobial drugs are available in donkeys. Many drugs commonly used by clinicians have not been evaluated in donkeys or mules. Thus, the possibility of adverse reactions or inappropriate response to treatment (considering that ineffective treatments could increase the risk of antimicrobial resistance) should be considered when extrapolating protocols from horses (Mendoza et al. 2019a).

In general, donkeys have faster metabolism for most studied antibiotics, with faster clearance rates and shorter half-lives compared to horses (Grosenbaugh et al. 2011). Thus, higher doses or shorter dosing intervals may be required in donkeys and mules (Table 1). Aminoglycosides (amikacin and gentamicin) are an exception, where dosing protocols are similar for these species.

Other considerations include the regulatory aspects that varies per country and the lack of information on drug accumulation in donkey byproducts (meat, milk) intended for human consumption.

Table 1 Recommended dose regimen for antimicrobial drugs used in donkeys and mules

Drug	Dose (mg/kg)	Route	Interval (h)	Observations	Donkey references
Amikacin	20–25	IV	24	Dose for foals	Horspool et al. (1994)
Ampicillin	10	IV	6-8-12		Horspool et al. (1992)
Amoxicillin	10–15	IV	8–12		Lavy et al. (1995)
Azithromycin[a]	10	PO	24	Use in foals only	
Ceftiofur[a]	2.2–4.4	IV	6–12	Higher doses in foals	
Cefquinome	1	IV	12		Widmer et al. (2009)
Clarithromycin[a]	7.5	PO	12	Use in foals only	
Chloramphenicol[a]	50–60	PO	6	Where legally allowed	
Doxycycline	10	PO	12		Chapuis et al. (2021)
Enrofloxacin	5	IV	24	Overdosing: lethargy or seizures	Sekkin et al. (2012)
Florfenicol	30	PO	24–48		Liu et al. (2024)
Gentamicin	2.2	IV	24		Welfare et al. (1996)
	6.6[a]	IV	24		
Metronidazole	Adult: 15–25	PO	8–12		El Meshad et al. (2013)
	Adult: 10[a]	IV	8–12		
	Foal: 10[a]	IV, PO	8–12	Assuming similar pharmacokinetics as equine foals	
Penicillin	20,000 IU/kg	IV	6-8-12		Oukessou et al. (1994)
Rifampin[a]	5–10	PO	12–24	Combined with macrolides for *Rhodococcus equi* or with sulfonamides	
Oxytetracycline	5–10	IV	12–24	Ideally, dilute in 250–500 ml of saline solution and administer over 20 min	Horspool and McKellar (1990)

(continued)

Table 1 (continued)

Drug	Dose (mg/kg)	Route	Interval (h)	Observations	Donkey references
Sulfonamide	12.5	IV	12		Peck et al. (2002)
	20–30[a]	PO	12		
Trimethoprim	2.5	IV	12–24		Peck et al. (2002)

IM intramuscular, *IV* intravenous, *PO* per os
[a]Dosage extrapolated from horses and based on clinical experience due to lack of donkey or mule specific pharmacokinetic information

2.2 Anti-inflammatory, Antipyretic and Antiendotoxic Drugs

The metabolism of non-steroidal anti-inflammatory drugs (NSAIDs) differ between donkeys and horses (Mendoza et al. 2019a). Most NSAIDs in donkeys have shorter half-lives and faster clearances than horses, with miniature donkeys having shorter dosing intervals compared to standard (mammoth) sized donkeys (Table 2) (Grosenbaugh et al. 2011). The risk of gastrointestinal or renal toxicity is higher in hypovolemic or dehydrated donkeys. Selective COX-2 inhibitors are used at similar dosages as horses although they are not approved for donkeys in most countries (Table 2).

Regarding corticosteroids, since no pharmacologic studies are available in donkeys or mules, doses for horses are currently used. Caution must be taken in obese or laminitic donkeys, and those diagnosed or suspected of endocrine disorders (PPID or ID) due to the risk of endocrinopathic laminitis (for additional information refer to the chapter on endocrinology).

Lidocaine as a continuous rate infusion (CRI) is used in donkeys and mules with acute gastrointestinal disorders, endotoxemia, and laminitis (Table 2). Although PK/PD studies have not been carried out in donkeys or mules, neurotoxicity has been reported in donkeys (Wood and Loomes, 2022). Clinical experience suggests that horse protocols work well in donkeys and mules.

Dipyrone (metamizole) has a shorter half-life in donkeys compared to horses, and ideally it should be administered at shorter intervals (Table 2). Its antipyretic, analgesic, and spasmolytic effects are better than its anti-inflammatory actions.

Paracetamol (acetaminophen) can be administered similar to horses to control fever and provide mild analgesia (Table 2) (Sullivan and Vogel 2023).

2.3 Antiparasitic Drugs

Most antiparasitic drugs are not approved for donkeys or mules but are used similar to horses (Table 3). The scarcity of species-specific information justifies close monitoring (Mendoza et al. 2018a). Donkeys are more sensitive to imidocarb

Table 2 Recommended doses for drugs with anti-inflammatory, antiendotoxic or antipyretic effects commonly used in donkeys and mules

Drug	Dose (mg/kg)	Route	Interval (h)	Observations	Donkey reference
Beclomethasone[a]	1–8 μg/kg	INH	12–24		
Budesonide[a]	1-3 μg/kg	INH	12		
Carprofen	0.7–1.3	IV, PO	24		Mealey et al. (2004)
Ciclesonide[a]	2744 μg 4116 μg	INH	12 for 5 days then every 24 for 5 days	Lower dose for 5 consecutive days and then higher dose for 5 days	
Cimicoxib	2–5	PO	24		Kim et al. (2014)
Dexamethasone[a]	0.05–0.2	IV, IM, PO	24		
Dimethyl sulfoxide (DMSO)[a]	1 mg/kg, >10% solution	IV	8–12	Causes hemolysis at concentrations higher than 10%	
Dipyrone (metamizole)	25	IV	12		Arcoverde et al. (2023)
Firocoxib	0.1–0.2	IV, PO	12–24		Matthews et al. (2009)
Flunixin meglumine	0.5–1.1	IV	8–24		Cheng et al. (1996)
Fluticasone[a]	2–5 μg/kg	INH	12–24		
Ketoprofen	1–3	IV	12		Anwer et al. (2012)
Lidocaine[a]	Loading: 1.3 Maintenance: 0.025–0.05	IV	IV: 20 min bolus CRI: mg/kg/minute	Some clinicians prefer use without bolus	
Meloxicam	0.6	IV	12–24	Fast clearance in donkeys	Mahmood et al. (2011)
Paracetamol (acetaminophen)	10[a]	IV	8–12		
	20–30	PO			Sullivan and Vogel (2023)

(continued)

Table 2 (continued)

Drug	Dose (mg/kg)	Route	Interval (h)	Observations	Donkey reference
Pentoxifylline[a]	10–15 mg/kg	IV, PO	8–12 h		
Phenylbutazone	2.2–4.4	IV, PO	12–24	Miniature donkeys: q 8 h	Matthews et al. (2001)
Prednisolone[a]	0.5–1	PO	12–24 h		

CRI constant rate infusion, IM intramuscular, *INH* inhaled, *IV* intravenous, *PO* per os
[a]Dosage extrapolated from horses and based on clinical experience due to lack of donkey or mule specific pharmacokinetic information

dipropionate toxicity than horses, so lower doses are recommended, and premedication with anticholinergic drugs such as scopolamine is recommended (Mendoza et al. 2024a). Macrocyclic lactones (ivermectin, moxidectin) should be used with caution in young donkeys and mules (less than 6 months of age) as well as in those with poor body condition due to the risk of neurotoxicity.

3 System-Specific Pharmacological and Therapeutic Considerations

3.1 Gastrointestinal Tract

Dental disorders are common in donkeys, principally in geriatric animals (Mendoza et al. 2018b). Treatment principles are discussed in detail elsewhere in this book (refer to the dental chapter).

Donkeys are more stoic than horses, and colic signs and severe pain can be overlooked, delaying diagnosis and treatment, and worsening the prognosis. Thus, it is recommended to determine the underlying cause of abdominal pain before pain relief is administered to avoid masking signs. Therapeutic options for medical colic (spasmodic, enteritis, gastric, ileal and large and small colon impactions) are similar to horses (Thiemann and Sullivan 2019), mostly based on fluid therapy, endotoxemia prevention and control (NSAIDs, polymyxin B), laxatives and cathartics (mineral oil, sodium sulfate, magnesium sulfate, dioctyl-sulfosuccinate sodium, Coca-Cola) (Table 4), laminitis prevention (cryotherapy, low-molecular weight heparin, soft hoof support), hyperlipemia prevention and control (see Chapter on liver function). Strangulating colic (volvulus, large colon displacements, intussusceptions, hernias) are managed similar to horses.

Donkey gastric ulcer syndrome (DGUS) affects both the squamous and glandular regions and is treated with proton pump inhibitors such as oral omeprazole (Table 4), although pharmacologic (PK/PD) studies have not been performed. Long-acting intramuscular omeprazole, or other proton pump inhibitors (pantoprazole,

Table 3 Recommended doses for most common anti-parasitic drugs used in donkeys and mules

Drug	Dose (mg/kg)	Route	Interval (h)	Protocol	Indication	Observations	Donkey references
Albendazole	10	PO	Once		Nematodes *D. arnfieldi*		Gokbulut et al. (2006)
	7.5	IM	Once				Li et al. (2024)
Buparvaquone	2–6	IM	Once		*Babesia caballi* *Theileria equi*		Mendoza et al. (2024a)
Closantel	20	PO	Once	Repeat at 8–10 weeks	*Fasciola hepatica* (mature)	Overdosing: blindness, anorexia, and ataxia	The Donkey Sanctuary (2018)
Diclazuril[a]	1	PO	Once	Weekly	EPM		
Diminazene aceturate	3.5	IM	Twice	Weekly	*B. caballi* and *T. equi*		Mendoza et al. (2024a)
Fenbendazole	5–10	PO	24	5 days	GI nematodes *D. arnfieldi*	Cyathostomins resistance has been observed Caution in early pregnant jennies (potential teratogenicity)	Grosenbaugh et al. (2011)
Imidocarb dipropionate	*B. caballi*: 2.2	IM	24	2 consecutive days	*Babesia caballi*	High risk of side effects (even death). To use lower dose needed	Mendoza et al. (2024a)
	T. equi: 2.2–4.4		48–72	3–4 doses	*Theileria equi*	Administer scopolamine IV prior to injection	
Ivermectin	0.2	PO	Once		GI nematodes *D. arnfieldi* Arthropods	Overdosing: blindness, staggering and seizures. No to use in younger than 6 months. Resistance observed in some countries	Gokbulut et al. (2005)
Mebendazole	10	PO	Once		Nematodes (Cyathostominae)		Gokbulut et al. (2016)

(continued)

Table 3 (continued)

Drug	Dose (mg/kg)	Route	Interval (h)	Protocol	Indication	Observations	Donkey references
Moxidectin	0.2	PO	Once		GI nematodes	Overdosing: staggering and seizures	Trawford et al. (2005)
	0.4	PO	Once		GI nematodes *D. arnfieldi* Arthropods Encysted cyathostomin larvae		The Donkey Sanctuary (2018)
Ponazuril[a]	5	PO	24	28 days	EPM		
Praziquantel	1–2.5	PO	Once		*A. perfoliata* and other cestodes		Getachew et al. (2013)
Pyrantel pamoate	19	PO	Once		GI nematodes and cestodes		Gokbulut et al. (2014)
	38	PO	Once		*A. perfoliata*		The Donkey Sanctuary (2018)
Secnidazole	10	PO	Once		*Balantidium coli*		Khan et al. (2013)
Triclabendazole	12–18	PO	Once		*Fasciola hepatica* (immature and adults)		Kinabo and Bogan (1989)

GI gastrointestinal, *EPM* equine protozoal myeloencephalitis, *IM* intramuscular, *PO* orally
[a]Dosage extrapolated from horses and based on clinical experience due to lack of donkey or mule specific pharmacokinetic information

Table 4 Recommended dose regimens for medications used for various disorders in donkeys and mules

Drugs	Dose	Route	Interval	Observations
Anticholinergics				
Hyoscine (scopolamine N-butyl bromide)[a]	0.3 mg/kg	IV	20–60 min	
Antivirals				
Acyclovir[a]	10 mg/kg 20 mg/kg	IV PO	12 h	Administer IV slowly
Ganciclovir[a]	2.5 mg/kg	IV	12 h	Administer slowly
Valacyclovir[a]	25–40 mg/kg	PO	8 h	
Bronchodilators				
Albuterol (salbutamol)[a]	0.5–2 μg/kg	INH	8–12-24 h	
Clenbuterol[a]	0.8–3.2 μg/kg	IV, PO	12–24 h	
Cathartics, absorbents, chelators and laxatives				
Coca-Cola®[a]	1 L/200 kg	IG	4 h	Without caffeine and sugar and remove gas prior to administration
Magnesium sulfate[a]	0.5–1 g/kg	IG	24 h	
Mineral oil[a]	1 L/200 kg	IG	24 h	
Sodium sulfate[a]	0.5–1 g/kg	IG	24 h	
Smectite (Biosponge®)[a]	0.5–1 g/kg	IG, PO	12–24 h	
Gastric ulcers				
Misoprostol[a]	1–5 μg/kg	PO	8–12 h	
Omeprazole[a]	2–4 mg/kg	PO	24 h	
Sucralfate[a]	20–40 mg/kg	PO	6–8 h	
Prokinetics				
Lidocaine	Loading dose: 1.3 mg/kg	IV	20 min bolus	Extrapyramidal signs (ataxia, circling, obnubilation, etc.) if toxicosis (commonly fast administration). Some clinicians prefer use without bolus
	Maintenance: 0.025–0.05 mg/kg/minute	IV	CRI	
Metoclopramide	0.1–0.5 mg/kg	IV, IM, SC, PO	4–8 h	Extrapyramidal signs (ataxia, circling, obnubilation, etc.) if toxicosis.
	0.04 mg/kg	IV	CRI	
Neostigmine	0.005–0.02 mg/kg	IV, IM, SC	4–6 h	Sialorrhea, tachycardia, abdominal discomfort

IG intragastric, *IM* intramuscular, *INH* inhaled, *IV* intravenous, *PO* per os, *SC* subcutaneous
[a]Dosage extrapolated from horses and based on clinical experience due to lack of donkey or mule specific pharmacokinetic information

esomeprazole) have not been studied in donkeys, although they are sporadically used. Misoprostol, a prostaglandin E synthetic analogue, is recommended using the equine dose (Table 4), although there are no studies in donkeys. This drug should not be used in pregnant jennies due to its abortive effect. Pregnant women should avoid contact with this drug.

Treatment of acute and chronic diarrhea follows similar considerations as horses, including medical support (fluids, correction of acid-base and electrolyte abnormalities, nutrition, endotoxemia control) (Mendoza et al. 2018b). Smectite has not been evaluated in donkeys, but it is used similar to horses (Table 4). Intravenous oxytetracycline (Table 1) is recommended in donkeys with diarrhea caused by *Neorickettsia risticii* (Potomac horse fever) or *Lawsonia intracellularis* (proliferative enteropathy). There is no evidence that antimicrobial administration is necessary to treat diarrhea caused by *Clostridium perfringens*, *Clostridioides difficile*, *Salmonella* spp., or viral pathogens (coronavirus, rotavirus).

Long-term oral prednisolone or dexamethasone administration (Table 2) are used to treat inflammatory bowel diseases (IBD) (see Chapters on gastrointestinal system for complementary reading). Routine fecal egg counts and deworming are highly recommended, in particular those with an unknown deworming history. Fecal microbiota transplant protocols described for horses can be used in donkeys with IBD and chronic diarrhea, although their efficacy remains to be determined. Alimentary allergy panels can be considered in animals with suspected nutritional intolerance to correct dietary management, although the validity of the results is unclear. A high fat diet can be considered in these animals.

Deworming should be based on regular fecal egg counts. Commercial dewormers are not marketed for donkeys but are used as horses. Formulations containing moxidectin and praziquantel are preferred compared to those based on ivermectin or moxidectin alone. Fenbendazole is a safe, cheap, and reliable deworming option in foals and adult donkeys (Table 3), particularly in regions where ivermectin resistance is a problem. Triclabendazole is the drug of choice against flukes (*Fasciola hepatica*) (Table 3).

3.2 Respiratory Tract

Most drugs intended to treat respiratory disturbances in horses can be used in donkeys. Caution is recommended due to the lack of species-specific information for most of them.

Antibiotic and anti-inflammatory drug recommendations and dosing protocols are similar to those described above (Tables 1 and 2). Although strangles (*Streptococcus equi* ssp. *equi*) and *Rhodococcus equi* infections appear to have low incidence in donkeys, antimicrobial considerations and biosecurity measures for horses apply to donkeys and mules. Therapeutic principles for pleuropneumonia are similar between species, including the use of broad-spectrum antimicrobials considering potential anaerobe infections, thoracocentesis (pleural fluid drainage), and treatment and prevention of complications (endotoxemia, hyperlipemia, fever, laminitis, thrombophlebitis).

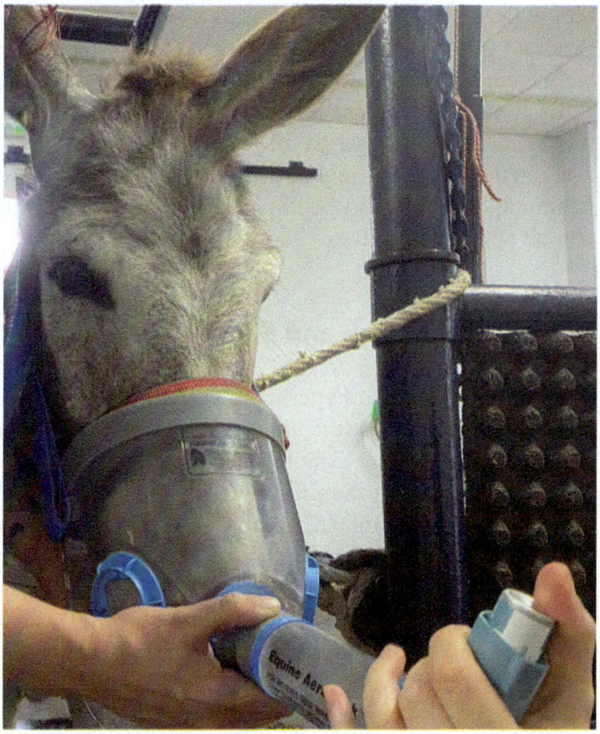

Fig. 1 Inhaled salbutamol administration using Equine AeroMask® in a donkey with acute respiratory distress due to interstitial pneumonia

The pharmacology of inhaled drugs used in horses have not been evaluated in donkeys or mules. Inhaled steroids such as fluticasone, beclomethasone, budesonide, or ciclesonide, as well as inhaled bronchodilators such as salbutamol and albuterol are currently administered similar to horses (Table 4). Devices such as AeroMask®, AeroHippus®, Equine Haler®, and Flexineb® can be used in donkeys (Fig. 1). Ciclesonide (where available) is effective at managing horses and donkeys with asthma. This drug is not absorbed into systemic circulation, which is an advantage over other glucocorticoids. Donkeys with recurrent endocrinopathic laminitis require additional attention, since steroid administration could exacerbate this condition.

Donkeys may develop interstitial pneumonia secondary to viral infections such as herpesvirus or influenza, or fibrosing interstitial pneumonia secondary to asinine herpesvirus infection (Mendoza et al. 2018c). Anti-viral drugs (acyclovir, valaciclovir or ganciclovir) have not been evaluated in this species (Table 4), and dosing protocols must be extrapolated from horses.

Lungworms (*Dictyocaulus arnfieldi*) are common in donkeys, which can act as carriers, hosting a high number of mature parasites with minimal to no clinical signs. They may represent a risk of disseminating lungworms to other equids (Rickards and Thiemann 2019). Macrocyclic lactones and benzimidazoles are effective against this parasite (Table 3). A specific parasite control program is recommended, which should include separated grazing areas for horses and donkeys in endemic areas, although this may not be necessary if animals are under a regular parasite control program (refer to the parasitology chapter for more information).

Vaccines against equine respiratory disorders (rhinopneumonitis [equine herpesvirus 1 and 4], influenza and viral arteritis) are not marketed for donkeys or mules. To the authors' knowledge, only the African Horse Sickness vaccine is specifically marketed for donkeys and mules, but it is only allowed in some countries (Australia, Ethiopia, South Africa). At this moment, vaccination protocols described for horses can be applied to donkeys and mules.

3.3 Metabolism and Endocrine System

Donkeys are prone to hyperlipemia due to a negative energy balance secondary to illnesses, stressful situations (transport, behavioral issues, environmental modifications), and physiological needs (pregnancy or lactation). Treatment is based on addressing the primary condition that includes halting lipid mobilization increasing caloric intake, increasing lipid clearance, and preventing potential complications from lipid infiltration (Mendoza et al. 2024b). Detailed information can be found in chapters on endocrinology and nutrition.

Pituitary pars intermedia dysfunction (PPID), insulin dysregulation (ID), and obesity (donkey metabolic syndrome, DMS) are the most common endocrine disorders in donkeys (Mendoza et al. 2019b). Daily oral pergolide administration, following equine recommendations, is the treatment of choice for PPID (Xue et al. 2023). Cabergoline, which is being used in horses as an alternative to pergolide, lacks information in donkeys. Weight control (increased physical activity, thyroxine administration, low carbohydrate-content diets), insulin sensitizing agents or secretagogues (metformin, glibencamide), and sodium-glucose transport protein 2 (SGLT2) inhibitors (e.g. canaglifozin, ertuglifozin) are therapeutical measures to consider in the management of ID and DMS (Mendoza et al. 2018b). The use of SGLT2 inhibitors in donkeys has been increasing because they reduce glucose and insulin concentrations and improve the signs of laminitis in horses and ponies. However, it is important to regularly measure triglycerides and liver enzymes in these animals because these drugs increase their concentrations.

Additional therapeutic options for endocrine disorders are described elsewhere in this book (see Chapter on endocrinology).

3.4 Cardiovascular System

Third-grade atrioventricular block is the most common rhythm disturbance in donkeys, and pacemaker devices have been placed in this species. Donkey-specific studies on anti-arrhythmic drugs have not been performed, thus doses must be extrapolated cautiously from horses. Other conditions such endocarditis (antimicrobials), myocarditis (corticosteroids), and pericarditis (drainage and antimicrobials) are treated similar to horses. No treatment for valvular dysfunction (regurgitation, stenosis) is available in donkeys. Ventricular septal defects or patent duct arteriosus could be surgically corrected like horses.

Inotropes, vasopressors, and angiotensin-converting enzyme inhibitors drugs have not been studied in donkeys, and equine dosing protocols can be used in donkeys.

Ionophores are cardiotoxic and should not be administered to donkeys.

3.5 Central Nervous System

Cervical stenotic myelopathy (Wobbler's syndrome) is rarely described in donkeys, which could be due to minimal use in sport activities, infrequent gait analysis, and infrequent veterinary care. Since donkeys tend to live longer than horses, degenerative joint disease of the cervical vertebrae may result in spinal cord compression. Medical treatment (rest and facet joint infiltration with corticosteroids) is similar to horses. Information on surgical correction is lacking, although similar techniques as horses could be used.

Supplementation with vitamin E and selenium could be beneficial in donkeys or mules with degenerative myeloencephalopathy or neuroaxonal dystrophy.

Several infectious pathogens can cause neurological disease in donkeys and mules, including viruses (Borna disease, West Nile encephalomyelitis, Eastern, Western, and Venezuelan equine encephalitis, Japanese encephalitis), bacteria (Borrelia spp. [Lyme's disease; neuroborreliosis]) and parasites (*Sarcocystis neurona* [equine protozoal myeloencephalitis, EPM], verminous myelitis) (Mendoza et al. 2018a). Due to the lack of species-specific studies on the treatment of these pathologies, equine recommendations could be used (Mendoza et al. 2018a). Briefly, treatment for viral pathologies consists of controlling inflammation, fluid therapy, nutritional support, medical support in recumbent animals (sling, bladder and rectal evacuation), and antimicrobials to control secondary infections. Anti-viral drugs have not been evaluated in donkeys but can be administered following equine dosing protocols (Table 4). Dimethyl sulfoxide (DMSO) can be administered similar to horses (Table 2). Oxytetracycline and doxycycline are recommended to treat neuroborreliosis (Table 1). Antibodies against *Sarcocystis neurona* have been detected in donkeys (Dubey et al. 2001), although the prevalence of EPM appears to be lower in donkeys and mules than horses. Ponazuril or diclazuril are the recommended treatments (Table 3).

Both tetanus and botulism affect donkeys and mules, with variable susceptibility depending on the country (developing versus developed), geographic location, and prophylactic vaccination (Kay and Knottenbelt 2007). Treatments and management are similar to horses. Reports of rabies in donkeys or mules are rare, and given the fatal nature of this condition, and the postmortem final diagnosis, euthanasia is mandatory.

Vaccination protocols for infectious agents follow similar principles as horses, since donkey and mule-specific vaccines have not been developed.

Vestibular disease, central or peripheral, is rarely reported in donkeys, and temporohyoid osteoarthropathy has not been reported in donkeys. Horner's syndrome may occur, although it is rare. Therapeutical management for these conditions is similar to horses.

3.6 Urinary System

Both enteral and parenteral fluid therapy principles (types of fluids, management of electrolytes and acid-base imbalances, fluid rates and calculations) described for horses to treat azotemia secondary to hypovolemia/dehydration (prerenal azotemia) or intrinsic renal azotemia (acute renal failure) can be applied to donkeys and mules. Post-renal azotemia is rarely documented and usually affects newborn donkey foals. Donkeys with chronic renal failure have the same clinical and hematologic abnormalities as horses including weight loss, anemia, azotemia, hypercalcemia, and normo- or hypophosphatemia (Mendoza et al. 2018c).

Non-selective NSAIDs (flunixin meglumine, phenylbutazone) should not be administered to hypovolemic donkeys until normovolemia is restored, or in donkeys with renal failure. Even when using COX-2 selective NSAIDs (meloxicam or firocoxib), care must be taken in these patients.

Furosemide or other diuretics have not been evaluated in donkeys, but if needed, they should be used as horses.

Although urinary tract infections are uncommon in donkeys, antimicrobials used for horses can be administered, considering that some may require different dosing protocols. Preferred antimicrobials are those excreted in urine in active form (penicillins, sulfonamides, or enrofloxacin).

Uroliths (bladder or urethra) and penile neoplasia are common causes of hematuria in donkeys, although incidence seems to be lower than horses.

3.7 Integumentary System

Masses and alopecia are the main reasons for clinical consultations, with ectoparasites, insect hypersensitivity and sarcoids being the most common conditions (White et al. 2019).

Sarcoid is the most common cutaneous neoplasm in donkeys and mules (Thiemann and Sullivan 2021). Surgical excision (conventional, laser, cryosurgery or electrosurgery) and intralesional cisplatin infiltration (1 mg/cm^3 mixed with chemical-grade sesame oil) are recommended treatments based on the type of sarcoid (Théon et al. 2007). Chemotherapy creams (imiquimod and 5-flurouracil) and immunotherapy (Bacillus Calmette-Guerin, and Mycobacterium cell wall extracts) can also be used but have lower success rates. Other therapeutical alternatives (electrochemotherapy, hyperthermia, photodynamic therapy or radiotherapy) can also be used in donkeys but are more expensive. Squamous cell carcinoma is most commonly observed in mucocutaneous junctions (penis, prepuce, eye and lips). Imiquimod or 5-flurouracil have better success rate than cisplatin in this type of neoplasia. Melanoma can also be observed in grey donkeys, although is less common than in grey horses. The Oncept® or Torigen® vaccines can be used in donkeys with melanoma, but scarce information is available about its efficacy.

Treatment of skin wounds, common in working donkeys is based on resting, clipping and wound care with diluted antiseptic solutions (chlorhexidine or povidone iodine). Ointments with antibiotics (nitrofurazone) can be used if secondary infection (pyoderma) is observed. Fly repellents are crucial to avoid secondary parasitic granulomas such as habronemiasis. This last condition is treated like in horses with debridement, topical creams containing ivermectin/moxidectin, dexamethasone, DMSO, and systemic anthelmintic.

Rain scald (*Dermatophilus congolensis* infection) can be observed in malnourished donkeys and due to exposure to wet or inclement weather. Treatment is based on clipping and cleaning affected areas, baths containing chlorhexidine or povidone iodine, and systemic administration of penicillin (Table 1). Other primary bacterial skin infections are uncommon in donkeys. Secondary pyoderma can be observed after skin injuries. Systemic antimicrobial (ceftiofur, doxycycline enrofloxacin, sulfonamides) can be administered in severe cases (Table 1).

Ringworm (dermatophytosis) is treated with chlorhexidine rinses and local shampoos or sprays containing antifungal drugs (e.g. ketoconazole, enilconazole, miconazole). Lime sulfur solution can also be used in some cases. Oral griseofulvin or intravenous iodide potassium can be administered but have potential side effects, including liver damage. It is also recommended to clean tack and harnesses.

Cutaneous parasitic lesions (fly granulomas, *Leishmania infantum* or *Besnoitia bennetti* nodules) are treated similar to horses. Sulfonamide/trimethoprim combinations in general are effective against *Besnoitia bennetti*. Treatment for *Leishmania infantum* is challenging given the resistance of the parasite to most antiprotozoal drugs, and it may be impractical or cost prohibitive. Treatment against mites, lice (chewing or suckling) and fleas is similar to horses, based on clipping, topical permethrins and oral ivermectin or moxidectin administration (Table 3), and local fipronil or lime sulfur solutions. Environmental tick control (acaricides, vegetation control, good hygiene) to reduce the incidence of tick-borne diseases (piroplasmosis, anaplasmosis, encephalitis) is important.

Insect hypersensitivity (*Culicoides* spp. [sweet itch] and *Stomoxys calcitrans* [stable fly]) are also observed in donkeys and mules (White et al. 2019). Sweet itch can have a different lesion pattern than stable fly hypersensitivity. Pruritus seems to be less frequently observed in donkeys, which could be explained either by a higher pruritic threshold compared to horses or it is nor observed by the owner. Treatment is like horses, based on insect repellents, blankets, and systemic corticosteroid administration (Table 1). Some animals may require topical treatment to reduce inflammation. In donkeys with skin allergies, skin or serological testing to identify allergens and develop allergen-specific immunotherapy (vaccines) can be done.

Urticaria is uncommon in donkeys, but corticosteroid administration is the main therapeutical option. Other uncommon disorders such as multisystemic eosinophil epitheliotropic disease (MEED) or pemphigus are treated with long-term corticosteroid administration (Table 1).

3.8 Musculoskeletal System

Laminitis is the most common musculoskeletal disorder in donkeys, usually associated with hyperinsulinemia (hyperinsulinemia-associated laminitis [HAL], recurrent endocrinopathic laminitis), or secondary to acute endotoxemia (e.g. colic, pleuropneumonia, colitis). Cryotherapy (mainly in the acute stage), rest, analgesia (NSAIDs, opioids, lidocaine CRI), and hoof care are central to laminitis management (Thiemann et al. 2022). Weight loss and low carbohydrate content diets are crucial to control HAL (See chapter on endocrinology). Regarding radiographic interpretation, use donkey-specific measurements to assess progression (Nocera et al. 2020), but also in the management (see chapter on endocrinology and nutrition). Pain relief drugs must be used carefully since excessive comfort could worsen sinking and/or rotation of the coffin bone. Deep flexor tendon tenotomy and digital nerves neurectomy are final and controversial alternatives for animals with chronic laminitis and severe pain.

Hoof abscesses can be more painful than laminitis, and sometimes be confused with colic since these animals are often recumbent. Treatment is similar to horses, based on drainage, soaking with antiseptic solutions, and hoof support (Mendoza et al. 2018c; McLean et al. 2023).

Tendon and ligament injuries are uncommon in companion donkeys due to their sedentary behavior; however emerging sport disciplines are gaining interest in donkeys and mules; thus, the prevalence of tendon and ligaments injuries are increasing, which are treated as horses. Therapeutic alternatives such as platelet-rich plasma (PRP), autologous conditioned plasma (Arthrex ACP), interleukin receptor antagonist protein (IRAP), autologous protein solution (Pro-Stride APS), α2-macroglogulin (A2M- Alpha2EQ), can also be used in donkeys and mules, but cost could be prohibitive. Septic arthritis management follows similar principles as in horses, including joint lavage, local and systemic antibiotics, NSAIDs, and support. Degenerative joint disease (arthrosis) is common in elder donkeys, and treatment consists of intra-articular corticosteroid injections (e.g. triamcinolone) and/or long-lasting NSAID administration (McLean et al. 2023). Depending on severity of pain, additional drugs such as opioids may be required. Polyacrylamide gel (Noltrex® VET) could also be used in donkeys and mules, mainly in those with PPID or laminitis.

3.9 Ophthalmology

There is no donkey-specific information on the treatment of ocular disorders and therapeutic protocols are extrapolated from horses. Topical antibiotics (neomycin, polymyxin B, gramicidin or gramicidin or bacitracin)), atropine, and systemic NSAIDs (flunixin meglumine) are the initial treatment of choice for uncomplicated corneal ulcers and uveitis. Topical antimicrobials must be modified upon bacterial culture sensitivity results (tobramycin, oxfloxacin, ciprofloxacin, gentamicin). Topical autologous serum, EDTA, and antifungals (voriconazole, itraconazole or miconazole) should be added in cases of mycotic ulcers and keratomalacia (melting ulcers). Recurrent uveitis is sporadically reported in donkeys, with similar treatment than horses. Ocular parasitic infections (*Thelazia* spp., *Onchocerca* spp., *Setaria* spp., *Besnoitia bennetti*) are treated with ivermectin/moxidectin and sulfonamides (Mendoza et al. 2018c). Ocular neoplasia treatment is discussed further elsewhere in this book (see chapters on dermatology and parasitology).

Topical atropine is used to induce mydriasis and prevent anterior synechia; however, it is important to monitor the contralateral eye pupil, gut motility, and fecal output due to its systemic effects as it can cause ileus and colic.

3.10 Miscellaneous: Infectious and Parasitic Diseases

Piroplasmosis due to *Babesia caballi* or *Theileria equi* infection is an important tick-borne disease of worldwide distribution in donkeys. Several protocols have been proposed for both parasites in donkeys and mules (Mendoza et al. 2024a). Most drugs improve clinical signs but are unable to completely eliminate *T. equi*, entailing a risk of relapse under stressful conditions. Imidocarb dipropionate (ID) is the most effective drug, but donkeys and mules are more prone to ID toxicity (liver and renal disturbances) and lower doses than horses should be used (Table 3). This drug should not be administered in febrile animals, and previous administration of flunixin meglumine, together with anticholinergic agents such as scopolamine (hyoscine) to prevent abdominal discomfort is recommended (Mendoza et al. 2024a)(Table 4). Buparvaquone and oxytetracycline combination have also been proposed as therapeutic alternatives against ID in donkeys and mules (Table 3). Diminazene diaceturate is another therapeutic option in some countries (Table 3).

Anaplasma phagocytophilum can cause infection in donkeys with typical (fever, edema, anemia) and atypical (muscular, neurological) signs (Aleman et al. 2024). Intravenous oxytetracycline or oral doxycycline are the treatments of choice (Table 1).

Infection with *Trypanosoma equiperdum* (dourine) and *T. evansi* (surra), two blood-borne parasites transmitted by biting flies, can be endemic in certain regions. The disease seems to be less severe in donkeys. Therapy with diminazene aceturate and quinapyramine sulfate can be administered in donkeys or mules (Table 3).

Although epidemiological information about leptospirosis in donkeys and mules is not available, it is assumed that recurrent uveitis, liver disease, renal failure, and abortion can also be observed in donkeys. Oxytetracycline and doxycycline are recommended antimicrobial options against this bacterium in donkeys (Table 1).

4 Conclusions

Scarce pharmacologic information is available for donkeys and mules; thus, clinicians must be judicious when extrapolating information from horses. Due to higher water content and metabolic rate, donkeys need shorter dosing interval and higher doses for commonly used drugs.

Questions for This Section

1. Which of the following statements is incorrect?
   (a) Donkeys have a faster clearance rates and shorter half-lives compared to horses for most antibiotics
   (b) Donkeys need a shorter dosing interval than horses for most drugs
   (c) Donkeys need a higher dose than horses for most drugs
   (d) **Donkeys have lesser side effects than horses for most drugs**
2. Which of the following drugs cannot be administered in donkeys with fever?
   (a) Paracetamol (acetaminophen)
   (b) Metamizole (dipyrone)
   (c) Meloxicam
   (d) **All of them can be used in pyretic donkeys**
3. What is the recommended treatment of donkeys with fluke?
   (a) Fenbendazole to high doses.
   (b) **Triclabendazole**
   (c) Moxidectin
   (d) Praziquantel
4. Which of the following drugs could administer in a donkey with suspect of interstitial pneumonia?
   (a) Oral dexamethasone
   (b) Inhaled budesonide
   (c) Oral clenbuterol
   (d) Inhaled salbutamol
   (e) **All of them**
5. Which of the following drugs used to treat piroplasmosis cause severe side effects in donkeys and mules?
   (a) **Imidocarb dipropionate**
   (b) Diminazene aceturate
   (c) Oxytetracycline
   (d) Buparvaquone

References

Aleman M, Vedavally U, Pusterla N, Wensley F, Berryhill E, Madigan JE (2024) Common and atypical presentations of Anaplasma phagocytophilum infection in equids with emphasis on neurologic and muscle disease. J Vet Intern Med 38:440–448. https://doi.org/10.1111/jvim.16964

Anwer MN, Rasheed MA, Ashraf M (2012) Pharmacokinetics of ketroprofen in healthy donkeys in Pakistan. J Anim Plant Sci 22:966–969

Arcoverde KN, Alves LSA, Cavalcante JM, Maranhao ACPM, Jurema AP, Araújo-Silva G, Urizar JTP, De Paula VV (2023) Pharmacokinetic monitoring of dypirone in Northeastern Brazilian donkeys (Equus asinus). Res Vet Sci 164:105034. https://doi.org/10.1016/j.rvsc.2023.105034

Chapuis RJJ, Smith JS, French HM, Toka FN, Peterson EW, Little EL (2021) Nonlinear mixed-effect pharmacokinetic modeling and distribution of doxycycline in healthy female donkeys after multiple intragastric dosing-preliminary investigation. Animals 11:2047. https://doi.org/10.3390/ani11072047

Cheng Z, McKellar Q, Nolan A, Lees P (1996) Preliminary pharmacokinetic and pharmacodynamic studies on flunixin meglumine in donkeys. Vet Res Commun 20:469–472

Dubey JP, Lindsay DS, Saville WJA, Reed SM, Granstrom DE, Speer CA (2001) A review of sarcocystis neurona and equine protozoal myeloencephalitis (EPM). Veterinary Parasitology 95:89–131

El Meshad AH, Ibrahim EM, Selim AM, Hamouda FK (2013) A new trend donkeys tetanus treatment using IgY. Research 5:23–29

Getachew AM, Innocent G, Proudman CJ, Trawford A, Feseha G, Reid SWJ, Faith B, Love S (2013) Field efficacy of praziquantel oral paste against naturally acquired equine cestodes in Ethiopia. Parasitol Res 112:141–146. https://doi.org/10.1007/s00436-012-3117-1

Gokbulut C, Boyacioglu M, Karademir U (2005) Plasma pharmacokinetics and faecal excretion of ivermectin (Eqvalan paste) and doramectin (Dectomax, 1%) following oral administration in donkeys. Res Vet Sci 79:233–238. https://doi.org/10.1016/j.rvsc.2005.01.002

Gokbulut C, Akar F, McKellar QA (2006) Plasma disposition and faecal excretion of oxfendazole, fenbendazole and albendazole following oral administration to donkeys. Vet J 172:166–172. https://doi.org/10.1016/j.tvjl.2005.02.022

Gokbulut C, Aksit D, Smaldone G, Mariani U, Veneziano V (2014) Plasma pharmacokinetics, faecal excretion and efficacy of pyrantel pamoate paste and granule formulations following per os administration in donkeys naturally infected with intestinal strongylidae. Vet Parasitol 205: 186–192. https://doi.org/10.1016/j.vetpar.2014.06.026

Gokbulut C, Aksit D, Santoro M, Roncoroni C, Mariani U, Buono F, Rufrano D, Fagiolo A, Veneziano V (2016) Plasma disposition, milk excretion and parasitological efficacy of mebendazole in donkeys naturally infected by Cyathostominae. Vet Parasitol 217:95–100. https://doi.org/10.1016/j.vetpar.2015.12.031

Grosenbaugh DA, Reinemeyer CR, Figueiredo MD (2011) Pharmacology and therapeutics in donkeys. Equine Vet Educ 23:523–530. https://doi.org/10.1111/j.2042-3292.2011.00291.x

Horspool LJI, McKellar QA (1990) Disposition of oxytetracycline in horses, ponies and donkeys after intravenous administration. Equine Vet J 22:284–285

Horspool LJ, Sarasola P, McKellar QA (1992) Disposition of ampicillin sodium in horses, ponies and donkeys after intravenous administration. Equine Vet J Suppl 24:59–61

Horspool LJI, Taylor DJ, McKellar QA (1994) Plasma disposition of amikacin and interaction with gastrointestinal microflora in Equidae following intravenous and oral-administration. J Vet Pharmacol Ther 17:291–298

Kay G, Knottenbelt DC (2007) Tetanus in equids: a report of 56 cases. Equine Veterinary Educacion 19:107–112. https://doi.org/10.2746/095777307X181320

Khan A, Khan MS, Avais M, Ijaz M, Abbas T (2013) Prevalence, hematology, and treatment of balantidiasis among donkeys in and around Lahore, Pakistan. Vet Parasitol 196:203–205. https://doi.org/10.1016/j.vetpar.2013.01.017

Kim T-W, Della Rocca G, Di Salvo A, Owen H, Sgorbini M, Giorgi M (2014) Pharmacokinetics of the novel cyclooxygenase 2 inhibitor cimicoxib in donkeys. J Equine Vet Sci 34:923–925. https://doi.org/10.1016/j.jevs.2014.02.004

Kinabo LD, Bogan JA (1989) Disposition of triclabendazole in horses, ponies and donkeys. Equine Vet J 21:305–307

Lavy E, Ziv G, Aroch I, Glickman A (1995) Pharmacokinetics, intramuscular bio-availability, and bioequivalence of amoxycillin in donkeys. J Veterinary Med Ser B 42:284–292

Li Z, Qin X, Wang C, Liu W (2024) Pharmacokinetics and tissue residues of albendazole sulphoxide and its metabolites in donkey after intramuscular injection. Vet Med Sci 10:e1393. https://doi.org/10.1002/vms3.1393

Liu S, Guo Y, Qu H, Dong Y, Zhao S, Fu T, Kang R, Cheng J, Huang S, Zhao L, Ma Q (2024) Pharmacokinetics of florfenicol and its metabolite florfenicol amine in the plasma, urine, and feces of fattening male donkeys following single oral administration. Front Vet Sci 10:1314029. https://doi.org/10.3389/fvets.2023.1314029

Lizarraga I, Sumano H, Brumbaugh GW (2004) Pharmacological and pharmacokinetic differences between donkeys and horses. Equine Vet Educ 16:102–112

Mahmood KT, Ashraf M (2011) Pharmacokinetics of meloxicam in healthy donkeys. Pak J Zool 43:897–901

Matthews NS, Peck KE, Taylor TS, Mealey KL (2001) Pharmacokinetics of phenylbutazone and its metabolite oxyphenbutazone in miniature donkeys. Am J Vet Res 62:673–675

Matthews NS, Grosenbaugh DA, Kvaternick V (2009) Pharmacokinetics and oral bioavailability of firocoxib in donkeys. 10th world congress of veterinary Anaesthesia, Glasgow, UK

Mclean AK, Falt T, Abdelfattah EM, Middlebrooks B, Gretler S, Spier S, Turoff D, Navas Gonzalez FJ, Knych HK (2023) Transdermal flunixin meglumine as a pain relief in donkeys: a pharmacokinetics pilot study. Metabolites 13:776. https://doi.org/10.3390/metabo13070776

Mealey KL, Matthews NS, Peck KE, Burchfield ML, Bennett BS, Taylor TS (2004) Pharmacokinetics of R(−) and S(+) carprofen after administration of racemic carprofen in donkeys and horses. Am J Vet Res 65:1479–1482. https://doi.org/10.2460/ajvr.2004.65.1479

Mendoza FJ, Toribio RE, Perez-Ecija A (2018a) Aspects of clinical relevance in donkeys. In: Reed SM, Bayly WM, Sellon DC (eds) Equine internal medicine, 4th edn. Elsevier, Amsterdam, pp 1513–1524

Mendoza FJ, Toribio RE, Perez-Ecija A (2018b) Donkey internal medicine – Part I: metabolic, endocrine, and alimentary tract disturbances. J Equine Vet Sci 65:66–74. https://doi.org/10.1016/j.jevs.2018.02.001

Mendoza FJ, Toribio RE, Perez-Ecija A (2018c) Donkey internal medicine – Part II: cardiovascular, respiratory, neurologic, urinary, ophthalmic, dermatology, and musculoskeletal disorders. J Equine Vet Sci 65:86–97. https://doi.org/10.1016/j.jevs.2018.02.025

Mendoza FJ, Perez-Ecija A, Toribio RE (2019a) Clinical pharmacology in donkeys and mules. Vet Clin North Am Equine Pract 35:589–606. https://doi.org/10.1016/j.cveq.2019.08.011

Mendoza FJ, Toribio RE, Perez-Ecija A (2019b) Metabolic and endocrine disorders in donkeys. Vet Clin North Am Equine Pract 35:399–417. https://doi.org/10.1016/j.cveq.2019.07.001

Mendoza FJ, Pérez-Écija A, Kappmeyer LS, Suarez CE, Bastos RG (2024a) New insights in the diagnosis and treatment of equine piroplasmosis: pitfalls, idiosyncrasies, and myths. Front Vet Sci 11:1459989. https://doi.org/10.3389/fvets.2024.1459989

Mendoza FJ, Toribio RE, Perez-Ecija A (2024b) Metabolic and endocrine insights in donkeys. Animals 14:590. https://doi.org/10.3390/ani14040590

Nocera I, Aliboni B, Ben David L, Gracia-Calvo LA, Sgorbini M, Citi S (2020) Radiographic and venographic appearance of healthy and laminitic feet in Amiata donkeys. Front Vet Sci 7: 601665. https://doi.org/10.3389/fvets.2020.601665

Oukessou M, Aissa M, Hmidouch A (1994) Pharmacokinetics study of benzylpenicillin and amoxicillin in the donkey (Equus asinus). Indian Vet J 71:1077–1081

Peck KE, Matthews NS, Taylor TS, Mealey KL (2002) Pharmacokinetics of sulfamethoxazole and trimethoprim in donkeys, mules, and horses. Am J Vet Res 63:349–353. https://doi.org/10.2460/ajvr.2002.63.349

Rickards KJ, Thiemann AK (2019) Respiratory disorders of the donkey. Vet Clin North Am Equine Pract 35:561–573. https://doi.org/10.1016/j.cveq.2019.08.009

Sekkin S, Gokbulut C, Kum C, Karademir U (2012) Plasma disposition of enrofloxacin following intravenous and intramuscular administration in donkeys. Vet Rec 171. https://doi.org/10.1136/vr.100653

Sullivan R, Vogel P (2023) Therapeutics in veterinary care of the donkey. UK-Vet Equine 7:32. https://doi.org/10.12968/ukve.2023.7.1.32

The Donkey Sanctuary (2018) In: Evans L, Crane M (eds) The clinical companion of the donkey, 1st edn. Matador

Théon AP, Wilson WD, Magdesian KG, Pusterla N, Snyder JR, Galuppo LD (2007) Long-term outcome associated with intratumoral chemotherapy with cisplatin for cutaneous tumors in equidae: 573 cases (1995–2004). J Am Vet Med Assoc 230:1506–1513. https://doi.org/10.2460/javma.230.10.1506

Thiemann AK, Sullivan RJE (2019) Gastrointestinal disorders of donkeys and mules. Vet Clin North Am Equine Pract 35:419–432. https://doi.org/10.1016/j.cveq.2019.08.001

Thiemann A, Sullivan R (2021) Guide to common skin disorders in donkeys. In Pract 43:318–327. https://doi.org/10.1002/inpr.80

Thiemann AK, Buil J, Rickards K, Sullivan RJ (2022) A review of laminitis in the donkey. Equine Vet Educ 34:553–560. https://doi.org/10.1111/eve.13533

Toutain PL, Ferran A, Bousquet-Melou A (2010) Species differences in pharmacokinetics and pharmacodynamics. In: Cunningham F, Elliott J, Lees P (eds) Comparative and veterinary pharmacology. Springer, Berlin, pp 19–48. https://doi.org/10.1007/978-3-642-10324-7_2

Trawford AF, Burden F, Hodgkinson JE (2005) Suspected moxidectin resistance in cyathostomin in two donkey herds at the donkey santuary, UK. In: 20th international conference of the world Association for the Advancement of Veterinary Parasitology. New Zealand

Welfare RE, Mealey KL, Matthews NS, Taylor TS (1996) Pharmacokinetics of gentamicin in donkeys. J Vet Pharmacol Ther 19:167–169

White SD, Bourdeau PJ, Brément T, Vandenabeele SI, Haspeslagh M, Bruet V, Sloet van Oldruitenborgh-Oosterbaan MM (2019) Skin disease in donkeys (Equus asinus): a retrospective study from four veterinary schools. Vet Dermatol 30:247–e276. https://doi.org/10.1111/vde.12733

Widmer A, Kummer M, Eser MW, Furst A (2009) Comparison of the clinical efficacy of cefquinome with the combination of penicillin G and gentamicin in equine patients. Equine Vet Educ 21:430–435. https://doi.org/10.2746/095777309x453137

Wood F, Loomes K (2022) Suspected lidocaine toxicity and the use of metoclopramide and neostigmine to treat post-operative small intestinal ileus in a donkey. Vet Rec Case Rep 10: e481. https://doi.org/10.1002/vrc2.481

Xue C, Davis J, Berghaus LJ, Hanafi A, Vaughn SA, Hart KA (2023) Pharmacokinetic properties of pergolide mesylate following single and multiple-dose administration in donkeys (Equus asinus). Equine Vet J 55:1078–1085. https://doi.org/10.1111/evj.13917

Anesthesia and Pain Management in Donkeys and Mules

Johannes P. A. M. van Loon

Abstract

This chapter describes the current knowledge on various anesthetic and analgesic protocols, drugs and techniques for donkeys and mules. Although the number of publications in this field is steadily growing, there is still much information that is not available about donkeys and mules. At these moments, the veterinary practitioner must rely on own personal experience and also on knowledge and experience from anesthesia and pain management in horses. With this chapter, we aim to describe the scientific evidence on anesthesia and pain management in donkeys and mules and try to merge this into the general knowledge that we can extrapolate from knowledge in horses. Where possible, we will compare and contrast the choice for drugs, protocols and techniques in donkeys and mules with those for horses. Like for equine anesthesia, there are inherent risks involved with the use of anesthetics for donkeys and mules too, but we can make the anesthetic procedures as safe as possible with sufficient theoretical backgrounds and appropriate and careful monitoring of our patients.

Keywords

Donkey · Mule · Anesthesia · Anaesthesia · Sedation · Analgesia · Pain

---

The original version of the chapter has been revised. A correction to this chapter can be found at https://doi.org/10.1007/978-3-032-18540-2_22

J. P. A. M. van Loon (✉)
Department of Clinical Sciences, Faculty of Veterinary Medicine, Utrecht University, Yalelaan, Utrecht, The Netherlands
e-mail: j.p.a.m.vanloon@uu.nl

A. K. McLean et al. (eds.), *Donkey and Mule Medicine: Clinical Medicine and Reproduction*, https://doi.org/10.1007/978-3-032-18540-2_14

1 Introduction

Veterinary care for donkeys and mules has received increased interest over the last decades. The number of donkeys and mules worldwide, serving as companions or working animals, is very high and the number of publications of welfare and pain management is steadily growing. There is also an increasing need for knowledge about anesthesia, options for both in-clinic situations and for field anesthesia procedures, and for locoregional techniques in donkeys.

In addition to the many similarities between horses and donkeys, there are also many important differences (Burden and Thiemann 2015), which is why this chapter will frequently compare the donkey to the horse. Anatomical, physiological and pharmacological differences between horses and donkeys will be discussed, as well as the use of injection anesthetics, inhalation anesthetics and the possibilities for pre- and postoperative analgesia.

Previous review articles (Matthews and van Loon 2013, 2019) have described very extensive evidence-based and empirically founded descriptions of anesthesia and pain management in the donkey and mule. Various publications have emerged recently about the pharmacological differences for anesthetics and analgesics between horses and donkeys (Coakley et al. 1999; Mealey et al. 1997; Sinclair et al. 2006; Taylor et al. 2008, Grosenbaugh et al. 2011).

The aim of this chapter is to further describe the important foundations for anesthesia and pain management in donkeys and mules and to integrate knowledge from older publications, with those from more recent studies.

2 Preanesthetic Evaluation and Case Preparation

Very similar to the pre-anesthetic screening that is common practice for horses that need to undergo anesthesia, the same approach can be used for donkeys and mules that need to undergo an anesthetic procedure. The physical status of the patient can be scored in the ASA classification system that is also used for horses or in the CHARIOT score, that has originally been described for horses by Hubbell et al. (2022) and has been further investigated by Brumund et al. (2024). In addition, an accurate estimation of the bodyweight is a key, since the bodyweight of donkeys is very often overestimated (Svendsen 2008). Hospitals that have a built-in scale for this are best prepared. The table from the Donkey Sanctuary that uses the height at the withers and the heart girth (Fig. 1) provides a suitable alternative when no scale is available. Alternatively, Kay et al. have published a similar scale that is suitable for estimating the live weight of mules (Fig. 2).

Many physical parameters and blood values may differ from those of horses (Svendsen 2008). Recent literature in horses (Barton et al. 2024) shows that preoperative fasting may bring some side-effects like decreased gastrointestinal motility and decreased water uptake, while not offering specific benefits for the anesthetized horse. For donkeys, there is no described advantage of having prolonged fasting times for anesthesia. Maney et al. (2019) described no improvements in gas

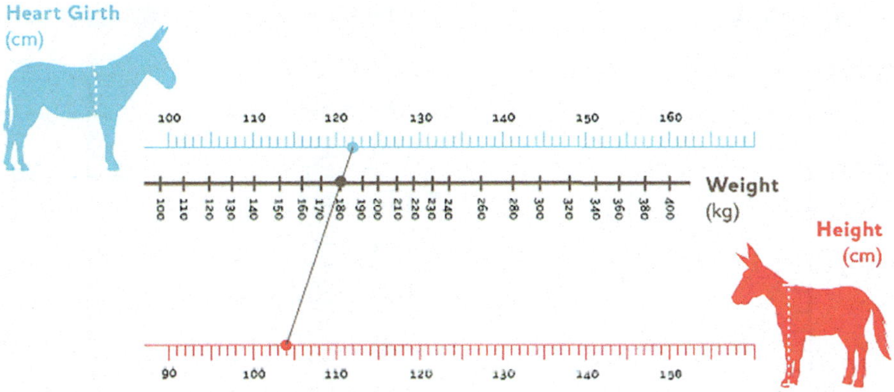

Fig. 1 Body weight estimation for donkeys, based on heart girth and height at withers. (Courtesy of Donkey Sanctuary UK)

exchange after 5 h of fasting in donkeys. At the same time, donkeys may be more prone to develop hyperlipemia after fasting and fasting times should therefore be limited very much for routine anesthesia procedures. For procedures that would require prolonged fasting (like abdominal laparoscopic procedures for abdominal cryptorchid stallions or mares suffering from an ovarian tumor), fasting should be performed very carefully with some support in the form of concentrates during the 24–36 h period of fasting to prevent hyperlipemia.

Jugular catheterization during premedication is very similar to the procedure in horses. Good restraint or mild sedation is sometimes required to perform this. Donkeys have a thicker cutanous coli muscle and thick skin that can make it more difficult to find the jugular vein and to insert a catheter. This can make it difficult to visualize the jugular vein and can complicate catheter placement.

3 Premedication

Premedication of donkeys and mules for general anesthesia or for standing surgical or diagnostic procedures is usually performed in a very similar way as in horses. This means that a combination of sedatives and analgesics is the typically the best approach. Throughout any anesthetic procedure, we aim to balance the effects of decreased consciousness (mental block), appropriate analgesia (sensory block, suitable degree of muscle relaxation (motor block) and autonomic stability (balanced tone of sympathetic and parasympathetic nervous system) as schematically shown in Fig. 3.

Sedation is best performed with an alpha2-agonist such as xylazine or detomidine (most commonly used alpha2-agonists in routine practice with donkeys and mules), and if applicable this can be combined with a low dose of acepromazine. To further deepen and stabilize the sedation, an opioid can be added, which also brings perioperative analgesia. Finally, for surgical procedures, an NSAID can be included

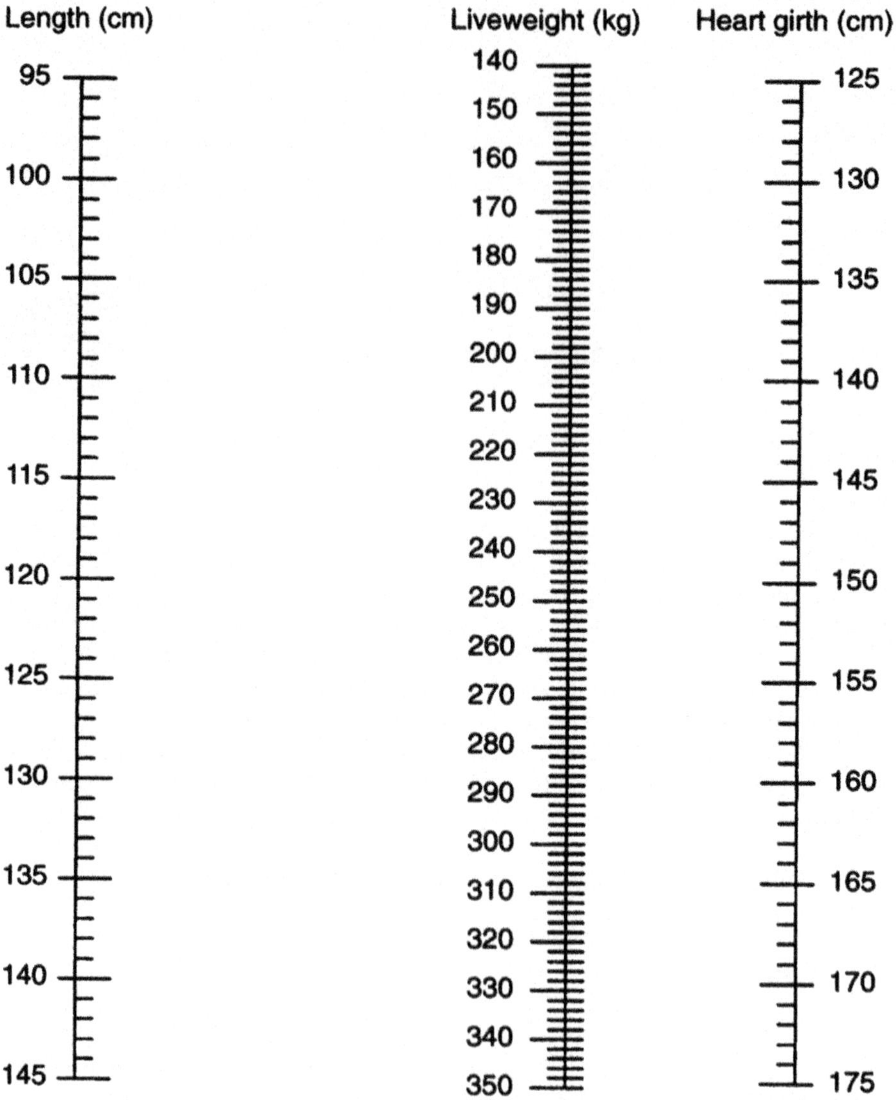

Fig. 2 Body weight estimation for mules, based on heart girth and length (measured from olecranon process of the elbow to the tuber ischii). (Courtesy of Kay et al. 2004)

in the premedication to prevent, as much as possible, the release of pro-inflammatory mediators.

The dosages for donkeys are very similar to what is used for horses and the depth and quality of sedation are best reached when individual titration of the drugs to the equid's response is performed. Several studies have described the use of detomidine, xylazine and romifidine in donkeys, and the advised dosages are shown in Table 1.

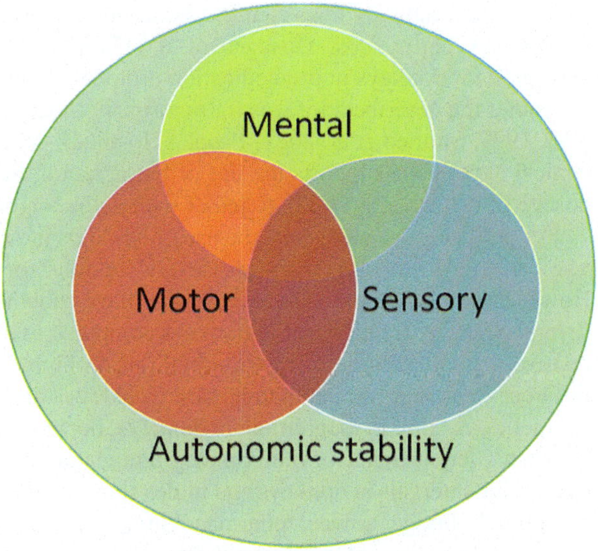

Fig. 3 Four pillars for building anesthesia protocols, with mental standing for mental block (sedation, loss of consciousness), sensory standing for sensory block (local anesthesia, analgesia), motor standing for motor block (muscle relaxation) and autonomic stability for stable sympathetic and parasympathetic nervous system tone

Table 1 Dosages of alpha2-agonists in premedication for horses, donkeys and mules

Sedative drug	Dose range Horse	Dose range Donkey	Dose range Mule
Xylazine	0.6–1 mg/kg IV	0.5–1 mg/kg IV 2–3 mg/kg IM	1.1–1.6 mg/kg IV
Detomidine	0.01–0.02 mg/kg IV	0.01–0.02 mg/kg IV	0.03 mg/kg IV
Romifidine	0.04–0.08 mg/kg IV	0.04–0.08 mg/kg IV	0.04–0.12 mg/kg IV

Based on: Brooke Hospital, Matthews et al. (1997), Mostafa et al. (1995), Alves et al. (1999)

For mules, there is clinical evidence that the required dosages for all alpha2-agonists are 40–50% higher, compared to horses and donkeys. Latzel (2008) described the pharmacokinetics and pharmacodynamics in mules. The half-life of xylazine is 15 min shorter in mules compared to horses, and the advised dose of xylazine was 50% higher compared to horses. Table 1 also shows the described dose ranges of sedatives in mules.

Acepromazine has been described for premedication in donkeys by Naddaf et al. (2015). The donkeys in this study were dosed with 0.04 mg/kg IV of acepromazine and showed significant levels of sedation. Lizarraga et al. (2014) described the sedative effects of 0.05 mg/kg IV Acepromazine in donkeys to be very limited. Although not extensively described in scientific studies, the combined use of acepromazine at low dosages (0.02–0.03 mg/kg) with alpha2-agonists has shown clinical efficacy. However, the occurrence of side-effects on blood pressure and thermoregulation due to peripheral vasodilation is very important, especially in younger and small-sized donkeys this could have a significant impact. No reports on the contra-indications for use of acepromazine in intact males have been published.

There are relatively few scientific studies on the use of opioids for premedication in donkeys and mules. Dosages of 0.025–0.05 mg/kg IV of butorphanol are described for donkeys in premedication protocols (combined with alpha2-agonists). For mules the described dosages for butorphanol are 0.033–0.044 mg/kg IV (Joubert et al. 1999, Straticò et al. 2021, Brooke Hospital). Apart from the scientific studies, clinical experience supports the use of butorphanol, but also that of morphine (at a dosage of 0.1–0.2 mg/kg IV, together with alpha2-agonists) in donkeys and mules.

For perioperative analgesia, the use of NSAIDs in premedication is also routinely advocated. To this aim, there are several options for use in donkeys and mules. However, several pharmacokinetic studies have shown that the pharmacokinetics of several NSAIDs are different in donkeys compared to that in horses. NSAIDs have a relatively small volume of distribution, due to high plasma protein binding and a different clearance compared to horses. Coakley et al. (1999) described the pharmacokinetics of flunixin meglumine in donkeys, horses and mules. Donkeys and mules have a higher clearance of flunixin meglumine than the horse. These results indicate that dosing intervals in donkeys and mules should be shorter than in horses. To arrive at optimal dosage advice, pharmacokinetic and pharmacodynamic studies with repeated doses should be carried out. A recent study by McLean et al. (2023) shows the transdermal application of flunixin meglumine in donkeys, with effective reductions in pro-inflammatory mediators over a time period of up to 72 h.

Sinclair et al. (2006) describe the comparative pharmacokinetics of meloxicam for the donkey and the horse. Meloxicam also shows a faster clearance in donkeys compared to horses. Based on this very short half live of meloxicam, the expected duration of analgesia for meloxicam in donkeys might be shorter compared to horses and therefore, the use of meloxicam in donkeys may not be very practical.

Mealey et al. (1997) described the pharmacokinetics of phenylbutazone in healthy horses and donkeys and also found increased clearance of this NSAID in donkeys compared to horses. All these studies indicate that the use of NSAIDs in donkeys and mules for premedication seems indicated and helpful, but that the dosing scheme of all these NSAIDs for postoperative analgesia should be adjusted. We recommend dosing flunixin meglumine and phenylbutazone in donkeys at the same dosage used in horses, but with a dosing scheme of 3 times daily.

4 Induction of General Anesthesia

General anesthesia is often induced with a combination of ketamine with or without a benzodiazepine, usually midazolam or diazepam. Various studies have described the differences in recumbency times between horses, donkeys and mules after premedication of an alpha2-agonist, with or without butorphanol, and induction with ketamine (Table 2). This table shows that mules show clearly shorter recumbency times after similar dosages of xylazine and ketamine, compared to horses and donkeys. This might be related to the differences in depth of sedation with these xylazine dosages, that are obtained in mules. The quality of induction was therefore inferior in these mules. A similar finding is shown for the induction with tiletamine/

Table 2 Recumbency times (min) with various injectable anesthetic combinations for horses, donkeys and mules

Drug combination	Horse	Donkey	Mule
Xylazine (1.1 mg/kg IV) Ketamine (2.2 mg/kg IV)	23 min	24 min	14 min
Xylazine (1.1 mg/kg IV) Butorphanol (0.04 mg/kg IV) Ketamine (2.2 mg/kg IV)	23.5 min	37 min	25.1 min
Xylazine (1.1 mg/kg IV) Tiletamine/Zolazepam (1.1 mg/kg IV)	30.7 min	43 min	21.1 min

Based on: Matthews et al. (1992, 1994)

zolazepam (Telazol in Table 2, Zoletil in Europe) after xylazine premedication. Both in donkeys and mules, the combined use of xylazine with butorphanol in premedication improved the quality of sedation and the subsequent recumbency time.

The author prefers to combine ketamine with a benzodiazepine, because of the superior quality of induction that is obtained with a benzodiazepine, as has been described in horses. In donkeys, especially miniatures, the dosages of diazepam or midazolam are usually around 0.1 mg/kg IV, combined with a dosage of ketamine of 2–2.5 mg/kg IV. Dar and Gupta (2016) describe the use of ketamine at 2.2 mg/kg IV with diazepam at a low dose of 0.03 mg/kg IV in mules after xylazine given at a dose of 1.3 mg/kg IV for premedication. The quality of induction was inferior with suboptimal muscle relaxation. Table 3 provides the dosages for midazolam/diazepam and ketamine over the range of bodyweights that can be applicable to donkeys ranging from miniature size to large sized donkeys like the Poitou.

Instead of the use of benzodiazepines, guaifenesin can also be used to improve the quality of induction. However, donkeys seem to be more susceptible to the sedative and muscle relaxant effects of guaifenesin, compared to horses. Matthews et al. (1997) described the pharmacokinetics of guaifenesin in horses and donkeys (Fig. 4) and described recumbency in donkeys at a lower dose of guaifenesin (125 mg/kg IV), compared to that of horses (211 mg/kg IV). The clearance of guaifenesin in donkeys was again faster compared to horses. The increased sensitivity to this muscle relaxant, together with the side-effects when overdosing (decreased MAP and minute ventilation) make this muscle relaxant a risky drug to use in donkeys, especially since donkeys are very often overestimated for bodyweight when no scale is available (personal experience). Therefore, we advocate to use benzodiazepines instead of guaifenesin in donkeys for muscle relaxation during induction of general anesthesia.

Apart from using ketamine in combination with a muscle relaxant, there are good descriptions of using barbiturates or non-barbiturates for induction in donkeys and mules as well. Thiopental has been described in several studies at dosages of 6 mg/kg IV for mules (after xylazine premedication) by Vullo et al. (2017) and at a dosage of 7 mg/kg IV in donkeys (after premedication with xylazine and acepromazine) by Emami et al. (2006). Abd-Almaseeh (2008) used a higher dose of thiopental (10 mg/

Table 3 Dosages for ketamine and benzodiazepines for induction of general anesthesia in donkeys and mules (ketamine (100 mg/ml) dosage 2.2 mg/kg, benzodiazepine (5 mg/ml) dosage based on metabolic weight (0.05–0.1 mg/kg)

Body weight	Dosage ketamine	Dosage midazolam/diazepam
50 kg	1.1 ml	1 ml
60 kg	1.3 ml	1.2 ml
70 kg	1.5 ml	1.3 ml
80 kg	1.8 ml	1.4 ml
90 kg	2.0 ml	1.6 ml
100 kg	2.2 ml	1.7 ml
110 kg	2.4 ml	1.8 ml
120 kg	2.6 ml	2.0 ml
130 kg	2.9 ml	2.1 ml
140 kg	3.1 ml	2.2 ml
150 kg	3.3 ml	2.3 ml
160 kg	3.5 ml	2.4 ml
170 kg	3.7 ml	2.5 ml
180 kg	4.0 ml	2.7 ml
190 kg	4.2 ml	2.8 ml
200 kg	4.4 ml	2.9 ml
250 kg	5.5 ml	3.4 ml
300 kg	6.6 ml	3.9 ml
350 kg	7.7 ml	4.4 ml
400 kg	8.8 ml	4.8 ml
450 kg	9.9 ml	5.3 ml
500 kg	11.0 ml	5.7 ml

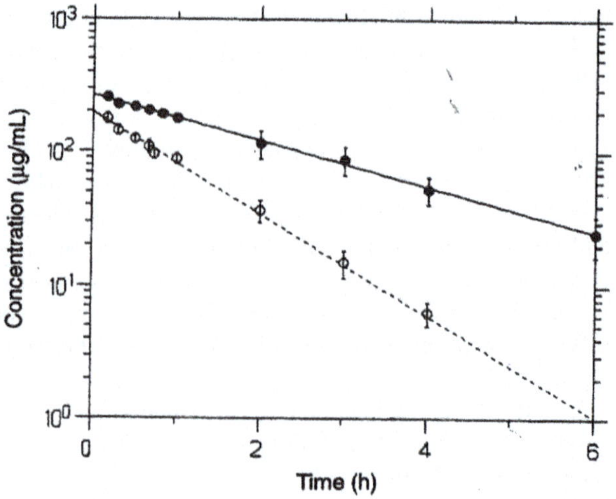

Fig. 4 Mean (±SD) plasma concentrations of guaifenesin in horses (•) and donkeys (°) after intravenous administration of guaifenesin of a mean dose of 125 mg/kg (donkeys) and 211 mg/kg (horses). (Courtesy of Matthews et al. 1997)

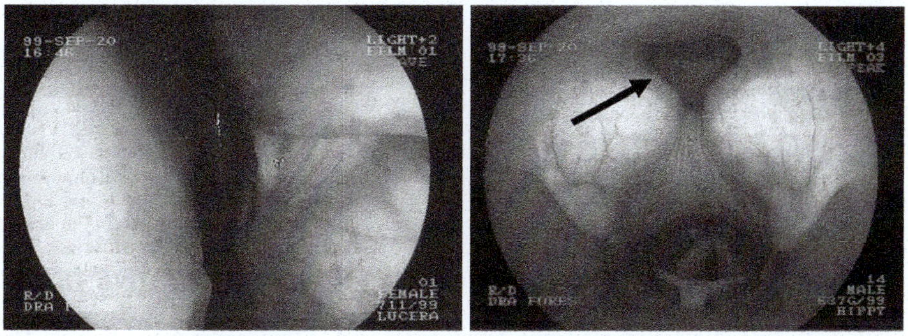

Fig. 5 Endoscopy of the upper airways in the donkey, with the narrow nasal cavities on the left and the deep pharyngeal recess (arrow) and the relatively small entrance between the vocal cords and the long and narrow epiglottis on the right (Courtesy of Forés et al. 2001)

kg IV) to induce donkeys after xylazine premedication. They observed very quick induction times (18 s) with more side-effects in terms of apnea, compared to induction protocols with propofol (at dosages of 2 mg/kg IV). Naddaf et al. (2015) described the use of propofol (2 mg/kg IV) to induce donkeys after acepromazine premedication (0.04 mg/kg IV). Molinaro Coelho et al. (2014) described the combined use of ketamine (1.5 mg/kg IV) and propofol (0.5 mg/kg IV) in donkeys after xylazine premedication. This ketofol combination of ketamine and propofol is also described in foals for induction of general anesthesia (Sun et al. 2021).

After induction of general anesthesia, donkeys and mules need to be intubated, especially when inhalant anesthesia is planned for maintenance. In field anesthesia procedures where intravenous maintenance is planned, intubation of the patient is often not performed and oxygen can be supplied by a nasally placed tube. Intubation, either by the orotracheal or nasotracheal route can be more challenging and complicated in donkeys compared to horses. This can be explained by the relatively narrow nasal passages, the relatively small entrance in between the vocal cords and the prominent pharyngeal recess. These structures are nicely described and shown in the study by Forés et al. (2001) (Fig. 5). Because of these anatomical circumstances that may make blind intubation more difficult in donkeys, we usually have a flexible endoscope ready to use for every donkey that needs general anesthesia to guide the tube during intubation. With this endoscope-assisted intubation technique, the procedure becomes much more predictable and quick.

5 Maintenance of General Anesthesia

For procedures that take longer than 60 min, inhalation anesthesia is recommended. Donkeys and mules have a similar MAC value for isoflurane (1.3%) and isoflurane can be administered in oxygen or in a mixture of oxygen in air. There are no publications on the use of constant rate infusions (CRI) to combine with an inhalant in donkeys, but the author uses CRI of detomidine (0.010 mg/kg/hr. IV) together

Table 4 Triple drip anesthesia in Donkeys

Drug combination	Horse	Donkey	Mule
Ketamine	2 mg/ml	2 mg/ml	2 mg/ml
Xylazine or Detomidine	1 mg/ml 0.02 mg/ml	0.5–1 mg/ml 0.02 mg/ml	1.5 mg/ml 0.03 mg/ml
Guaifenesin	100 mg/ml	50 mg/ml	50 mg/ml
CRI speed	1–2 ml/kg/h	1–2 ml/kg/h	1–2 ml/kg/h

These dosages are based on personal experience of the author and on the findings of Taylor et al. (2008) and Molinaro Coelho et al. (2014)

with isoflurane in oxygen in a similar way compared to this technique in horses. The use of other alpha2-agonists, ketamine or lidocaine for CRIs with isoflurane is advised with similar dosages compared to those in horses (personal experience). Monitoring of these patients can routinely be performed with a multiparameter monitor, using ECG, capnography for End Tidal CO_2 and isoflurane concentrations, invasive blood pressure, spirometry if available and arterial blood gas analysis. Invasive blood pressure can be measured from the facial, transverse facial or auricular arteries. Fluid therapy can be performed by means of isotonic crystalloid infusions like Ringers or Lactated Ringers at a speed of 5–10 ml/kg/hr.

For Total Intravenous Anesthesia (TIVA), various combinations of alpha2-agonists, ketamine and muscle relaxants can be used. Because of the increased sensitivity of donkeys to guaifenesin, the dosages and concentrations of the different triple drip solutions are preferably adapted to this. Vullo et al. (2017) described the use of triple drip in mules for field castration. Taylor et al. (2008) and Molinaro Coelho et al. (2014) described this technique for maintenance in donkeys. In both studies, guaifenesin was used at a concentration of 50 mg/ml, whereas triple drip solutions for horses usually contain guaifenesin at 100 mg/ml. Table 4 shows several options to perform triple drip anesthesia in donkeys and mules using the adapted 5% concentrations of guaifenesin, in combination with different alpha2-agonists.

6 Recovery

Donkeys usually show less flight responses during recovery from general anesthesia and take more time to stand compared to horses. Nevertheless, assisting them during recovery usually improves the quality of the recovery. This can easily be performed by manual assistance (Fig. 6), both in field or hospital settings. To support donkeys during recovery, it is advised to administer oxygen by means of a nasally placed tube and to place a nasopharyngeal tube to secure the upper airway when donkeys exhibit respiratory noises after extubation.

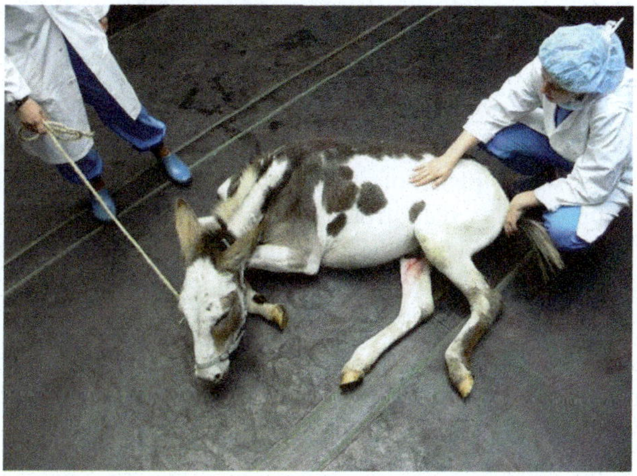

Fig. 6 Hand-assisted recovery in a young donkey after surgical castration

7 Standing Surgery in Donkeys and Mules

For standing surgical procedures in donkeys or mules, the premedications that have been described for donkeys and mules previously in this chapter can be used as a starting point. Because of the risk of worsening hypothermia, especially during longer lasting standing procedures, the authors don't routinely include acepromazine in the premedication protocol, as hypothermia is a more common complication during standing procedures. The use of convective temperature management systems, blankets and warm infusions can help to prevent hypothermia as much as possible (Florczyk et al. 2020). After that, the sedation could be prolonged by either top-up dosages or the use of CRIs with an alpha2-agonist. Although it has not been described in the literature, the use of a detomidine, xylazine or romifidine CRI can be used to maintain a steady state of sedation. For the authors, the use of detomidine CRI with a starting dose of 0.01 mg/kg/hr. and subsequent individual tailoring to the sedation depth of the patient works well. For surgical procedures, proper locoregional techniques are necessary to provide optimal surgical conditions.

8 Locoregional Anesthetic Techniques

Both for procedures in general anesthesia and in the standing sedated donkey or mule, the use of locoregional anesthetic techniques are a valuable and essential element of the anesthetic protocol (Lamont 2008).

8.1 Intrafunicular Use of Local Anesthetics for Castration

Suriano et al. (2014) described the use of intrafunicular lidocaine administration in isoflurane-anesthetized donkeys. The donkeys that received intrafunicular lidocaine showed lower heart rates and needed lower End-Tidal isoflurane concentrations to be in a stable plane of anesthesia during surgical castration. These findings are supported by data from studies in horses, that also show improved surgical conditions (Haga et al. 2006) and improved postoperative analgesia (Abass et al. 2018) in horses that were receiving intratesticular and intrafunicular local anesthetics for surgical castration.

8.2 Epidural Anesthesia

Epidural anesthesia has already been described in the donkey by Shoukry et al. (1975), but is still also described in more recent papers like for instance in the publication of Hamed et al. (2017), that compares the use of epidural xylazine and dexmedetomidine and Marzok and El-Khodery (2015), that described the use of tramadol and lidocaine after epidural administration in donkeys.

Indications for epidural anesthesia in the donkey and mule are widespread. The use of epidurally administered local anesthetics and alpha2-agonists can be used when surgical corrections are required for rectal or vaginal prolapses or for surgical removal of melanomas in the tail and perineal region. Epidurally administered opioids like morphine or methadone (with or without alpha2-agonists) can be used during abdominal surgery. Aziz et al. (2008) describe laparoscopic ovariectomy in standing sedated donkeys. For procedures like these, the epidural administration of morphine (0.1–0.2 mg/kg diluted to a volume of 15–20 ml), as described in horses by van Hoogmoed and Galuppo (2005), is a very good addition to the protocol. The use of epidural morphine will improve the surgical conditions and will help to bolster the postoperative analgesic protocol. Epidurally administered morphine can also be used for long-term analgesia after hindlimb surgery or with painful conditions in the hindlimb such as septic arthritis or Einschuss, when administered through an epidural catheter. Burnham (2008) described the anatomy of the sacral and coccygeal vertebrae in donkeys, with the first intercoccygeal space in the donkey being narrower than the second. Table 5 shows dosages for different anesthetics and analgesics that can be used in donkeys, with dosages that have been described in donkeys or that have been extrapolated from studies in horses.

Local anesthetics and alpha2-agonists are administered without dilution, for the administration of morphine (with or without an alpha2-agonist), the volume of morphine can be increased by adding saline for a more cranial spreading of the analgesic effect (for instance for applying analgesia in patients that need hindleg or caudal abdominal analgesia.

Table 5 Epidural administration of anesthetics and analgesics in donkeys and mules

Drug	Dosage (mg/kg)	Onset of action	Duration of effect
Lidocaine (2%)	0.22–0.35	5–15 min	60–90 min
Lidocaine (2%) + adrenaline	0.22–0.35	5–15 min	120–180 min
Mepivacaine (2%)	0.22–0.35	5–15 min	90–120 min
Bupivacaine (0.5%)	0.06	10–30 min	120–240 min
Xylazine	0.10–0.20	10–30 min	90–150 min
Detomidine	0.01–0.02	10–15 min	120–180 min
Dexmedetomidine	0.005	5–10 min	120–180 min
Morphine	0.1–0.2	30–60 min	18–24 hr

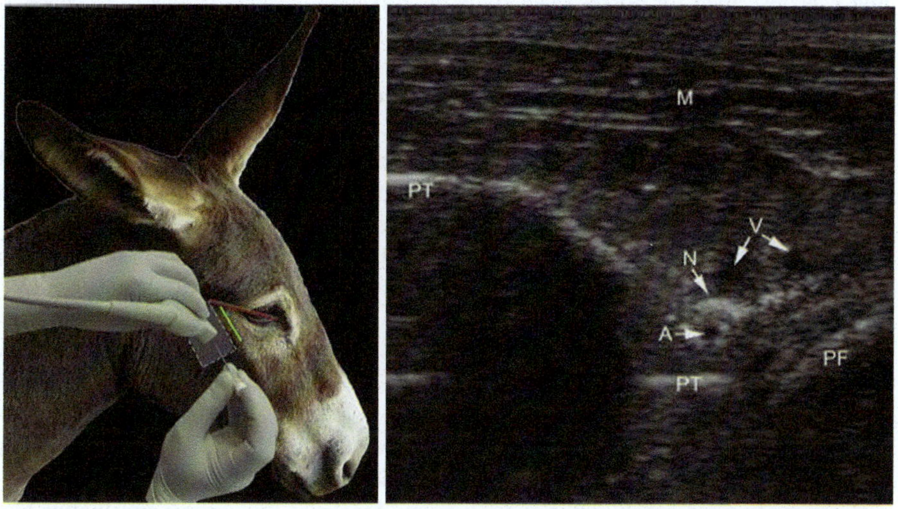

Fig. 7 The left side shows the position of the ultrasound probe and the needle for maxillary nerve block. The right side shows the ultrasound image of the relevant structures: the masseter muscle (M), the maxillary nerve (N) related to branches of the deep facial vein (V), the infraorbital artery (A) and the perpendicular plate of the palatine bone (PT) and fascia of the periorbital cone (PF). (Courtesy of Veterinary Anaesthesia and Analgesia Hagag and Nahas 2022)

8.3 Local Anesthetic Techniques of the Head

These techniques will enable the performance of dental and ophthalmic procedures in standing sedated donkeys and will also help to improve protocols for general anesthesia, where they can be combined with inhalant or intravenous anesthesia. The reviews of Tremaine (2007) and Hermans et al. (2019) provide very useful visual anatomical landmarks and descriptions of ultrasound guided techniques in horses. Many of these techniques can also be used in donkeys. Hagag and Nahas (2022) described blind versus ultrasound guided maxillary nerve blocks in donkeys and provide very nice visual descriptions of the location of the probe and the needle and of the ultrasound images (Fig. 7). Sobhy et al. (2019) described the ultrasound-guided inferior alveolar nerve block (mandibular nerve) in donkeys.

The study was started with dissection of the lower jaw of donkeys and showed a lot of similarities between the inferior alveolar anatomy in donkeys compared to horses. The color doppler function of the ultrasound offered very good assistance to localize the blood vessel that accompanies the inferior alveolar nerve. Hagag and Nahas (2022) described the ultrasound-guided retrobulbar nerve block in donkey cadavers in a very similar approach to that described in horses (Hermans et al. 2019) and described adequate infiltration of the intraconal space.

For surgery of the ears, for instance in case of removal of sarcoids, the caudal auricular and auriculopalpebral branches of the facial nerve can be desensitized. In a case report by McCluskie and Tremaine (2009), these nerves were also desensitized in a recumbent donkey for surgical removal of an auricular sarcoid with a technique that was described for horses (McCoy et al. 2007).

8.4 Pudendal Nerve Block for Urogenital Surgery

El-Sherif et al. (2017) described the technique of ultrasound-guided blocking of the pudendal nerves in male donkeys. They first performed dissection studies and then used the technique in sedated donkeys to successfully anesthetize the anus, perineum, penis, prepuce and glans penis. This technique enables urogenital surgery in the standing sedated male donkey.

9 Conclusions

Donkeys demonstrate many differences compared to horses regarding their behavior and their physiology. To formulate adequate anesthetic and analgesic protocols, the differences in pharmacokinetics and pharmacodynamics for various anesthetics like sedative drugs, NSAIDs, muscle relaxants and dissociative anesthetics must be considered. Many publications are now available with detailed descriptions of various anesthetic and analgesic drug protocols that have led to improved quality and safety of anesthetic procedures in donkeys. Although there is a need for further studies exploring these differences, allowing us to further improve our knowledge on donkey anesthesia and pain management, we hope that this chapter will provide assistance in performing good quality anesthesia and pain management in donkeys and mules.

Diagnostic and Treatment Considerations: Anesthesia, Sedation, and Pain Management in the Field

A Case Study Comparing Horses, Donkeys, and Mules

Fernanda Castro Dosal

Introduction

Anesthesia has been defined in various ways over the years, with Bailey's first definition (1751) as "a defect in sensation." The word anesthesia became synonymous with "unconsciousness that provides insensitivity to pain" (Muir and Hubbell 2009). This definition was used for the following 50 years (Muir and Hubbell 2009).

In 1957, Woodbridge included four specific components described as sensory blockade (analgesia), motor blockade (muscle relaxation), mental blockade (unconsciousness), and blockade of undesired effects on respiratory and cardiac systems. Woodbridge calls this balanced anesthesia (Muir and Hubbell 2009). The objectives of anesthesia are analgesia, relaxation, unconsciousness, reversible, and absence of memories.

Animal anesthesia began with Hickman (1824) using carbon dioxide to induce unconsciousness. However, the practical advantages of anesthesia and its benefits in surgery were realized by G.H. Dadd (1847), during a time when physical restraint methods with harnesses and ropes were used to tie the limbs. The first alpha-2 adrenergic was xylazine, introduced between 1960 and 1970, which was quickly adopted for use in combination with ketamine for short anesthetic procedures.

For veterinarians who practice clinical work with domestic equids (donkeys and horses) in fieldwork, anesthesia and sedation are working tools that must be used in practice, acknowledging that different field conditions can cause differences in the effect of anesthetics.

The physiological parameters of each equid must always be considered, which includes heart rate, respiratory rate, ocular reflexes, mucous membrane color, hematocrit, and protein concentration. Environmental conditions such as ambient temperature are also important. Induction, maintenance, and recovery are critical phases of anesthesia, both for the equid and for the doctors managing it.

To perform field anesthesia, there is no ideal drug, rather the protocol to use is the one with which one is most familiar, to ensure the shortest duration necessary for the procedure, followed by a gentle recovery.

Under general anesthesia, the equid is expected to remain within its physiological parameters. For example, if the normal heart rate of a horse is 28–40 bpm or a donkey is 31–53 bpm, we expect that when under the effect of a sedative or anesthetic, the equid stays within these parameters.

When performing a protocol with an alpha-2 adrenergic sedative like xylazine and a dissociative anesthetic like ketamine, the equid will experience analgesia, generalized muscle relaxation, and deep sedation. The equid lowers its head, relaxes the lips and eyelids, positions limbs widely apart, relaxes body joints and becomes ataxic, which are signs of moderate to deep sedation. When sedation is very deep, muscle spasms and involuntary head contractions can also be seen (Valverde 2010; Lerche 2013).

Xylazine is considered the least potent of the alpha-2 drugs, which also include detomidine, romifidine, and medetomidine. Xylazine is known to produce less sedation in excited or nervous equids, and in many cases, can be completely ineffective in causing sedation in these animals. This is due to the selectivity for alpha 2 receptors being lower in xylazine than in medetomidine, romifidine, and detomidine (Muir and Hubbell 2009).

The cardiovascular effects that xylazine exerts are significant after intravenous (IV) administration. One minute after IV administration, arterial pressure increases in response to peripheral vasoconstriction and is followed by the onset of bradycardia and often accompanied by a second-degree atrioventricular block. However, it is reported that xylazine causes these changes to a lesser extent (by about 30%) compared to detomidine or romifidine, which are considered more potent alpha-2 agonists. Over the course of about 10–15 min after xylazine administration, the initial peripheral vasoconstriction changes to vasodilation and decreased blood pressure (Muir and Hubbell 2009; Lerche 2013).

Respiratory rates can decrease below the normal range during sedation with xylazine because it increases airflow resistance through the nasal passages, pharynx, and larynx. The degree to which the respiratory rate is affected can vary. The arterial oxygen pressure tends to decrease significantly within 5–30 min of administration.

Xylazine is a pre-anesthetic widely used prior to induction with ketamine. Ketamine is considered a dissociative anesthetic that induces analgesia and immobility in healthy equids without depressing cardiovascular function, as it counteracts the cardiovascular effects of xylazine and increases heart rate, blood pressure, and cardiac output. The respiratory effects of ketamine are also commonly considered beneficial, as it has a bronchodilator effect.

When ketamine is administered without sedation or in prolonged infusions (1–2 h), or as IV boluses over time, horses will show excitation effects, causing rigidity, muscle stiffness, mydriasis, sweating, and ataxia. To avoid these signs, a bolus of xylazine can be administered.

The intravenous use of combined sedatives and anesthetics is known as total intravenous anesthesia (TIVA). Its benefits include being more accessible to use, not requiring an oxygen machine, and recovery is smoother and more predictable (Muir and Hubbell 2009; Valverde 2010; Doherty and Valverde 2006). However, its disadvantages can include drug accumulation causing rough or long recovery, and it is completely metabolized by the liver, which is a disadvantage when you have an equid with hepatic failure who needs surgery.

Anesthesia Case Study

An anesthetic protocol was performed with 30% xylazine (300 mg/ml) and ketamine 2 g (200 mg/ml), measuring times, heart rate, respiratory rate, as well as induction and anesthetic recovery in different geographical areas serviced by the Donkey Sanctuary-Universidad National Autonomous of Mexico (UNAM) Program Field Service.

Objective

To determine the total time of anesthesia, its effect on heart and respiratory rates, as well as the type of induction and recovery, using a protocol with 30% xylazine (300 mg/ml) at doses of (1.1 mg/kg IV) in horses and (1.3 mg/kg IV) in donkeys; ketamine at (2.2 mg/kg PV IV) in horses and (2.3 mg/kg PV IV) in donkeys; in different geographical areas of Veracruz state.

Materials and Methods

Materials

- Anesthesia tables
- Thoracic diameter tape
- Stethoscope
- Micro-centrifuge
- Refractometer
- Pulse oximeter
- Xylazine 300 mg/ml
- Ketamine 200 mg/ml
- 5 ml syringes
- 10 ml syringes
- #20 needles
- Alcohol swabs
- Capillary tubes

Methods

Animals

A total of 24 male equids (16 horses, 8 donkeys) that presented for bilateral orchidectomy surgery were included in this study. The average age was 5 years (2–14 years), based on incisor (Easley et al. 2005). The average weight was 333 kg for horses and 109 kg for donkeys (112–520 kg), as determined using the Carroll and

Huntington thoracic tape formula. The average body condition score was 2.7 (1.5–3) based on a 5 point scoring system (Svendsen 2008).

Baseline ranges for RR, HR, and mucous membrane color were obtained through respiratory auscultation at the level of the trachea and cardiac auscultation. Pulmonary fields and cardiac sounds were also auscultated. Hematocrit and total plasma proteins were also measured. The blood was left inside tubes with EDTA and waited for them to settle, and then measured with a tape, and proteins were assessed with a refractometer.

All equids received a xylazine bolus (horses 1.1 mg/kg, donkeys 1.3 mg/kg) by direct intravenous injection, followed by a ketamine induction bolus 3 min later (horses 2.2 mg/kg, donkeys 2.4 mg/kg).

During anesthesia, eyes were covered with a drape and monitoring was performed every 3 min for RR, HR, and mucosa color. Ocular reflexes and sensitivity to touch stimulus on ear and lips were observed.

At recovery, a halter or rope headstall was placed, allowing the equid to get into sternal recumbency and one person with knowledge of how to provide head, assisted the equid to get up. This was evaluated using a 5-point scale (Tables A4 and A5).

Before anesthetizing a patient, the following were considered:

- Age (estimated by dentition)
- Weight (using thoracic diameter tape formula by Carroll and Huntington) (Muir and Hubbell 2009; Sanchez 2003)
- General physical examination, with the circulatory, respiratory, nervous, and locomotor systems being prioritized

The weight calculation method was based on the Carroll and Huntington formula:

$$\text{Weight} = \text{PTr}^2 \times \text{LPHTI}/11900$$

(PTr = thoracic perimeter in cm; LPHTI = Length from point of shoulder to point of ischial tuberosity)

General Physical Examination:

Circulatory System:

- Cardiac Auscultation:
  - 3rd to 6th intercostal space on left side (pulmonary, aortic, and mitral valves)
  - 3rd and 4th intercostal space on right side (tricuspid valve)
- Heart rate was assessed and categorized based on ranges listed in Table A1.

Respiratory System:

- Tracheal auscultation:

  Air should pass freely through the trachea, auscultating from cranial to caudal.
- Landmarks Used for Pulmonary Auscultation:

  An imaginary line was drawn from the caudal angle of the scapula to the olecranon tuberosity, and semi circularly toward the proximal portions of the

Table A1 Physical parameters during anesthesia comparing horses and donkeys

Parameter	Horse	Donkey
Stage	Adult	Adult
Heart rate (bpm)	28–40	31–53
Respiratory rate (rpm)	8–16	13–31
CRT	2 s	2 s
Mucous membrane color	Pink	Pink
Hematocrit%	32–52%	25–38%
Plasma total protein	6–8 g/dl	5.8–8.2 g/dl
Temperature °C/°F	37.5–38.5 99.5–101.3	36.5–37.7 97.7–99.8

sixth, eleventh, and sixteenth ribs, to pass over the transverse processes of the thoracic vertebrae to the caudal angle of the scapula.

- Respiratory rate, capillary refill time (CRT), and mucosal membrane appearance were assessed and categorized using ranges described in Tables A2 and A3.

Nervous System assessment:

- Ocular reflexes
- Patient attitude/demeanor
- Posture, head position
- Proprioception, postural responses
- Locomotion

Anesthesia and Monitoring Protocol

Anesthetic drugs:

- Horses: xylazine (1.1 mg/kg IV) and ketamine (2.2 mg/kg IV)
- Donkeys: xylazine (1.3 mg/kg IV) and ketamine (2.4 mg/kg IV)

Parameters evaluated:

- HR, RR, weight (thoracic tape), age (estimated by dentition)
- Hematocrit and total protein, O_2 saturation
- Time from induction bolus to recumbency
- Total anesthesia time (sedation to end time)
- Induction and recovery scale quality

Table A2 Horses

#	Time of induction bolus (ketamine) to recumbency	Total anesthesia time	Total time from xylazine to the end of anesthesia	C.R Bpm	R.R Rpm	Hematocrit %	Protein g/dl	Induction scale	Recover scale
1	60 s	24 min	33 min	28	9	32	7.5	3	3
2	80 s	12 min	21 min	27	11	32	7.5	4	5
3	60 s	16 min	20 min	21	12	32	7.2	4	5
4	60 s	9 min	13 min	24	21	32	7	4	4
5	49 s	10 min	15 min	38	25	32	7.5	3	4
6	57 s	18 min	22 min	31	9	33	6.8	4	5
7	60 s	16 min	20 min	30	13	35	7.6	3	4
8	54 s	13 min	17 min	40	17	35	8	5	5
9	80 s	11 min	15 min	40	9	33	6.8	5	5
10	60 s	18 min	22 min	32	11	55	7	3	2
11	40 s	15 min	19 min	37	8	38	6.8	4	4
12	60 s	20 min	24 min	34	8	41	8	5	5
13	90 s	16 min	35 min	28	15	37	6.8	4	5
14	90 s	13 min	18 min	33	18	50	8	5	5
15	55 s	14 min	26 min	27	12	37	7.5	5	4
16	60 s	16 min	21 min	39	15	46	7.5	5	5
Average	63.4 s	14.4 min	21.3 min	32	13	37.5	7.3	4.1	4.3

C.R cardiac range, *R.R* respiratory range

Table A3 Donkeys

#	Time of induction bolus (ketamine) to recumbency	Total anesthesia time	Total time from xylazine to the first movement	C.R Bpm	R.R Rpm	Hematocrit %	Protein g/dl	Induction scale	Recover scale
1	56 s	5 min	8 min	50	23	32	7.5	5	5
2	60 s	7 min	14min	41	20	35	7.5	4	5
3	50 s	13 min	17 min	40	19	38	7.5	4	5
4	52 s	14 min	35 min	40	21	42	7.5	4	5
5	35 s	7 min	10 min	39	30	38	7.5	3	3
6	41 s	12 min	16 min	60	30	43	6.8	3	3
7	40 s	6 min	11 min	55	31	50	8	3	3
8	50 s	14 min	18 min	43	20	45	7	5	5
Average	48 s	9.7 min	16.1 min	46	24	40.5	7.4	3.8	4.2

C.R cardiac range, *R.R* respiratory range

Results

The following parameters were measured: induction bolus (ketamine) to recumbency time, total anesthesia time, total time from sedation to end of anesthesia, heart rate, respiratory rate, hematocrit and proteins, scale for induction quality, and scale for recovery quality (see Tables A2 and A3).

The results obtained from the 24 anesthetized equids were:

- *Induction bolus (ketamine) to recumbency time*: Horses showed an average time of 63.4 s, with the shortest time being 40 s and the longest time 90 s, and donkeys showed an average time of 48 s, with the shortest time being 35 seconds and the longest time 60 s.
- *Total anesthesia time*: Horses showed an average time of 14.4 min, with the shortest time being 9 min and the longest time 24 min, and donkeys showed an average time of 9.7 min, with the shortest time being 5 min and the longest time 14 min.
- *Total time from sedation to end of anesthesia*: Horses showed an average time of 21.3 min, with the shortest time being 13 min and the longest time 35 min, and donkeys showed an average time of 16.1 min, with the shortest time being 8 min and the longest time 35 min.
- *Heart rate during anesthesia*: Horses showed an average heart rate of 32 beats per minute, with the lowest rate being 21 beats per minute and the highest rate being 40 beats per minute, and for donkeys, an average heart rate of 46 beats per minute with the lowest being 39 beats per minute and the highest being 60 beats per minute.
- *Respiratory rate during anesthesia*: Horses showed an average respiratory rate of 13 respirations per minute, with the lowest being 8 respirations per minute and highest of 25 respirations per minute. In donkeys, an average respiratory rate of 24 respirations per minute was obtained, with the lowest being 19 respirations per minute and the highest being 31 respirations per minute.
- *Hematocrit and plasma proteins*: Horses showed an average hematocrit of 37.5%, with lowest being 32% and highest 52%, and average plasma proteins of 7.3 g/dl, with lowest being 6.8 g/dl and highest 8 g/dl. In donkeys, an average hematocrit of 40.3% was obtained, with lowest being 32% and highest 50%, and average plasma proteins of 7.3 g/dl, with lowest being 6.8 g/dl and highest 8 g/dl.
- *Induction quality scale*: Based on a scale of 5 to 1, an average induction score of 4.1 was obtained in horses and 3.8 in donkeys (Table A4).
  Here 5 represents smooth induction with good muscle relaxation and 1 represents vigorous limb movements, paddling, and increased muscle activity during lateral recumbency transition.
- *Recovery quality scale*: Based on a scale of 5 to 1, an average recovery score of 4.3 was obtained in horses and 4.2 in donkeys (Table A5).
  Here 5 represents a single, coordinated effort to stand up with slight or no ataxia and 1 represents many uncoordinated attempts to stand that cause major injuries or involve life-threatening situations (fractures).

Table A4 Induction scale

Scale	Induction
5	Soft induction, with good muscle relaxation
4	Soft transition to lateral recumbency with few limb and facial movement
3	Slower lateral recumbency with more muscular stiffness and limb movements
2	Increased muscle activity, before and during the transition to lateral recumbent
1	Vigorous limb movements, kicking, and increased muscle activity during the transition to recumbent

Scale taken from mama et al. (1992) and Garcia (2001)

Table A5 Recovery

Scale	Recovery
5	A single, coordinated effort to get up with slight or no ataxia
4	Single attempt with slight ataxia
3	Smooth recovery with more than one attempt to get up
2	Several uncoordinated attempts that cause minor injuries (lacerations on limbs or head)
1	Many uncoordinated attempts to get up that cause major or life-threatening injuries (fractures)

Scale taken from mama et al. (1992) and Garcia (2001)

Discussion

The equid anesthesia protocols utilized in this study are well described in the literature, where it is stated that physiological parameters should remain within or slightly below range compared to those same parameters measured before anesthesia. In addition, these protocols should result in recumbency between 30 and 90 s after the ketamine induction bolus is given IV. The total anesthesia time associated with administration of a single xylazine-ketamine bolus has an average of 20–30 min, and the ketamine induction bolus has a time of 5–15 min until the first movement (total anesthesia time) (Muir and Hubbell 2009; Lerche 2013; Svendsen 2008).

The results of this study show that anesthesia times for both horses and donkeys were within those reported in the literature. No problems with induction, maintenance, or recovery were observed.

When comparing between species, we observed that donkeys have shorter anesthesia times, as already described in the literature, likely due to a higher hepatic metabolism than horses, allowing them to metabolize and eliminate anesthetics more quickly (Muir and Hubbell 2009; Lerche 2013).

The induction with xylazine-ketamine boluses was smooth with both species, and without danger to the animal or personnel. The ataxia cited in the literature corresponding to each species was considered (Muir and Hubbell 2009; Valverde 2010; Lerche 2013; Svendsen 2008); however, it was maintained as a satisfactory method for inducing total intravenous anesthesia in horses and donkeys. It was observed that the eye position, mucous membrane color, capillary refill time, and cardiac and respiratory rates remained within the baseline range measured before anesthesia.

With the combination, nystagmus was detected while the animal was under anesthesia. However, certain voluntary movement of the ears, lips, and head; muscle fasciculations; and groaning were recorded in the first moments of anesthesia, which in some donkeys was at 5 min.

Conclusion

The use of xylazine (300 mg/ml) in combination with ketamine (200 mg/ml) works for field anesthesia of equids. In both species, by 3 min after xylazine administration, sedation was sufficient to proceed with induction. Additionally, in horses, recumbency was achieved at 63 s with a total anesthesia of 21.3 min, and in donkeys, recumbency at 48 s with a total anesthesia of 16.1 min, which are sufficient for an experienced doctor to perform bilateral orchidectomy.

This study utilized a small number of standard sized donkeys, so caution must be taken in applying the findings to other donkey populations. While both donkeys and horses in our study responded to this anesthetic combination with adequate time and appropriate maintenance of physiological parameters, extreme care must be taken with donkeys, as the personnel handling them must know that the anesthesia time will be shorter.

Questions

1. *Describe the difference in recumbency times in horses compared to donkeys in this study?*
   A. You can see faster effects in donkeys, all the sedation signs like lips, ears, and eyelid relaxation comes first in donkeys, and they have shorter times to recumbency.
2. *When comparing induction rates were they the same in horses and donkeys?*
   A. For induction you need to use higher doses in donkeys and shorter times between bolus for maintenance during anesthesia.
3. *Would we expect similar clearance in mules?*
   A. Mules show completely different, anatomic differences. For example, to sedate a mule, we need to use higher doses of xylazine (1.6 mg/kg) and ketamine (2.6 mg/kg), but it will depend on the temper of the mule. In some cases, when the mule is too aggressive or nervous, you need to use 1.8 mg/kg of xylazine and 2.8 mg/kg of ketamine.
   They have different recumbency times they go down faster because of their metabolism, the signs of sedation they show are more subtle, they are always alert, and the I.V. anesthesia times are shorter than donkeys.

4. *Was the temperature, pulse, and respiration the same during anesthesia in horses and donkeys?*
   A. Under anesthesia you need to maintain the same parameters they have before the induction, but between equids I find some differences in the cardiac and respiratory range, it was higher in donkeys, and I find out that some donkeys vocalize during the anesthesia.

References

Auer JA, Stick JA (2012) Equine surgery, 4th edn. Elsevier Saunders

Bernal B (2013) Estudio comparativo de dos tecnicas de anesthesia total intravenosa de corta duracion en hibridos equinos del estado de tlaxcala, Mexico. Tesis de licenciatura. Mexico, D.F. Universidad nacional autonoma de Mexico.

Doherty T, Valverde A (2006) Manual of equine anesthesia and analgesia. Blackwell Publishing.

Donkey sanctuary Mexico, world horse welfare, Universidad Nacional Autonoma de Mexico, ciudad de Mexico, Distrito Federal: c2009 DS-WHW-UNAM. http//www.fmvz.unam.mx/fmvz/donkey/acerca.html

Easley J, Dixon PM, Schumacher J (2005) Section 2. Morphology: aging. In: Equine dentistry, 3rd edn

Fernando JA (2006) Anormalidades y patologias dentales graves en equidos rurales de seis estados de Mexico. Tesis de licenciatura. Mexico D.F. Universidad nacional autonoma de Mexico

Knottenbelt DC (2006) Equine formulary. Elsevier Saunders

Lerche P (2013) Total intravenous anesthesia in horses. Department of Veterinay Clinic Sciences, the Ohio State University

Muir WW, Hubbell JAE (2009) Equine anesthesia monitoring and emergency therapy, 2nd edn. Elsevier Saunders

Reed SM, Bagly WM, Sellon DC (2004) Equine internal medicine, 2nd edn. Saunders

Sanchez F (2003) Evaluacion de cinco metodos para calculary indirectamente el peso corporal de caballos de la policia montada en la ciuda de mexico. Tesis de licenciatura. Mexico, D.F. Universidad nacional autonoma de Mexico

Svendsen E (2008) The professional handbook of the donkey, 4th edn. Whittet Books

Valverde A (2010) Alpha-2 Agonist as pain therapy in horses. Department of clinical studies, Ontario veterinary collage, University of Guelph

References

Abass M, Picek S, Garzón JFG, Kühnle C, Zaghlou A, Bettschart-Wolfensberger R (2018) Local mepivacaine before castration of horses under medetomidine isoflurane balanced anaesthesia is effective to reduce perioperative nociception and cytokine release. Equine Vet J 50:733–738. https://doi.org/10.1111/evj.12947. Epub 2018 May 16

Abd-Almaseeh ZT (2008) Comparative anesthetic protocols: propofol and thiopental in xylazine premedicated donkeys. J Anim Vet Adv 7:1563–1567

Alves GES, Faleiros RR, Gheller VA, Vieria MM (1999) Sedative effect of romifidine in untamed mules. Cienc Rural 29:51–55

Aziz DM, Al-Badrany MS, Taha MB (2008) Laparoscopic ovariectomy in standing donkeys by using a new instrument. Anim Reprod Sci 107:107–114

Barton CK, Hector RC, Hendrickson DA, Kawcak CE, Nelson BB, Goodrich LR (2024) Fasting horses perioperatively decreases manure production and increases time to manure output postoperatively: a controlled randomized trial. J Am Vet Med Assoc 262:1201–1208. https://doi.org/10.2460/javma.24.04.0235

Brumund L, Wittenberg-Voges L, Rohn K, Kästner SBR (2024) Risk assessment in equine anesthesia: a first evaluation of the usability, utility and predictivity of the two-part CHARIOT. Front Vet Sci 11. https://doi.org/10.3389/fvets.2024.1384525

Burden FA, Thiemann A (2015) Donkeys are different. J Equine Vet Sci 35:376–382

Burnham SL (2008) Anatomical differences of the donkey and mule. Proc Am Assoc Equine Pract 48:102–109

Coakley ME, Peck K, Taylor TS, Matthews NS, Mealey KL (1999) Pharmacokinetics of flunixin meglumine in donkeys, mules, and horses. Am J Vet Med Res 60:1441–1444

Molinaro Coelho CM, Duque Moreno JC, Goulart DS, Caetano LB, Soares LK, Coutinho GH, Es Alves G, da Silva LF (2014) Evaluation of cardiorespiratory and biochemical effects of ketamine-propofol and guaifenesin-ketamine-xylazine anesthesia in donkeys (Equus asinus). Vet Anaesth Analg 41:602–612

Dar KH, Gupta AK (2016) Total intravenous anaesthesia in adult mules. Vet Anaesth Analg 43: 204–208

Emami MR, Seifi H, Tavakoli Z (2006) Effects of totally intravenous thiopental anesthesia on cardiopulmonary and thermoregulatory system in donkeys. J Appl Anim Res 29:13–16

El-Sherif M, El-Khamary A, Nazih M, Senosy W (2017) Ultrasound-guided pudendal nerve block in male donkeys. Assiut Vet Med J 63:81–85

Forés P, López J, Rodríguez A, Harrán R (2001) Endoscopy of the upper airways and the proximal digestive tract in the donkey (Equus Asinus). J Equine Vet Sci 21:17–20

Grosenbaugh DA, Reinemeyer CR, Figueiredo MD (2011) Pharmacology and therapeutics in donkeys. Equine Vet Educ 23:523–530

Haga H, Lykkjen S, Revold T, Ranheim B (2006) Effect of intratesticular injection of lidocaine on cardiovascular responses to castration in isoflurane-anesthetized stallions. Am J Vet Res 67: 403–408

Hagag U, Nahas AEL (2022) Ultrasound-guided retrobulbar nerve block in donkey cadavers. Vet Anaesth Analg 49:197–202. https://doi.org/10.1016/j.vaa.2022.01.002. Epub 2022 Jan 19

Hamed MA, Abouelnasr KS, Ibrahim HMM, El-khodery SA (2017) Comparative, sedative, and analgesic effects of epidural Dexmedetomidine and Xylazine in donkeys (Equus asinus). J Equine Vet Sci 59:104–109

Hermans H, Veraa S, Wolschrijn CF, van Loon JPAM (2019) Local anaesthetic techniques for the equine head, towards guided techniques and new applications. Equine Vet Educ 31:432–440. https://doi.org/10.1111/eve.12757

Hubbell JAE, Muir WW, Hopster K (2022) Rethinking equine anaesthetic risk: development of a novel Combined Horse Anaesthetic Risk Identification and Optimisation tool (CHARIOT). Equine Vet Educ 34:134–140

Joubert KE, Briggs P, Gerber D, Gottschalk RG (1999) The sedative and analgesic effects of detomidine-butorphanol and detomidine alone in donkeys. J S Afr Vet Assoc 70:112–118. https://doi.org/10.4102/jsava.v70i3.769

Kay G, Pearson RA, Ouassat M (2004) Estimation of the liveweight of working mules in Morocco from their body measurements. Vet Rec 154:85–88. https://doi.org/10.1136/vr.154.3.85

Lamont LA (2008) Multimodal pain management in veterinary medicine: the physiologic basis of pharmacologic therapies. Vet Clin North Am Small Anim Pract 38:1173–1186

Latzel ST (2008) Clinical and pharmacological studies on elimination kinetics of Xylazine (RompunR/Bayer) in Mules. Dissertation (thesis), Ludwig-Maximillians-University, Faculty of Veterinary Medicine, Munich, 1st edn, Verlag Dr Hut, Munich. ISBN 978-3-89963-894-3

Lizarraga I, Castillo-Alcala F, Varner KM, Robinson LS (2014) Sedation and mechanical antinociception after intravenous administration of detomidine in donkeys: a dosage–effect study. Vet Rec. https://doi.org/10.1136/vr.102569

Florczyk A, Simhofer H, Rosser J (2020) Hypothermia prevention in long-standing equine dental procedures. Equine Vet Educ 32:368–371. https://doi.org/10.1111/eve.12981

Maney J, Little E, Dzikiti T (2019) Effect of a short pre-anesthetic fast on arterial blood gas values in isoflurane-anesthetized donkeys. Can Vet J 60:1057–1059

Marzok MA, El-khodery SA (2015) Comparative analgesic and sedative effects of tramadol, tramadol-lidocaine and lidocaine for caudal epidural analgesia in donkeys (Equus asinus). Vet Anaesth Analg 42:215–219

Matthews NS, Taylor TS, Hartsfield SM, Williams JD (1992) A comparison of injectable anaesthetic regimens in mammoth asses. Equine Vet J Suppl 11:37–40

Matthews NS, Taylor TS, Hartsfield SM, Hayton WL, Jones DH (1994) Pharmacokinetics of ketamine in mules and mammoth asses premedicated with xylazine. Equine Vet J 26:241–243

Matthews NS, Peck KE, Mealey KL, Taylor TS, Ray AC (1997) Pharmacokinetics and cardiopulmonary effects of guaifenesin in donkeys. J Vet Pharmacol Ther 20:442–446

Matthews N, van Loon JPAM (2013) Anaesthesia and analgesia of the donkey and the mule. Equine Vet Educ 25:47–51. https://doi.org/10.1111/j.2042-3292.2011.00312.x

Matthews N, van Loon JPAM (2019) Anesthesia, sedation, and pain management of donkeys and mules. Vet Clin North Am Equine Pract 35:515–527. https://doi.org/10.1016/j.cveq.2019.08.007. Epub 2019 Oct 3

McLean AK, Falt T, Abdelfattah EM, Middlebrooks B, Gretler S, Spier S, Turoff D, Javier Navas Gonzalez F, Knych HK (2023) Transdermal Flunixin Meglumine as a pain relief in donkeys: a pharmacokinetics pilot study. Metabolites 13:776. https://doi.org/10.3390/metabo13070776

McCluskie LK, Tremaine WH (2009) Surgical removal of an aural sarcoid in a donkey using ultrasonic shears. Vet Rec 164:561–563

McCoy AM, Schaefer E, Malone E (2007) How to perform effective blocks of the equine ear. Proc Am Assoc Equine Pract 53:397–398

Mealey KL, Matthews NS, Peck KE, Ray AC, Taylor TS (1997) Comparative pharmacokinetics of phenylbutazone and its metabolite oxyphenbutazone in clinically normal horses and donkeys. Am J Vet Res 58:53–55

Mostafa MB, Farag KA, Zomor E, Bashandy MM (1995) The sedative and analgesic effects of detomidine (domosedan) in donkeys. J Vet Med A 42:351–356

Naddaf H, Baniadam A, Rasekh A, Arasteh A, Sabiza S (2015) Cardiopulmonary effects during anaesthesia induced and maintained with propofol in acepromazine pre-medicated donkeys. Vet Anaesth Analg 42:83–87

Shoukry M, Saleh M, Fouad K (1975) Epidural anaesthesia in donkeys. Vet Rec 97:450–452

Sinclair MD, Mealy KL, Matthews NS, Peck KE, Taylor TS, Bennett BS (2006) Comparative pharmacokinetics of meloxicam in clinically normal horses and donkeys. Am J Vet Res 67:1082–1085

Sobhy AM, Nomir AG, El-khamary AN, Abo-rawash AA, Shamaa A, Kassem MM (2019) Ultrasound-guided inferior alveolar nerve block in donkeys. Damanhour J Vet Sci 2:14–18

Straticò P, Carluccio A, Varasano V, Guerri G, Suriano R, Robbe D, Cerasoli I, Petrizzi L (2021) Analgesic effect of Butorphanol during castration in donkeys under total intravenous Anaesthesia. Animals 11(8):2346. https://doi.org/10.3390/ani11082346

Sun TC, Bellei M, Robinson A, Martinez-Taboada F (2021) Ketofol: a ketamine-propofol admixture for induction of anaesthesia in two foals. Vet Rec Case Rep 9:e136. https://doi.org/10.1002/vrc2.136

Suriano R, Varasano V, Robbe D, Carluccio A, Straticò P, Contri A, Petrizzi L (2014) Intraoperative analgesic effect of Intrafunicular lidocaine injection during orchiectomy in isoflurane-anesthetized Martina Franca donkeys. J Equine Vet Sci 34:793–798

Svendsen ED (2008) The professional handbook of the Donkey, 4th edn. Whittet Books Ltd. ISBN-10: 1873580681

Taylor EV, Baetge CL, Matthews NS, Taylor TS, Barling KS (2008) Guaifenesin-ketamine-xylazine infusions to provide anesthesia in donkeys. J Equine Vet Sci 28:295–300

Tremaine WH (2007) Local analgesic techniques for the equine head. Equine Vet Educ 19:495–503

Van Hoogmoed LM, Galuppo LD (2005) Laparoscopic ovariectomy using the endo-GI stapling device and endo-catch pouches and evaluation of analgesic efficacy of epidural morphine sulfate in 10 mares. Vet Surg 34:646–650

Vullo C, Carluccio A, Robbe D, Meligrana M, Petrucci L (2017) Guaiphenesin-ketamine-xylazine infusion to maintain anesthesia in mules undergoing field castration. Acta Vet Scand 59:67. https://doi.org/10.1186/s13028-017-0335-7

Mules and Donkeys as Surgical Patients; Special Considerations

Claude A. Ragle

Abstract

As surgical patients, donkeys and mules have distinct differences from horses (Taylor and Matthews 2002). These differences stem primarily from specific physiologic and behavioral attributes (McLean and Wang 2013; McLean et al. 2019). That is not to say that mules and donkeys do not suffer many of the same diseases or injuries of the horse! Although the need for surgical treatment of colic in the mule seems rarer than it is in horses and surgical treatment of colic in the donkey appears to have poorer outcomes and more complications (Merridale-Punter et al. 2023) as compared to the horse, we still require more data to drive improvements in treatments and outcomes. With few exceptions mules and donkeys are afflicted with the same maladies as the horse. There are however some surgical conditions that are either more or less prevalent in the mule or donkey as compared to the horse. Two examples of surgical issues that mules all most never suffer from are, left laryngeal hemiplegia and cryptorchism. It is more difficult to prove this claim with scientific analysis due to lack of large population comparisons. As there are disease differences recognized within breeds of horses, there is a variation within breeds of donkeys and crosses of mules. As humans have certain diseases that are related to occupations, so does the equine. A study of mules removed from use as pack mules in the Germany military was most commonly due to, degenerative hock disease considered related to the job and terrain where the mules performed and was similar to the findings of the Halflinger horses doing similar work (Brunner et al., 2002). All factors being equal it is this authors experience that the "use" and "lifestyle" of a horse, donkey or mule will have a greater influence on disease patterns than specie differences.

C. A. Ragle (✉)
Department Veterinary Clinical Sciences, College of Veterinary Medicine, Washington State University, Pullman, WA, USA
e-mail: ragle@wsu.edu

A. K. McLean et al. (eds.), *Donkey and Mule Medicine: Clinical Medicine and Reproduction*, https://doi.org/10.1007/978-3-032-18540-2_15

For successful treatment of donkeys and mules the veterinarian must appreciate; the medical, behavioral and use differences, within and between the breeds, as well as the vast similarities of the Equus genus.

Keywords

Donkey · Mule · Surgery · Castration · Ovariectomy · Laparoscopy

1 Castration

Castration is likely one of the oldest and most common surgical procedure performed on the male equine. Castration complications are also among the more common issues leading to malpractice claims against veterinarians (Meagher 2005). Complication rates related to castrations, depending on how defined and classified have been reported between 10–60%. (Baldwin 2023). The mule and donkey warrant special consideration when it comes to castration. Although the number of peer reviewed studies are limited, a reoccurring theme is that castration of mules and donkeys are more prone to complications, especially related to hemorrhage and inadequate anesthesia (Racine et al. 2018; Moll et al. 1995).

Successful outcome of a surgical procedure is often related to proper case selection, patient management, surgical technique and client communication. The physical and behavior attributes of mules and donkeys vary enough from horses that that the uninitiated handler may find themselves over challenged (McLean et al. 2019). This entire text is dedicated to the intricacies of the mule and donkey, and refering to chapters dedicated to behavior, anatomic, physiologic and pharmacological specifics will aid the readers ability to apply the surgical techniques described in this chapter.

1.1 General Considerations

1.1.1 Age at Castration

There has been no consensus as to the ideal age for castration (Cognie 2023) and ages reported at castration range from several days old to animals in theirs 20 s. The author's recommendation, when possible, is to castrate several weeks prior to weaning. The rationale to this approach is, the patient still has maternal antibodies, the testicle and pedicle are small, and the patient can return to its mother after the operation, reducing stress and suffering. Unfortunately, veterinarians treat the animals as presented by their owners and must be prepared for a wide range of ages.

1.1.2 Behavior Considerations

Mules and donkeys have a strong sense of self preservation. This characteristic must be appreciated and accommodated by the veterinarian for optimal outcomes. Mules are endowed with hybrid vigor which seems to heighten cognition as well (Proops

2008). In the authors experience it is important to sedate the animal prior to any dramatic change of routine on the day of surgery. I generally sedate with intramuscular acepromazine (see chapter "Gastrointestinal Disorders of Donkeys and Mules" for doses) while the patient is in the stall and then they are left undisturbed for 30 minutes. If ever the subsequent pre-induction sedation (typically xylazine IV) does not provide the expected level of sedation, it is better to delay the operation to another day, than increasing doses of drugs hoping for a better response. It is important that only calm and relaxed animals are anesthetized for elective surgical procedures. How the animals are managed prior to induction of anesthesia is equally important as the management of anesthesia after induction. Other than the surgical technique of castration used, the negative physiological effects from being an overly frightened or struggling patient, can in my experience, increase operative and postoperative complications. They key is to listen to the mule or donkey and adapt as needed to foster a quiet and relaxed transition into anesthesia. This requires slow and confident handling of the animal giving them time to process what is asked of them and use of negative and positive reinforcement to foster acceptance of the required preoperative procedures.

1.1.3 Pre-surgery Considerations

All patients require a through physical exam prior to the operation. Special attention to scrotal palpation to confirm normal anatomy, when possible, if safety concerns prevent presurgical palpation, the owner should be informed it will be completed under general anesthesia and the findings could alter the surgical plan. Depending on the age and history of the animal additional diagnostics (blood work etc.) may be indicated. The patient should be current on deworming and vaccinations with particular attention to having had proper vaccination for tetanus.

True abdominal cryptorchidism is very rare in the mule, to the point that if a mule is examined and only 1 or no testicles are present a complete and full workup is warranted due to the extreme rarity of this condition in mules. The few I found in the literature were due to a testicular teratoma and not a routine retained testicle (Sanders 1903). I don't remember ever operating a cryptorchid mule and got similar experiences from the colleagues I surveyed for this chapter, representing well over 100 years of surgical experience in several university and private practices in the usa and abroad. (personal communication: Dr. Cliff Honnas—none, Dr. Jim Schumacher—maybe one but no more). This is an interesting observation worthy of further study, till then perhaps it is a result of heterosis (Baranwal et al. 2012). Donkeys on the other hand can present as a true abdominal or inguinal cryptorchid, especially in the miniature breeds (Sprayson and Thielmann 2007) A survey of a herd of 274 donkeys in Nigeria demonstrated 30% with cryptorchidism, thought due to inbreeding (Kumi-Diaka et al. 1981). It seems even more remarkable that both sides of the family of mules suffer the condition of cryptorchidism, but the mule is spared!

If the animal is especially tractable a presurgical scrub of the surgical site can be performed and surgical scrub applied for adequate contact time, which will help reduce the time required under anesthesia which benefits the patient. It is not generally needed to withhold feed prior to anesthesia, especially if on a primarily hay or grass diet, grain should be withheld as well, free access to water should be provided. Castration properly performed is classified as a "clean" procedure and as such antibiotic treatment is not indicated. If any extenuating circumstances increase the risk of infection in a patient, perioperative antibiotics are used and must be administered such that a systemic therapeutic level is present in the patient at the time of skin incision. In all cases the use of NSAIDs is indicated, starting before surgery and for a minimum of 3 days. Daily careful examination of physical parameters and behavior of the patient can determine if a longer duration is indicated. (See chapter "Endocrines Disorders of Donkeys and Mules" for antibiotic and NSAID administration) (see chapter "Gastrointestinal Disorders of Donkeys and Mules" for anesthesia administration).

1.1.4 Technique considerations

I will describe in detail my preferred technique for castration of mules and donkeys, but first I want to address some of the confusion surrounding castration nomenclature. The traditional nomenclature has used overly simplified terminology such as, open, closed, half-closed, primary closure etc. These terms are not useful as they attempt to reduce a multistep operation with several variable options to a single word. This has made the value of retrospective studies on castration outcomes limited due difficulty in standardized recording of techniques and complication classification (Rodden 2024). A proper surgical report should record the details of the castration procedure, as is required in all operations performed in veterinary medicine. Castrations of equines can be done with the patient standing or recumbent. In the authors experience castration of the donkey and mule are best performed with the patient in dorsal recumbency under anesthesia. The technique of castration of mules and donkeys should ensure immaculate hemostasis and secure closure of the vaginal tunic, thus preventing post operative hemorrhage and eventration. The former should be the expectation and standard for castration of any member of the Equus genus.

1.1.5 Surgical Technique

After the patient is quietly anesthetized and placed in dorsal recumbency the important steps of the procedure begin. First the scrotum is wiped with alcohol and each testicle is injected with local anesthetic (2% lidocaine HCl or mepivacaine HCl) using an 18-gauge, 1 ½ inch needle. The injection volume (~10–40 mls) should be such that the testicle become obviously rigid, this ensures the pressure gradient will favor migration up the testicular pedicle. This step is important in that it is part of a balanced anesthesia approach, reducing pain both during and immediately after the surgical procedure. The injected local anesthetic will also lead to paralysis of the cremaster muscle, preventing contraction, which could create unwanted tension during transection. The surgical site is the cleaned and sterilely prepped and draped for surgery. Two parallel skin incisions are made about 2 cm on either side of the median raphe, no longer than the length of the testicles. One testicle is grasped and brought up to the corresponding skin incision which is continued to the depth of the vaginal tunic, an incision is made through the vaginal tunic at the level of the cranial pole of the testicle, large enough to accommodate the surgeons index finger (Fig. 1).

An index finger is placed into the vaginal tunic and enough traction is applied expose the vaginal tunic in preparation to strip it free of any overlying fascia or adipose tissue. A pair of Ochner/Kocher forceps are applied grasping the vaginal tunic and tail of the epididymis and proper ligament and held in the palm of the same hand with the index finger inside the vaginal tunic (Figs. 2 and 3).

This step is important in that it allows adequate traction to be applied for stretching the cord without failure and tearing of the vaginal tunic, which can complicate that process. Traction is applied to the vaginal tunic with the hand holding the forceps and tunic while the opposite hand is pushing an encircling gauze to remove excess connective tissue from the length of the exposed tunic (Fig. 4).

Next a pair of Kelly forceps are used to separate the laterally attached cremaster muscle from the vaginal tunic, it wraps around the tunic in a c shape, be sure to separate all the fibers and take care not to penetrate or damage the integrity of the vaginal tunic (Fig. 5).

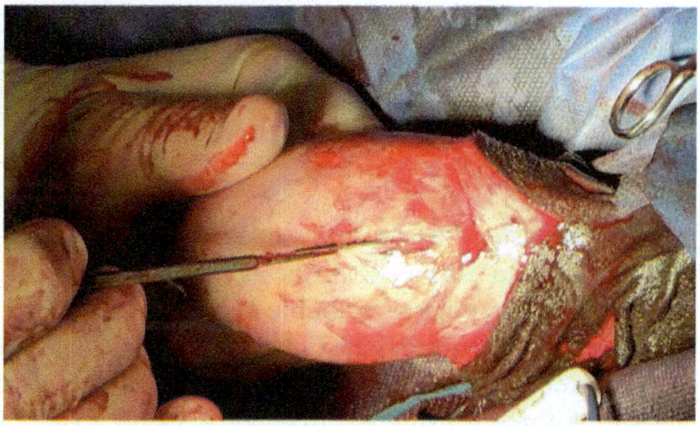

Fig. 1 Incision into vaginal tunic at cranial pole of testicle

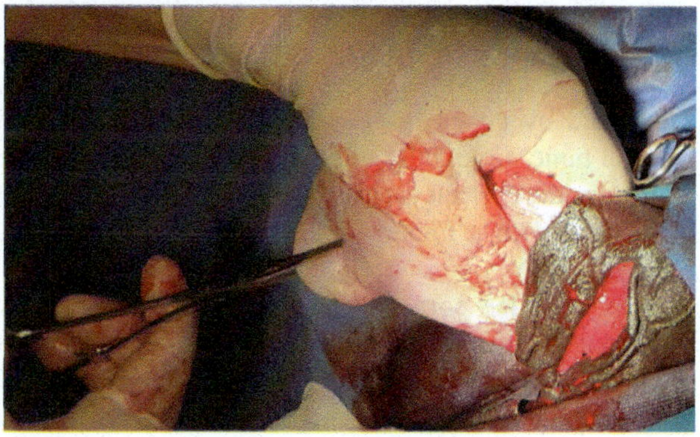

Fig. 2 Ochner/Kocher forceps applied to testicle

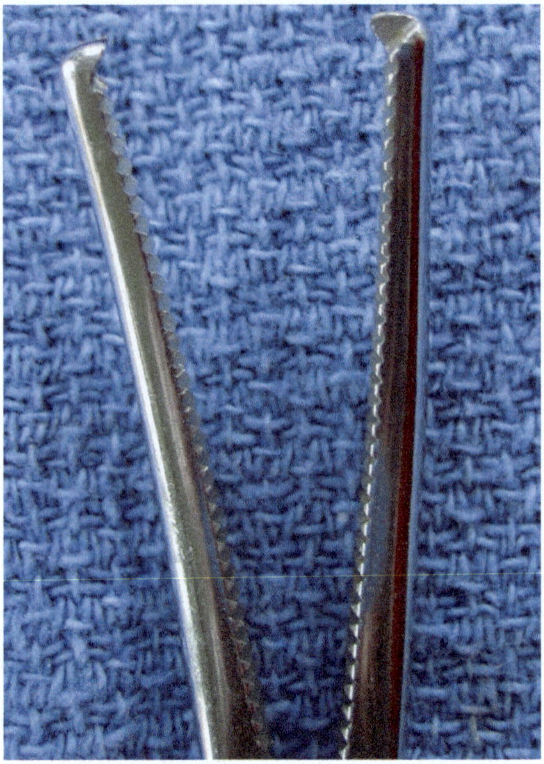

Fig. 3 Tip of Ochner/Kocher Forceps (only to be used on tissue being discarded)

The cremaster muscle is now to be double clamped, using a Ferguson angiotribe forceps, one can use an angiotribe as the proximal clamp and use a Carmalt clamp on the distal aspect, wherever double clamping is used. The function of the extra distal clamp is reducing retrograde bleeding from the excised tissues (Fig. 6a, b).

The cremaster muscle is now transected just distal to the proximally placed angiotribe forceps using electrocautery, cautery is applied to the cut surface of the muscle protruding from the proximal clamp, with special attention to any vessels contained in the clamp. If performing this surgery in the field, a disposable or reusable electrocautery pen provides acceptable cautery (for examples see Bovie® Cautery). The proximal clamp on the cremaster muscle is released and hemostasis is confirmed by visual inspection, Scissors are used to lengthen the opening in the vaginal tunic proximally to expose the pedicle of the testicle (Fig. 7).

The pedicle is elevated from the vaginal tunic and the mesorchium is bluntly penetrated with a finger, dividing the pedicle into a vascular and fibrous portion (Fig. 8).

The vascular portion is double clamped using a Ferguson agiotribe as the proximal clamp. The vascular portion is double ligated proximally to the clamp using an absorbable suture material (example #1or #2 polyglatin 910 or polydioxanone). Each of the two sutures are place as a double encircling ligature,

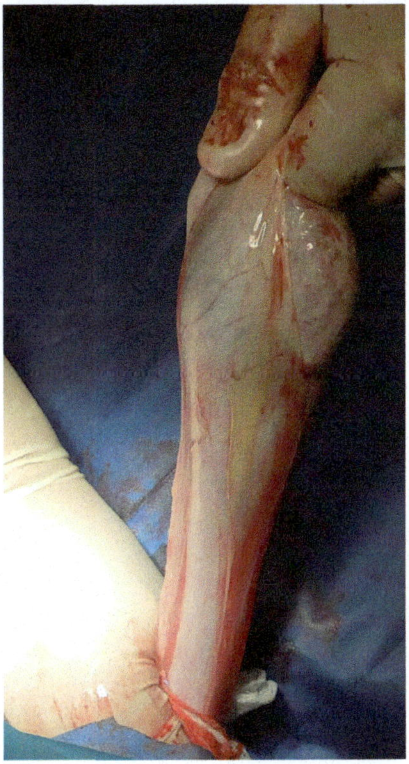

Fig. 4 Stripping fascia from the testicular cord, which clearly exposes the vaginal tunic and cremaster muscle

where the vascular pedicle is encircled once and a surgeon's knot followed by the appropriate number of throws for the suture material (Silver et al. 2016), followed by the tails of the suture encircling the pedicle 180 degrees and another surgeon knot followed by square knots for a total of 5 knots (Fig. 9).

The vascular pedicle is then divided using electrocautery and the proximal stump is carefully inspected and cautery applied to the ends of the exposed vessels (Fig. 10).

The proximal clamp on the vascular pedicle is removed and hemostasis confirmed by inspection. The double ligated vascular pedicle is replaced within the vaginal tunic and the vaginal tunic is then double clamped distal to the location of the vascular pedicle stump (Fig. 11). The vaginal tunic and remaining fibrous portions of the testicular pedicle (e.g. vas deferens) are ligated in the same manner as the vascular portion and divided with electrocautery, with additional cautery applied to the stump (Fig. 12).

The proximal clamp is removed, hemostasis is confirmed as is the integrity of the vaginal tunic and the same steps are repeated on the remaining testicle. After the final testicle has been removed, each incision and the dead space in the scrotum is carefully inspected for bleeding and electrocautery applied as needed and any of the stripped tissue from the cord that can protrude out the incisions is trimmed using cautery (Fig. 13).

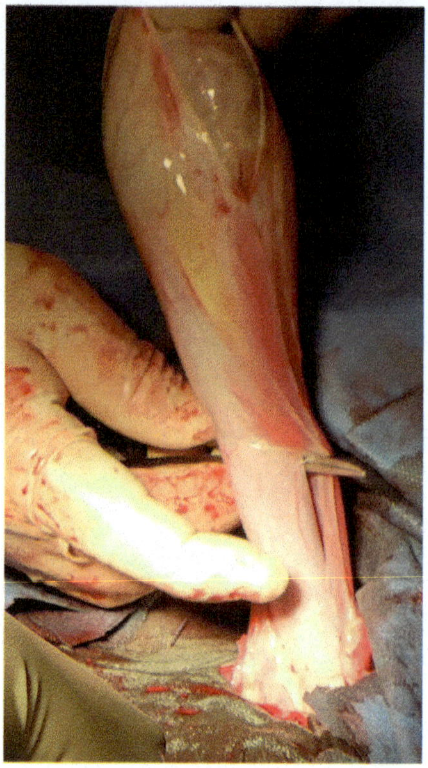

Fig. 5 Isolating cremaster muscle from vaginal tunic

Even the slightest hint of bleeding is to be addressed as the surgical positioning and tendency of anesthesia to lower blood pressure will underrepresent what bleeding could be present in the standing and awake patient. Immaculate hemostasis is required to be able to appose the skin of the scrotum without risk of hematoma formation. I prefer to invert the redundant skin of the scrotum and apply cyanoacrylate glue, which is then covered with an elastic adhesive bandage as the method of protecting the surgical wound (Figs. 14 and 15). This will allow protected healing of the wound in the expected case but also in the rare case of a bleeder being missed allows drainage of the scrotum without the need to remove any sutures.

Refer to the table below for additional information on induction and recumbency in donkeys and mules (refer to Table 1).

1.1.6 Castration Aftercare

NSAIDs in an age and specie appropriate dose and rate is the only postop drugs required, typically for 3–5 days (see chapter "Endocrines Disorders of Donkeys and Mules") The aftercare may vary slightly depending on the age of the patient. Animals still with their mothers are returned to with them, but if in a herd situation, they are kept separate for several days to ensure normal progression of healing. If after several days all is normal return to normal housing is possible but depending on

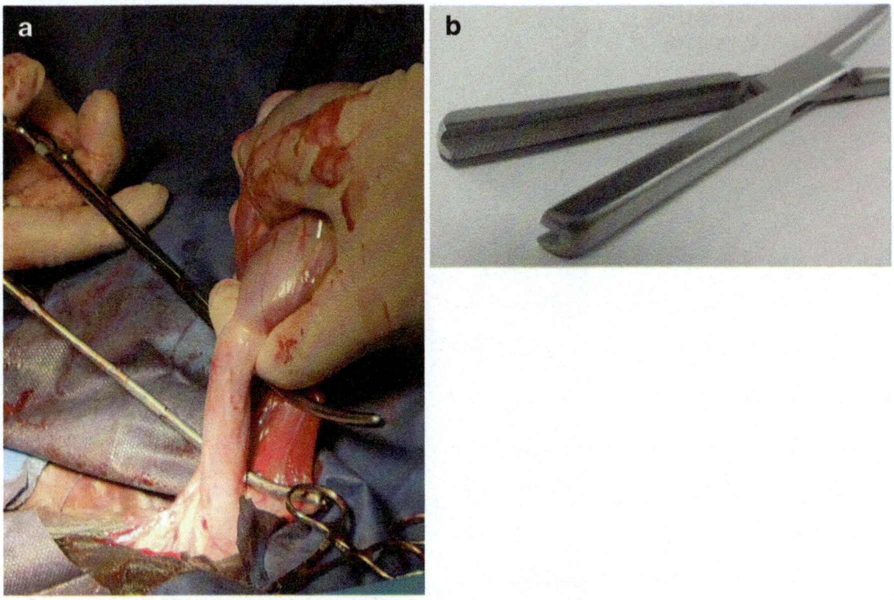

Fig. 6 (**a**) Double clamping of the cremaster muscle. (**b**) Tip of Ferguson Angiotribe forceps

the specifics of the housing and the herd dynamics, there is not a real downside to giving the foal a couple of weeks of TLC before returning to a robust lifestyle. The older horse should be restricted to a quiet stall or pen for a least a day or two, they can be hand walked for general health and attitude reasons but not directly needed because of the operation. Adults should be provided a couple of weeks to heal before starting back to a more normal lifestyle. The bandage is left to loosen on its own, which generally happens within 5–10 days. Inspection of a well healed surgical site and a patient in good health and spirit are the best guide for increasing activity. Of course, postoperative care on the whole must be dictated by the individual animal and cannot be based solely by the procedure performed, a general starting plan for the average is just the initial goal, with daily monitoring guiding any modifications needed. A final mention of potential unwanted sexual behavior after castration is warranted. It is reported in the horse that up to 30% of gelding will show some signs of unwanted sexual behavior (Line et al. 1985). Although no studies exist specifically related to the mule it is the author' experience that the castrated john mule seems to be similar and if anything, it may be to a greater degree. It is not all that uncommon to have john mules achieve erection and mount and penetrate when penned with a mare in estrus. Not all johns are so inclined but, in my opinion, the later degree of sexual behavior is more common in johns than geldings. That behavior can be frustrating to owners and veterinarians, but options are limited other than preventing cohabitation with cycling mares.

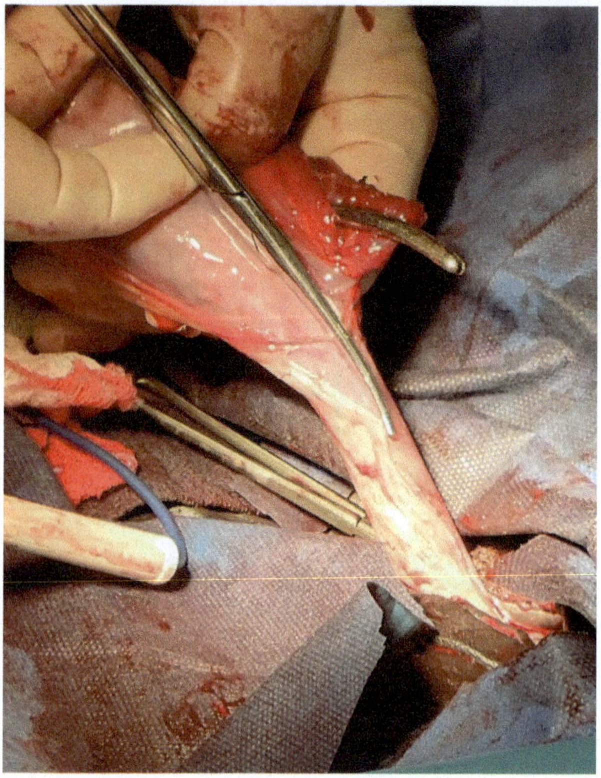

Fig. 7 Extending opening proximally in the vaginal tunic

2 Ovariectomy of the Female Mule

2.1 Introduction

Ovariectomy of female equines has been implemented on a very minor scale compared to castration of the male equine. In spite of the stark differences in the frequency of application between the equine genders, the major goals of the operations are similar: permanent behavior modification and elimination of reproduction. Two obvious differences between the genders, behavior and anatomy, account for the disparity in the operations. Intact females are not as difficult or dangerous to house and handle as intact males, and the external location of testicles permit much easier access for removal. Few colts and no young mules are kept intact as breeding stallions, whereas nearly all fillies and mollies remain intact. This is in part due to the previous lack of minimally invasive surgical options for ovariectomy. Until the application of laparoscopy to ovariectomy, many owners and veterinarians considered the available operative techniques too invasive and flawed for use except when surgery was a last resort to salvage a problem animal (Jorgensen 1996). Of 800 equine practitioners that responded to a survey regarding the effects of the

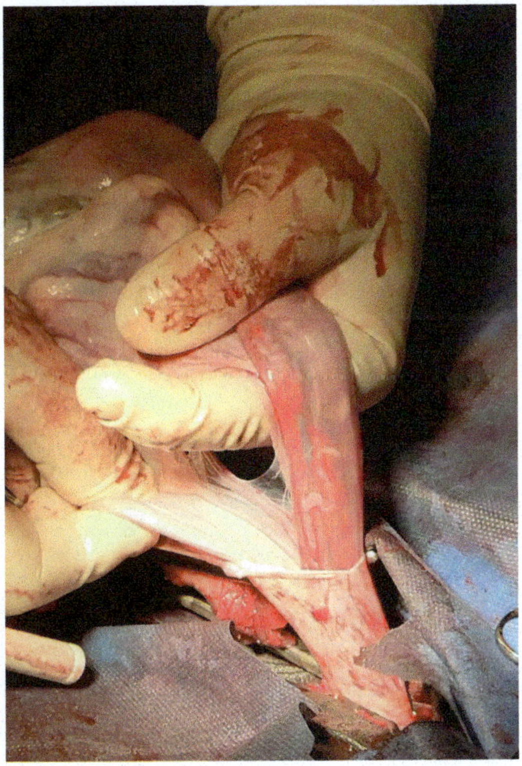

Fig. 8 Puncture of the mesorchium, isolating the vascular pedicle

estrous cycle on athletic performance, 89% had treated mares for variation in performance during estrus (Jorgensen et al. 1996).

When a mule is presented for operation with the complaint of behavior problems, confirming the behavior's association with the estrous cycle is important. Publications that examine the effects of the estrous cycle on the athletic mare are useful for reviewing and sharing to aid owners in making informed decisions regarding their mare (Jorgensen et al. 1996; Pryor and Tibary 2005). Mule behavior issues can be a complex problem with factors involved other than the estrous cycle. Mental attitude problems of molly unrelated to estrus are not improved by ovariectomy.

Mules are most often are presented for surgery not due to abnormal behavior but for the intent to promote a more "gelding-like" animal (Heaton et al. 2018; Petrizzi et al. 2020). There are owners who elect to spay mares intended solely for a riding career from the belief that it will provide a more docile horse that is easier to manage around other horses. This practice is common in some regions and in the United States can be traced back to at least the 1900s (Hobday 1900).

It should be appreciated that the female equine differs from other domestic species in that an ovarian source of estrogen does not seem to be required for at least some degree of estrous behavior. The most determining hormonal factor in

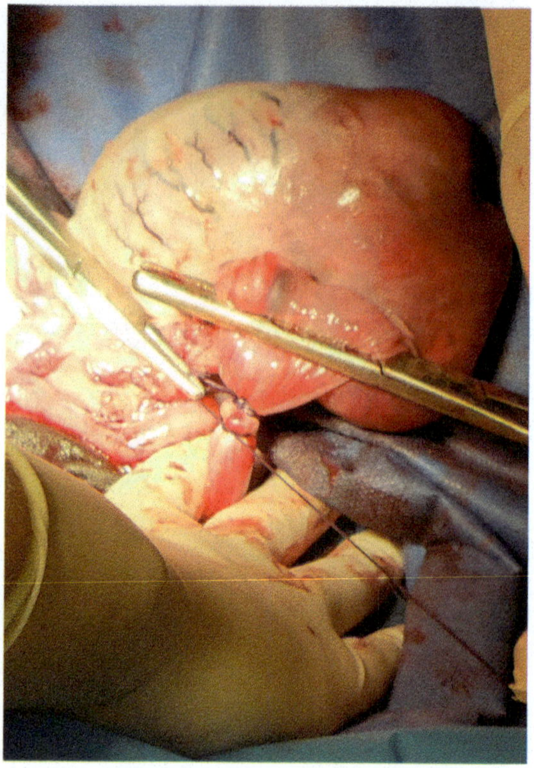

Fig. 9 Double ligation of the vascular pedicle

estrous behavior in the mare seems to be the absence of progesterone (Asa et al. 1980a, b). Ovariectomy removes the ovarian source of progesterone and estrogen. It is common for ovariectomized mares to be used as jump mares for collecting stallions with or without supplementation of estrogen (Brinsko et al. 2010). Considering these observations, it is important to note that published reports of postovariectomy mares operated for behavior problems related to estrus indicate that only about 30% of mares are noticed to manifest some signs of estrus (Hooper et al. 1993; Collar et al. 2021). There are no corresponding studies on the ovariectomized mule to know if it differs significantly from that of the mare horse. It is worth noting that after castration, approximately 20% to 30% of geldings also display masculine behavior toward mares and aggression toward other horses (Line et al. 1985). It has been stated that when used for semen collection, ovariectomized mares rarely need exogenous hormonal therapy to display behavioral signs of estrus (Brinsko et al. 2010). Phytoestrogens are found in certain plants and are structurally and/or functionally similar to mammalian estrogens and their active metabolites (Samocha 2017; Ferreira-Dias et al. 2013) Common supplements fed to horses can contain phytoestrogens (e.g., soybean meal) and in the authors experience may contribute to unwanted signs of estrus in the ovariectomized equine (Adkin 2018). It is also important to note that ovariectomized mares, as well as intact mares, can exhibit "paradoxical estrous" behavior during seasonal anestrus that is associated

Fig. 10 Cautery of the stump of the divided vascular pedicle for increased hemostasis

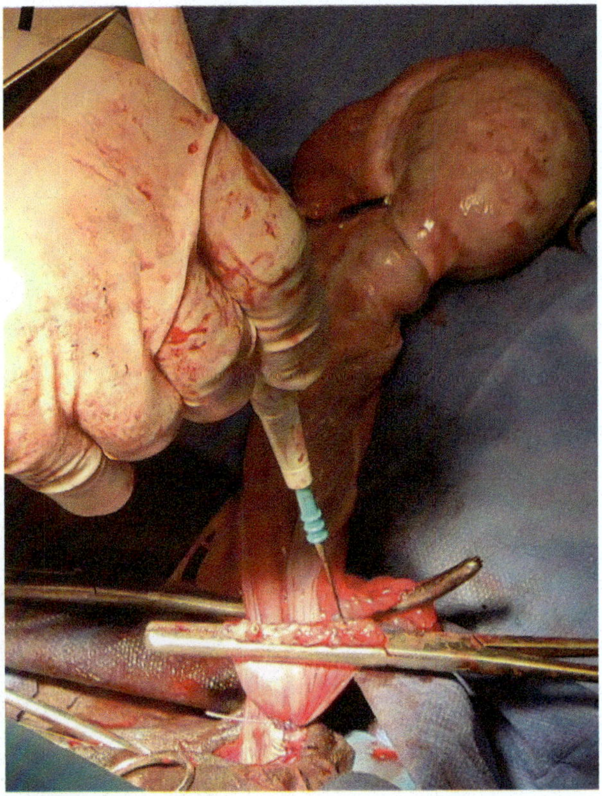

with hormone secretion from the adrenal cortex (Crowell-Davis 2007). Despite these apparent contradicting facts, owner satisfaction surveys of mares ovariectomized for "behavior problems during estrus" report that approximately 80% are satisfied with the outcome of the operation (Hooper et al. 1993; Hanson and Galuppo 1998; Kamm and Hendrickson 2007; Smith and Mair 2008; Devick et al. 2020).

The problem behaviors associated with estrus that most improve following ovariectomy are aggressive behavior and a generalized disagreeable demeanor; excitability, kicking, and biting; and problems during training (Kamm and Hendrickson 2007). The time frame during which improvement is noted ranges from 3 to 18 months.

Laparoscopic bilateral ovariectomy offers significant advantages over previous techniques via colpotomy (vaginal celiotomy), laparotomy, and ventral celiotomy (Kamm and Hendrickson 2007). Inherent in the laparoscopic approach is providing clear viewing of the target tissues, ability to manipulate the ovary without undue tension on the pedicle, excellent hemostasis, and small incisions. This combination of technical advantages is not possible with the nonlaparoscopic techniques of ovariectomy (Colbath et al. 2017; Devick and Hendrickson 2019). Laparoscopic ovariectomy performed in the standing mule or donkey presents technical advantages over the technique performed in the recumbent animal. Foremost is

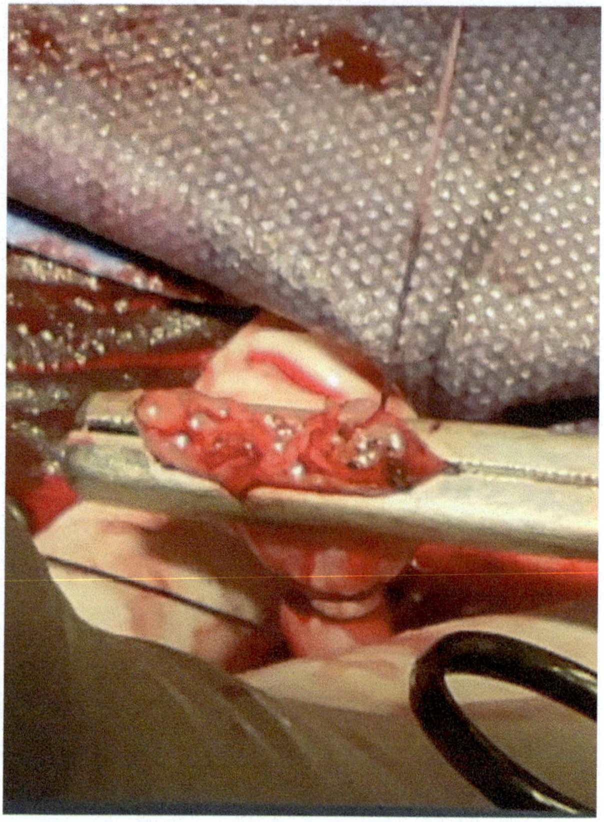

Fig. 11 Double ligation of the vaginal tunic and remaining fibrous portion of the testicular pedicle

that the mesovarium and ovary as suspended in the standing molly or jenny are much easier to locate and manipulate as compared to the passively positioned reproductive tract of the female positioned in dorsal recumbency. The dorsal positioning requires specific instrument manipulation to elevate the uterine horn to localize the ovary and requires a second instrument for grasping it and for applying traction during sectioning of the pedicle (Ragle and Schneider 1995).

Contraindications for standing elective bilateral laparoscopic ovariectomy are few and similar to any elective laparoscopic procedure performed on the standing mule. The mule should be of correct size and training to allow safe restraint in standing stocks (Fig. 16a, b).

It is preferable to perform the operation when the molly is in diestrus or anestrus. Performing the operation while the molly is in estrus should be avoided. Although it seems intuitive that ovariectomy of mares with "behavior problems during estrus" would be more successful in younger mares, at least one study indicates that the age at ovariectomy did not significantly affect the outcome (Kamm and Hendrickson 2007). One can surmise that the same may be true in the mule but further studies are needed.

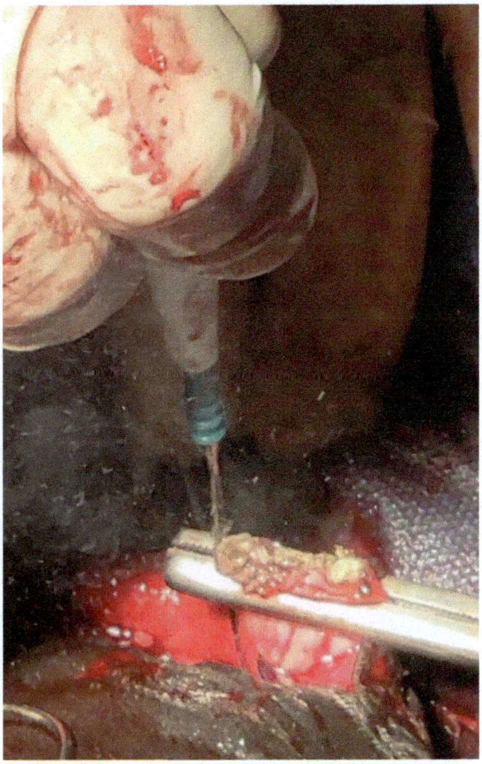

Fig. 12 Cautery applied to stump of double ligated and transected vaginal tunic and fibrous remains of testicular pedicle

2.2 Anesthesia, Positioning, and Surgical Preparation

Normal preoperative screening prior to surgery is standard. Mules that present with no significant current or past illness receive a physical exam, including a rebreathing exam and an electrocardiogram. Complete blood count with packed cell volume (PCV) and total protein (TP) is usually performed, although a PCV and TP are adequate in a young healthy horse. In the majority of horses, it is advisable to reduce the intestinal contents prior to surgery to improve intraoperative viewing. The preoperative diet is modified by feed withdrawal or feeding a high energy, low bulk diet. Water is not restricted prior to surgery. A 450-kg horse will have a digestive tract volume capacity of about 200 L of ingesta (Kararli 1995). The weight of the gastrointestinal contents will range from 5% to 10% of the body weight depending on the diet composition (5% with concentrate and 10% with roughage) (Harris et al. 2006). Intestinal transit time, or mean retention time, of feedstuffs in horses varies in studies from 18 to 60 hours depending on methodology of measurement, diet, level of exercise, and so on (Van Weyenberg et al. 2006). For clinical application, it is considered that 45–72 hours are required for food to completely pass through the digestive tract of the equine (Swinker 2002). The goal is that prior to surgery, the mule should be concave in the paralumbar fossae. Observing the

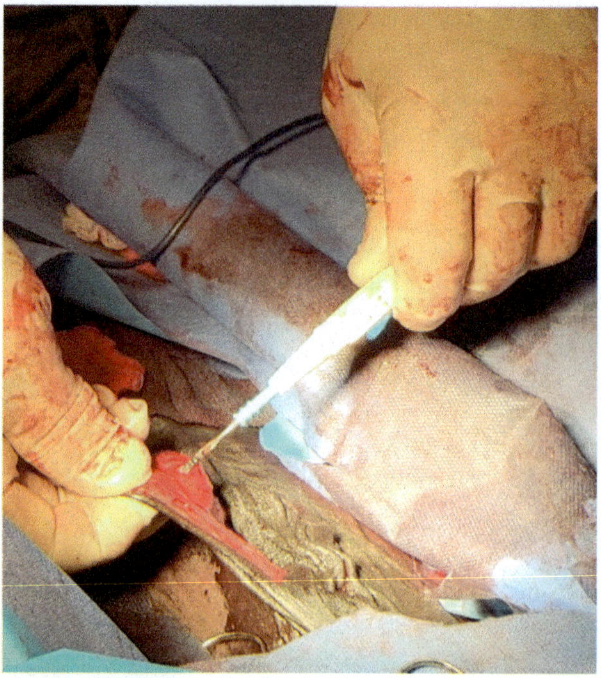

Fig. 13 Through search and cautery of surgical area to prevent hematoma formation

concave shape in the flanks and a low intestinal bulk detected during transrectal exam are the best indicators that the operation can proceed. Depending on the mule, the time to achieve this can vary from 12 to 72 hours. The alternative to complete feed withdrawal is to restrict feed to a low bulk, high energy diet (e.g., 1 cup of soaked complete pelleted feed with ¼ cup corn oil four times a day) until there is a concaved contour to the paralumbar fossae (~48 hours).

The greater the internal body fat, the more important it is to reduce the contents in the gastrointestinal tract. Excess internal body fat has a significant negative impact on the size of the laparoscopic visual field and in the ease of establishing and maneuvering laparoscopic portals. Although the body condition score may be one indicator of possible excess internal body fat, some horses seem to have greater internal body fat (especially in the retroperitoneal area of the paralumbar fossae, mesocolon, and mesovarium) than their amount of subcutaneous fat would suggest. Transrectal examination is performed prior to surgery to evacuate the rectum, to confirm the reproductive tract status, and to verify that the paralumbar fossa area is clear of any obstructions that would complicate portal placement.

Mollies that are anticipated to be nervous in the stocks prior to surgery are premedicated with acepromazine at about 0.03–0.06 mg/kg IM 30 minutes prior to loading into the stocks. Before laparoscopy, the mule is administered penicillin G potassium (22,000 U/kg [10,000 U/lb], IV), gentamicin sulfate (6.6 mg/kg [3 mg/lb], IV), and flunixin meglumine (1.1 mg/kg [0.045 mg/lb], IV).

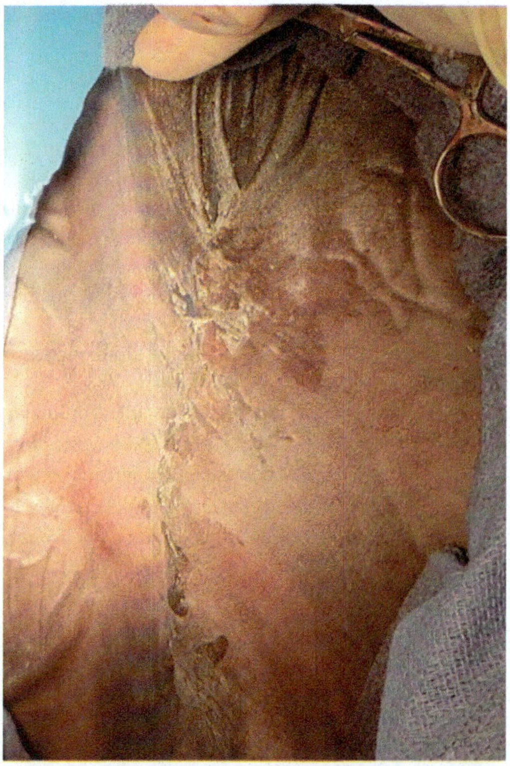

Fig. 14 Scrotal skin inverted and apposed using cyanoacrylate glue

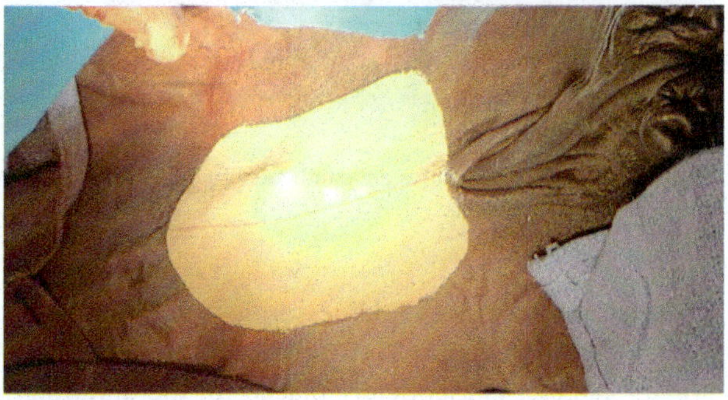

Fig. 15 Elastic adhesive bandage over the surgical site

The mule is led into a standing operating room, restrained in stocks, and positioned to gain access to the paralumbar fossae. When needed extra stall mats can be stacked at the rear of the stocks to achieve a 10- to 12-cm elevation of the pelvis to aid viewing of the caudal abdomen. Detomidine hydrochloride (0.015 mg/kg [0.0068 mg/lb], IV) and butorphanol tartrate (0.025 mg/kg [0.011 mg/lb], IV) are administered. Sedation is maintained with detomidine hydrochloride continuous rate infusion (CRI) and adjusted as needed (0.1 mg/kg/min). The hair over the caudal thorax, paralumbar fossae, and caudal flanks are clipped, and the sites are prepared

Table 1 After mule/donkey is induced and positioned in dorsal recumbency

1. Wipe scrotal injection sites with alcohol-prior to surgical prep.
2. Inject each testicle until full with local anesthetic.
3. Aseptic preparation of surgical site and surgeon.
4. Make two parallel scrotal incisions, each about 2 cm from median raphe.
5. Incise into the cranial portion of the vaginal tunic of one testicle, place index finger into the tunic and place kocher forceps firmly over the tunic and tail of the epididymis.
6. Apply firm tension of the tunic/testicle and with the other hand strip away the connective tissue and fat covering the tunic and cremaster muscle.
7. Divide cremaster from vaginal tunic double clamp with angiotribes and section with electrosurgery.
8. Expose pedicle from inside tunic, Blunt puncture mesorchium to separate into vascular and non-vascular segments of the pedicle.
9. Double lamp and double ligate vascular portion and section with electrosurgery.
10. Replace ligated vascular portion inside tunic.
11. Double clamp and double ligate tunic and section with electrosurgery.
12. Repeat steps 5–11 on remaining testicle.
13. SubQ closure of scrotal incisions (and/or) invert scrotum and glue.

Must have immaculate hemostasis for skin incision closure

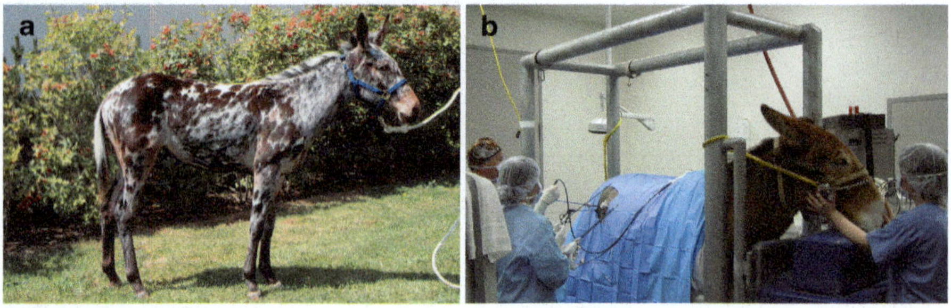

Fig. 16 (**a, b**) Female mules are candidates for standing laparoscopic ovariectomy when accepting to be restrained in stocks

for surgery with chlorhexidine surgical scrub solution alternated with sterile gauze soaked in saline (0.9% NaCl) solution. The rectum is manually evacuated and a urinary catheter is aseptically placed for the duration of the operation.

2.3 Instrumentation

1	57-cm 30° laparoscopic telescope
1	Video camera, monitor, light source, insufflator
1	5-mm, 32-cm, 18-ga tip laparoscopic injection needle
2	12-mm × 11.5-cm laparoscopic cannula assemblies
2	10-mm laparoscopic cannula assemblies

(continued)

1	57-cm 30° laparoscopic telescope
1	Video camera, monitor, light source, insufflator
1	5-mm, 32-cm, 18-ga tip laparoscopic injection needle
2	12-mm × 11.5-cm laparoscopic cannula assemblies
2	10-mm laparoscopic cannula assemblies
1	5- or 10-mm traumatic grasping forceps
1	10-mm vessel-sealing device (LigaSure)
1	Morcellator

Items to have available

- Ligating loops
- Hemoclips
- Staples
- Semm claw forceps
- Specimen bag
- 10-mm × 20-cm laparoscopic cannula assemblies
- 5-mm laparoscopic scissors

2.4 Surgery

The primary task of the operation is to achieve excellent hemostasis during transection of the ovarian pedicle (Fig. 17) and removal of the ovary from the abdominal cavity. Practically every method available for use in operations has been applied to ovariectomy of the equine. An excellent review of these has been published (Lee and Hendrickson 2008; Shettko and Hendrickson 2008). This surgical description will focus on a two-portal technique utilizing vessel-sealing technology and morcellation (Lund et al. 2014).

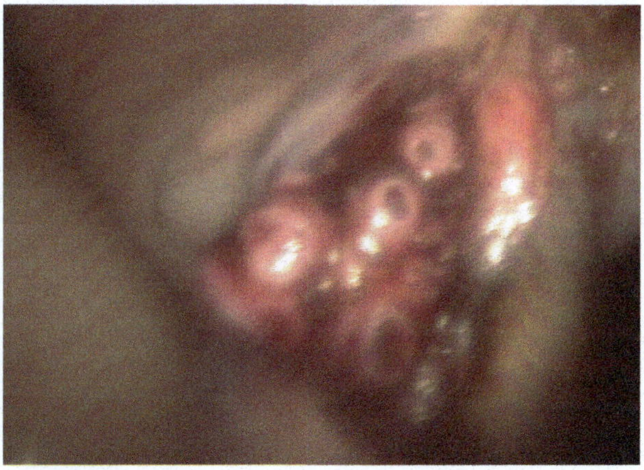

Fig. 17 The ovarian artery is tortuous and may be cleaved several times when transected

After draping the surgical areas, 2% mepivacaine hydrochloride (20–40 mL) is infiltrated throughout the full thickness of the body wall with special attention to the skin at each of the four portal sites. Intradermal injection of the local anesthetic is performed, creating a 3- to 5-cm wheal over each portal site (Fig. 18). An additional intradermal site can be injected at the ventral aspect of the drape fenestration to permit towel clamp application.

On the molly's left side, the first portal (laparoscope) is located at the level of the ventral tuber coxae and between the last two ribs. To avoid possible penetration of the pleural space, it is important that the intercostal portal is *not* located dorsal to a horizontal line from the most ventral aspect of the tuber coxae (Fig. 19). The second portal (instruments) is located 5 cm below the dorsal margin of the internal abdominal oblique muscle midway between the last rib and tuber coxae. On the right side, the first portal is located just dorsal to the internal abdominal oblique muscle and just caudal to the last rib. The second portal is located 5 cm distal and caudal to the first portal. A 15-mm skin incision is made over the first portal location on the left side. A 10- to 12-mm portal sheath with conical obturator is placed into the abdominal cavity. The obturator is removed and replaced with a 50-cm-long, 10-mm-diameter, 30° angle viewing laparoscope.

Fig. 18 Local anesthetic injection creating a ~ 5-cm intradermal wheal followed by deep infiltration to the peritoneum for each of the portal locations

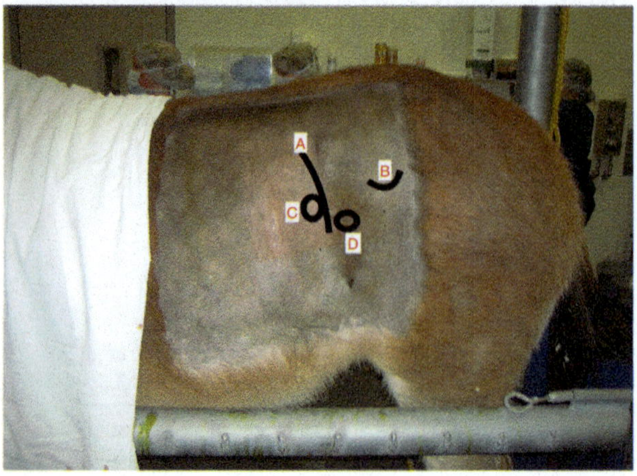

Fig. 19 (**a**) location of last rib (**b**) Ventral border of Tuber coxae (**c**) Locations of first portal, (**d**) Location of second portal

If the field of view is not adequate, insufflation is used. The amount of insufflation should be kept to the minimum amount to permit viewing. Not every mule will require the use of an insufflator for adequate viewing of the ovarian pedicles. When needed, the abdomen is insufflated with carbon dioxide (CO_2) (7 L/min) for ~4 minutes with adjustments as needed and then turned off, the goal being to allow clear viewing of the ovarian area and to provide an intra-abdominal pressure below 10 mmHg.

Once a field of view is established on the left side, the laparoscope is passed under the descending colon at the level of the bladder and then directed dorso-cranial, providing a view of the right paralumbar fossa area. A 15-mm skin incision is made over the first portal site, and a 5- to 12-mm sheath and obturator is inserted using the laparoscopic view to confirm placement into the abdominal cavity. The second portal on the right is placed using an identical technique. The laparoscope is withdrawn from under the descending colon and focused on the intra-abdominal location of the paralumbar fossa for the second portal on the left side; it again is placed under laparoscopic guidance to confirm cleanly entering the abdominal cavity.

After all four portal sheaths are established, a long injection needle is placed in the newly established port and 30–60 mL of 2% lidocaine is infiltrated along the mesovarium of the ovary (Fig. 20).

The needle is withdrawn from the portal and a vessel-sealing device (LigaSure, Covidien, Medtronics Minneapolis, Mn) is inserted to begin the removal of the ovary. The instrument is capable of both vessel sealing and transection of the mesovarium with each placement of the jaws. The cutting function is activated by a trigger on the instrument handle and should only be used when vessel sealing has been successful, as indicated by visual assessment of tissue reaction to the energy source and the correct audible signal from the unit. The maximum length of tissue that the jaws will accommodate is 20 mm. Transection can begin at the cranial or caudal aspect of the ovary. It is technically easier to begin at the cranial aspect as the

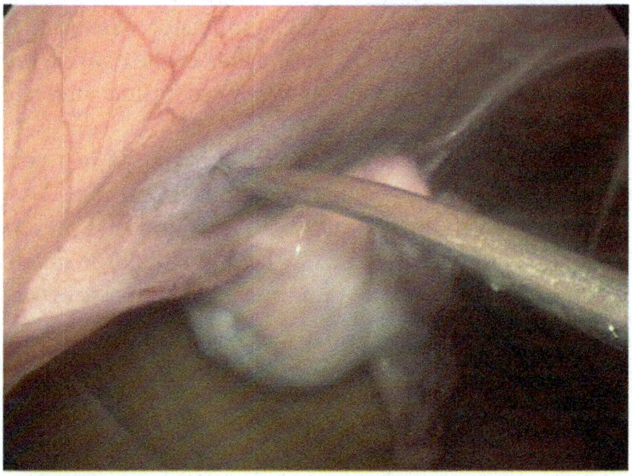

Fig. 20 Injection of local anesthetic into the ovarian pedicle

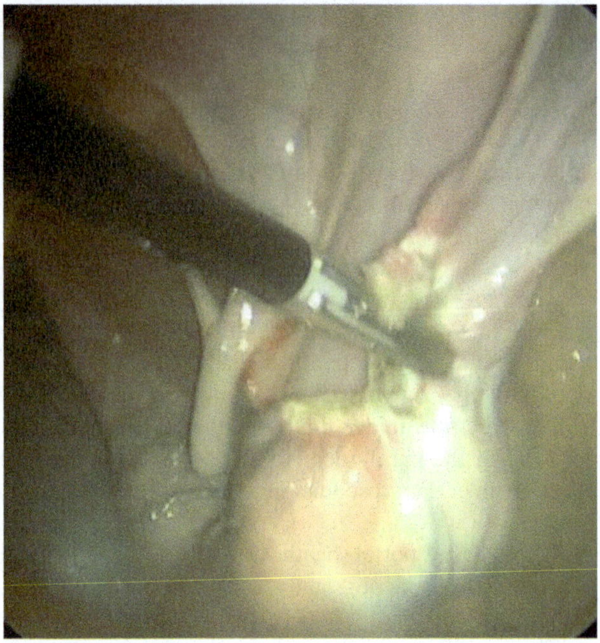

Fig. 21 Vessel sealing devices provide excellent hemostasis

angle of approach is straight and the mesovarium is a single-layer structure with a free leading margin at that point as compared to the double-layer structure (proper ligament and mesosalpinx) at the caudal attachments to the uterine horn. If the mare has excessive adipose tissue in the mesovarium, extra caution should be taken to assure hemostasis. When the pedicle becomes too thick due to fat to allow secure closure of the handle, the pedicle should be split sagittally using the jaws of the vessel-sealing instrument. This will reduce the thickness of the tissue placed between the jaws, allow transection of the pedicle to progress, and improve hemostasis. Although use of vessel sealing technology is mildly more inflammatory than ligating loops it is likely not of clinical significance. (Seabaugh et al. 2014). It is also possible to perform a double seal over the area of the ovarian artery by placing the first grasps of the jaws more dorsal on the mesovarium and releasing the jaws after sealing without triggering the scissors. (Fig. 21).

The next placement of the jaws should be just ventral to the last seal, this time completing another sealing cycle and transecting. As the transection continues, a 5-mm band of attached mesovarium should be preserved to suspend the ovary for morcellation (Lund et al. 2014). Care should be taken to transect the robust proper ligament of the ovary and minimize using that structure for suspending the ovary for morcellation. There is risk of damage to the uterine horn by pulling it into the morcellator if the proper ligament of the ovary remains attached to the ovary. If at anytime during the operation an additional laparoscopic instrument would augment the viewing and dissection of the ovary, a 5-mm portal can be established. Depending on the individual anatomy of the mare, the extra portal can be located

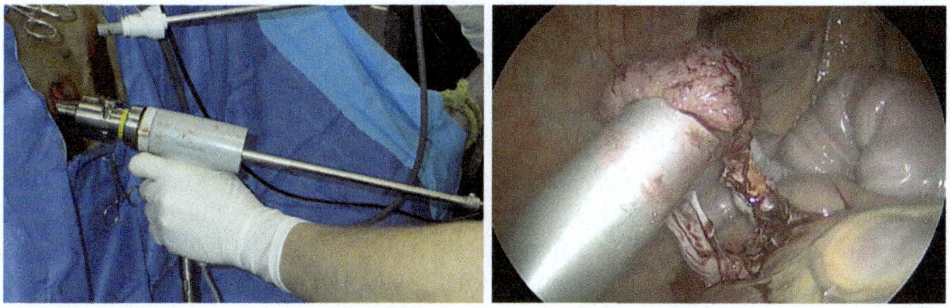

Fig. 22 Morcellation allows removal of the ovaries through a 20 mm portal

approximately 10 cm ventral or dorsal to the existing portal in the paralumbar fossae. If the extra 5-mm portal is used, then the ovary can be suspended from the 5-mm forceps for morcellation. Once the ovary is freed from all but the small band of mesovarium, the vessel-sealing instrument and associated portal are removed. The skin incision is slightly enlarged to accommodate the 20-mm diameter of the morcellator. The morcellator consists of a cable, handpiece, blade, cannula, obturator and power unit. The obturator is inserted in the 19-cm cannula, and placed into the abdomen. The obturator is removed and the blade (attached to the handpiece) is inserted in the cannula. A pair of 5- or 10-mm laparoscopic tenaculum forceps is placed through the portal of the morcellator to grasp a leading edge of the ovary. The ovary is drawn to the morcellator, and the morcellator is activated either by foot pedal or handpiece, depending on the surgeon's preference (Fig. 22).

The morcellator has a 20 mm blade consisting of a rigid inner tube that rotates within an outer cannula. The handpiece connects to an electric motor drive unit. When activated, the blade extends 2–4 mm out beyond the tube except in a small guarded area that extends from the outer tube to permit cutting via a peeling action. The morcellator can also be set where the cutting blade is extended beyond the guard if a core cut is needed. For morcellation of the ovary, the adjustable guard is used and set in the 6 o'clock position so it can be rotated as needed to aid in the process of morcellation. It is important that excellent viewing of the morcellation is maintained. If the view becomes obscured, the morcellator blade should be immediately disengaged until the view is reestablished. The ovary is extracted in peeled sections and, depending on the size, consistency, and configuration of the ovary, it can be removed in typically one to seven segments (see morcellation video). The small band of mesovarium that suspends the ovary is cut by the blade of the morcellator as the final portion of suspended ovary is pulled into the portal and the portal is angled to transect the band as it is stretched over the blade edge. Alternatively, the remaining piece of suspended ovary can be grasped with 5-mm tenaculum forceps, and a pair of 5-mm scissors can be placed alongside to cut the band. Any fragments of ovary that may become separated and free in the abdomen are located and removed through the morcellator portal. Although not a common practice in the horse, free transected ovaries have been left in the abdomen with little apparent negative effect (Shoemaker et al. 2004; Devick and Hendrickson 2021). At the conclusion of the operation on the left side, a final check of the ovarian pedicle is performed to confirm hemostasis. The laparoscope and morcellator are then removed, and the operation is repeated on the right side, or the transected right ovary can be passed to the left side for morcellation.

A note on morcellator units, due to legal and medical complications with morcellation in human surgery, manufacture of morcellators have been very limited for the last decade (FDA 2014) Pre-owned units have been availble during that time period for veterinary use but that supply is dwendling. With the FDA update of 2020 and approved use of contained morcellation, production of new units may increase. Morcellators with a 20 mm blade are the most efficient for use on the equine ovary (Bekhit et al. 2014).

When the ovariectomy is completed, remaining gas in the abdominal cavity is decompressed. Suction can be used to aid in the removal of the gas to more completely empty the abdomen to reduce the occurrence of subcutaneous emphysema (Devick et al. 2018). The surgical site is cleaned of any signs of blood, and the skin incisions are apposed using skin staples or size 2-0 or 0 monofilament nonabsorbable suture. Typically, there is no need to suture the aponeurosis of the external abdominal oblique muscle. Adherent bandages are placed over the incisions before the horse is returned to its stall (Fig. 23).

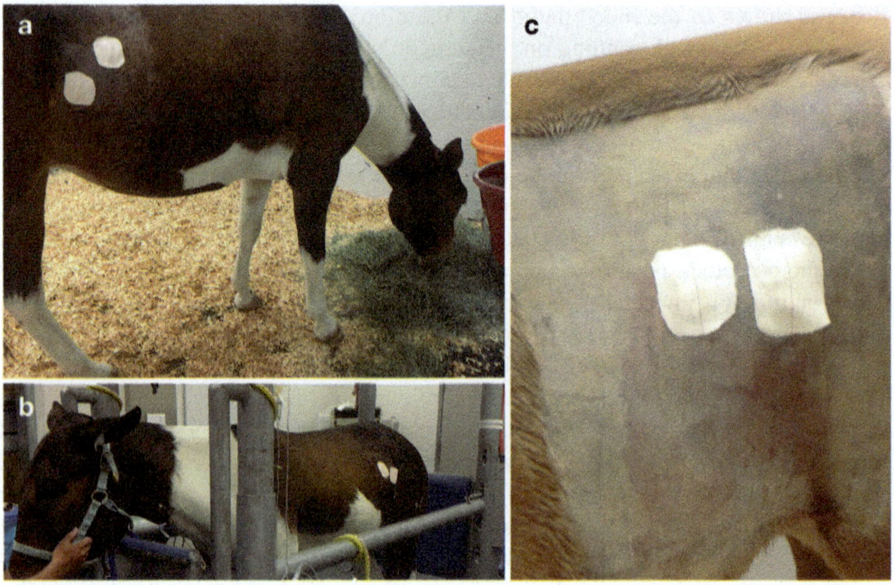

Fig. 23 Postoperative bandage of laparoscopic portals

2.5 Postoperative Management

Postoperatively, the animal is returned to its stall and allowed food and water several hours after surgery. Antibiotics are continued for 12 hours postoperatively. The day after surgery, transrectal palpation can be performed to sweep the resected ovarian pedicles and to free any tissue that might have become adhered. Non-steroidal anti-inflammatories are administered for 3–7 days after surgery. Body temperature should be monitored at least once per day for 3–4 days. The surgical incisions are kept bandaged and are monitored daily. Mild reaction to palpation over the portal sites during the first 3–5 days is to be expected. If pressurized insufflation was used during the operation, the paralumbar area should be monitored for subcutaneous crepitus. Owners should be instructed on observation of the mule after release from the hospital and on how to recognize subcutaneous crepitus as it is common when insufflation is used. Mules are confined to a stall for 24 hours, and then hand-walked or allowed small paddock turnout. Light riding can begin after staple or suture removal at 2 weeks and then progress to full work at 3–4 weeks. Owners should be cautioned about any exposure of their mule to exogenous estrogen and potential phytoestrogens in the diet. This is because after ovariectomy removing the source of progesterone the animal will be extremely sensitive to any exposure of exogenous estrogen or estrogen-like compounds.

2.6 Complications

There are several technical challenges in performing laparoscopic ovariectomy that can lead to complications. At the top of the list would be the challenge of establishing the laparoscopic portals. The risk of damage to internal organs (e.g., spleen, kidney, cecum, mesovarium, mesocolon, and descending colon) or failure to enter the peritoneal cavity are the primary complications (Desmaizie et al. 2003). Recognizing these concerns, cannula manufacturers have devised special laparoscopic cannulas to aid in placement. The common feature of these ports is that they permit placement while the laparoscope is within the cannula for viewing the tissue planes as the trocar is advanced (Vázquez et al. 2023). The peritoneum overlying the paralumbar fossae of the horse is resistant to penetration and is easily displaced by a blunt obturator. This can lead to the common length (15 cm) cannula being too short to enter the abdomen, creating a retroperitoneal space that degrades the field of view. If a sharp trocar or a longer cannula is used, the peritoneum is penetrated, but the risk to underlying viscera during this blind placement is increased. The technique of placing a cannula with a blunt obturator in the 17th intercostal space on the left is a method to easily penetrate the peritoneum without displacing it. This is possible because the peritoneum is tightly adhered to the ribs in this location. Although portals can be established in the 17th intercostal space of the right side, the peritoneum is not adhered here as it is on the left. Even in using the left intercostal or optical trocar techniques care must be taken to only insert the obturator until it just penetrates the peritoneum; over-advancing can cause blunt damage to the spleen or kidney. During placement of the cannula, the obturator should be removed every few centimeters advanced to check for the aspiration of air from the negative intra-abdominal pressure. Once the intercostal portal is established, it permits the remaining portals to be placed under direct viewing. If during direct viewing there is trouble with the obturator or the trocar penetrating the peritoneum, the "chopstick" maneuver can be employed. The chopstick maneuver involves using an ipsilateral instrument or a laparoscope to alternately cross over the dorsal and ventral aspects of the tip of the advancing cannula, thus pushing the peritoneum over the cannula from inside the abdomen.

Although there are a multitude of technical errors that can occur during laparoscopic ovariectomy (e.g., dropping a freed ovary, thermal or sharp damage to the adjacent viscera, inadequate hemostasis, poor field of view, and inability to efficiently manipulate instruments), it has still been demonstrated to have fewer complications than previous techniques (Rocken et al. 2011; Easley et al. 2018).

3 Ovariectomy Conclusion

Bilateral laparoscopic ovariectomy offers many advantages over the previous techniques of bilateral ovariectomy in the mule. Continued improvements in technique and instrumentation will only add to the efficacy, affordability, and operative benefits. The use of ovariectomy can canimprove the consistency of performance as riding, driving and packing mules.

There is a dearth of research related to the long-term effects of bilateral ovariectomy on the overall behavior and health of the mare and molly. There has been little to no research effort evaluating the in-depth normal estrous behavior of the mare mule (Gonzalez et al. 2015; Heaton et al. 2018) Understanding what is normal both for a population and for an individual is a prerequisite for being able to compare the effects of ovariectomy. Without doubt, it is a challenging area of study but one that could drastically improve the decision-making process for laparoscopic bilateral ovariectomy in the mule.

Questions

1. **Why do mules and donkeys have a higher reported risk of castration-related complications compared with horses, particularly hemorrhage and anesthetic failure?**
   A. Mules and donkeys appear more prone to castration complications due to a combination of anatomical, physiological, and behavioral factors. Compared with horses, they often have stronger cremaster muscle tone, thicker spermatic cords, and a heightened stress response, all of which may increase the risk of hemorrhage if meticulous hemostasis is not achieved. Inadequate anesthesia is also reported more frequently, likely reflecting species differences in drug metabolism, stress-induced catecholamine release, and underestimation of sedative and

anesthetic requirements when extrapolating from horse protocols. These factors underscore the need for species-specific anesthetic planning and strict surgical technique.

2. **How do the behavioral and cognitive characteristics of mules and donkeys influence preoperative handling, sedation strategies, and overall surgical success?**
   A. Mules and donkeys exhibit strong self-preservation behaviors and heightened situational awareness, which can be misinterpreted as resistance or noncompliance. Abrupt changes in routine, rushed handling, or escalating restraint can increase fear and physiological stress, negatively affecting anesthetic induction and recovery. Effective management relies on early, low-stress sedation, calm and deliberate handling, and allowing sufficient time for the animal to process new stimuli. Surgical success is closely tied to achieving a calm, relaxed patient prior to anesthesia, as excessive fear and struggling can contribute to anesthetic complications and poorer postoperative outcomes.
3. **What factors should guide age selection for castration in donkeys and mules, and how do management realities alter ideal recommendations?**
   A. There is no universally accepted ideal age for castration in equids, and donkeys and mules are presented for surgery across a wide age range. When possible, castration prior to weaning is advantageous due to smaller testicular size, reduced surgical trauma, passive transfer of maternal antibodies, and the ability to return the patient to the dam, minimizing stress. However, veterinarians must adapt to the age, temperament, and management circumstances of each patient, emphasizing appropriate anesthetic planning, hemostasis, and postoperative care regardless of age at presentation.
4. **Given the rarity of cryptorchidism in mules and its prevalence in donkeys, what biological or genetic mechanisms might explain this difference, and how should it influence diagnostic and surgical planning?**
   A. True abdominal cryptorchidism appears to be exceedingly rare in mules, in contrast to its recognized occurrence in donkeys, particularly miniature breeds. This disparity may reflect genetic factors associated with heterosis, though definitive mechanisms remain unproven due to limited population-level studies. Clinically, absence of palpable testicles in a mule should prompt thorough diagnostic evaluation rather than assumptions of cryptorchidism. In donkeys, retained testicles should be anticipated more readily, and surgical planning should account for both inguinal and abdominal presentations.

References

Adkin A (2018) The characterization of phytoestrogens in equine feeds and serum. PhD Dissertation. University of Florida, Gainesville

Asa CS, Goldfoot DA, Garcia MC et al (1980a) Sexual behavior in ovariectomized and seasonally anovulatory pony mares (*Equus caballus*). Horm Behav 14:46–54

Asa CS, Goldfoot DA, Garcia MC et al (1980b) Dexamethasone suppression of sexual behavior in the ovariectomized mare. Horm Behav 14:55–64

Baldwin CM (2023) A review of prevention and management of castration complications. Equine Veterinary Education 36:97–106. https://doi.org/10.1111/eve.13880
Baranwal VK, Mikkilineni V, Zehr UB, Tyagi AK et al (2012) Heterosis: emerging ideas about hybrid vigour. J Exp Bot 63–18:6309–6314. https://doi.org/10.1093/jxb/ers291.Epub
Bekhit M, Jackson T, Advincula AP (2014) Simulation study comparing the effectiveness of the Gynecare Morcellex®, the MOREsolution™, and the Rotocut™ G1 tissue Morcellators. Surg Technol Int 24:237–242
Brinsko SP, Blanchard TL, Varner DD et al (2010) Manual of equine reproduction, 3rd edn. Mosby, St. Louis
Brunner H, Gerhards H, von Rennenkampff F (2002) Radiographic screening examination in mules and haflinger horses exclusively used as pack animals in mountain areas. Pferdeheilkd Equine Med 18:359
Colbath AC, Hackett ES, Lesser CS, Hendrickson DA (2017) Left paralumbar laparoscopic bilateral ovariectomy in mares. Vet Surg 46(4):574–579
Collar EM, Duesterdieck-Zellmer KF, Huber MJ et al (2021) Outcome of bilateral equid laparoscopic ovariectomies. Vet Surg 50(5):975–983
Cognie J, Freret S, Lansade L, Parias C, Barriere P, Gesbert A, Reigner F, & Deleuze S (2023) Early castration in foals: Consequences on physical and behavioural development. Equine veterinary journal, 55(2), 214–221.https://doi.org/10.1111/evj.13580
Crowell-Davis SL (2007) Sexual behavior of mares. Horm Behav 52:12–17
Desmaizie LM, Martinot S, Lepage OM et al (2003) Complications associated with cannula insertion techniques used for laparoscopy in standing horses. Vet Surg 32:501–506
Devick IF, Hendrickson DA (2019) Left paralumbar fossa approach combined with mesocolon fenestration for bilateral equine ovariectomy. Vet Surg 48(5):735–741
Devick IF, Hendrickson DA (2021) Effects of leaving amputated ovaries intra-abdominally during elective bilateral standing laparoscopic ovariectomy in equids. Animals (Basel) 11(1)
Devick IF, Leise BS, Rao S et al (2018) Evaluation of post-operative pain after active desufflation at completion of laparoscopy in mares undergoing ovariectomy. Can Vet J 59(3):261–266
Devick IF, Leise BS, McCue PM et al (2020) Ovarian histopathology, pre- and post-operative endocrinological analysis and behavior alterations in 27 mares undergoing bilateral standing laparoscopic ovariectomy. Can Vet J 61(2):181–186
Easley JT, McGilvray KC, Hendrickson DA et al (2018) Vessel sealer and divider instrument temperature during laparoscopic ovariectomy in horses. Vet Surg 47(S1):O26–O31
Ferreira-Dias G, Botelho M, Zagrajczuk A et al (2013) Coumestrol and its metabolite in mares' plasma after ingestion of phytoestrogen-rich plants: potent endocrine disruptors inducing infertility. Theriogenology 80(6):684–692
Gonzalez S, Gomes R, Souza A et al (2015) Evidences of regular estrous cycles in mules and successful use of these animals a recipients for donkey embryos. J Equine Vet Sci 35:869–872
Hanson CA, Galuppo LD (1998) Bilateral laparoscopic ovariectomy in standing mares. In: Proceedings of the American Association of Equine Practitioners, 44, pp 300–301
Harris PA, Coenen M, Frape D et al (2006) Equine nutrition and metabolic diseases. In: Higgins AJ, Snyder JR (eds) The Equine Manual, 2nd edn. Saunders, London, p 164
Heaton K, Ragle C, Godderidge MT et al (2018) Estrous behavior in mules—an owner's perspective. J Equine Vet Sci 60:109–112
Hobday FF (1900) Oophorectomy or ovariotomy of the mare. Vet J 1:84–88
Hooper RN, Taylore TS, Varner DD et al (1993) Effects of bilateral ovariectomy via colpotomy in mares: 23 cases (1984–1990). J Am Vet Med Assoc 203:1043–1046
Jorgensen JS (1996) Effects of the estrous cycle on performance in athletic mares. Compend Contin Educ Pract Vet 18:692–699
Jorgensen JS, Vivrette S, Correa M et al (1996) Significance of the estrous cycle on athletic performance in mares. In: 1996 Proceedings, 42nd Annual Meeting of the American Association of Equine Practitioners. American Association of Equine Practitioners, Denver, pp 98–100

Kamm JL, Hendrickson DA (2007) Clients' perspectives on the effects of laparoscopic ovariectomy on equine behavior and medical problems. J Equine Vet Sci 27:435–438

Kararli TT (1995) Comparison of the gastrointestinal anatomy, physiology, and biochemistry of humans and commonly used laboratory animals. Biopharm Drug Dispos 16:366

Kumi-Diaka J, Kwanashie GG, Ogwu D 1981 Cryptorchidism and malformations of the scrotum and testes of donkeys.Theriogenology. Mar;15(3):241-3. https://doi.org/10.1016/0093-691x(81)90045-5.PMID: 16725583.

Lee M, Hendrickson D (2008) A review of equine standing laparoscopic ovariectomy. J Equine Vet Sci 28:105–111

Line SW, Hart BL, Sanders L (1985) Effect of prepubertal versus postpubertal castration on sexual and aggressive behavior in male horses. J Am Vet Med Assoc 186(3):249–251

Lund CM, Ragle CA, Lutter JD et al (2014) Use of a motorized morcellator for elective bilateral laparoscopic ovariectomy in standing equids: 30 cases (2007-2013). J Am Vet Med Assoc 244(10):1191–1197

Mclean A, Wang W (2013) Pilot study comparing hematologic and serum biochemical parameters in healthy horses (Equus caballus) and mules. Journal of Equine Veterinary Science 33:353–354. https://doi.org/10.1016/j.jevs.2013.03.079

Mclean A, Navas González F, Canisso I (2019) Donkey and mule behavior. Vet Clin North Am Equine Pract 35:575–588. https://doi.org/10.1016/j.cveq.2019.08.010

Meagher DM, Brokken TD (2005) A review of equine malpractice claims. https://www.semanticscholar.org/paper/A-reviewof-equine-malpractice-claims. https://api.semanticscholar.org/CorpusID:69023195

Merridale-Punter MS, Prutton JSW, Stefanovski D, Worsman FCF, Payne RJ, Wylie CE, Archer DC, Menzies-Gow NJ, Coles L, McGovern KF, Parker RA (2023) Outcome following emergency laparotomy in 33 UK donkeys: a retrospective multicentre study. Equine Vet J 55(2): 222–229. https://doi.org/10.1111/evj.13578. Epub 2022 Jun 21. PMID: 35478419

Moll HD, Pelzer KD, Pleasant RS, Modransky PD, May KA (1995) A survey of equine castration complications. Journal of Equine Veterinary Science 15(12):522–526

Petrizzi L, Guerri G, Straticò P et al (2020) Laparoscopic ovariectomy in standing mule mares. J Equine Vet Sci 84:102857

Proops L, Burden F, & Osthaus B (2008) Mule cognition: a case of hybrid vigour?. Animal cognition, 12(1), 75–84. https://doi.org/10.1007/s10071-008-0172-1

Pryor P, Tibary A (2005) Management of estrus in the performance mare. Clin Tech Equine Pract 4: 197–209

Racine J, Vidondo B, Ramseyer A, Koch C (2018) Complications associated with closed castration using the Henderson equine castration instrument in 300 standing equids. Veterinary surgery. VS 2448(1):21–28. https://doi.org/10.1111/vsu.12960

Ragle CA, Schneider RK (1995) Ventral abdominal approach for laparascopic ovariectomy in horses. Vet Surg 24:492–497

Rocken M, Mosel G, Seyrek-Intas K et al (2011) Unilateral and bilateral laparoscopic ovariectomy in 157 mares: a retrospective multicenter study. Vet Surg 40(8):1009–1014

Rodden EBK, Suthers JM, Busschers E, Burford JH, Freeman SL 2024 A scoping review on intraoperative and postoperative surgical castration complications in domesticated equids. Equine Vet J. Nov;56(6):1115-1128. https://doi.org/10.1111/evj.14122 Epub 2024 Jul 12. PMID: 38993145

Samocha MB (2017) The effects of a high-phytoestrogen diet on equine estrous cycles and fertility using domestic mares as models for captive exotic ungulates. University of California, Davis

Sanders JR (1903) Three anomalies I have met with in castration. J Comp Med Vet Arch 24(3):164-166

Seabaugh KA, Goodrich LR, Bohn AA et al (2014) A comparison of peritoneal fluid values in mares following bilateral laparoscopic ovariectomy using a vessel sealing and dividing device versus placement of two ligating loops. Vet J 202(2):297–302

Shettko DL, Hendrickson DA (2008) Principles of laparoscopic hemostasis: ligation techniques. J Equine Vet Sci 28(1):46–50

Shoemaker RW, Read EK, Duke T et al (2004) In situ coagulation and transaction of the ovarian pedicle: an alternative to laparoscopic ovariectomy in juvenile horses. Can J Vet Res 68:27–32

Silver E, Wu R, Grady J, Song L 2016 Knot Security- How is it Affected by Suture Technique, Material, Size, and Number of Throws? J Oral Maxillofac Surg. Jul;74(7):1304-12. https://doi.org/10.1016/j.joms.2016.02.004. Epub 2016 Feb 18. PMID: 26979258

Smith LJ, Mair TS (2008) Unilateral and bilateral laparoscopic ovariectomy of mares by electrocautery. Vet Rec 163(10):297–300

Sprayson T, Thiemann A (2007) Clinical approach to castration in the donkey. In Practice 29:526-531

Swinker A (2002) The anatomy and function of the horse's gastrointestinal tract. http://www.das.psu.edu/research-extension/equine/news/the-anatomy-and-function-of-the-horse2019s

Taylor TS, Matthews NS (2002) Donkey and mule scenarios: When to stop, think, read or call. Proc Am Assoc Equine Pract 48:115–116

U.S. Food and Drug Administration (2014) UPDATED laparoscopic uterine power morcellation in hysterectomy and myomectomy: FDA safety communication [archived]. FDA, Silver Spring. Available at: https://wayback.archiveit.org/7993/20170404182209/https:/www.fda.gov/MedicalDevices/Safety/AlertsandNotices/ucm424443.htm

U.S. Food and Drug Administration (2020) Update: perform only contained morcellation when laparoscopic power morcellation is appropriate. FDA Safety Communication. FDA, Silver Spring. Available at: https://www.fda.gov/medical-devices/safety-communications/update-perform-only-contained-morcellation-when-laparoscopic-power-morcellation-appropriate-fda

Van Weyenberg S, Sales J, Janssens GPJ (2006) Passage rate of digesta through the equine gastrointestinal tract: a review. Livest Sci 99(1):3–12

Vázquez FJ, Vitoria A, Gómez-Arrue J et al (2023) Complications in laparoscopic access in standing horses using cannula and trocar units developed for human medicine. Vet Sci 10(1)

Part III

Reproduction and Advanced Breeding

Male Reproductive Health: JACKASS REPRODUCTION

J. Miró and J. Catalán

Abstract

A donkey is not a horse with long ears. There are important reproductive differences between both species. Jackasses have, on average, a larger penis and testicles than horses. The accessory sex glands are also larger in the donkey except for the vesicular glands. Jackasses are territorial. When confronted with a jenny in estrous they display a long precopular phase, with an intense interaction, and a short copular phase. In general, jackasses ejaculate a good quality semen, with high sperm concentration and reduced sperm abnormalities. Donkey spermatozoa move fast, having an active metabolism and producing important amounts of lactic acid and ROS. Seminal plasma (SP) is a complex fluid produced by sexual accessory glands. More than 2000 proteins are identified in donkey SP, and only a fraction are known. Some of them are antioxidant enzymes or some are involved in the jenny post-insemination endometrial reaction. Jackass SP has a lot of metabolites different than stallions with important sperm metabolism differences. However, SP has a deleterious effect on donkey sperm preservation and must be eliminated in the cryopreservation process. Post-thaw donkey sperm survival results are good (40–60%) and are similar or better than obtained results for stallion frozen/thawed semen. However, low conception rates are obtained when jennies are inseminated. Further studies are needed, taking into account recent scientific updates, in order to improve fertility results in donkeys by artificial insemination.

Keywords

Anatomy · Sperm · Behavior · Seminal plasma · Metabolics · Semen preservation

J. Miró (✉) · J. Catalán
Equine Reproduction Service, Department of Animal Medicine and Surgery, Veterinary Faculty, Universitat Autònoma de Barcelona, Bellaterra (Cerdanyola del Vallès), Spain
e-mail: Jordi.miro@uab.cat

A. K. McLean et al. (eds.), *Donkey and Mule Medicine: Clinical Medicine and Reproduction*, https://doi.org/10.1007/978-3-032-18540-2_16

1 Anatomy

1.1 Penis and Prepuce

The penis is the organ of copulation and consists of 3 parts: (1) The root or bulb, attached to the tuber ischia, (2) The body or shaft, which is the main portion of the penis; and (3) The glans, the enlarged and free end of the penis. When erect, the penis doubles in length and thickness and the glans increases by 4 times. The jack's erect penis is larger than the stallion's penis (Fig. 1), and the glans presents a more pronounced dilatation (Miragaya et al. 2018).

The prepuce or sheath is a double invagination of the skin that covers the distal portion of the penis when not erect. Externally, near the preputial orifice, the jackass has 2 nipples, one on each side. The nipples are significant prominences while in the stallion they are smaller and in the inguinal position (Fig. 2).

1.2 Testicles and Epididymes

Donkey testes are ovoid structures contained in the scrotum. On average, donkey testicles are larger than horses of similar body weight (Table 1). The testicle is comprised of seminiferous tubules in which spermatogenesis occurs and interstitial tissues of which the Leydig cells produce testosterone. Produced spermatozoa are drained to the epididymis. The epididymis has 3 parts: head, body and tail. The head of the epididymis is closely attached to the testicle's anterior dorsal aspect. The body of the epididymis continues along the testicle's back side. The head and body of the epididymis are involved in spermatozoa maturation. Next is the epididymis tail, attached to the caudal part of the testicle and involved in the final spermatozoa maturation and storage. The donkey epididymis tail is significantly larger than that of the stallion. It is possible that the larger donkey testicular volume is related to the higher spermatozoa production and the larger epididymis tail is related to the

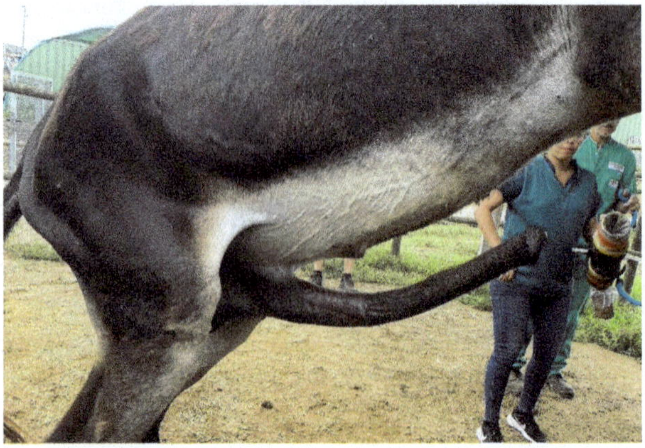

Fig. 1 Large erect penis of a jack

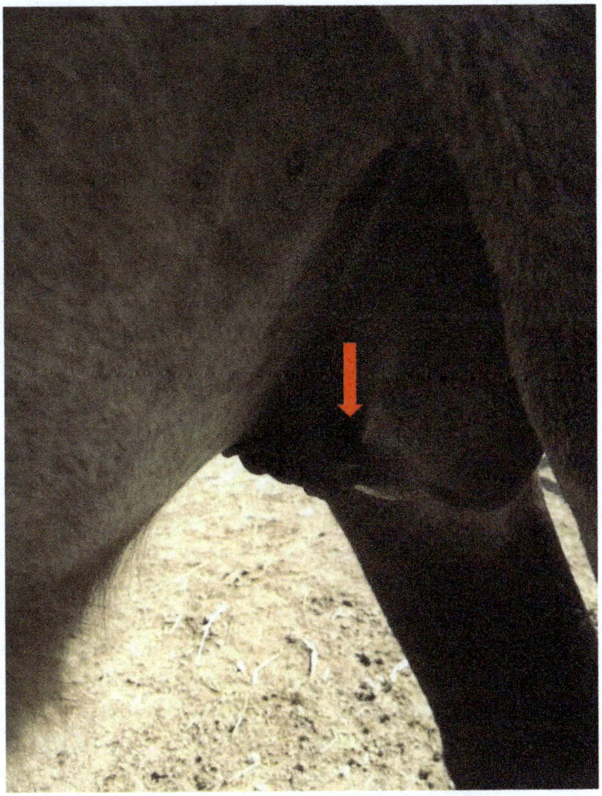

Fig. 2 Nipple in the prepuce of a jack (red arrow)

TESTICLES AND EPIDIDYMES

Table 1 Testicles sizes in Catalan and Ragusano donkeys compared to horses

	Catalan donkey (unpublished data of author)	Ragusano donkey (Quartucio et al. 2011)	Martina Franca donkey (Carluccio et al. 2013)	Stallion (unpublished data of author)
Long (cm)	10–12	8–11	9.7–10	8–12
High (cm)	7.5–8.5	5.8–8	6.9–7.3	6–7
Width (cm)	6.5–8	5.1–6.5	5.9–6.2	5

increased capacity for spermatozoa storage in the jack. Quartucio et al. (2011) observed a poor correlation between testicle size and the daily sperm output in 8 Ragusano donkeys. However, these authors suggest that increasing the number of males in the study might allow a correlation to be verified, as in other domestic species.

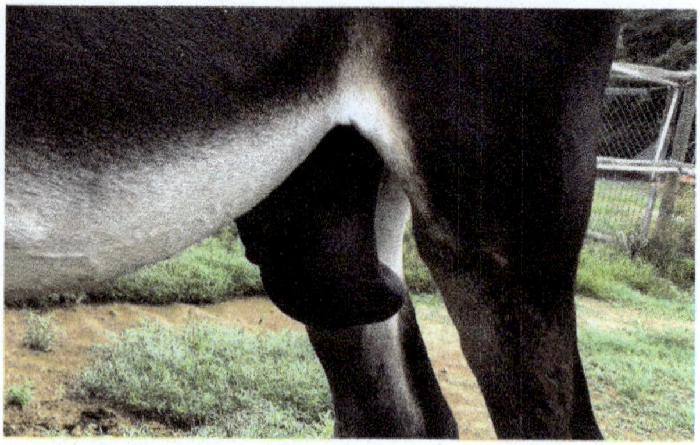

Fig. 3 Donkey pendulous testicles

The scrotum is an outpouching of skin in the inguinal area. It has a longitudinal midline fusing two sacs, one for each testis. The normal testicles should be freely movable within the scrotal sac. The scrotum's skin is soft and is very important in thermoregulation with a large amount of sweat glands. The donkey's scrotum is more pendulous than a stallion's but is not as pendulous as a bull's (Fig. 3). Within the scrotum, the testicles have a horizontal position or sometimes a slight cranio-caudal inclination. The larger donkey epididymis tail is visible on external examination of the caudal part of the scrotum (Fig. 4). This can help us to easily diagnose a 180° testicular torsion when it occurs in the donkey.

The ultrasonographic appearance of the normal testicle is similar in the stallion. The parenchyma is homogeneous, gray, and granular. A small amount (<5 mm) of anechoic fluid can be seen in the vaginal cavity between the visceral and parietal tunicae like in the stallion. The central vein of the testicle is usually visible and larger than the stallion's. The central vein crosses the center of the testicle as an anechoic duct (Fig. 5 and refer to Table 2).

The caudal part of the testicle is very easy to explore by ultrasonography due to the large tail of the epididymis in the jack. Because the epididymis is composed of a highly convoluted duct, its ultrasonographic appearance is very heterogeneous, representing multiple cross sections of ducts (fluid-filled lumen and epididymal wall) (Fig. 6).

1.3 Spermatic Cord

The spermatic cord extends from the abdominal inguinal ring to its attachment on the testis. It acts as passageway for the deferent duct, nerves and blood vessels. The cremaster muscle is lateral but is not part of the cord.

Blood is supplied to the testicle by the testicular artery. The testicular artery is a very tortuous vessel. The veins draining the testis form an intimate network around

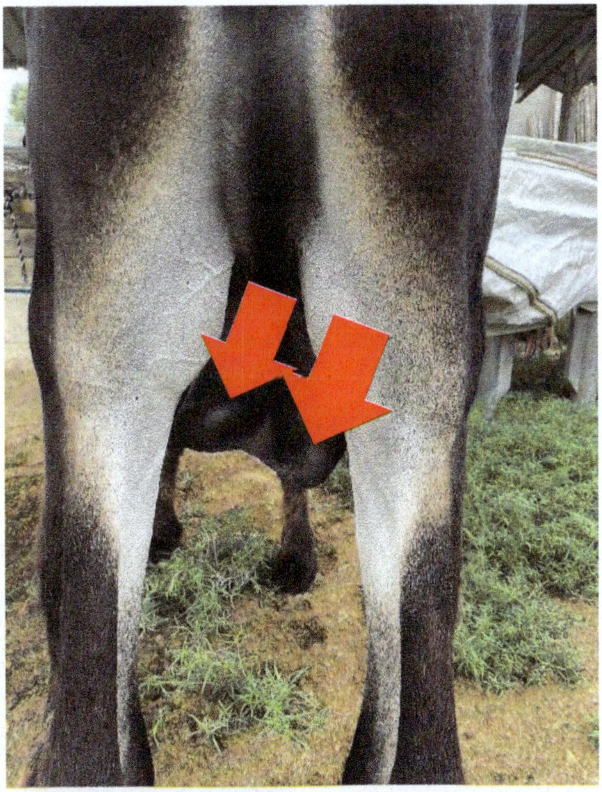

Fig. 4 The tails of the epididymis visible from the caudal view of the jack (red arrows)

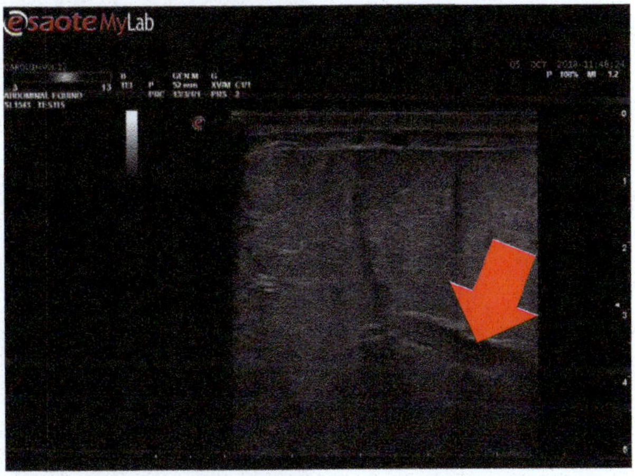

Fig. 5 Testicular parenchyma with central vein as seen on ultrasonographic examination of the testicle of a jack (red arrow)

the artery called the pampiniform plexus. This plexus serves as a countercurrent heat exchange system, decreasing the testicular arterial blood supply by 4–5 °C before it reaches the testicle. Due to this special vascularization system, the spermatic cord shows a very heterogeneous ultrasonographic image (Fig. 7).

Table 2 Different letters (a, b) in the same line indicates a significant difference ($P < 0.05$) (Gacem et al. 2020)

	Pampiniform Plexus Proximal A	Pampiniform Plexus Supratesticular B	Testis Periphery C
Resistive Index (RI)	2.91 ± 0.48^a	2.37 ± 0.57^b	1.27 ± 0.29^c
Pulsatility Index (PI)	0.87 ± 0.05^a	0.83 ± 0.05^b	0.67 ± 0.09^c
Total Arterial Blood Flow (TABF)	0.55 ± 0.02^a	0.46 ± 0.02^b	0.23 ± 0.01^c
Peak Systolic Velocity (PSV, cm/s)	72.79 ± 14.85^a	56.78 ± 18.3^b	27.51 ± 5.7^c
End Diastolic Velocity (EDV, cm/s)	7.74 ± 2.23^a	8.06 ± 2.39^a	8.17 ± 2.47^a

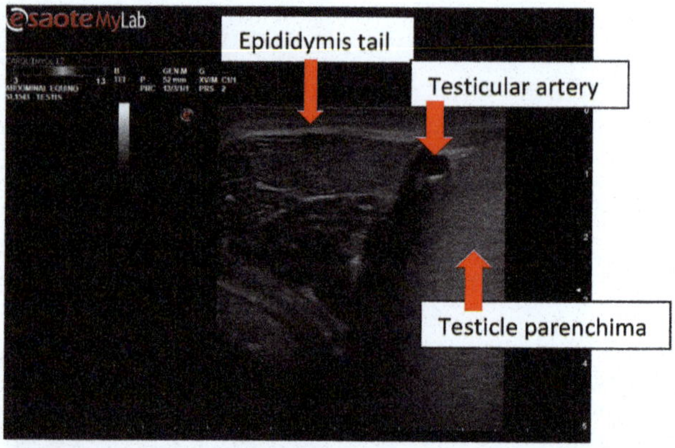

Fig. 6 Donkey tail of the epididymis on ultrasonographic examination. Labeled arrows point to the epididimis tail, testicular artery, and testicle parenchyma

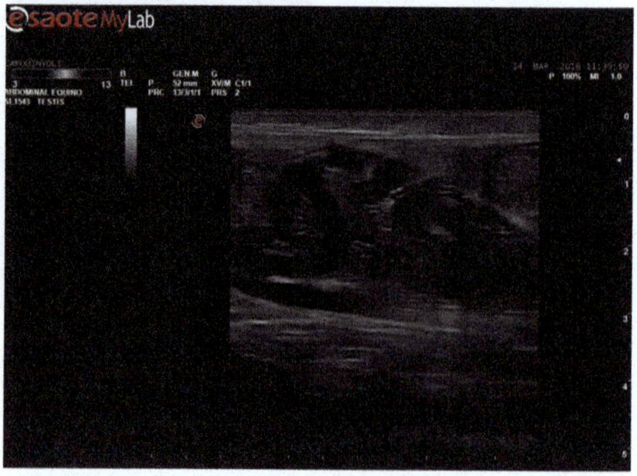

Fig. 7 Ultrasonographic image of the spermatic cord

Pulse Doppler ultrasound is a good technique to evaluate the vascular perfusion of the testicles in stallions and also as a tool for the diagnosis of chronic testicular dysfunction (Ortiz-Rodríguez et al. 2017). Gacem et al. (2020), evaluated 3 different locations of the jackass testicular artery: At the pampiniform plexus proximal (A), pampiniform plexus supra-testicular (B) and the testis periphery (C). The Doppler parameters evaluated were Resistive Index (RI), Pulsatility index (PI), Total Arterial Blood Flow (TABF), Peak Systolic Velocity (PSV) and End Diastolic Velocity (EDV). RI, PI, TABF and PSV decreased significantly to the testicle and resulted in a testis temperature and oxygenation decrease (Fig. 8). On the contrary, donkey

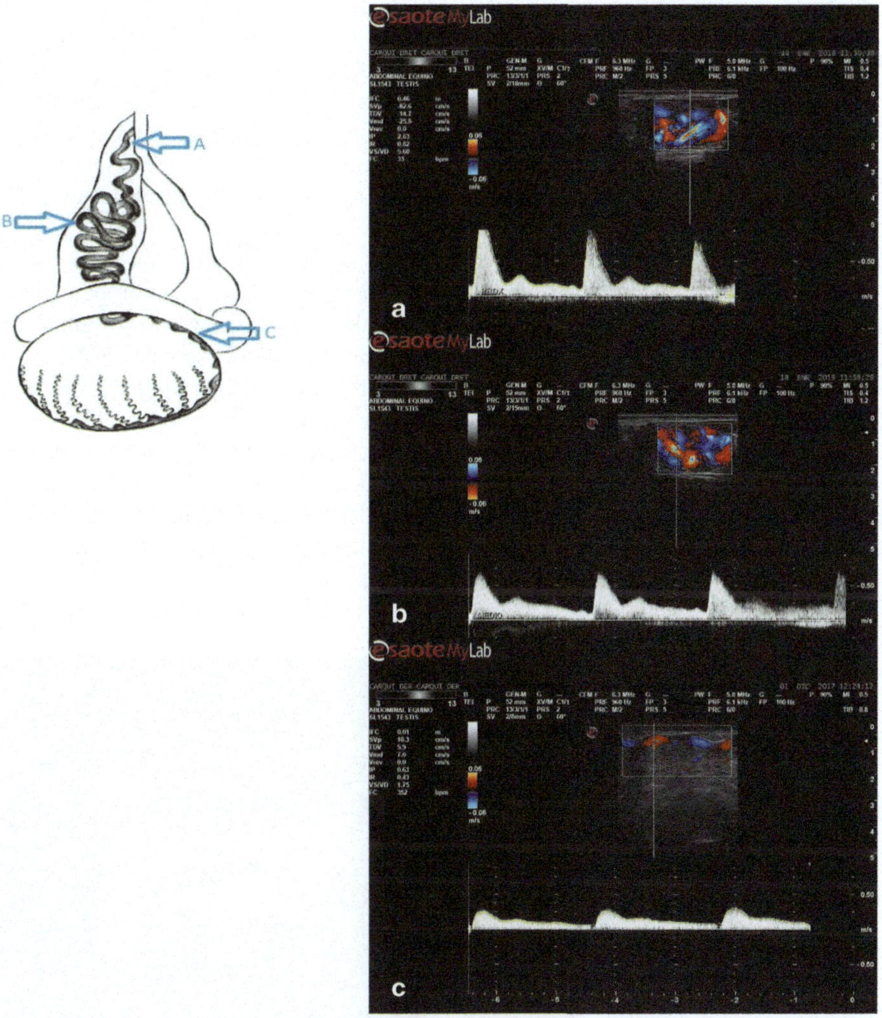

Fig. 8 Pulse Doppler ultrasound of the testicular artery. The left side shows an anatomical diagram with labeled parts: A, B, and C. The right side includes three ultrasound images labeled a, b, and c, displaying Doppler flow patterns and waveforms. Each ultrasound image shows color-coded blood flow and corresponding waveforms, indicating different flow velocities

testicular arterial blood flow was higher than that of the stallion, perhaps related to the relatively higher volume of testicular tissue in the jack. However, the value of PI was negatively correlated to sperm number and velocity. PI could be a good predictor of sperm motility and concentration in donkeys.

1.4 Deferent Duct

The deferent duct is a continuation of the epididymis and extends from the cauda epididymis through the spermatic cord to the pelvic urethra. The deferent duct of the donkey, as in the stallion, has an extremely thick wall. It can be palpated within the spermatic cord readily through the scrotal skin.

In its approach to the pelvic urethra, the deferent ducts enlarge significantly in diameter and are called ampullae. The ampulla of the paired deferent duct of the donkey are finger-shaped. Donkey ductus deferens ampullae are significantly higher than stallion's. The ampullae are directed caudally at the ventral aspect of the seminal vesicles and the prostate, and dorsal to the neck of the urinary bladder (Fig. 9).

Ampullae can be checked easily trasresctally by palpation or using ultrasonography just cranial to the prostate, showing two important differences compared to the stallion: size and ultrasound appearance. Donkey ampullae have a significantly larger diameter (25–35 mm *vs* 8–15 mm) (Table 3). However, the walls of the stallion ampullae show a grey homogeneous ultrasonographic image, while in the jack the walls of the ampullae appear irregular, with anechogenic zones, like a gland (Figs. 9 and 10). Abou-Elhamd et al. (2012) described the histology of the ampula of the deferent duct of the donkey and concluded that they consisted of a glandular enlargement of the terminal part of each duct. Therefore, ampullae are considered accessory sex glands (along with seminal vesicles, prostate and bulbouretral glans) producing the species-specific donkey seminal plasma.

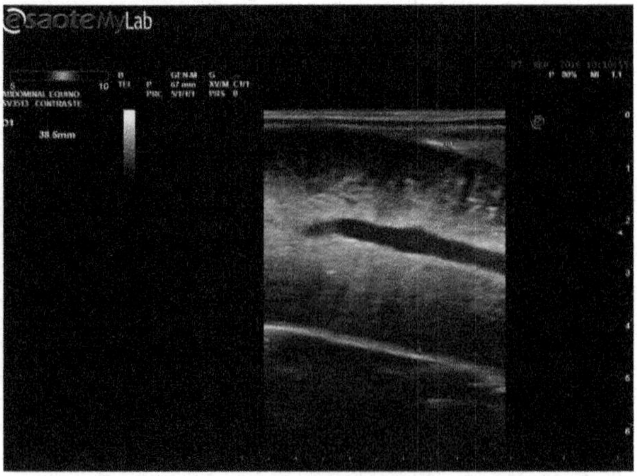

Fig. 9 Ultrasonographic image of the donkey ductus deferens ampulla

Table 3 Accessory sex gland size: Donkey *vs* horse

	Bulbourethal Glands Length × height, mm	Prostate Center height mm	Vesicular Glands Diameter mm	Ductus Deferens Ampulae Diameter mm
Donkey (Catalan donkey) (Gacem et al. 2020)	18.5 × 36.67	28.28	8.46	31.9
Horse (Pozor and McDonnell 2002)	12.2 × 14.3	14.1	34.7	9.1

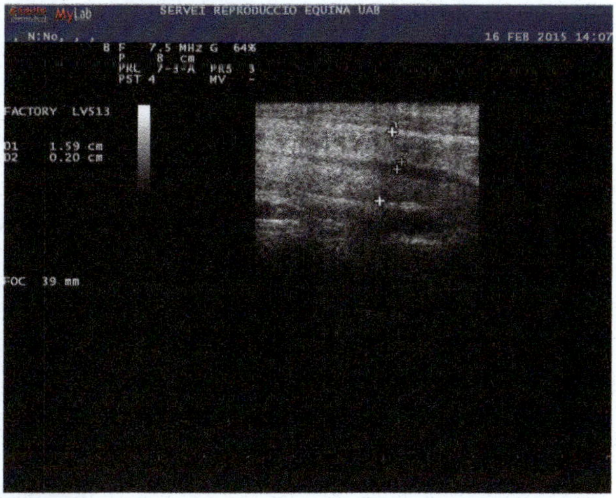

Fig. 10 Ultrasonographic image of the stallion ductus deferens ampulla

1.5 Accessory Glands

The vesicular glans (or seminal vesicles), prostate gland and bulbourethral glands are collectively called the accessory sex glands (Fig. 11). However, as explained before, donkey ductus deferens ampulla could also be included in this group. Together, they are involved in the seminal plasma production. These glands can be rectally palpated. The vesicular glands are the most difficult to palpate because of the great size of the adjacent ampulla and they collapse prior to teasing with a mare.

Rectally we can explore all accessory sex glands. Bulbourethral glands are the first glands moving in the caudo-cranial direction, just after the anus. The donkey bulbourethral glands, a paired gland (right and left), are larger and elliptic in shape in donkeys compared to those in horses which are round (Table 3) (Fig. 12).

The jackass prostate is on the midline just caudal to the ampullae (Fig. 13). It is larger than a stallion's (Table 3) with large anechogenic spaces without sexual stimulation (Fig. 14). The prostate gland of the donkey appears more active during the spring. This reactivity is decreased during other seasons of the year. Minimal reactivity was recorded during the winter (Abou-Elhamd et al. 2013).

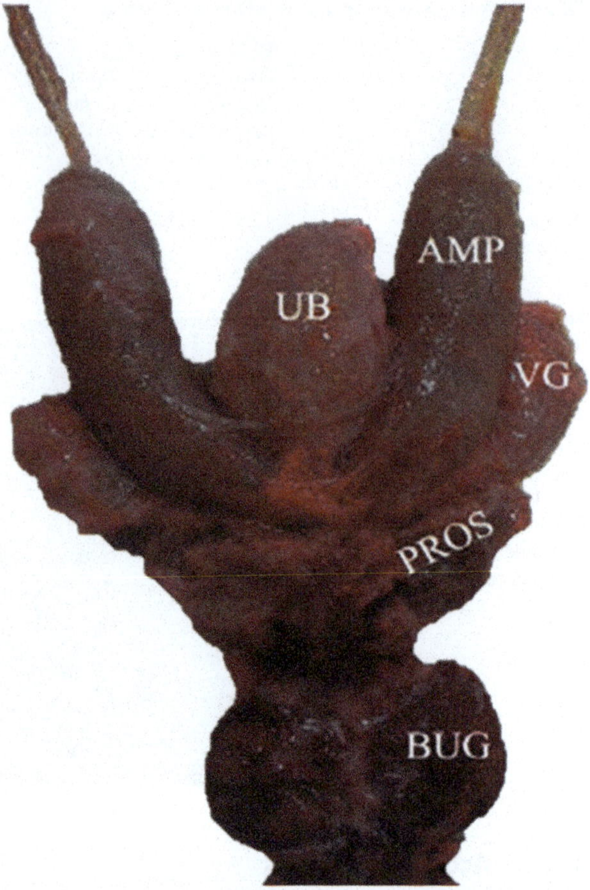

Fig. 11 Donkey accessory sex glands (UB urine bladder, AMP ductus deferens ampulla, VG vesicular gland, PROS prostate, BUG bulbourethral gland)

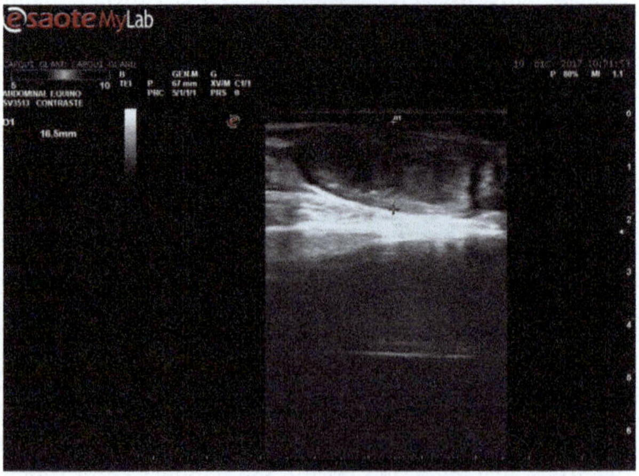

Fig. 12 An ultrasonographic image of the bulbourethral gland in a jack

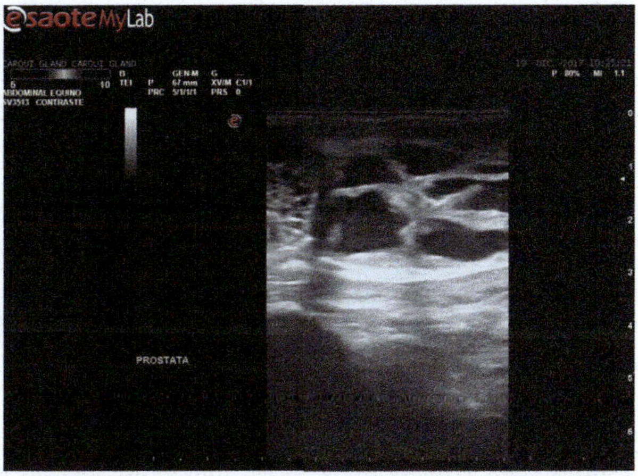

Fig. 13 Ultrasonographic image of the donkey prostate

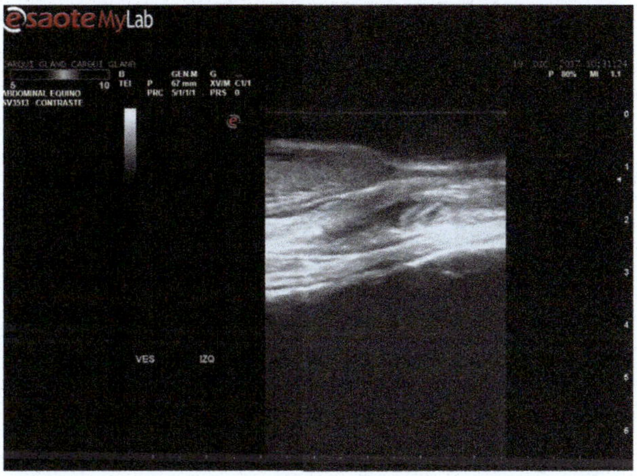

Fig. 14 Ultrasonographic image of the vesicular gland just under the end of the ampulla, in a jack

The donkey vesicular glands (or seminal vesicles) are paired, located slightly caudal and lateral to the ampullae (Fig. 10). They are thinner than those of the stallion. Without sexual stimulation they are usually homogeneous (Fig. 15), however with sexual stimulation the fluid in the lumen will increase.

2 Physiology & Behavior

In the wild, jacks are territorial and remain in the same territory if there is enough food and water (McDonnell 1998). Jacks defend their territory ferociously, particularly against other jacks. Occasionally, a jack may have another subordinate who can help in defending the territory (McLean et al. 2019).

Jennies are migratory animals in small groups (3–4 females, sometimes with offspring). They show very apparent and particular signs of heat. Mouth clapping

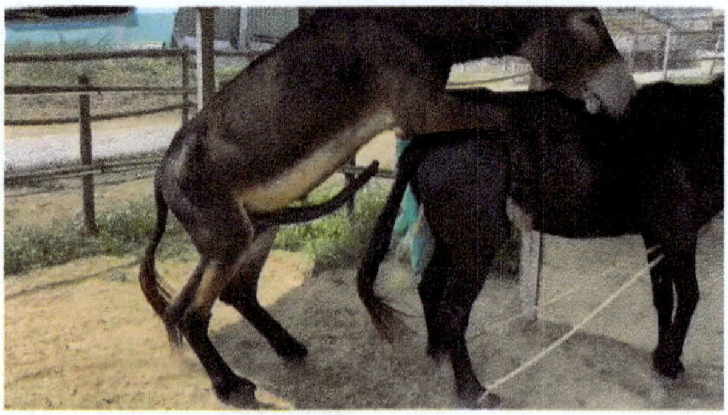

Fig. 15 Jackass breeding a mare

with sialorrhea and ears depressed against the extended neck is the most typical estrous sign. Usually, they show immobility and no aggression towards the jack. Occasionally, jennies show tail raising, urinating and contraction and relaxation of the labia vulva with eversion of the clitoris. But jennies in oestrous, within their group of females, also show heterotypical sexual behaviour, including chasing one another, mounting, the Flehmen response and sniffing. This type of behaviour is common among cows in oestrus, but not in mares (Taberner et al. 2008). Then, in the wild, jennies attract the attention of the jack and are bred by the male of the territory they cross. This seems like an interesting species strategy to ensure genetic variability.

The jackass natural breeding has 2 phases: a long precopular phase (5 to more than 30 min) with important and unique interactions with the jenny (much different than between stallion and mare) and a short copular phase ($\leq$ 1 min) (much quicker than stallion).

The precopular phase is initially intense. The male vocalizes, mounts without erection and bites the neck and ears. He descends and sniffs the perineal area and shows the Flehmen response and bites the flank, and extremities. Then the jack can mount the jenny 3–4 times without erection but with decreasing intensity. Between mounts the jack seems to lose interest in the female, remains distant and grazing (Taberner et al. 2008; McLean et al. 2019). Finally the jack raises the tail and exposes the penis, initially the external prepuce invagination and then the glans and the internal prepuce invagination. In erection the male approaches the female to copulate. But the complete erection of the glans will only occur inside the jenny's vagina.

In a controlled breeding situation and to avoid the intense precopular interaction, jacks can be stimulated near a teaser or a female in estrus, kept at a distance until erection and then conducted to mount the female or dummy.

During the copular phase, after 4–5 lumbar thrusts, the jackass ejaculates, then becomes relaxed, lowers the tail and dismounts slowly, showing a mucous secretion on the urethral orifice.

Naturally, jackasses show more attraction to jennies than mares. Only a reduced number of mares (31–41%) display estrus to the jack and the estrus signs are different (Canisso et al. 2010). In general, for donkeys to accept breeding mares, it is best that they do not breed jennies and that they live with mares from a young age (Fig. 15). Be careful because some mares, even in heat, can be aggressive towards the jack. To avoid accidents, it is best to resort to artificial insemination .

The jackass has efficient sperm production, with a short spermatogenesis (47.5 days) and good semen quality (Quartuccio et al. 2011).

3 Semen Collection

The collection of semen in the jackass is similar to that of the horse. However, the time spent is usually much higher because, as explained, jackasses have a long preovulatory phase. Young males are usually faster than adults (Canisso et al. 2019).

When we collect semen in a place where another male has previously been, the collection time can be lengthened. The jack displays the flehmen response, especially if there are feces from previous donkeys. In this case, it is normal for him to defecate on top the existing feces to mark his territory. Possibly this is the result of the territorial behaviour explained before.

Jackasses can be easily trained to collect on a dummy. If they are not trained or breed jennies naturally, we can use a jenny in estrous or not. For donkeys accustomed to breeding mares we can also use a mare. But we can also easily collect semen via ground of a stimulated jack in front of a jenny.

The same types of artificial vaginas used for stallions are useful for donkeys.

Semen collection in a jackass can last up to 90 min. The PGF2α analogue cloprostenol sodium (125 μg i.m.) given immediately upon exposure to a jenny in estrous hastens the time to collect semen (9.4 ± 2.0 min) with no effects on semen quality (Panzani et al. 2020).

4 Spermatozoon

In general donkey semen has good quality (Table 4). On average, semen volume is similar to that for a stallion. But donkey sperm concentration and sperm motility are higher, while sperm abnormalities are lower than for stallions.

L-lactate production indicates the metabolic status of spermatozoa. Donkey spermatozoa show a high velocity, with high metabolic activity. L-lactate production is a good predictor of donkey semen quality, strongly correlated with other important variables such as sperm count, HOST (hypoosmotic swelling test) or motility parameter evaluated by a computerized system (CASA) (Miró et al. 2005).

Evaluating sperm morphometry by a computerized system (ASMA), the donkey spermatozoid head is smaller than that in stallions, while the midpiece is significantly larger in donkey spermatozoid (Miró et al. 2008) (Table 5). Spermatozoa midpiece contains the mitochondria, and donkey shows higher mitochondrial activity than horses evaluated by flow cytometry (Yeste et al. 2015; Miró and Papas 2018). As a result of this and as explained before, donkey spermatozoa move faster

Table 4 General characteristics of Catalonian donkey semen (Miró et al. 2005)

	Mean ± S.E.	Confidence interval (95%)
Filtered volume (mL)	68.40 ± 16.59	55.68 ± 57.55
Sperm count (_10^6 /mL)	280.88 ± 228.94	271.12 ± 290.63
Total motility, VAP > 10 mm/s (%)	68.40 ± 16.59	67.73 ± 69.06
Ph	7.77 ± 0.35	7.75 ± 7.78
Sperm viability (%)	70.06 ± 16.46	69.40 ± 70.72
Sperm immature tail (%)	10.22 ± 4.80	10.03 ± 10.41
Sperm coiled-tail (%)	1.22 ± 1.16	1.17 ± 1.26
Sperm head abnormality (%)	2.5 ± 2.37	2.41 ± 2.60
Tailless spermatozoon (%)	2.95 ± 2.17	2.86 ± 3.04
Immature sperm with proximal cytoplasmic droplet (%)	1.36 ± 1.42	1.31 ± 1.41
Immature sperm with distal cytoplasmic droplet (%)	0.39 ± 0.55	0.37 ± 0.42
Total abnormalities	18.99 ± 8.62	18.30 ± 18.99
L-lactate production (mmol/mg de protein _ 60 min)	0.85 ± 1.32	0.74 ± 0.95
HOST test (%) (hypoosmotic swelling test)	24.19 ± 19.26	22.64 ± 25.74
Sperm curvilinear velocity (VCL, μm/s)	80.20 ± 51.72	78.21 ± 82.20
Sperm linear velocity (VSL, μm/s)	49.80 ± 43.18	48.13 ± 51.46
Mean velocity (VAP, μm/s)	59.39 ± 43.04	57.74 ± 61.05
Linear coefficient (LIN, %)	60.50 ± 28.79	61.60 ± 59.39
Straightness coefficient (STR, %)	79.21 ± 24.50	78.28 ± 80.14
Wobble coefficient (WOB, %)	71.97 ± 20.27	71.20 ± 72.74
Mean lateral head displacement (mean ALH, μm)	2.36 ± 2.05	2.28 ± 2.44
Frequency of head displacement (BCF, Hz)	12.71 ± _ 5.20	12.51 ± 12.91
Minor harmonic oscillation of the head (HLO, μm)	0.49 ± 1.30	0.44 ± 0.54

Table 5 Computerized sperm morphometry. Donkey *vs* Horse (Miró et al. 2008)

	Donkey	Horse
LH (head length, μm)	5.45 ± 0.53	5.67 ± 0.36
WH (head width, μm)	2.67 ± 0.21	2.85 ± 0.31
AH (head area, $μm^2$)	11.56 ± 1.32	13.42 ± 1.72
WMP (midpiece width, μm)	1.19 ± 1.09	0.81 ± 0.27
AMP (midpiece area, $μm^2$)	3.21 ± 0.99	1.63 ± 0.56
AHMP (angle head-midpiece, °)	8.76 ± 17.46	6.28 ± 6.34

than horse spermatozoa, but produce higher Reactive Oxygen Species (ROS) which are capable of inducing cellular damage (Miró and Papas 2018).

HOST test is a good method to evaluate the spermatozoa membrane integrity. This test shows that donkey spermatozoa are more sensitive to a hypoosmotic medium than are stallion spermatozoa (Caiza de la Cueva et al. 1997; Miró et al.

2005). Rota et al. (2010) observed that the simple incubation for 5 or 45 min at 37 °C of one part of semen with three parts of bi-distilled water is an applicable HOS-test in the semen analysis of donkey spermatozoa.

Several studies analyzed the seasonal influences on jenny reproductive activity (Ginther et al. 1987; Blanchard et al. 1999; Carluccio et al. 2003; Taberner et al. 2008), showing less influence than in mares. However, the literature pertaining to seasonal influences on jackass reproduction is limited. Carluccio et al. (2013) in Martina Franca jackasses observed no differences in testicular morphometric characteristics during the year, but a certain influence on reaction time and some seminal parameters. A lower reaction time was observed in the spring and summer compared with autumn and winter. On the other hand, they observed that the total volume was significantly higher in winter compared with all the other seasons; gel-free volume was higher in winter compared with summer and autumn. Mean sperm concentration was significantly lower in winter compared with spring and summer. Total and progressive motility and membrane integrity did not show any significant difference between seasons. A lower average path velocity was observed in autumn compared with spring and summer, and in summer a higher straight-line velocity was detected compared with spring and autumn, and a lower curvilinear velocity compared with winter. No differences in amplitude of lateral head displacement, beat cross frequency, straightness and linearity were observed between seasons. Seminal pH did not show any seasonal difference; also sperm morphology did not show any significant difference during the whole year (Table 6).

The computerized analysis of sperm motion (CASA) and the statistical analysis of obtained data identified different sperm subpopulations with different motility characteristics in several mammalian species (Quintero Moreno et al. 2003). Donkeys, like other mammals, have four subpopulations of spermatozoa with different motility patterns (Miró et al. 2005) (Table 7). The characteristics of donkey sperm motility subpopulations are very different in comparison with stallion sperm motility subpopulations (Quintero-Moreno et al. 2003). This means that donkey spermatozoa are faster but also have a different kind of motility compared to horse spermatozoa.

Flores et al. (2008) observed that freezing/thawing induced motility changes, but did not change the existence of 4 motile subpopulations in two phylogenetically different mammalian species: boar or donkey ejaculates. The characteristics of sperm motility patterns of the different sperm subpopulations could play some role in the maintenance of the overall function of mammalian ejaculates and, possibly, provide a good overall indicator of semen quality.

5 Seminal Plasma

Seminal plasma is a complex fluid produced by sexual accessory glands and ductus deferens ampulla as previously described. Usually, semen analysis includes only the cells (spermatozoa) analysis; however, the role of seminal plasma in the fertilization process is very important. Each species has its own reproductive strategy, and the

Table 6 Values (mean ± SD) of semen parameters in the seven considered Martina Franca jackasses in the different seasons (Carluccio et al. 2013)

	Winter	Spring	Summer	Autumn
Total volume (mL)	99.7 ± 31.35[a]	86.1 ± 23.50[b,c]	82.5 ± 19.77[c]	80.5 ± 18.29[c]
Gel-free volume (mL)	77.3 ± 21.55[a]	69.6 ± 18.90[b]	65.1 ± 18.82[b, c]	64.3 ± 15.59[c]
Concentration (10^6/mL)	285.9 ± 69.32[a]	319 ± 73.22[b]	329.3 ± 66.07[b,c]	310.8 ± 64.67
pH	7.25 ± 0.06	7.25 ± 0.06	7.25 ± 0.05	7.24 ± 0.05
Total motility (%)	82.5 ± 6.05	81.9 ± 5.45	81.2 ± 5.61	81.4 ± 5.42
Progressive motility (%)	73.3 ± 6.16	73.5 ± 6.76	72.8 ± 6.73	73 ± 6.28
VAP (mm/s)	123.7 ± 15.48[b]	122.3 ± 13.83[b]	124.5 ± 13.11[b]	116.8 ± 15.65[a]
VSL (mm/s	101.6 ± 6.3[b]	100.1 ± 7.01[b]	102.9 ± 5.09[a]	99.6 ± 5.52[b]
VCL (mm/s)	198.3 ± 20.38[b]	194.2 ± 22.36[b]	188.1 ± 18.75[a]	194.7 ± 18.73[b]
ALH (mm)	6.6 ± 0.73	6.6 ± 0.68	6.8 ± 0.70	6.4 ± 1.46
BCF (Hz)	40.9 ± 2.5	40.5 ± 2.83	39.7 ± 2.81	40.5 ± 2.97
STR (%)	86.4 ± 3.26	85.9 ± 3.9	84.9 ± 4.92	85.9 ± 4.53
LIN (%)	57.4 ± 4.69	58.6 ± 5.81	59.4 ± 5	58.1 ± 4.80
Membrane integrity (%)	81.3 ± 5.8	80.4 ± 5.71	80.3 ± 5.47	80.4 ± 5.37
Total abnormalities (%)	3.72 ± 0.73	3.70 ± 0.71	3.69 ± 0.72	3.71 ± 0.61
Primary abnormalities (%)	1.27 ± 0.45	1.25 ± 0.42	1.26 ± 0.38	1.23 ± 0.37
Secondary abnormalities (%)	2.45 ± 0.47	2.45 ± 0.48	2.42 ± 0.46	2.48 ± 0.41

Letters indicate: a versus b, $P < 0.05$; a versus b, c, $P < 0.01$; a versus c, $P < 0.001$
Abbreviations: *ALH* amplitude of lateral head displacement, *BCF* beat cross frequency, *LIN* linearity, *STR* straightness, *VAP* average path velocity, *VCL* curvilinear velocity, *VSL* straight line velocity

role of seminal plasma is different. The interest in this field is increasing, however much still remains unknown.

Detailed analysis of donkey seminal plasma compounds is lacking. The main parts are proteins, plus sugars like fructose or antioxidants like melatonin, ascorbic acid or glutathione.

The total amount of proteins in donkey seminal plasma is significantly higher than in horses (48.00 ± 28.49 *vs* 8.70 ± 4.81 mg/mL). Some proteins are involved in the control of jenny endometrial reaction postbreeding or insemination (Miró et al. 2018). Other proteins act as antioxidants but the main role of proteins remains unknown.

The jenny postbreeding endometrial reaction results in a large number of polymorphonuclear neutrophils (PMNs) in the uterine lumen (Miró et al. 2013; Vilés et al. 2013a, b). These PMNs are activated by seminal plasma, degranulating and releasing the granules which contain, Elastase, Mieloperoxidase and Catepsine.

Table 7 Sperm subpopulations and motility descriptors in Catalonian donkey semen

Sperm motility descriptors	Sperm subpopulations			
	1	2	3	4
N	845	845	447	818
(%)	31.8	20.6	16.8	30.8
VAP (μm/s)	114.96 ± 0.74^a	25.37 ± 0.98^c	65.03 ± 1.05^b	25.73 ± 0.84^c
LIN (%)	86.85 ± 0.61^a	29.84 ± 0.80^c	28.38 ± 0.87^c	69.70 ± 0.70^b
ALHmed (μm/s)	1.94 ± 0.06^c	2.64 ± 0.08^b	4.94 ± 0.08^a	1.36 ± 0.07^d
MADalg (μm)	-1.68 ± 0.64^c	-28.40 ± 0.82^d	12.85 ± 0.89^a	0.87 ± 0.70^b
BCF (Hz)	13.54 ± 0.20^a	13.27 ± 0.25^a	11.93 ± 0.27^b	11.01 ± 0.21^c
HLO (μm)	0.66 ± 0.05^a	0.40 ± 0.06^b	0.81 ± 0.07^a	0.23 ± 0.05^b

Abbreviations: *VAP* average path velocity, *LIN* linearity, *ALMed* mean lateral head displacement, *MADalg* algebraig angular mean displacement, *BCF* beat cross frequency, *HLO* minor harmonic oscillation of the head

Different superscripts (a, b, c and d) in the same row indicate significant differences ($P < 0.05$). Results are expressed as means ± S.E. from 78 semen samples from four Catalonian donkeys. The total number of motile spermatozoa analyzed was 2556

Table 8 Spermatozoa intracellular ROS (peroxides levels, H2DCFDA/PI)

Without seminal plasma	27.46 ± 16.65
With seminal plasma	8.86 ± 3.27

These enzymes induce the extrusion of their DNA, which together with nuclear and cytoplasmic proteins form web-like structures outside the cell called neutrophil extracellular traps (NETs) (Mateo-Otero et al. 2021; Miró et al. 2021). Most of the sperm are caught for these traps and most of them remain alive with energetic tail movements. And some, after 3 h, are released with excellent motility patterns (Miró et al. 2020). A DNAse present in seminal plasma can promote the digestion of NETs (Alghamdi and Foster 2004). This mechanism appears to be important in sperm selection and, therefore, in the reproductive strategy of donkeys.

Sperm movement metabolism produces reactive oxygen species (ROS), but NETs produce 4–5 times more ROS.

Seminal plasma antioxidants control the oxidative stress on spermatozoa induced by ROS produced by the spermatozoa and by the endometrial reaction post AI (Table 8).

As explained, seminal plasma has non-enzymatic and enzymatic mechanisms to control the ROS. Melatonin is usually associated with reproduction seasonality in several species; however the presence of melatonin in seminal plasma is not associated with this seasonality. Donkeys have seminal plasma melatonin levels lower than dogs or boars and significantly lower than bulls (Table 9). The main role of melatonin in seminal plasma may be to protect spermatozoa from oxidative damage (González-Arto et al. 2016).

Table 9 Concentration of melatonin (pg/mL) in seminal plasma produced by the donkey, stallion, boar, bull and dog (González-Arto et al. 2016)

	Mean ± SEM (pg/mL)	Range (pg/mL)
Donkey (n = 4)	2.82 ± 0.38[b]	0.68–6.47
Stallion (n = 4)	2.48 ± 0.14[b]	1.24–3.39
Boar (n = 6)	9.34 ± 1.38[b]	1.07–26.71
Bull (n = 6)	19.10 ± 7.37[a]	0.74–88.03
Dog (n = 7)	6.06 ± 2.28[b]	0.74–46.22

n number of analyzed males. Different letters (a, b) account for significant differences between species ($p < 0.05$)

Table 10 MT1 and MT2 receptor distribution in the donkey, stallion, boar, bull, deer and dog spermatozoa, assessed by indirect immunofluorescence (González-Arto 2016)

	Head			Flagellum		
	Acrosome	Equatorial band	Postacrosome	Neck	Midpiece	Tail
Donkey	1,2	[1], 2	2	2	2	2
Stallion	1,2	1	1, [2]	1, [2]	1, [2]	1, [2]
Boar		1		1,2	1	
Bull	[1]	[1]	[1]	[1], 2	1	1
Deer	[1]	[1]	[1]	[1], 2	1	1
Dog	[1]	[1]	[1]	1	1, 2	

Brackets indicate that in that location, the receptor was not detected in all the spermatozoa of the sperm sample. The absence of a number indicates that there was no immunostaining in that sperm region

Table 11 Total enzymatic activity

	SOD (U/mL)	GPx (U/L)	GR (U/L)	CAT (U/L)
Donkey	1845.15 ± 1368.25	667.40 ± 625.81	79.60 ± 29.02	11740.89 ± 2700.04
Horse	223.05 ± 67.71	249.80 ± 11.23	17.40 ± 8.64	621.46 ± 1561.94

Two kinds of melatonin receptors (MT1 and MT2) were observed in spermatozoa from several species; however, the presence and distribution of these receptors is very different between species (Table 10) (González-Arto et al. 2016).

The antioxidant enzyme defense system, comprising superoxide dismutase (SOD), glutathione reductase (GR), glutathione peroxidase (GPx) and catalase (CAT), has been described in sperm of several species (Alvarez and Storey 1989; Beconi et al. 1993; Marti et al. 2003). Donkeys and horses showed high SP plasma enzymatic activity (Table 11) compared to other species like sheep (Marti et al. 2003). Donkeys show a significantly higher enzymatic activity than horses that seems to be related to higher spz activity and ROS production (Miró and Papas 2018). However, when SP enzymatic activity is evaluated in relation to total SP protein, the role of each enzyme is different between both species. GPx and GR are relatively more important in horse SP, while SOD and CAT is in donkeys (Table 12). The ROS control seems to be species-specific.

Jackass Seminal Plasma proteomics isolated more than 2000 different proteins (unpublished data), but proteins between 30 and 100 kDa, rather than entire seminal

Table 12 Enzymatic activity in relation to seminal plasma protein

	Total Protein (mg/mL)	SOD (U/mg)	GPx (U/g)	GR (U/g)	CAT (U/g)
Donkey	48.00 ± 28.49	49.65 ± 26.56	14.02 ± 4.08	0.91 ± 0.67	194.84 ± 184.83
Horse	8.70 ± 4.81	25.34 ± 10.20	25.06 ± 4.77	1.69 ± 1.20	76.25 ± 75.83

plasma, appear to be involved in the control of the jenny postbreeding endometritis (Miró et al. 2021). Included in this protein group there are known proteins, like SOD, CAT, GPx, GSR, DNAse, LDH (lactate deshydrogenase), and a lot of unknown proteins.

Metabolomics of donkey seminal plasma was evaluated by means of nuclear magnetic resonance. Important differences between horse and donkey were detected. Lactate concentrations are much higher in horses, but many amino acids or derivatives are higher in donkeys. Horse spermatozoa is dependent on seminal plasma glucose levels but donkey spermatozoa are not, indicating that the sources used by the sperm are different between both species. These findings may be important in the design of extenders for donkey cooled or frozen semen (Catalán et al. 2023).

On the other hand, recently donkey seminal plasma microvesicles were isolated and characterized (Catalán et al. 2025). Seminal microvesicles are secreted by functional cells of the male genital tract. These vesicles are nano-sized (30–1000 nm) lipid bilayer-enclosed particles that transport bioactive molecules, including proteins, lipids, and nucleic acids and play a key role in modulating reproductive processes, including sperm function and immune response in the female genital tract. Further studies are needed in order to understand the specific role of these microvesicles in the donkey reproductive strategy.

6 Sperm Preservation

6.1 Semen Cooling

As we mentioned previously seminal plasma contains a lot of proteins. Most of them are beneficial. However, there are also proteins with detrimental effects on spermatozoa. These proteins make contact with the spermatozoa during ejaculation, interact with the spermatozoid membrane, and influence different characteristics, including motility. These undesirable effects can be reduced by Jackass semen dilution or the total or partial removal of seminal plasma for cooled semen (Miró et al. 2009).

In practice it is common to use horse protocols, a 1:3 dilution in horse extenders when preparing cooled donkey semen for shipment in 20 mL per dose. Or, alternatively, semen can be extended 1:5 with a skim milk-based extender, centrifuged (600 g for 15 min at 20 °C) and the pellet resuspended in a horse extender to 50 to 100 million/mL in 20 mL per dose (Canisso et al. 2019). Donkey semen doses are cooled at 4–7 °C and shipped in commercial styrofoam semen boxes.

Horse semen extenders are normally also used for donkeys. However, as explained before, there are important differences between both species. The amount of proteins, and specially of antioxidants, are significantly higher in donkey seminal plasma (Papas et al. 2019). Horse spermatozoa is dependent on seminal plasma glucose levels, which is not the case in donkeys. This means that donkey spermatozoa are depending on other sources of energy (Catalán et al. 2023). The impact of ROS on donkey sperm functionality is different than on horse sperm. Donkey sperm tolerate high Reduced Glutathione (GSH) concentrations (8 mM and 10 mM) but horse sperm suffer significant damage with more than 4 mM (Yánez-Ortiz et al. 2021). Further studies are needed in order to improve donkey cooled semen preservation and the artificial insemination results.

6.2 Semen Freezing

Protocols and extenders used to process donkey semen for freezing are also adapted from stallions with the same problem, jackass sperm morphology, motility, metabolism and physiology are different than in stallions.

In our lab (Autonomous University of Barcelona, Spain) donkey semen is processed as described above. Firstly, a good quality ejaculate should be obtained and analysed for sperm concentration by an haemocytometer or an spectrophotometer, for motility by a CASA system, and for sperm viability and abnormalities by an eosin-nigrosin stain. Semen is rapidly diluted 1:5 (sperm:extender) by a skim-milk extender and allowed in 50 mL Corning tubes. A high dilution factor will protect spermatozoa from lactic acid and ROS, as well as some deleterious seminal plasma proteins. In addition, using a 1:5 sperm:extender dilution ratio when eliminating seminal plasma by centrifugation seems to provide better results than those reported using a 1:1 ratio. The sperm damage decreases and the recovery efficacy increases (Miró et al. 2005).

Diluted semen samples are placed in a water bath at 20 °C during semen evaluation and handling. In order to eliminate the seminal plasma, diluted semen is centrifuged at 600 g 15 min at 20 °C. The supernatant is discarded and the resultant pellet is resuspended in a small volume of freezing semen extender (2–4 mL depending on pellet volume). The pellet is carefully homogenized by a micropipette and then analyzed for sperm concentration and viability. Then the sperm concentration is adjusted at 200×10^6 spz/mL. Finally, semen samples are packaged in 0.5 mL straws, refrigerated, and frozen using an automatic freezer. Once the temperature reaches −120 °C, the straws are plunged into liquid nitrogen and stored in tanks until their further use.

Post-thaw donkey sperm survival results are good (40–60%). These rates are similar or better than those obtained for stallion frozen/thawed semen (Sabatini et al. 2014; Flores et al. 2008). Donkey cryopreserved spermatozoa are able to penetrate zona pellucida-free bovine oocytes *in vitro* (Taberner et al. 2010), and it's also able to fertilize mares by artificial insemination to produce mules. However, low conception rates are obtained when jennies are inseminated with jackass frozen/thawed semen (Vidament et al. 2009; Oliveira et al. 2012; Rota et al. 2012).

There are few studies trying to increase the conception rate by artificial insemination in donkeys, some try to modify the freezing media, others the sperm concentration or add seminal plasma (Oliveira et al. 2012; Rota et al. 2012). More studies are needed to improve obtained results taking into account the recent scientific discoveries on donkey reproductive strategy, sperm metabolism and physiology, seminal plasma role, proteomics, metabolomics, etc., and jenny post-artificial insemination endometrial reaction or preovulatory follicular fluid role.

Study Questions

1. Which of the accessory sex glands is smaller in the donkey than in the horse?
   (a) Prostate
   (b) Vesicular glands
   (c) Bulbouretral glands
   (d) Ductus deferens ampullaes
2. The jackass precopular phase
   (a) is shorter than the copular phase
   (b) is characterized by an intense interaction with the jenny
   (c) can be increased by prostaglandins administration
   (d) is longer in spring
3. The seminal plasma enzymatic antioxidants control
   (a) the reactive oxygen species (ROS) produced by sperm
   (b) the ROS produced by the jenny endometrial reaction post-AI
   (c) booth (a and b)
   (d) the oxydative stress produced by oocyte fertilization
4. Donkey sperm metabolism
   (a) is depending on glucose
   (b) produce lactic acid
   (c) is reduced after ejaculation
   (d) produce ROS
5. During donkey semen freezing process
   (a) Seminal Plasma is eliminated
   (b) Semen is maintained at 25 °C for 24 h.
   (c) Protocols are significantly different than for stallion
   (d) Spermatozopa are rapidly vitrified

References

Abou-Elhamd A, Salem A, Selim A (2012) Histological and histochemical studies on the ampulla of the deferent duct of donkey (*Equus asinus*). J Adv Vet Res 2:261–270

Abou-Elhamd A, Ahmed O, Salem A, Selim A (2013) Histomorphological studies on the prostate gland of donkey Equus Asinus during different seasons. J Hist. https://doi.org/10.1155/2013/643287

Alghamdi AS, Foster DN (2004) Seminal DNAse frees spermatozoa entangled in neutrophil extracellular traps. Biol Reprod 73:1174–1181

Alvarez J, Storey B (1989) Role of glutathione peroxidase in protecting mammalian spermatozoa from loss of motility caused by spontaneous lipid peroxidation. Gamete Res 23:79–90

Beconi M, Francia C, Mora N, Affranchino M (1993) Effect of natural antioxidants on frozen bovine semen preservation. Theriogenology 40:841–851

Blanchard TL, Taylor TS, Love CL (1999) Estrous cycle characteristics and response to estrus synchronization in mammoth asses (*equus asinus americanus*). Theriogenology, 52(5): 827–834. https://doi.org/10.1016/S0093-691X(99)00175-2

Caiza de la Cueva F, Pujol MR, Rigau T, Bonet S, Miró J, Briz M, Rodríguez-Gil JE (1997) Resistance to osmotic stress of horse spermatozoa: the role of ionic pumps and their relationship to cryopreservation process. Theriogenology 48:947–968

Canisso IF, Carvalho G, Morel MD (2010) Sexual behaviour and ejaculate characteristics in Pêga donkeys (*Equus asinus*) mounting estrous horse mares (*Equus caballus*). Theriogenology 72: 56–63

Canisso IF, Panzani D, Miró J, Ellerbrock RE (2019) Key aspects of donkey and mule reproduction. Vet Clin Equine 35:607–642

Carluccio A, Pau S, Picci GF, Tosi U (2003) Ciclo estrale nell'asina di Martina Franca: osservazioni comportamentali ed ecografiche. Veterinaria Pratica.

Carluccio A, Panzani S, Contri A, Bronzo V, Robbe D, Veronesi MC (2013) Influence of season on testicular morphometry and semen characteristics in Martina Franca jackasses. Theriogenology 79:502–507

Catalán J, Yánez I, Martínez-Rodero I, Mateo Y, Nolis P, Yeste M, Miró J (2023) Comparison of the metabolite profile of donkey and horse seminal plasma and its relationship with sperm viability and motility. Res Vet Sci 165:105046

Catalán J, Martínez-Díaz P, Parra A, Yeste M, Roca J, Barranco I, Miró J (2025) Isolation and characterization of extracellular vesicle subsets in donkey seminal plasma. https://doi.org/10.2139/ssrn.5117870

Flores E, Taberner E, Rivera MM, Peña A, Rigau T, Miró J, Rodríguez-Gil JE (2008) Effects of freezing/thawing on motile sperm subpopulations of boar and donkey ejaculates. Theriogenology 70:936–945

Gacem S, Papas M, Catalán J, Miró J (2020) Examination of jackass (*Equus Asinus)* accessory sex glands by B-mode ultrasound and of testicular artery blood flow by colour pulsed-wave Doppler ultrasound: correlations with semen production. Reprod Domest Anim 55:1–8. https://doi.org/10.1111/rda.13604

Ginther OJ, Scraba ST, Bergfelt DR (1987) Reproductive seasonality of the jenney. Theriogenology, 27(4):587–592. https://doi.org/10.1016/0093-691X(87)90053-7

González-Arto M, Vicente-Carrillo A, Martínez-Pastor F, Fernández-Alegre E, Roca J, Miró J, Rigau T, Rodríguez-Gil JE, Pérez-Pe R, Muiño-Blanco T, Cebrián-Pérez JA, Casao A (2016) Melatonin receptors MT1 and MT2 are expressed in spermatozoa from several seasonal and non seasonal breeder species. Theriogenology 86:1958–1968

Marti J, Marti E, Cebrian-Perez J, Muiño-Blanco T (2003) Survival rate and antioxidant enzyme activity of ram spermatozoa after dilution with different extenders or selection by a dextran swim-up procedure. Theriogenology 60:1025–1037

Mateo-Otero Y, Zambrano F, Catalán J, Sánchez R, Yeste M, Miró J, Fernández-Fuertes B (2021) Seminal plasma, and not sperm, induces time and concentration-dependent neutrophil extracellular trap release in donkeys. Equine Vet J 54:1–12

McDonnell SM (1998) Reproductive behaviour of donkeys (*Equus asinus*). Appl Anim Behav Sci 60:277–282

McLean AK, Navas González FJ, Canisso IF (2019) Donkey and mule behaviour. Vet Clin North Am Equine Pract 35:575–588

Miragaya M, Neild D, Alonso A (2018) A review of reproductive biology and biotechnologies in donkeys. J Equine Vet Sci 65:55–61

Miró J, Taberner E, Rivera M, Peña A, Medrano A, Rigau T, Peñalba A (2009) Effects of dilution and centrifugation on the survival of spermatozoa and the structure of motile sperm cell subpopulations in refrigerated Catalonian donkey semen. Theriogenology 72:1017–1022

Miró J, Papas M (2018) Improvement of cryopreservation protocol in both purebred horses including Spanish horses. Span J Agric Res. 16(4):e0406.

Miró J, Marín H, Catalán J, Papas M, Gazem S, Yeste M (2020) Seminal plasma, sperm concentration, and sperm-PMN interaction in the donkey: An In Vitro model to study endometrial inflammation at post-insemination. Int J Mol Sci 21: 3478. https://doi.org/10.3390/ijms21103478

Miró J, Papas M (2018) Post–artificial insemination endometrial inflammation and its control in donkeys. J Equine Vet Sci 65:38–43

Miró J, Lobo V, Quintero-Moreno A, Medrano A, Peña A, Rigau T (2005) Sperm motility patterns and metabolism in Catalonian donkey semen. Theriogenology 63:1706–1716

Miró J, Taberner E, Ocaña M, Flotats A, Peña A (2008) Morphometry and motility of donkey spermatozoa vs horse spermatozoa. Reprod Domest Anim 43(Suppl 4):75

Miró J, Vilés K, garcía W, Jordana J, yeste M (2013) Effect of donkey seminal plasma on sperm movement and sperm-polymorphonuclear neutrophils attachment *in vitro*. Anim Reprod Sci 140:164–172

Miró J, Catalán J, Marín H, Yánez-Ortiz I, Yeste M (2021) Specific seminal plasma fractions are responsible for the modulation of sperm-PMN binding in the donkey. Animals 11:1388

Oliveira JV, Papa FO, Melo-Ocaña CM, Monteiro GA, Puoli-Filho JNP, Alvarenga MA (2012) New procedures to freeze donkey semen and its influence on mares and jennies fertility. J Equine Vet Sci 32:503–504

Ortiz-Rodríguez JM, Anel-López L, Martín-Muñoz P, Alvarez M, Gaiotskell-Phillips G, Anel L et al (2017) Pulse Doppler ultrasound as a tool for the diagnosis of chronic testicular dysfunction in stallions. Plus One 12(5):e0175878

Panzani D, Quaresma M, Fanelli D, Camilloo F, Moroni R, Rota A, Martins-Bessa, Nóvoa M, Catalán J, Caniso IF, Conte G, Miró J (2020) Hastening time to ejaculation in donkey jacks treated with the PGF2α analog, Cloprostenol Sodium. Animals 10:2231. https://doi.org/10.3390/ani10122231

Papas M, Arroyo L, Bassols A, Catalán J, Bonilla-Correal S, Gacem S, Yeste M, Miró J (2019) Activities of antioxidant seminal plasma enzymes (SOD, CAT, GPX and GSR) are higher in jackasses than in stallions and are correlated with sperm motility in jackasses. Theriogenology 140:180–187

Pozor MA, McDonnell SM (2002) Ultrasonographic measurements of accessory sex glands, ampullae, and urethra of normal stallions of various size types. Theriogenology 58:1425–1433

Quartuccio M, Marino G, Zanghì A, Garufi G, Cristarella S (2011) Testicular volume and daily sperm output in Ragusano donkeys. J Equine Vet Sci 31:143–146

Quintero-Moreno A, Miró J, Rigau T, Rodriguez-Gil JE (2003) Identification of sperm subpopulations with specific motility characteristics in stallion ejaculates. Theriogenology 58: 1973–1990

Rota A, Bastinacci C, Magelli D, Panzani D, Camilo F (2010) Evaluation of plasma membrane integrity of donkey spermatozoa. Reprod Domest Anim 45:228–232

Rota A, Panzani D, Sabatini C, Camilo F (2012) Donkey jack (*Equus asinus*) semen cryopreservation: studies of seminal parameters, post breeding inflammatory response, and fertility in donkey jennies. Theriogenology 78:1846–1854

Sabatini C, Mari G, Mislei B, Love C, Panzani D, Camilo F, Rota A (2014) Effect of post-thaw addition of seminal plasma on motility, viability and chromatin integrity of cryopreserved donkey jack (*Equus asinus*) spermatozoa. Reprod Domest Anim 14:989–994

Taberner E, Medrano A, Peña A, Rigau T, Miró J (2008) Oestrus cycle characteristics and prediction of ovulation in Catalonian jennies. Theriogenology 70:1489–1497

Taberner E, Morató R, Mogas T, Miró J (2010) Ability of Catalonian donkey sperm to penetrate zona pellucida-free oocytes matured in vitro. Anim Reprod Sci 118:354–361

Vidament M, Vincent P, Martin FX, Magistrini F, Blesbois E (2009) Differences in the ability of jennies and mares to conceive with cooled and frozen semen containing glycerol or not. Anim Reprod Sci 112:22–35

Vilés K, Rabanal R, Rodriguez-Prado M, Miró J (2013a) Effect of ketoprofen on the uterine inflammatory response after AI of jennies with frozen semen. Theriogenology 79:1019–1026

Vilés K, Rabanal R, Rodriguez-Prado M, Miró J (2013b) Influence of seminal plasma on leukocyte migration and amount of COX-2 protein in the jenny endometrium after insemination with frozen-thawed semen. Anim Reprod Sci 143:57–63

Yánez-Ortiz I, Catalán J, Delgado-Bermúdez A, Carluccio A, Miró J, Yeste M (2021) Addition of reduced glutathione (GSH) to freezing medium reduces intracellular ROS levels in donkey sperm. Vet Sci 8:302

Yeste M, Estrada E, Rocha LG, Marín H, Rodríguez-Gil JE, Miró J (2015) Cryotolerance is related to ROS production and mitochondrial membrane potential rather than to the integrity of sperm nucleus. Andrology 3:395–407

Updates on Embryo Technologies in the Donkey and Mule

Duccio Panzani

Abstract

Embryo technologies in donkeys (*Equus asinus*) are being developed to conserve endangered equid species and improve reproduction. Although donkey populations have declined in industrialized countries, embryo transfer (ET) from donor jennies to recipients like other donkeys, horses, or mules has shown promise. Despite early challenges, recent advances have led to successful pregnancies and births. Jennies' reproductive anatomy is similar to mares with some differences. They breed year-round, and ovulation can be induced hormonally. Artificial insemination (AI) with fresh or chilled semen has varying success, while frozen-thawed AI is less effective. Sterile techniques are vital for embryo recovery, with factors like ovulation count and timing affecting success rates.

Synchronizing estrus cycles between donors and recipients is critical for ET. Healthy, similarly sized recipients, ensure successful pregnancy and foaling, with mixed success using horses and mules as recipients. Efforts in embryo cryopreservation face challenges due to donkey embryos' unique structure, but studies show potential for future use. Continued research and funding are needed to advance these technologies for conserving endangered breeds and improving reproductive outcomes.

Keywords

Embryo · Endangered equids · Embryo transfer · Jennies · Ovulation

D. Panzani (✉)
Dipartimento di Scienze Veterinarie, Università di Pisa, Pisa, Italy
e-mail: duccio.panzani@unipi.it

A. K. McLean et al. (eds.), *Donkey and Mule Medicine: Clinical Medicine and Reproduction*, https://doi.org/10.1007/978-3-032-18540-2_17

1 Introduction

The donkey species, scientifically known as *Equus asinus*, has its origins traced back to two primary ancestors: the Nubian donkey (*E. asinus africanus*) and the Somali donkey (*E. asinus somaliensis*) (Beja-Pereira et al. 2004; Kimura et al. 2013). Historical evidence suggests that donkeys played a significant role in human expansion, migrating from central Africa to Europe and Asia over 10,000 years ago (Kimura et al. 2013). This is supported by numerous references to donkeys in various religious texts, traditions, and artworks throughout history (Bough 2011).

In 2017, the Food and Agriculture Organization (FAO) reported that there are approximately 46 million donkeys worldwide. The majority of these donkeys, over 52%, are found in Africa, while 33% reside in Asia. Central and South America account for 14% of the donkey population, and only 1% are found in Europe, North America, and Oceania combined (http://www.fao.org/faostat/en/#data/QA).

The distribution of donkeys is largely influenced by their role as working animals. In regions where donkeys are still utilized for labor, their populations remain stable (Carneiro et al. 2018). However, in areas where donkeys have lost their traditional social and economic roles, local breeds are at risk of extinction. In Europe, for example, 27 out of 28 reported donkey breeds are classified as critically endangered or at risk, with their populations having declined by approximately 50% over the past 70 years (Camillo et al. 2018). Following the Convention on Biological Diversity and the Nagoya Protocol, the European Commission's Council Regulation (EC) 870/2004 has actively promoted the conservation of biological diversity both in situ and ex situ, particularly focusing on the germplasm of endangered donkey breeds. This regulatory framework has been instrumental in safeguarding the genetic diversity of donkeys, ensuring their survival and sustainability.

In recent years, there has been a growing interest in donkeys, largely due to the beneficial properties of donkey milk for human consumption and its applications in the cosmetic industry (Martini et al. 2015; Ragona et al. 2016; Bibbiani et al. 2017). Studies have highlighted that donkey milk, especially from jennies, is an excellent substitute for human milk, particularly for children who are intolerant to cow's milk protein. This has led to an increased demand for donkey milk and its products.

However, the unsustainable and indiscriminate slaughter of donkeys and the trade of their skins to produce Ejao, a traditional Chinese medicine, have posed significant threats to donkey populations. This situation has accelerated the development of assisted reproductive techniques for donkeys, aimed at improving their fertility and economic value. These efforts are also crucial for ensuring the ethical sustainability of donkeys as a "production species" (Lesté-Lasserre 2019).

Organizations like the Donkey Sanctuary, one of the most prominent charities dedicated to donkey welfare, have been advocating for these measures. They emphasize the importance of ethical practices in the treatment and use of donkeys, striving to protect and preserve these animals for future generations.

2 Anatomy and Physiology of Reproduction in the Jenny

The reproductive system of a jenny, or female donkey, is quite similar to that of a mare, or female horse, with a few notable differences. One of the main distinctions is the jenny's cervix, which is longer and narrower compared to that of a mare. The ovaries of a jenny are similar in shape and structure to those of a mare, featuring an ovarian cortex on the outside, a medulla on the inside, and an ovulation fossa located at the free margin of the ovary (Guintard et al. 1998; Renner Martin et al. 2009).

Donkeys are non-seasonal breeders, meaning they exhibit reproductive activity throughout the entire year (Ginther et al. 1987; Carluccio et al. 2003; Taberner et al. 2008). However, some studies have noted instances of seasonality in donkeys, which are likely influenced by factors such as negative energy balance rather than geographical latitude (Lemma et al. 2006a; Quaresma et al. 2014, 2015; Quaresma and Payan-Carreira 2015). Jennies typically reach puberty between their first and second year of life, but they are generally not bred until they are at least 3 years old. The reproductive cycle of a jenny lasts between 24 and 26 days, with the diestrus phase spanning 15–19 days and the estrus phase lasting 4–10 days. Ovulation usually occurs within 24–48 h before the end of the estrus phase (Henry et al. 1987, 1998; Lodi et al. 1995).

During estrus, the uterine edema in jennies is less pronounced compared to mares, making it necessary to use a teaser male for accurate estrus detection (Lemma et al. 2006b). The presence of a corpus luteum, observed through ultrasonographic evaluation, using B-Mode and Color Doppler, confirms ovulation (Miró et al. 2015a; Panzani et al. 2018c). Jennies typically exhibit a single follicular wave (Derar and Hussein 2011), with follicle deviation occurring 8–9 days before ovulation when the follicle reaches a diameter of approximately 20 mm (Vandeplassche et al. 1981; Kebede et al. 2012; Quaresma and Payan-Carreira 2015). The pre-ovulatory follicle diameter in jennies ranges from 30 to 48 mm, and the incidence of double ovulations per cycle varies significantly among different breeds, ranging from 0% to 50%. (Meira et al. 1995, 1998; Blanchard et al. 1999; Contri et al. 2014; Quaresma and Payan-Carreira 2015; Perez-Marin et al. 2016; Díaz-Duran et al. 2017).

3 Uterine Embryo Recovery

Embryo transfer (ET) has been previously utilized as a conservation tool for endangered equid species, such as Przewalski's Horses (*Equus Przewalskii*; Summers et al. 1987). This technique, which allows for the potential production of a foal from each donor ovulation, holds significant promise for the preservation of endangered donkey breeds. Additionally, it can accelerate the selection of desirable genetic traits in donkeys used for production purposes.

However, there are several practical challenges in developing ET for donkeys. These challenges primarily stem from the anatomical and physiological differences between donkeys and horses, the latter being the species from which the ET technique was originally adapted. While ET has been a common practice in equine reproduction since the last decade of the twentieth century, its application in donkeys remains relatively uncommon. This is despite the fact that the initial studies on donkey ET were conducted only a few years after the first successful horse ET (Oguri and Tsutsumi 1974; Allen et al. 1985).

The successful implementation of ET in donkeys could play a crucial role in both conserving endangered breeds and enhancing the genetic quality of production animals. Overcoming the existing challenges in this area would significantly benefit the preservation efforts and the overall sustainability of donkey populations.

As with all the other mammals, for a jenny to be a viable embryo donor, she must be capable of ovulating, conceiving, and allowing the embryo to develop and survive until it is extracted from the uterus. The induction of ovulation in a jenny in estrus has been successfully achieved within 24–48 h following an injection of 1500–2500 IU of hCG administered intravenously, or a GnRH analogue, such as 100 mg of Lecirelin given intravenously, 400 μg of Buserelin administered subcutaneously, or 0.8 mg of Deslorelin injected intramuscularly (Carluccio et al. 2007; Camillo et al. 2014; Canisso et al. 2019).

When compared to conception rates after natural mating, the results with fresh and cooled semen artificial insemination (AI) in jennies are comparable to those observed in mares (Canisso et al. 2019). However, the outcomes with frozen-thawed semen AI have been less satisfactory. The highest pregnancy rates with frozen-thawed AI in jennies were achieved using insemination doses of 500 million spermatozoa, which were resuspended in homologous seminal plasma after thawing, resulting in a 67% pregnancy rate (8 out of 13). In contrast, without post-thaw dilution in seminal plasma, the pregnancy rate dropped to 23.1% (6 out of 26) in the same study (Rota et al. 2012).

Similar to mares, jennies experience a transient post-breeding endometritis, which is often less detectable through ultrasonography compared to horses. This condition occurs more frequently when frozen-thawed semen is used and is considered abnormal if it persists for more than 2 days after insemination (Rota et al. 2012). However, there is a lack of sufficient studies to determine the impact of post-breeding endometritis on subsequent fertility in donkeys. Embryo recovery has been successfully carried out between 5.5 and 9 days post-ovulation in jenny donors (Camillo et al. 2010; Fanelli et al. 2020; Panzani et al. 2020). The recovery can be performed using a one-way tubing system without a filter or a Lactated Ringer solution with a custom-built "Y" tubing system that includes an embryo transfer filter (Camillo et al. 2010; Panzani et al. 2012). For both methods, a cuffed polyethylene or silicon catheter, ranging from 20 to 32 French, is inserted through the cervix and approximately 5 cm into the uterine body. The varying body sizes of different jenny breeds and the unique structure of the jenny cervix present additional challenges during catheterization. These factors must be carefully considered to ensure successful embryo recovery (Camillo et al. 2010; Panzani et al. 2012).

Typically, the flushing catheter is inserted into the vagina using a hand covered by a long sterile glove. The cervix is then grasped and gently pulled back to facilitate the insertion of the catheter. Once the catheter is positioned within the uterus, its cuff is inflated with 50 mL of air. This inflation ensures that the catheter is securely placed against the internal os of the cervix, preventing the flushing medium from flowing back into the vagina.

The flushing medium is then introduced into the uterus and subsequently collected in a cylinder or through an in-line filter with the assistance of transrectal uterine massage. This process is repeated three to four times, with each flush using between 0.5 and 1 liter of medium, depending on the uterine volume assessed through transrectal palpation during the procedure. After the final flush, the cuff is deflated, and the catheter is carefully withdrawn from the uterus.

It is crucial to monitor the volume of the recovered fluid closely (Camillo et al. 2010). As with this procedure in mares, approximately 90% of the fluid introduced should be recovered at the end of the flushing process to minimize the risk of leaving the embryo within the uterus. Immediately following the uterine flushing, a luteolytic dose of prostaglandin analogue is administered intramuscularly. This includes options such as 5 mg of Dinoprost, 3 mg of Alphaprostol, 0.075 mg of cloprostenol, or 7.5 mg of luprostiol (Carluccio et al. 2008; Panzani et al. 2018b; Canisso et al. 2019). This administration helps prevent an unwanted pregnancy and accelerates the onset of a new estrus cycle, thereby promoting physiological uterine clearance. Embryo recovery rates following artificial insemination with either fresh or chilled semen have shown a variation between 30% and 75%, with no significant impact from the photoperiod. In one particular study, 83 estrus cycles of five Pantesca jennies were subjected to artificial insemination with fresh semen over a span of 2 years. The embryo recovery rate, which measures the number of embryos retrieved per cycle, and the positive flushing rate, indicating cycles where at least one embryo was recovered, were influenced by the number of ovulations and the timing of embryo recovery in relation to the donor's ovulation (Camillo et al. 2010).

In a more recent investigation, 59 embryo recoveries were conducted on 17 Spanish breed jennies, including Andalusian and Zamorano-Leones, between 6.5- and 8.5-days post-ovulation. The findings revealed that the embryo recovery rate was affected by the quality of the semen used, but not by the specific day of flushing (whether 6.5, 7.5, or 8.5 days), the age of the donor (ranging from 3 to 19 years old), or the cellularity of the recovered uterine fluid, which varied from no cellular debris to high levels of cellularity (Pérez-Marín et al. 2017).

4 The Donkey Embryo

The donkey embryo, much like the horse embryo, produces a mucin-like glycoprotein capsule that replaces the zona pellucida during the blastocyst stage (Oriol et al. 1993). This capsule is crucial for the embryo's development and protection.

In horses, the viability of embryos is typically assessed based on their size, developmental stage, and overall quality at the time of recovery. Since 1988, the clinical evaluation of horse embryos has been based on the morphological criteria proposed by McKinnon and Squires (1988). These criteria include parameters such as shape, color, cell number and compactness, size of the perivitelline space, number of extruded and degenerated cells, and the developmental stage relative to the embryo's age. This quality scoring system, ranging from 1 (excellent) to 5 (degenerate or dead), has been used to qualitatively assess embryos in all studies related to donkey embryo transfer. Most donkey embryos recovered in these studies have been reported to be of excellent or good quality according to these parameters.

The extensive evaluation of equine embryos in scientific literature has established the physiological dimensions and developmental stages of recovered embryos (McKinnon and Squires 1988; Carnevale et al. 2000). However, in donkeys, fewer than 100 embryos have been analyzed post-recovery, leading to significant variations among studies, particularly concerning the diameters of day 8 embryos (Camillo et al. 2010; Panzani et al. 2012, 2018a; Fanelli et al. 2020).

Embryo recovery in jennies conducted between days 6 and 7 post-ovulation has resulted in embryos with diameters of less than 300 micrometers, which are smaller compared to those in horses. By day 8, this size discrepancy between donkey and horse embryos becomes less pronounced (Vendramini et al. 1997; Camillo et al. 2010; Panzani et al. 2012, 2018a; Pérez-Marín et al. 2017; Bottrel et al. 2019; Dorado et al. 2020; Fanelli et al. 2020).

Only three studies have reported the average diameters of donkey embryos at various developmental stages, but unfortunately, they did not specify the exact ages of the embryos. By comparing the reported diameters, it can be inferred that donkey embryos transition from the morula stage to early blastocyst around day 6 and reach the blastocyst stage by approximately day 7.5 (Pérez-Marín et al. 2017; Bottrel et al. 2019).

5 Transfer of the Embryo

For successful embryo transfer (ET) in donkeys, it is crucial that the recipient jennies are young, healthy, and of a similar size to the donor. These recipients must be capable of maintaining the pregnancy, giving birth, and raising the foal until it is weaned. In the early studies on donkey ET, the selection of recipients was often limited due to resource constraints, which likely contributed to the less than satisfactory results observed.

Currently, there are no specific studies that have evaluated the optimal synchronization of ovulation between donor jennies and their recipients. Generally, donkey embryos that are 7–8 days old are transferred into recipient jennies between 5 and 8 days after ovulation, following practices similar to those used in equine ET (Vendramini et al. 1997; Camillo et al. 2010; Panzani et al. 2012, 2018a; Bottrel et al. 2017; Pérez-Marín et al. 2017; Dorado et al. 2020). To ensure successful transfer, it is essential to accurately identify the recipient's estrus phase and ovulation day. This often requires daily transrectal palpations and ultrasound examinations, especially during estrus when a dominant follicle is detected. It is also recommended to perform an ultrasound examination just before the transfer to confirm the presence of a functional corpus luteum (CL) using B-mode or Color Doppler, and to ensure there is no edema or fluid accumulation in the uterine lumen (Miró et al. 2015b; Panzani et al. 2018b).

When planning a donkey ET program, it is beneficial to have a consistent herd of recipients to achieve spontaneous synchronization, similar to practices in equine ET. Alternatively, synchronization of estrus between donors and recipients can be planned. Various estrus synchronization protocols have been tested in jennies, primarily to optimize artificial insemination (AI) management and to improve synchronization between donors and recipients for ET (Fanelli et al. 2019a, b).

As an alternative approach, horses and mules have been utilized as surrogate mothers for donkey embryos. Studies have shown that the pregnancy rates up to 40 days in donkey-to-horse embryo transfer (ET) are comparable to those observed in horse-to-horse ET. However, early research indicated that the formation of endometrial cups and the implantation of the fetus did not occur, resulting in the termination of all pregnancies between 80 and 100 days. Additionally, there was a 15% loss rate before the end of the pregnancy (Allen 1982; Allen et al. 1985; Enders et al. 1996).

In more recent developments, a significant breakthrough was achieved when an acyclic recipient mare was treated with long-acting progesterone. The treatment involved administering 300 mg of progesterone every 15 days from day 15 to day 120 of the pregnancy (Peña-Alfaro et al. 2014). This regimen successfully led to the birth of a healthy donkey foal at full term. This advancement highlights the potential for using horses as surrogate mothers for donkey embryos, offering a viable alternative for reproductive strategies in equine species.

Recently the birth of 3 healthy mule foals after embryo transfer in 3 mule recipients has been reported in the same year, one in Argentina and one in Italy, respectively (Fanelli et al. 2022). The Argentinian mule recipients were acyclic and their pregnancies were supported with long acting progesterone up to day 100, after 3 days of pharmacologic estrus induction with exogenous estradiol 17ß. The Italian mule was cyclic instead and did not receive any hormone supportive treatment.

To achieve optimal results in embryo recovery, it is crucial to maintain a highly sterile environment. This involves using techniques, instruments, and media that do not compromise the viability of the embryos. Historically, embryo recovery in jennies has been conducted through transcervical uterine flushing. To prevent uterine contamination during this process, it is essential to thoroughly wash and dry the donor's perineum, similar to the procedure used for mares.

The initial studies on donkey-to-donkey embryo transfer (ET) in endangered breeds such as Poitou and Pantesca yielded disappointing pregnancy rates. When horse protocols were applied, the pregnancy rates at 14 days post-transfer were only 16.7% (3 out of 18) and 22.4% (13 out of 58). These figures are significantly lower compared to the 63% success rate for surgical transfer of donkey embryos to mares and the 67% for horse embryos to donkeys. Typically, horse-to-horse ET results in pregnancy rates ranging from 50% to 75%.

Interestingly, the use of Dulbecco's Phosphate-Buffered Saline (DPBS) as an embryo-holding medium resulted in a mere 7.1% pregnancy rate (1 out of 14), whereas Ringer Lactate and EmCare Holding Solution (Bio98 Srl, Milano, Italy) achieved higher rates of 27.2% (9 out of 33) and 27.2% (3 out of 11), respectively. This suggests that phosphate may have a detrimental effect on donkey embryo development, similar to observations in hamsters, where it was attributed to the "Crabtree effect." This effect involves enhanced glycolysis in culture media containing phosphate, which inhibits respiratory activity and oxidative metabolism of embryos.

Further studies investigated the causes of the different outcomes of ET between horse and donkey species. The first hypothesis proposed that the flushing and holding media might negatively impact the quality of donkey embryos. However, this hypothesis was disproven as the percentage of dead cells in donkey embryos, assessed using DAPI staining, was less than 2%, which is significantly below the 20% threshold for viable horse embryos.

The second hypothesis suggested that the challenges in catheterizing the jenny cervix during ET could trigger the release of PGF2α, leading to luteolysis in recipient jennies, a phenomenon previously observed in mares. This hypothesis was also refuted, as there was no significant increase in PGF-metabolites in jennies 5 days post-ovulation following sham non-surgical transfers. Additionally, the pregnancy rates were identical for donkey-to-donkey and donkey-to-horse embryo transfers, both at 50% (6/12 and 3/6, respectively), while horse-to-donkey ETs resulted in a 100% pregnancy rate (5/5). The findings suggest that the low pregnancy rates in the Pantesca donkey ET study might be attributed to the lack of recipient selection and the inability to exclude recipients after repeated pregnancy failures, a common practice in commercial horse ET. Conversely, a recent ET study on Pega donkeys in Brazil, which used Ringer's lactate and a non-phosphate-based embryo holding media, achieved a 45% pregnancy rate (4/11) after transfer.

In conclusion, while more extensive ET studies are necessary to confirm these results, it appears feasible to attain satisfactory pregnancy rates by transferring fresh donkey embryos into donkey recipients.

Recovered embryos are typically washed in a holding medium such as DPBS, Ringer Lactate, or a commercial zwitterionic embryo holding medium. After evaluating the quality of the embryos, they are gently aspirated into a 0.25–0.5 mL French straw. This straw is prepared with a small amount of holding solution and an air bubble both before and after the embryo. The French straw is then inserted into a bovine embryo transfer (ET) French gun, and the embryos are non-surgically transferred to synchronized recipients.

For the embryo transfer procedure, recipients are usually brought to a stock where their perineum is thoroughly washed, rinsed, and dried, following the same protocol used for mares. Some operators administer acepromazine to the embryo recipients before the transfer to relax the animal, facilitate cervix manipulation, and prevent the release of prostaglandin F2 alpha (PGF2α).

The operator inserts the guarded gun, which is protected by a sanitary sheath, through the vagina. The vaginal part of the cervix is grasped with three fingers and gently pulled backward. The tip of the gun is then blindly inserted into the cervical os, the sanitary sheath is broken, and the cervix is manipulated to aid the insertion and progression of the gun. The embryos are usually released into the uterine lumen of the recipients without the need for transrectal manipulation.

6 Embryo Cryopreservation

Given the poor fertility outcomes with frozen donkey semen, researchers have focused on cryopreserving donkey embryos with minimal damage to establish both in situ and ex situ donkey germplasm banks. Cryopreservation extends the viability of embryos before embryo transfer (ET) and can be performed either by cooling at 4 °C or by freezing/vitrification at −196 °C. The cooling method, commonly used in the horse industry, allows embryos to be transferred within 24 h without significantly reducing pregnancy rates compared to fresh embryo ET. This method enables the collection of embryos on a farm and their subsequent transfer to an ET center via express courier. Typically, equine embryos are placed in a 5 mL plastic or glass tube filled with a warmed embryo holding medium, securely capped and sealed with parafilm. This tube is then placed inside a closed and parafilm-sealed 50 mL centrifuge tube filled with flushing media, secured in a passive cooling device.

Although there are no specific studies on the transfer of cooled donkey embryos, it has been reported that when held in Ringer Lactate and Equitainer for 24 h, they exhibited only 2% dead cells, comparable to the rate observed in freshly recovered donkey embryos. The presence of a glycoprotein capsule surrounding the embryos of mares and jennies has been suggested as a reason for the poor results after conventional freezing and vitrification of embryos larger than 300 micrometers in diameter (Panzani et al. 2018a).

Building on this concept, a limited number of studies have examined the vitrification outcomes of donkey embryos larger than 300 μm. In vitro evaluations of vitrified/thawed embryos showed only a slight increase in cell death compared to fresh embryos. The only in vivo study resulted in the birth of two healthy donkey foals after transferring 11 high-quality embryos. In this study, the pregnancy rates after transferring vitrified donkey-to-donkey embryos smaller than 300 micrometers (4 out of 11) and horse-to-horse embryos (4 out of 10) were very similar (Fanelli et al. 2020).

Despite the limited literature on donkey embryo cryopreservation, these early studies highlight the potential of this technique and its possible applications in donkey reproduction.

7 Oocytes Technologies and In Vitro Produced Embryo

Donkey oocytes have been successfully retrieved from ovaries collected at abattoirs through methods such as scraping or follicle aspiration, as well as from living jennies using transvaginal ultrasound-guided follicle aspiration. Scraping yielded a higher recovery rate of oocytes per ovary from slaughtered jennies, with an average of 4.86 ± 0.67 COCs per ovary, compared to 3.14 ± 0.36 COCs per ovary using an 18-G needle and 0.61 ± 0.16 COCs per ovary with a 20-G needles (Abdoon et al. 2014).

Early studies on transvaginal ultrasound-guided follicle aspiration, employing a 12-G double-lumen needle as used in horses, resulted in a 48–57% oocyte recovery rate (Goudet et al. 2016; Deleuze et al. 2018) The maturation rate of oocytes increased significantly from 39% to 77% when cultured in media such as TCM199, DMEM-F12, or CR1aa (Goudet et al. 1997, 2016; Zhao et al. 2011; Abdoon et al. 2014; Deleuze et al. 2018). However, the initial attempt at in vitro fertilization (IVF) of in vivo collected donkey oocytes was not successful, with only 15% of oocytes forming pronuclei (Goudet et al. 1997).

A single pregnancy has recently been reported following the transfer of an in vitro produced donkey embryo (Dobbie et al. 2024). Out of 61 oocytes matured for 46–48 h and fertilized via ICSI, eight developed into blastocysts and were subsequently vitrified. Three of these eight blastocysts were transferred into three jennies, two on day 4 and one on day 5 post-ovulation. Only the day 5 transfer resulted in a pregnancy, which was detectable until day 12 but was lost before day 14.

No foals born after transfer of in vitro produced oocytes or donkey clones have been reported yet.

8 Conclusion

Recent advancements in embryo technologies for donkeys represent a significant step forward in both species conservation and reproductive management. Despite the anatomical and physiological challenges unique to jennies, progress in artificial insemination, embryo recovery, and transfer techniques has led to increasingly successful pregnancy outcomes. The use of alternative recipients—such as horses and mules—combined with hormone-supported protocols, has broadened the scope for effective embryo transfer.

Although cryopreservation and in vitro embryo production in donkeys remain in early stages, initial results are promising and suggest strong potential for future application. Notably, only a limited number of research groups worldwide are currently focused on donkey reproduction. Yet, their encouraging findings indicate that endangered donkey breeds could greatly benefit from the adaptation of equine biotechnologies—especially if supported by funding from international charities or governmental bodies.

To fully realize the potential of these technologies, sustained research, technical refinement, and institutional backing are essential. These efforts will be critical in ensuring the long-term sustainability of donkey breeding programs and the preservation of genetic diversity within this historically and economically important species.

Questions

1. **Which of the following is a key anatomical difference between donkeys (jennies) and horses (mares) that impacts assisted reproductive technologies?**
   A. Donkeys have a smaller ovary size and longer uterine body
   B. Donkeys have a larger uterine body and higher cervical resistance ✓
   C. Donkeys have shorter estrous cycles than mares
   D. Donkeys have more pronounced seasonal anestrus
2. **In embryo transfer (ET) in donkeys, why is synchronization between donor and recipient estrous cycles critical?**
   A. To reduce stress on the donor animal
   B. To increase the chance of successful embryo development and implantation ✓
   C. To prevent twin pregnancies
   D. To reduce costs of hormonal treatments
3. **What has been a major limitation in applying in vitro fertilization (IVF) techniques in donkeys compared to horses?**
   A. Lack of funding for donkey research
   B. Low fertilization and embryo development rates ✓
   C. Donkey sperm cannot be cryopreserved
   D. Donkey embryos are too large for manipulation
4. **Which reproductive technology has shown potential for conservation of endangered donkey breeds?**
   A. Artificial insemination alone
   B. Embryo cryopreservation ✓
   C. Use of cross-species surrogates without synchronization
   D. Natural service breeding only
5. **Why is research into donkey embryo technologies particularly important?**
   A. Donkeys are no longer widely used in agriculture
   B. Donkeys have identical reproductive physiology to horses
   C. Donkeys are under economic and conservation pressures, and embryo technologies can support both farming sustainability and breed preservation ✓
   D. Donkeys reproduce more efficiently than horses and require no intervention

References

Abdoon ASS, Abdel-Rahman HA, Shawki SM, Kandil OM, Fathalla SI (2014) Influence of follicle size, methods of retrieval on oocytes yield and morphology in Egyptian Jennies ovaries with special reference to maturation rate in vitro. Vet Res Commun 38:287e95. https://doi.org/10.1007/s11259-014-9617-y

Allen WR (1982) Immunological aspects of the endometrial cup reaction and the effect of xenogeneic pregnancy in horses and donkeys. J Reprod Fertil Suppl 31:57–94

Allen WR, Kydd J, Boyle MS, Antczak DF (1985) Between-species transfer of horse and donkey embryos: a valuable research tool. Equine Vet J 17:53–62. https://doi.org/10.1111/j.2042-3306.1985.tb04594.x

Beja-Pereira A, England PR, Ferrand N et al (2004) African origins of the domestic donkey. Science (1979) 304:1781. https://doi.org/10.1126/science.1096008

Bibbiani C, Biagini P, Salari F, Martini M (2017) Dairy donkey: an alternative building layout. J Agric Eng 48:1–8. https://doi.org/10.4081/jae.2017.637

Blanchard TL, Taylor TS, Love CL (1999) Estrous cycle characteristics and response to estrus synchronization in mammoth asses (Equus asinus americanus). Theriogenology 52:827–834

Bottrel M, Fortes T, Ortiz I et al (2017) Establishment and maintenance of donkey-in-mule pregnancy after embryo transfer in a non-cycling mule treated with oestradiol benzoate and long-acting progesterone. Span J Agric Res 15. https://doi.org/10.5424/sjar/2017154-11653

Bottrel M, Ortiz I, Pereira B et al (2019) Cryopreservation of donkey embryos by the cryotop method: effect of developmental stage, embryo quality, diameter and age of embryos. Theriogenology 125. https://doi.org/10.1016/j.theriogenology.2018.11.011

Bough J (2011) The mirror has two faces: contradictory reflections of donkeys in western literature from lucius to Balthazar. Animals 1:56–68. https://doi.org/10.3390/ani1010056

Camillo F, Panzani D, Scollo C et al (2010) Embryo recovery rate and recipients' pregnancy rate after nonsurgical embryo transfer in donkeys. Theriogenology 73. https://doi.org/10.1016/j.theriogenology.2009.11.024

Camillo F, Vannozzi I, Tesi M et al (2014) Induction of ovulation with buserelin in jennies: in search of the minimum effective dose. Anim Reprod Sci 151. https://doi.org/10.1016/j.anireprosci.2014.09.011

Camillo F, Rota A, Biagini L et al (2018) The current situation and trend of donkey industry in Europe. J Equine Vet Sci 65:44–49

Canisso IF, Panzani D, Miró J, Ellerbrock RE (2019) Key aspects of donkey and mule reproduction. Vet Clin North Am Equine Pract 35. https://doi.org/10.1016/j.cveq.2019.08.014

Carluccio A, Pau S, Picci GF, Tosi U (2003) Ciclo estrale nell'asina di Martina Franca: osservazioni comportamentali ed ecografiche. Veterinaria Pratica …

Carluccio A, Panzani S, Tosi U et al (2007) Efficacy of hCG and GnRH for inducing ovulation in the jenny. Theriogenology 68. https://doi.org/10.1016/j.theriogenology.2007.07.005

Carluccio A, Panzani S, Contri A et al (2008) Luteal function in jennies following PGF2α treatment 3 days after ovulation. Theriogenology 70. https://doi.org/10.1016/j.theriogenology.2008.03.010

Carneiro GF, Cavalcante Lucena JE, de Oliveira Barros L (2018) The current situation and trend of the donkey industry in South America. J Equine Vet Sci 65:106–110

Carnevale EM, Ramirez RJ, Squires EL et al (2000) Factors affecting pregnancy rates and early embryonic death after equine embryo transfer. Theriogenology 54. https://doi.org/10.1016/S0093-691X(00)00405-2

Contri A, De Amicis I, Veronesi MC, Carluccio A (2014) Effect of the season on some aspects of the estrous cycle in Martina Franca donkey. Theriogenology 81:657–661

Deleuze S, Douet C, Couty I, Moros-Nicolas C, Barriere P, Blard T et al (2018) Ovum pick up and in vitro maturation of jennies oocytes toward the setting up of efficient in vitro fertilization and in vitro embryos culture procedures in donkey (Equus asinus). J Equine Vet Sci 65:111e7. https://doi.org/10.1016/j.jevs.2018.03.004

Derar RI, Hussein HA (2011) Ovarian follicular dynamics during the estrous cycle in jennies in Upper Egypt. Vet Med Int 2011:1–6

Díaz-Duran M, Zarco L, Boeta AM (2017) Ovarian dynamics and estrous cycle length in the donkey (Equus asinus). Theriogenology 103:1–8

Dobbie T, Felix MR, Gleason K, Hinrichs K (2024) Transvaginal aspiration, oocyte maturation and production of blastocysts and a pregnancy via ICSI in an endangered breed: Baudet du Poitou donkey. Theriogenology 224:34–40

Dorado J, Bottrel M, Ortiz I et al (2020) Factors affecting embryo recovery rate, quality, and diameter in andalusian Donkey jennies. Animals 10. https://doi.org/10.3390/ani10111967

Enders AC, Meadows S, Stewart F, Allen WR (1996) Failure of endometrial cup development in the donkey-in-horse model of equine abortion. J Anat 188(Pt 3):575

Fanelli D, Tesi M, Bagnato G et al (2019a) New simplified protocols for timed artificial insemination (TAI) in milk-producing donkeys. Theriogenology 139. https://doi.org/10.1016/j.theriogenology.2019.08.005

Fanelli D, Tesi M, Rota A et al (2019b) Studies on the use of prostaglandin F2α and gonadotropin-releasing hormone analogs for timed artificial insemination in jennies. J Equine Vet Sci 74. https://doi.org/10.1016/j.jevs.2018.12.001

Fanelli D, Panzani D, Rota A et al (2020) Cryopreservation of donkey embryos: comparison of embryo survival rate after in vitro culture between conventional freezing and vitrification. Theriogenology 154. https://doi.org/10.1016/j.theriogenology.2020.05.020

Fanelli D, Losinno L, Castañeira C, Alonso C, Bragulat AF, Panzani D, Bocci C, Degl'Innocenti A, Moroni R, Camillo F, Wilsher S, Allen WR (2022) First report of mule-in-mule pregnancies with live births following embryo transfer. J Equine Vet Sci 113. https://doi.org/10.1016/j.jevs.2022.103999

Ginther OJ, Scraba ST, Bergfelt DR (1987) Reproductive seasonality of the jenney. Theriogenology 27:587–592

Goudet G, Bezard J, Duchamp G (1997) Equine oocyte competence for nuclear and cytoplasmic in vitro maturation: effect of follicle size and hormonal environment. Biol Reprod 57:232e45. https://doi.org/10.1095/biolreprod57.2.232

Goudet G, Douet C, Kaabouba-Escurier A, Couty I, Moros-Nicolas C, Barriere P et al (2016) Establishment of conditions for ovum pick up and IVM of jennies oocytes toward the setting up of efficient IVF and in vitro embryos culture procedures in donkey (Equus asinus). Theriogenology 86:528e35. https://doi.org/10.1016/j.theriogenology.2016.02.004

Guintard C, Moreau J, Tainturier D et al (1998) Cervix conformation: a first anatomical approach in baudet du poitou jenny asses. Anim Sci 66:741–744

Henry M, Figueiredo AE, Palhares MS, Coryn M (1987) Clinical and endocrine aspects of the oestrous cycle in donkeys (Equus asinus). J Reprod Fertil Suppl 35:297–303

Henry M, Lodi LD, Gastal MMFO (1998) Sexual behaviour of domesticated donkeys (Equus asinus) breeding under controlled or free range management systems. Appl Anim Behav Sci 60: 263–276. https://doi.org/10.1016/S0168-1591(98)00175-0

Kebede H, Lemma A, Negussie H (2012) Ultrasonographic studies on ovarian dynamics and associated estrus manifestations of jennies under controlled management, Ethiopia. Trop Anim Health Prod 44:1965–1970

Kimura B, Marshall F, Beja-Pereira A, Mulligan C (2013) Donkey domestication. Afr Archaeol Rev 30:83–95. https://doi.org/10.1007/s10437-012-9126-8

Lemma A, Bekana M, Schwartz HJ, Hildebrandt T (2006a) The effect of body condition on ovarian activity of free ranging tropical jennies (Equus asinus). J Veterinary Med Ser A 53:1–4

Lemma A, Schwartz HJ, Bekana M (2006b) Application of ultrasonography in the study of the reproductive system of tropical jennies (shape Equus asinus). Trop Anim Health Prod 38:267–274

Lesté-Lasserre C (2019) Donkeys face worldwide existential threat. Science (1979) 366:1294–1295. https://doi.org/10.1126/science.366.6471.1294

Lodi LD, Henry M, Paranhos da Costa MJR (1995) Behavior of donkey jacks (Equus asinus) breeding horse mares (Equus caballus) at pasture. Biol Reprod 52:591–598

Martini M, Altomonte I, Manica E, Salari F (2015) Changes in donkey milk lipids in relation to season and lactation. J Food Compos Anal 41:30–34

McKinnon AO, Squires EL (1988) Equine embryo transfer. Vet Clin North Am Equine Pract 4: 305–333. https://doi.org/10.1016/S0749-0739(17)30643-0

Meira C, Ferreira JCP, Papa FO et al (1995) Study of the estrous cycle in donkeys (Equus asinus) using ultrasonography and plasma progesterone concentrations 1. Biol Reprod 52:403–410

Meira C, Ferreira JC, Papa FO, Henry M (1998) Ovarian activity and plasma concentrations of progesterone and estradiol during pregnancy in jennies. Theriogenology 49:1465–1473

Miró J, Vilés K, Anglada O et al (2015a) Color Doppler provides a reliable and rapid means of monitoring luteolysis in female donkeys. Theriogenology 83:485–490

Miró J, Vilés K, Anglada O et al (2015b) Color Doppler provides a reliable and rapid means of monitoring luteolysis in female donkeys. Theriogenology 83. https://doi.org/10.1016/j.theriogenology.2014.10.007

Oguri N, Tsutsumi Y (1974) Non-surgical egg transfer in mares. J Reprod Fertil 41:313–320. https://doi.org/10.1530/jrf.0.0410313

Oriol JG, Betteridge KJ, Hardy J, Sharom FJ (1993) Structural and developmental relationship between capsular glycoproteins of the horse (Equus caballus) and the donkey (Equus asinus). Equine Vet J 25:14–18. https://doi.org/10.1111/j.2042-3306.1993.tb04816.x

Panzani D, Rota A, Crisci A et al (2012) Embryo quality and transcervical technique are not the limiting factors in donkey embryo transfer outcome. Theriogenology 77. https://doi.org/10.1016/j.theriogenology.2011.08.032

Panzani D, Rota A, Tesi M et al (2018a) Update on donkey embryo transfer and cryopreservation. J Equine Vet Sci 65. https://doi.org/10.1016/j.jevs.2017.11.012

Panzani D, Tardella M, Govoni N et al (2018b) Effect of the administration of alfaprostol 3 or 6 days after ovulation in jennies: ultrasonographic characteristic of corpora lutea and serum progesterone concentration. Theriogenology 121. https://doi.org/10.1016/j.theriogenology.2018.08.014

Panzani D, Tardella M, Govoni N et al (2018c) Effect of the administration of alfaprostol 3 or 6 days after ovulation in jennies: ultrasonographic characteristic and plasma progesterone concentration of corpora lutea. Theriogenology 121:175–180

Panzani D, Fanelli D, Camillo F, Rota A (2020) Embryo technologies in donkeys (Equus asinus). Theriogenology 156. https://doi.org/10.1016/j.theriogenology.2020.06.041

Peña-Alfaro C, Barros L, Carneiro G, Gastal M, Gastal E (2014) Embryo transfer in Pega donkeys (*Equus asinus*) in Brazil. J Equine Vet Sci 1:185

Perez-Marin CC, Galisteo I, Perez-Rico A, Galisteo JJ (2016) Effects of breed, age, season, and multiple ovulations on cyclic, PGF2α-induced, and postpartum estrus characteristics in Spanish jennies. Theriogenology 85:1045–1052

Pérez-Marín CC, Vizuete G, Galisteo JJ (2017) Embryo recovery results in Hispano-Arabe horse and Spanish donkey breeds. Livest Sci 206. https://doi.org/10.1016/j.livsci.2017.10.011

Quaresma M, Payan-Carreira R (2015) Characterization of the estrous cycle of Asinina de Miranda jennies (Equus asinus). Theriogenology 83:616–624

Quaresma M, Martins AMF, Rodriguez JS et al (2014) Pedigree and herd characterization of a donkey breed vulnerable to extinction. Animal 8:354–359

Quaresma M, Silva SR, Payan-Carreira R (2015) Reproductive patterns in the non-breeding season in Asinina de Miranda jennies. Reprod Domest Anim 50:784–792

Ragona G, Corrias F, Benedetti M et al (2016) Amiata donkey milk chain: animal health evaluation and milk quality. Ital J Food Saf 5:5951

Renner Martin TFP, Martin TFPR, Forstenpointner G et al (2009) Gross anatomy of the female genital organs of the domestic donkey (Equus asinus Linné, 1758). Anat Histol Embryol 38: 133–138

Rota A, Panzani D, Sabatini C, Camillo F (2012) Donkey jack (Equus asinus) semen cryopreservation: studies of seminal parameters, post breeding inflammatory response, and fertility in donkey jennies. Theriogenology 78. https://doi.org/10.1016/j.theriogenology.2012.07.015

Summers PM, Shephard AM, Hodges JK et al (1987) Successful transfer of the embryos of Przewalski's horses (Equus przewalskii) and Grant's zebra (E. burchelli) to domestic mares (E. caballus). J Reprod Fertil 80:13–20

Taberner E, Medrano A, Peña A et al (2008) Oestrus cycle characteristics and prediction of ovulation in Catalonian jennies. Theriogenology 70:1489–1497

Vandeplassche GM, Wesson JA, Ginther OJ (1981) Behavioral, follicular and gonadotropin changes during the estrous cycle in donkeys. Theriogenology 16:239–249. https://doi.org/10.1016/0093-691X(81)90106-0

Vendramini OM, Bruyas JF, Fieni F et al (1997) Embryo transfer in poitou donkeys, preliminary results. Theriogenology 47. https://doi.org/10.1016/s0093-691x(97)82536-8

Zhao G, Wu K, Cui L, Zhao L, Liu Y, Tan X et al (2011) In vitro maturation and artificial activation of donkey oocytes. Theriogenology 76:700e4. https://doi.org/10.1016/j.theriogenology.2010.10.039

Mule and Hinny Reproduction: From Sterility to Fertility in Hybrids

Amy K. McLean

Abstract

Mules and hinnies are hybrid offspring of donkeys (*Equus asinus*) and horses (*Equus caballus*), with hinnies representing the reciprocal cross. Scientific knowledge about these hybrids is limited, especially regarding hinnies. While they are generally considered sterile, historical records, including references by Homer, describe fertile mules. Rare cases worldwide document mules and hinnies producing live offspring when mated with a jack.Hybrid sterility is primarily attributed to differences in chromosome number and structure, yet many hybrids still exhibit typical reproductive behaviors. Male mules, often castrated, may display aggression or produce spermatids if intact. Female mules and hinnies may show estrus behaviors, sometimes requiring ovariectomy to manage these behaviors. Beyond fertility, female hybrids have also been studied as recipients for equine embryo transfers, particularly for donkey embryos. Reciprocal crosses present additional reproductive challenges. Mares carrying mule foals may produce fewer endometrial cups, lower levels of equine chorionic gonadotropin (eCG), and reduced progesterone, all critical for maintaining pregnancy. Jennies mated to stallions may experience poor conception rates, potentially due to uterine immune responses to stallion versus jack semen. After birth, neonatal concerns include the risk of isoerythrolysis. With the growing interest in conserving donkey breeds and breeding mules for recreational or performance purposes, understanding the reproductive physiology, pregnancy management, and control of reproductive behaviors in these hybrids has become increasingly important.

A. K. McLean (✉)
Department of Animal Science, University of California Davis, California, USA
e-mail: acmclean@ucdavis.edu

A. K. McLean et al. (eds.), *Donkey and Mule Medicine: Clinical Medicine and Reproduction*, https://doi.org/10.1007/978-3-032-18540-2_18

Keywords

Fertile mules · Spermatogenesis · Oocytogenesis · Recipients · Embryo development · eCG · Castration · Ovariectomy

1 Introduction

Participatory surveys we have conducted (McLean et al. 2014, 2016) with mule and hinny breeders worldwide reveal clear preferences for one hybrid over the other, often influenced by availability and intended use. In many countries, mules are considered the ideal choice for work, though they are not always accessible. In Central and South America, mules remain important for draught purposes, including farming and transporting goods such as coffee. In parts of Mexico, dealers import mules from the U.S., with prices ranging from $4–5000 depending on size. Some southern Mexican Cattlemen Associations import larger jacks for breeding, allowing members to mate for free. Government-supported mule-breeding programs exist in countries such as India, Argentina, Chile, and Mexico, supplying mules to farmers for hauling bricks, goods, or other farm work. In South America, including Brazil, Colombia, and Argentina, mules are highly valued and preferred for work on cattle ranches. Breeders in areas without formal programs often offer jacks at stud to help produce mules locally. In regions where mares are scarce, hinnies are more commonly raised. Geographic availability also varies in Mali, West Africa, only donkeys and a few horses are present, with no mules, while countries like Morocco and Ethiopia have both mules and hinnies widely used for draught work (Fig. 1).

In Baja, Mexico cattle ranchers prefer hinnies to mules due to the very rough and desert type terrain. The ranchers believe that hinnies perform and work better in this dry, desert climate and can withstand longer periods without water compared to a

Fig. 1 A hinny from Miranda, Portugal an area where there are many hinnies used in the communities for ploughing potatoes and helping harvest grapes. Although, most of the hinnies are quite old averaging 24.3+ years and soon there will be less hinnies in this area and less people using hinnies

mule. This attribute more closely resembles characteristics of a donkey versus a horse (Corazon Vaquero 2008). Other places like Portugal, Colombia, and Mexico, hinnies are found based on the availability of female donkeys and mares. In some communities it's easier to find mares to breed with jacks (male donkeys) and sometimes in more resource poor communities' donkeys may only be available, therefore, a female donkey is then mated with a stallion to produce a hinny. In some cases, the hinny has even been referred to as the "poor man's mule" (Fig. 2).

Breeding for Mules and Hinnies Hinny breeders in many countries have all shared challenges with raising hinnies. One such challenge is behavior, convincing a stallion to breed a female donkey but the same can be true when asking a jack to breed a mare horse. In Colombia like in the US through participatory interviews in October of 2014 we found the perception has generally been that hinnies are smaller and somewhat inferior to mules. However, many breeders have begun using artificial insemination and to improve the size and quality of hinnies being produced (Fig. 3).

Fig. 2 Hinny being used for cow working competition at the World Mule Show in Bishop, CA (World Champion Cow Mule (Hinny), LeMoan and Walter Nunn, Bryan, Texas)

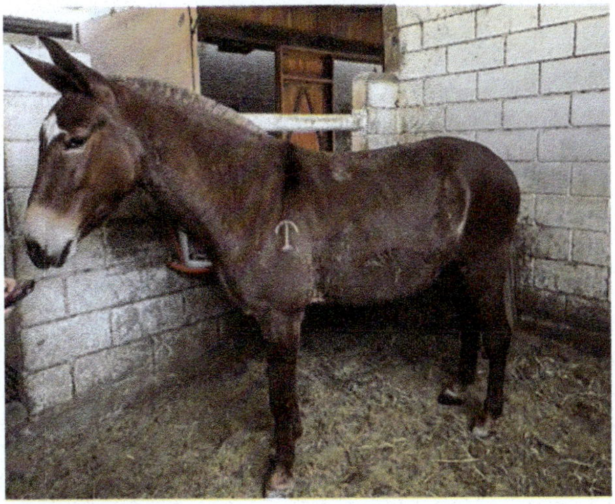

Fig. 3 Monarch is a hinny that was bred in Guadalajara, Mexico using frozen semen from a famous American Quarter Horse stallion bred to a jenny. Extremely nice quality equine as shown in the photograph

Fig. 4 A stallion in the state of Veracruz, Mexico used for breeding to jennies (female donkeys) for hinny production

During our survey work we found some breeders have reported difficulties in having jennies settle or conceive when breed to a horse. Several theories have been proposed that the decreased conception rates are due to acrosomal differences on the acrosome of the spermatozoa of the stallion compared to the donkey and other theories have been based on ideas that the intrauterine environment is somewhat different in a jenny compared to a mare. Generally, once a stallion is found to breed jennies and produce live foals, that stallion will continue to be used for hinny production (refer to Fig. 4).

2 Are All Mules and Hinnies Sterile?

There have been several reports of hinnies and mules having foals. A hinny stallion imported from Texas in the late 60s was reported to produce live and mature spermatozoa's (Trujillo et al. 1969). However, there was much controversy over the fertile animal but it's possible. Another case of a fertile mule in China during the late 80s later turned out to be a fertile hinny that produced a live foal named Dragon Foal (Rong et al. 1985, see Fig. 5). Figure 6 refers to the schematics and genetics of hinnies and mules that produce live foals and if they are mated with a horse or jack and what the outcome would be. More and likely there are more fertile mules and hinnies than we know of, but most people do not attempt to breed them. Depending on where they are born, and the beliefs associated with mules and hinnies having foals there's probably more live foals born in developing countries than we know of. In some cases, the live foal of a mule or hinny maybe seen or considered a blessing and in other cases a curse. Recent reports of live foals from mules have come from Colorado and Morocco.

So, how and why do we believe mules and hinnies are sterile? A donkey has 31 pairs of chromosomes, and a horse has 32 resulting in 63 pairs for a mule (see

Fig. 5 Mule mother produced a live foal named Dragon Foal in China in 1981 (Rong et al. 1985)

Fig. 6 Fertile hinny and mule schematics when bred with a horse or donkey

C. J. Zhao *et al.*

Table 1 Maternal chromosome-passing pattern in hybrids of horse and donkey[a]

	Mule (♀, 2n = 63)	Hinny (♀, 2n = 63)
Horse (♂, 2n = 64)	Horse (♀ or ♂, 2n = 64)	Hinny (♀ or ♂, 2n = 63)
Donkey (♂, 2n = 62)	Mule (♀ or ♂, 2n = 63)	Donkey (♀ or♂, 2n = 62)

[a]The first row and column are maternal and paternal parents, respectively, and the rest are the offspring from the corresponding parents in the first row and column.

Fig. 7). When the two are mated, the chromosomes must correspond in information and pair. The chromosome complements between these two species are not only numerically different but also morphologically distinct, which impairs proper homologous pairing during meiosis, leading to sterility in most hybrids (Trujillo et al. 1962). Based on the pairing of chromosomes researchers have suggested its more common female mules or hinnies can reproduce compared to males.

Studies have suggested that male mules when undergoing spermatogenesis, sperm cells will form but maturation will not continue past stage 2 or spermatid formation (Neves et al. 2002). Neves et al., compared donkey and mule testes morphometry and seminiferous epithelium length. Male mules have smaller testes compared to their sire the jack but interestingly this study found that despite having smaller testes, the seminiferous tubules contained 95% germ cells (spermatocytes), similar Sertoli cells but fewer Leydig cells in mules compared to jacks. The study found that spermatogenesis did not advance past the secondary phase (8 in total) to form mature spermatozoa, but the morphology of mule testes could be used as potential modeals for intraspecies transplant of germ cells (e.g. horse or donkey).

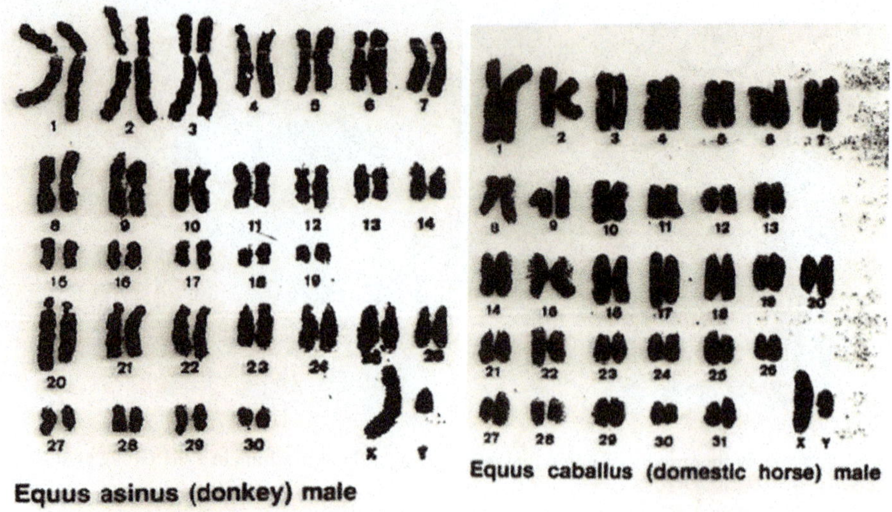

Fig. 7 Chromosomes from a donkey (31) and horse (32). The chromosomes must match and correspond to share information even the sex chromosomes are different (Trujillo et al. 1962)

Fig. 8 Miracle mule, mule who gave birth to a foal sired by a jack in Fez, Morrocco at the American Fonduk Vet Clinic and attending veterinarian Dr. Gigi Kay and director of the Fonduk carried out karyotyping to confirm the miracle at the vet school in Rabat

Yet the delay in spermatogenesis is the reason for the suspected assumption that mules are sterile. Another study compared three equine hybrids including a hinny, mule and Przewalski horse/domestic horse cross and found that abnormal pairing was seen in both the mule and hinny during the meiotic phase of spermatogenesis likely halting the progression of cell formation (Chandley et al. 1975).

Similar findings in female hybrids where oocytogenesis does not complete maturation as in the male hinny or mule (Han et al. 2016). However, in some cases this has proven false! The first documented case of a female mule producing a live offspring after being mated with a jack was in 1898 in Morocco. Many years later in 2002 another mule, a 14 year old mare mule gave birth to a foal at the American Fonduk Vet Hospital in Fez, Morrocco. Karotyping was carried out at a lab in Rabat (http://news.bbc.co.uk/2/hi/science/nature/2290491.stm) (see Fig. 8).

There have been scientific findings and proof of fertile female hinnies and mules (Rong et al. 1985). Different parental testing involving cytogenic testing have proven offspring of various fertile female mule and hinnies are in fact their offspring (Zhao et al. 2006; Ryder et al. 1985).

3 Hybrid Challenges from the Embryo to Neonate- How to Prevent Embryonic and Neonate Loss

Despite the fact that mules and hinnies may occasionally reproduce, such hybrid pregnancies present distinct physiological challenges. One significant issue is the early embryonic loss frequently observed in mares carrying mule foals. Fetal genotypes may affect placental development in equids specifically the chronic girdle (Allen and Short 1997). This appears to stem from a failure to develop secondary or accessory corpora lutea around days 45–60 of gestation, which is pivotal for sustained progesterone production. Without this luteal support, the mare's endocrine system may mistakenly recognize the pregnancy as not viable, leading to embryo resorption or loss (therio.vetmed.lsu.eduScienceDirect).

Furthermore, mares carrying hybrid foals tend to form fewer endometrial cups, structures responsible for secreting equine chorionic gonadotropin (eCG) which helps maintain pregnancy (Wilsher and Hoogewijs 2024). Serum eCG concentrations haven been reported to be lower in jennies carrying hinnies and mares carrying mule embryos compared to horses carrying horses or jennies carrying donkey embryos (Wilsher and Hoogewijs 2024). A recommendation for mares or jennies who have previously experienced early embryonic loss from low eCG is to provide active immune support by transferring donkey lymphocytes (Wilsher and Hoogewijs 2024). Documenting endometrial cup formation by ultrasounding can be indicative of development and if luteal support is warranted in early pregnancy. In other reports, accessory or secondary corpus lutea formation fail to develop in females carrying hybrid embryos and augmenting progesterone levels during early pregnancy. Higher pregnancy loss in mule pregnancies, with endocrine differences tracked. A comparative study monitored mares carrying mule vs. horse embryos by measuring progesterone and eCG in serum and monitoring embryonic vesicle growth and size. Boeta and Zarco found like other studies that pregnancy loss was significantly higher in mule pregnancies and mares inseminated with donkey semen had significantly (p 0. < 0.05) less progesterone and eCG compared to mares carrying embryos from horses (Boeta and Zarco 2005). Therefore, proactive progesterone supplementation is often advised in these cases to bridge the hormonal gap until endometrial cups fully develop and sustain luteal function naturally. When cups *do* develop in extraspecies pregnancies, eCG/progesterone can be high. Reports from embryo-transfer programs note that donkeys carrying mule embryos can show large cups and very high eCG, with corresponding high progesterone, reinforcing the central role of the cups/eCG → accessory CL axis. (clinicaltheriogenology.net). These conditions should be considered when performing breeding for hybrids and or performing interspecies embryo transfer from one species to the next (e.g. donkeys to horses or mules, or mules to horses, horses to mules).

When a jenny or mare carries their foal to term the next concern is to make sure the foal (hinny or mule) can nurse the colostrum from its dam (the mother). The problem lies with a difference in red blood antigens found in both donkeys and mules but not in horses (McClure et al. 1994). Further research found that not all donkeys carry this "unique" red blood cell (Traub-Dargatz et al. 1995). This has been described by some as "donkey factor." Over 10% of the mule foals are considered to be affected by this factor. Screening prior to the mare foaling should be conducted by testing her blood for the producing an anti-red blood cell antibody to donkeys. Colostrum can be tested by mixing the mare's blood with the colostrum (further information can be found in the immunology chapters and neonatal chapter).

4 Practical Management Suggestions

1. **Close early monitoring (days 14–120):**
   - Schedule frequent ultrasonography plus serum progesterone (and where available, eCG) checks from day ~35 onward, when the cups would normally begin forming. Escalate monitoring around days 45–100, the risk window highlighted in hybrid pregnancies. (ScienceDirect, ResearchGate, AGRIS, Periodikos)
2. **Progestin support when indicated:**
   - If progesterone is suboptimal or there's concern about absent/insufficient accessory CL, consider exogenous progestin supplementation (e.g., altrenogest/synthetic progesterone) through the window when eCG/accessory CL would normally sustain luteal function (often to ~day 120, tailored to labs and scans). This strategy aims to "bridge" the endocrine gap documented in hybrid pregnancies. (Periodikos, rep.bioscientifica.com, Equine-Reproduction.com, LLC)
3. **Document endometrial cup status:**
   - Use ultrasound and clinical timing to infer cup formation; if cups appear reduced/failed (as in donkey-in-horse analogs), maintain or intensify luteal support and surveillance, given the association with pregnancy loss. (PubMed, PMC)
4. **Risk-aware breeding calendar & general management:**
   - Manage body condition, stress, and seasonality (seasonal lack of supplementary CL can add risk). Optimize nutrition and disease prevention to reduce compounding factors for early loss. (ScienceDirect)

Protocol checklist you can adapt for breeders managing mares carrying mule or jennies carrying hinny pregnancies.

5 Hybrid Pregnancy Monitoring & Support Checklist

(Mares carrying mule or Jennies carrying hinny embryos)

1. **Early Pregnancy Confirmation**
   - Day 14–16: Ultrasound for pregnancy detection.
   - Confirm embryo location & viability.
2. **Critical Monitoring Window (Day 35–120)**
   Why? Endometrial cups should form (~Day 37–42), secrete eCG, induce accessory CLs → maintain progesterone. Failures are common in hybrid pregnancies.
   - Ultrasound scans every 2–3 weeks (focus on cups, CLs, placental development).
   - Serum progesterone check at each scan.
   - If available: eCG assay to assess cup function.
3. **Hormonal Support**
   - If progesterone <4 ng/mL (or downward trend):
     - Start altrenogest (Regumate®) or synthetic progesterone.
     - Continue supplementation until Day 120 (when placenta normally assumes role).
   - Adjust dose based on labs + ultrasound findings.
4. **Risk Period Alerts**
   - Day 45–100 = Highest risk for loss.
     - Watch for resorption/abortion signs: vaginal discharge, abnormal ultrasound findings, declining hormone levels.
     - Increase monitoring frequency if abnormalities suspected.
5. **General Mare Management**
   - Maintain body condition score 5–6/9.
   - Minimize stress (avoid transport, herd aggression).
   - Ensure balanced nutrition (adequate protein, trace minerals, Vit E).
   - Vaccinations & parasite control up to date.
   - Limit seasonal risks (heat stress, low supplementary CL formation in off-season).
6. **Communication & Documentation**
   - Record all scan results, hormone levels, treatments, and dates.
   - Educate owners about higher embryonic loss rates in mule pregnancies.
   - Discuss cost/benefit of intensive monitoring + hormone support upfront.

Goal *Maximize pregnancy maintenance until Day 120, when placenta reliably produces enough progesterone.*

Evidence Basis *Reports of increased embryonic loss in mule pregnancies linked to insufficient endometrial cup formation, low eCG, and poor accessory CL development. Exogenous progesterone shown to support luteal phase when natural support is inadequate.*

6 Mule Moms

Despite if mare mules are fertile or not studies have suggested that mare mules make excellent mothers (González et al. 2015; Gambini et al. 2025). Reports suggest that mare or molly mules are popular choices as recipient mares for embryo transfer specifically for carrying donkey embryos (see Fig. 9 and refer to chapter "Advanced In Vitro Breeding Technologies for Donkeys and Mules" on ART for more information about embryo transfer in mules). González et al. (2015) reported out of 72 mare mules 61.1% had ovaries that were cyclic (n = 70), 45.8% had corpus luteaus, and 11 oocytes were found from antral follicles. Mules in one study carrying donkey embryos carried the fetuses to terms 372 to 379 days. The study reported the mare mules had excellent maternal ability of and produce large amounts of milk. Martini et al. 2024 reported mule milk qualitative characteristics and found their milk to be similar to horse and donkey milk. The researchers were able to monitor milk quality throughout the lactation period and did note a change in the composition which likely reflected a change in the growth of the foal and nutritional needs.

7 Horse'n Around- Reproductive Behavior in Male and Female Hybrids

To not repeat what has already been covered in the behavior and the surgery chapters but surgical procedures for both male and female mules and hinnies are often practiced shifting reproductive behavior such as decreasing aggression or estrus behavior. Male mules in parts of the United States are referred to as "horse mules" due to their playful yet aggressive nature with one another also referred to as "horsing around." In some parts of the world mules are not castrated yet reports of aggression such as pawing, biting and kicking have been seen in intact male mules. Generally, speaking the number one reason for castration is to alter behavior. Mule owners should consider castration options at an early age before learned behaviors develop. Female mules and hinnies may display signs of estrus or may not (see

Fig. 9 Mule mom with donkey foal resulting in an embryo transfer. (Photo credit, Dr. Luis Arryos, 2018, University of Cordoba, Argentina)

Fig. 10 Estrus behavior in mules appears similar to estrus in donkeys with head and neck extended forward, mouth chewing or clapping, other female mules' express interest and mount in some cases

Fig. 10). Heaton et al. (2018) conducted a study with over 100 mule owners seeking information about estrous behavior in mules from an owner's perspective. From the owners who were surveyed 71% recognized estrous behavior in mules and 24% asked for veterinary advice due to changes in behavior. Most of the mules were used for riding and performance purposes. Treatments suggested included hormonal treatment (application of synthetic progesterone, Regumate®) or owners considered or used ovariectomy as a permanent solution. So, regardless of if mules are sterile or not they still may show reproductive tendencies and considering how to manage this behaviors should be taken into account from castrations to hormonal treatment.

8 Conclusion

Although mules and hinnies are generally assumed to be sterile, this is not universally true. Further research is needed, including controlled breeding studies with larger numbers of male and female hybrids, to better understand the occurrence of fertility. Most male mules and hinnies are castrated at a young age, yet it remains unclear whether spermatogenesis is truly prevented in all intact males. Similarly, no large-scale studies have been conducted with female hybrids to systematically evaluate their reproductive potential. Existing reports indicate that some cyclic female mules can carry embryos to term, and rare cases of fertile hinnies and mare-mules have been documented.

Hybrid pregnancies can be complicated by placental development issues, and introducing semen into a jenny's uterus may pose additional challenges. Despite these obstacles, mules make excellent recipients for embryo transfer, offering good milk quality and producing foals with strong survival instincts. Reproductive management—including castration or ovariectomy—may still be necessary to manage behaviors, even if hybrids are presumed sterile.

Advances in assisted reproductive technologies, breed conservation, and embryo transfer techniques are helping to optimize hybrid breeding protocols. By understanding and managing reproduction from insemination through foaling, we can improve efficiency and success in breeding mules and hinnies.

Questions and Answers

1. Are all mules and hinnies sterile?
   True or **False**
2. Female mules and hinnies can display signs of estrous.
   True or False
3. Male mules should only be castrated if their behavior is aggressive.
   True or **False**
4. Why is there early embryonic loss in hybrid equine?
   (a) To much progesterone because the embryo is a hybrid
   (b) **Not enough progesterone and eCG produced**
   (c) Because it's a hybrid
   (d) There is no difference in early embryonic loss across equid species and hybrids
5. Can mare mules carry embryos and produce milk
   True or False

References

Allen WR, Short RV (1997) Interspecific and extraspecific pregnancies in equids: anything goes. Journal of Heredity 88(5):384–392

Boeta M, Zarco L (2005) Progesterone and equine chorionic gonadotropin concentrations around the time of pregnancy loss in mares impregnated by donkeys or stallions. Journal of Equine Veterinary Science 25(12):531–536

Chandley AC, Short RV, Allen WR (1975) Cytogenetic studies of three equine hybrids. J Reprod Fertil Suppl 23:356–370. https://pubmed.ncbi.nlm.nih.gov/1060807

Corazon Vaquero DVD (2008) www.corazonvaquero.com

Gambini A (2025) Current and Emerging Advanced Techniques for Breeding Donkeys and Mules. Animals 15(7):990. https://doi.org/10.3390/ani15070990

González SM, Gomes RG, Souza AK, Silva CB, Silva-Santos KC, Seneda MM (2015) Evidences of regular estrous cycles in mules and successful use of these animals as recipients for donkey embryos. J Equine Vet Sci 35:869–872

Grondin TM, Dewitt SF (2014) Chapter 106: Normal Haematology of the horse and donkey. In: Schalm's veterinary Haematology, 6th edn. Wiley-Blackwell, Hobodken, pp 821–828

Han H, Wang A, Liu L, Zhao G, Su J, Wang B, Li Y, Zhang J, Wu B, Sun W, Hu S, Li S, Zhao L, Li X (2016) Testicular characteristics and the block to spermatogenesis in mature hinny. Asian-Australas J Anim Sci 29(6):793–800. https://doi.org/10.5713/ajas.15.0670

Heaton K, Ragle C, Godderidge MT (2018) Estrous behavior in mules—an owner's perspective. J Equine Vet Sci 60:109–112

Martini M, Degl'Innocenti A, Altomonte I, Sodi I, Bocci C, Fanelli D, Moroni R, Panzani D, Camillo F, Salari F (2024) Report on a milking mule: milk qualitative characteristics during lactation. Animals 14(11):1585. https://doi.org/10.3390/ani14111585

McClure JJ, Koch C, Traub-Dargatz J (1994) Characterization of a red blood cell antigen in donkeys and mules associated with neonatal Isoerythrolysis. Anim Genet 25(2):89–94. https://doi.org/10.1111/j.1365-2052.1994.tb00091.x

McLean AK, Wang W, Navas-Gonzalez FJ, Rodrigues JB (2014) Haematological and serum biochemical parameters in healthy working horses, donkeys, mules and hinnies in Portugal and Spain. In: Proceedings 7th international colloquium on working equid. Royal Holloway, University of London, London, p 201

McLean AK (2016) Comparing the physiological and biochemical parameters of mules and hinnies to horses and donkeys. In: Proceedings of the Donkey and Mule Welfare Symposium, Hydra Donkey and Mule Conference, Greece

Neves ES, Chiarini-Garcia H, França LR (2002) Comparative testis morphometry and seminiferous epithelium cycle length in donkeys and mules. Biol Reprod 67(1):247–255. https://doi.org/10.1095/biolreprod67.1.247

Rong R, Yang X, Cai H, Wei J (1985) Fertile mule in China and her unusual foal. J R Soc Med 78: 821–825

Ryder OA, Chemnick LG, Bowling AT, Benirschke K (1985) Male mule foal qualifies as the offspring of a female mule and Jack donkey. J Hered 76(5):379–381. https://pubmed.ncbi.nlm.nih.gov/4056372

Traub-Dargatz JL, McClure JJ, Koch C, Schlipf JW (1995) Neonatal isoerythrolysis in mule foals. JAVMA 206: 1 Clincal Report:67–70

Trujillo JM, Stenius C, Christian LC, Ohno S (1962) Chromosomes of the horse, the donkey, and the mule. Chromosoma (Berl) 13:243–248

Trujillo JM, Ohno S, Jardine JH, Atkins NB (1969) Spermatogenesis in a male hinny: histological and cytological studies. Jrnl Hered 60(2):79–84

Wilsher S, Hoogewijs M (2024) What has embryo recovery and transfer taught us about equine reproductive physiology? Clin Theriogenol. https://doi.org/10.58292/CT.v16.10602

Zhao CJ, Qin YH, Lee XH, Wu C (2006) Molecular and cytogenetic paternity testing of a male offspring of a hinny. J Anim Breed Genet 123(6):403–405. https://doi.org/10.1111/j.1439-0388.2006.00615.x

Neonatal Care and Considerations

Carla Bargi Belli, Natalia Dinoá Leite de Vasconcelos,
and Claudia Barbosa Fernandes

Abstract

Neonatology in donkeys and mules presents unique challenges for veterinarians and professionals working with these animals. While many conditions occur and should be managed similarly to other equine neonates, there are distinct physiological differences that must be considered in these species. Available literature on neonatal care in donkeys and mules is limited, particularly for mules, and is nearly nonexistent for hinnies. Most publications consist of case reports or surveillance data that include donkeys and/or mules alongside horses, often without sufficient detail for direct comparison. Comprehensive resources specifically dedicated to these animals remain scarce, with only few books and isolated chapters addressing their unique characteristics. In this chapter, we provide an overview of the key differences in neonatal care for donkeys and mules, emphasizing aspects of the immediate postpartum period and discussing the primary diseases affecting neonates of these species.

Keywords

Neonatology · Behavior · Diseases · Mule · Donkey

C. B. Belli
Department of Internal Medicine, School of Veterinary Medicine and Animal Science, University of São Paulo, São Paulo, SP, Brazil

N. D. L. de Vasconcelos · C. B. Fernandes (✉)
Department of Animal Reproduction, School of Veterinary Medicine and Animal Science, University of São Paulo, São Paulo, Brazil

A. K. McLean et al. (eds.), *Donkey and Mule Medicine: Clinical Medicine and Reproduction*, https://doi.org/10.1007/978-3-032-18540-2_19

1 Peripartum

1.1 General Considerations

The increasing demand for the birth of healthy animals with high economic potential and superior genetics affects both horses and donkeys, and their hybrid offspring are no exception (Fig. 1). Effective monitoring of gestation and parturition, along with neonatal evaluations, is essential to safeguard both mare and neonate's health and welfare in equid breeding practices. Gestation length is one of the primary parameters used to estimate the expected foaling date and to ensure proper monitoring.

Jennies pregnant with donkeys have been reported to have an average gestation length of 362.0 ± 15.3 days (range: 331–421 days) (Galisteo and Perez-Marin 2010), with another study showing a duration of 374 days (Fielding 1988). However, shorter gestation lengths have also been observed, such as 353.4 ± 13.0 days (range: 339–370 days) (Crisci et al. 2014). A study conducted by Boakari et al. (2019) examined differences in gestation length between mares bred to stallions and those bred to donkeys. Contrary to anecdotal reports suggesting that pregnancies carrying mule fetuses last longer than horse pregnancies, the study found that the stallion-bred group had a gestation length 6.1 days longer. No significant differences were observed concerning fetal sex.

Gestation intervals for mares carrying mule fetuses, as reported in the literature, range from 316.9 to 365.3 days (Boakari et al. 2019), 342.3 ± 10.8 days (Paolucci et al. 2012), and 343.1 days (range: 315–369 days) (Giger et al. 1997). The chapter authors hypothesize that these variations are likely influenced more by the maternal average gestational days than by fetal factors.

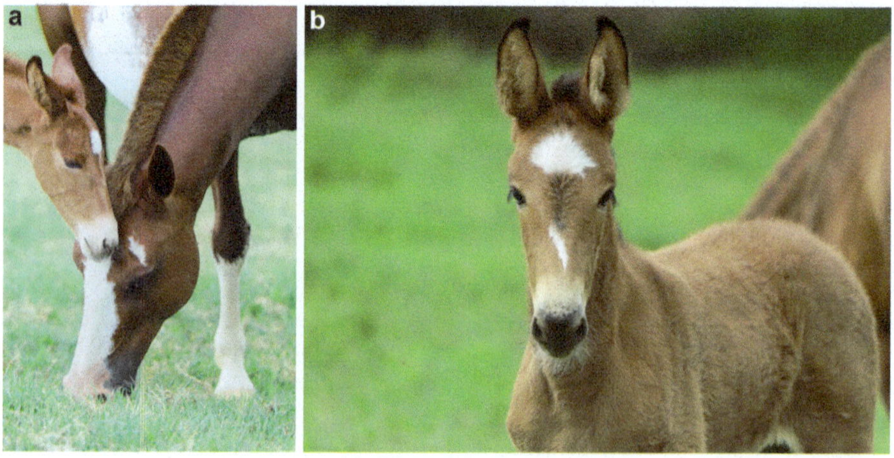

Fig. 1 Mule (*Equus asinus* × *Equus caballus*) neonate. (Source: DVM PhD Maria Augusta Alonso)

Regarding foaling prediction, one study assessed physical indicators (vulvar relaxation and sacrosciatic ligament relaxation) and mammary secretion parameters (pH, BRIX index, and electrolyte concentration), comparing mares carrying mule versus horse fetuses to evaluate the effectiveness of these markers in determining the day of birth. The study concluded that, in all groups, vulvar and sacrosciatic ligament relaxation increased as parturition approached. However, mares carrying mule fetuses exhibited greater relaxation on the day of birth. No significant differences were observed in the other parameters among the groups (Octaviano et al. 2024). These findings suggest that, despite some specific differences, mule pregnancies closely resemble equine pregnancies.

Regarding placental evaluation, although no macroscopic differences in mule versus horse placental morphology were observed, Tinel et al. (2023) demonstrated that the allantochorion membrane of mule pregnancies exhibited a significantly wider base of microcotyledons in the nongravid uterine horn and showed a tendency toward increased microvilli tissue composition within the allantoic mesoderm of the uterine body. These findings provide evidence of structural differences in the term allantochorion membranes between interspecies and horse pregnancies (Tinel et al. 2023).

Vaccine schedules, deworming, colostrum quality testing, balanced nutrition for mares carrying mule fetuses and pregnant jennies, and all other aspects of peripartum care require the same level of attention as those commonly practiced in horse breeding.

2 The Immediate Postpartum

2.1 General Considerations

Knowing the normal timing and behaviors of the newborn after birth is extremely important to quickly recognize when there is a problem and intervene when possible and necessary. It is also important to avoid thinking there is a problem when in fact there isn't. For horse neonates, these parameters are already well known and described in the literature, even including small breed variations. Such information is less available and known when it comes to donkeys and especially mules. And it is nonexistent for hinnies.

2.2 Perinatal Parameters

Key Points

- **Mule** and **donkey** neonates have higher Apgar scores and stand up and nurse earlier than horse neonates.
- **Mule** neonates take longer to eliminate meconium than horses.

The most important information and considerations are the same regarding horse neonates: the presence of a strong suckle reflex, the time to get into sternal decubitus prior to standing, the ability to nurse, the production of urine and passing of meconium, the findings from a modified Apgar system, etc.

For mule and donkey foals, Apgar scores of 7 and 8 are also considered normal, but with higher averages than in horse foals (Sgorbini et al. 2009; Bonelli et al. 2020; Alonso et al. 2023).

A study (Bindi et al. 2023) showed that for mules, the times recorded were: 1 min to lift the head, 2 min for sternal decubitus, 20 min for the suckling reflex, 38.7 ± 13.4 min to stand up, 72.3 ± 30.5 min to nurse, and 416.3 ± 401.8 min to eliminate meconium. This indicates that mules' foals take less time to stand up and nurse compared to what is known for horses but take longer to eliminate meconium.

There is a study that compared these parameters between horse and mule neonates from the same property, with all the mares being of Mangalarga, a breed from Brazil (Alonso et al. 2023) and found the same conclusions. In this study, mule foals exhibited higher Apgar scores in all evaluation times compared to horse foals, with a slight advantage in females. Additionally, mule foals began nursing faster than horse foals, indicating a quicker adaptation to extrauterine life. On the other hand, mule neonates took longer to pass meconium, but all were healthy and did not require intervention. Also, the time for placental expulsion was shorter in the third phase of parturition for mares that foaled mules compared to those that gave birth to horse foals. Although this study only examined one breed and compared horses and mules from the same farm (Alonso et al. 2023), it is expected that the overall behavior is the same for other horse breeds used to produce mules, even though the results may vary slightly.

For Amiata donkey foals (Bonelli et al. 2020), in animals with an Apgar score of 7 or 8, the times recorded after birth were: 2 min for sternal recumbency, up to 20 min for the suckling reflex, 40 ± 8.9 min (Apgar 8) and 40 ± 18.8 min (Apgar 7) to stand, 120 ± 10.4 min (Apgar 7) and 180 ± 32.8 min (Apgar 8) to nurse, with no differences between males and females. In another study (Veronesi et al. 2014a) with a different breed, Martina Franca donkey foals, times recorded were 60.4 ± 9.9 min to stand and 99.6 ± 11.55 min to nurse. Hence, there is some variation depending on the breed considered.

In donkeys, similar to horses and different from mules, the time for meconium elimination is up to 3 h after birth (Mendoza and Toribio 2024). Asinine foals are not very resilient and require more attention in the first weeks (Burton et al. 2014).

Although there is not much specific research, the findings in cases of prematurity or dysmaturity are similar to those seen in horses (Burton et al. 2014; Evans and Crane 2018), such as weakness, low birth weight, difficulty standing up and maintaining a standing position, a more domed head, drooping ears, furrier coat than normal for the species, delayed hoof hardening, failure of cuboidal bone ossification, tendon laxity, pulmonary atelectasis, feeding intolerance, failure to maintain blood glucose levels, and laboratory abnormalities (such as a lower neutrophil/lymphocyte ratio). However, it is important to remember that the expected gestational period for donkeys and mules differ from horses and that twin pregnancies are more common in donkeys (Mendoza and Toribio 2024).

3 Parameters of the Neonate Examination

3.1 Physical Exam Parameters

Key Points
- **Donkey** foals and **mules** are more likely to develop hypothermia than horses.

Just like horses, donkey foals also tend to have higher rectal temperature, heart rate and respiratory rate than adult donkeys, with external factors also playing a role (Zakari et al. 2018). A study (Bonelli et al. 2020) showed that, for Amiata donkey neonates, in the first 5 min of life, heart rate was 120 ± 7.5 bpm (Apgar 8) and 100 ± 10.9 bpm (Apgar 7), respiratory rate was 64 ± 1.8 breaths per minute (Apgar 8) and 36 ± 4.3 breaths per minute (Apgar 7), and rectal temperature was 38.3 ± 0.1 °C (Apgar 8) and 38 ± 0.2 °C (Apgar 7).

After this initial period, in asinine foals, normal values are considered to be: temperature of 37.5–38.5 °C, a heart rate of 80–120 bpm, and a respiratory rate of 30–40 breaths per minute after 12 h of life (Barrio et al. 2019).

It is noteworthy that donkeys and mules, both adults and foals, are more likely to develop hypothermia in cold climates. Even though they are born with a denser coat, donkey neonates, regardless of breed or health status, can develop hypothermia more easily than horses, and it is not true that they are more tolerant to low temperatures (Mendoza and Toribio 2024). If they get wet from rain, they can develop hypothermia very quickly, which can progress to death. In cases of foals struggling to maintain body temperature in the first hours after birth, it is important to check for hypothyroidism (Burton et al. 2014).

In addition to physiological parameters, equid neonatal examination should evaluate behavioral changes, congenital abnormalities, etc. For example, in both donkeys and mules, umbilical hernias have also been reported (Moustafa et al. 2022).

Another interesting point to note during the examination is that donkeys are less sensitive to the cough reflex than horses (Barrio et al. 2019).

Healthy donkey neonates, in the first week of life, nurse 5–6 times per hour and consume about 20% of their body weight, and weight gain varies according to the breed (Mendoza and Toribio 2024).

3.2 Laboratory Findings and Complementary Exams

Key Points
- In both **donkeys** and **mules**, red blood cells, hematocrit, and hemoglobin decrease in the first days of life.
- **Asinine** neonates show an increase in hematocrit only at higher percentages of dehydration.
- Sick **asinine** neonates are more likely to develop increased triglycerides than horses.

- **Mule** neonates have more immature carpal bones at birth compared to horses.
- In **asinine** neonates, the involution of umbilical structures is slower than in horses.

When collecting blood, it is important to consider some differences, such as the fact that asinine foals have thicker skin than horses (Mendoza and Toribio 2024).

Some important differences regarding laboratory tests between donkeys and horses are that hematocrit tends to be higher in young donkeys, and in general, donkeys do not show an increase in hematocrit until significant dehydration occurs (greater than 12%). The serum of asinine also tends to be clearer than that of horses (Burton et al. 2014). Additionally, hematocrit and hemoglobin tend to be higher in donkey foals than in horses (Mendoza and Toribio 2024). A summary of donkey and mule neonates' laboratory values found in different studies is presented in Tables 1 and 2.

Regarding the neonate examination in the immediate postpartum period, a study conducted in Brazil with horse and mule foals from Mangalarga mares (Boakari et al. 2021) showed that glucose and lactate concentrations were different between the two groups, but with similar patterns: lower glucose at birth, increasing after nursing then stabilizing, and higher lactate after birth with a gradual decrease over time. However, even though the pattern was the same, the concentrations differed: mule foals had lower blood glucose immediately after birth when compared to horse foals and this parameter reversed at 12–24 h postpartum. On the other hand, mean blood lactate concentration in the neonatal period was lower in mule foals compared to horses. One explanation for these differences is that mule foals may have more efficient adaptability after birth, demonstrated by higher APGAR scores and different behavior (Alonso et al. 2023), and a possible increased oxygen uptake, that would decrease the anaerobic metabolism and subsequently result in lower lactate concentrations with less intracellular acidosis (Boakari et al. 2021). In summary, regardless of the concentrations, it is important to monitor trends in the individual neonate. In this study, analyses were performed both in the laboratory and with a portable device showing different concentrations obtained by each method, but with similar patterns over time. The same profile with higher lactate levels and more acidotic at birth were also found in donkey (Veronesi et al. 2014b; Carluccio et al. 2017) and mule neonates from different breeds (Bindi et al. 2023).

Regarding hematological values, the trend for Martina Franca donkey foals is that red blood cell count, hematocrit, and hemoglobin levels are higher at birth and decrease during the first 3–4 days after birth. Meanwhile, leukocytes remain stable during the first 3 days and then increase significantly, with lymphocyte levels rising on days 6 and 7 (Veronesi et al. 2014b).

Regarding serum biochemical evaluation, studies on Martina Franca donkey neonates (Veronesi et al. 2014b) showed that ALT concentrations increased (higher in females than in males) 24 h after birth and returned to previous levels by day 7. CK levels were higher in the first 12 h of life and decreased between days 3 and 10. ALP was also higher in the first 24 h, as were urea and creatinine concentrations. As expected, total protein (but not albumin) was lower at birth and increased after

12 h. Bilirubin was slightly lower at birth, increased after the first and second feedings, and then decreased over time. Other parameters showed some fluctuations, but with significant individual variations. A similar biochemical pattern was also observed in Amiata donkeys, though with some differences in values (Sgorbini et al. 2013).

Very high creatinine concentrations and/or concentrations that remain elevated for longer than expected after birth may indicate placental insufficiency, hypovolemia, or renal impairment (Mendoza and Toribio 2024).

Regarding mules (Italian Trotter mares bred with Amiata donkeys), a study that monitored laboratory values in the first 48 h of life observed a decrease in creatinine and an increase in magnesium (Mg) during this period (Bindi et al. 2023). Another study (Vernaccini et al. 2024) evaluated mules (Italian Trotter mares bred with Amiata donkeys) from birth to 2 months of age and reported a decrease in red blood cell count, hematocrit, and hemoglobin levels from the first 24 h to 1 week of age. There was also a decrease in mean corpuscular volume and mean corpuscular hemoglobin during the first month. Additionally, the study observed an increase in leukocyte count from birth until the seventh day of life, an increase in AST, and a decrease in ALP in the first month, along with a reduction in lactate, urea, and creatinine in the first 24 h.

Although not routinely measured, a study showed differences between horse and mule neonates in oxidant and antioxidant parameters (Riccio et al. 2023). In this study, mules exhibited lower levels of lipid peroxidation along with a higher concentration of glutathione peroxidase, as well as lower bilirubin concentrations, indicating species differences in adaptation to the extrauterine environment. Low effectiveness of the antioxidant system at birth was identified in donkey foals (Sgorbini et al. 2021).

The use of serum amyloid A (SAA) measurement has increased in equine medicine. A study evaluated the values in female donkeys and their foals 48 h and 30 days after birth, all without alterations (Bazzano et al. 2022). The SAA value for these foals was 37.44 ± 19.75 μg/ml at 48 h of life, and 16.04 ± 18.14 μg/ml at 30 days after birth.

Triglycerides increase from birth until the 7th day of life and then remain relatively stable into adulthood. Triglyceride concentrations are slightly higher in asinine neonates receiving milk replacers than in others, but there is no difference in blood glucose, cholesterol, or non-esterified fatty acids (Mendoza and Toribio 2024). Hyperlipemia is considered common in sick asinine foals (Burton et al. 2014; Mendoza and Toribio 2024), different from what is observed in horses.

As with horse neonates, blood glucose evaluation is very important, as these animals frequently and rapidly develop hypoglycemia. Other electrolyte disorders, such as hypocalcemia and hypomagnesemia, can also occur in sick animals and should be checked, as well as hypokalemia, hypomagnesemia, and hypophosphatemia in neonates with hyperlipemia and/or receiving parenteral nutrition (Mendoza and Toribio 2024).

As used in horses, cytological evaluation and culture of tracheal lavage can also be used as complementary respiratory system tests for donkey and mule neonates (Carneiro et al. 2017a).

Radiographic examinations can also be highly useful in these animals, but there are few studies comparing findings to what is known in horses. One study (Nocera et al. 2023) evaluated the radiographic development of the distal portion of the thoracic limbs and carpus of mule neonates (from Standardbred mares and Amiata donkeys) from birth to 3 months of life, showing morphological characteristics intermediate between horses and donkeys. An important finding is that, at birth, these animals have more immature carpal bones than horses, with rounded margins on the cuboid bones and greater spacing between them. This is highly relevant to avoid misdiagnosis of prematurity or dysmaturity in these animals.

Similarly, ultrasonographic examination can be used in neonates of all equids. Regarding umbilical evaluation, one study (Vitale et al. 2021) found no differences in the appearance or measurements of umbilical structures in the first week of life between asinine foals and horses, despite donkey foals having a lower birth weight than horses. However, donkey foals exhibit a slower involution of these structures compared to horses. During the first 7 days of life, no significant differences in structure size were observed, resembling what is known for calves and differing from horses, in which a 20% reduction in umbilical structure size is expected within the first week of life.

4 Diseases

4.1 General Considerations

The examination of neonates depends on knowledge of conditions as well as a thorough clinical examination. Knowing how to approach the animal without stressing the mother and safely and calmly restraining the foals help a lot in the assessment. It is recommended that neonates be examined immediately after birth or as soon as possible, even if no evident conditions are present, to detect any congenital or other abnormalities and initiate early treatment if necessary.

Regular and early handling of the animals should be encouraged. This will help neonatal donkeys, mules, and hinnies develop better interaction and trust with humans, making any procedures progressively easier and facilitating future handling (such as ear access, halter placement, temperature measurement, etc.) (McLean et al. 2019).

Although in some parts of the world, mules and donkeys receive the same quality care as horses, this is not the case in many others. Proper nutrition, deworming, vaccinations, and basic veterinary care can make a significant difference in the occurrence of neonatal conditions and in the prognosis regarding treatment.

4.2 Congenital Disorders

Most congenital disorders don't have a known inherited genetic pattern. There are few known genetic diseases in donkeys. Another factor to be considered is the higher degree of inbreeding found in some asinine breeds, which may increase the number of congenital abnormalities in these populations (Robbe et al. 2012).

An important factor to consider in mule cases is the genetic predisposition that may exist from the base breed in the mare, but it is not known how it will manifest itself in the hybrid offspring.

Congenital alterations can be benign, with no major consequences for the animal's life, such as a case of congenital oral papilloma (Fig. 2), which is easily resolved with a simple surgical procedure. However, some can hinder the quality of life or result in mortality. Malformations of the limbs that cannot be corrected and prevent the animal from standing (Fig. 3) is one such example. Other alterations that

Fig. 2 Asinine (*Equus asinus*) neonate with congenital papillomatous formation on the lower lip. (Source: FMVZ-USP Veterinary Hospital)

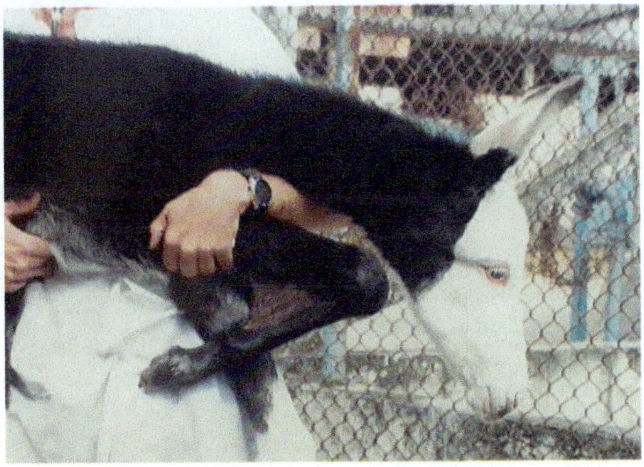

Fig. 3 Mule neonate. Note the malformation in the right thoracic limb with the formation of a skin "membrane" between the proximal and distal portions of the limb, preventing it from fully extending. Tendons, vessels and nerves passed through this membrane, preventing surgical correction. The animal was euthanized during surgery. (Source: FMVZ-USP Veterinary Hospital)

are incompatible with extrauterine life, such as hydranencephaly, has been reported in a mule by Chacur et al. (2014).

Congenital abnormalities can appear in an isolated form or as multiple defects, such as the reported case of musculoskeletal alterations associated with prognathism in a neonatal mule (Nocera et al. 2024), which can sometimes complicate treatment or worsen the prognosis for animals with multiple associated abnormalities.

Some congenital diseases already reported in donkeys: epidermolysis bullosa (hereditary); epitheliogenesis imperfecta; entropion (Grove 2008); digit agenesis (Robbe et al. 2012), atresia ani and hypospadias (Nelson et al. 2015), rectal atresia (Teixeira et al. 2010), cleft palate (Shaw et al. 2015), choanal atresia and laryngeal deformities (Barton and Ohnesorge 2009), osteopetrosis with dental eruption abnormalities, brachygnathia, and pathological fractures due to the absence of osteoclasts (Williamson et al. 2016), lateral patellar luxation (Abu-Seida and Shamaa 2022), and persistent ductus arteriosus (Lowe 1972). In mules, strabismus has also been reported (Lassaline and Latimer 2018). One author (Knottenbelt 2008) mentions a case of skin fragility in a donkey, similar to skin asthenia seen in other species, including horses.

Common alterations are flexural deformities, which may appear independently or as part of the findings in premature or dysmature animals (Fig. 4). Some simpler cases may correct themselves spontaneously, while others will require splints (Fig. 5), corrective hoof trimming, and even surgical treatment, similar to what is recommended for equine neonates.

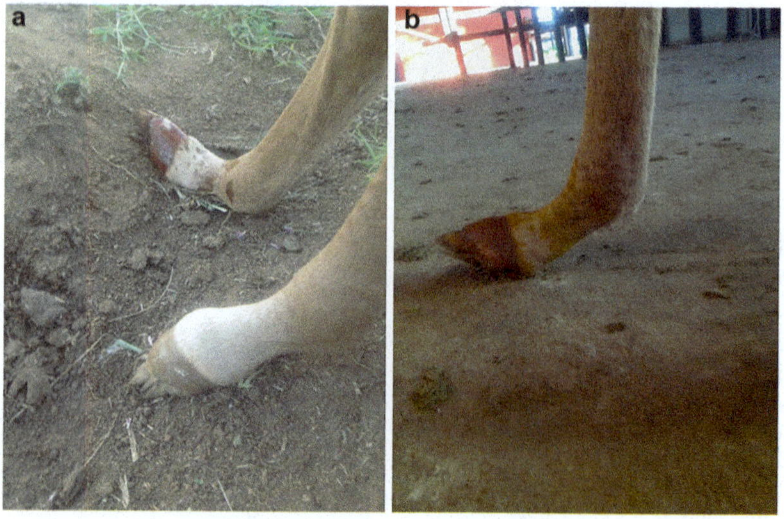

Fig. 4 Mule neonate with flexural deformity in limb's distal portion. (Source: DVM PhD Maria Augusta Alonso)

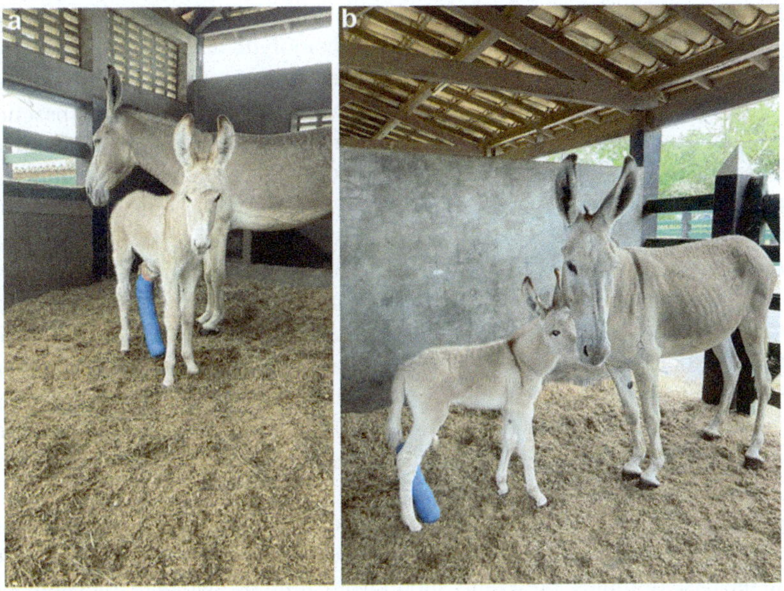

Fig. 5 Asinine neonate undergoing treatment with orthopedic splints. (Source: DVM MsC Fernanda Timbó)

4.3 Neonatal Isoerythrolysis (NI)

Key Points

- Isoerythrolysis is more common in **mule** neonates

Isoerythrolysis occurs due to incompatibility between the blood type of the female and the male that will produce the neonate, with the need for the female to be previously sensitized to the foal's blood type (through pregnancy or transfusion of previous blood components). After birth, the foal ingests colostral antibodies against its own red blood cell antigens, resulting in hemolysis and/or agglutination.

In the case of mules, although there are still few conclusive studies on the subject, theoretically all pregnancies are considered incompatible (Johnson 2006), due to the presence of a different blood type in donkeys (donkey factor). This "donkey factor" is a different antigen, not found in red blood cells in horses and will therefore always initiate the occurrence of antibodies against this factor in horses exposed to it (by transfusion or pregnancy). But apart from this information, little is known about blood types in donkeys and mules.

Therefore, the occurrence of NI is likely higher in mule foals than in other equine neonates. It is not yet known whether the occurrence in mule neonates is actually higher than reported due to subclinical cases going undiagnosed and only those cases with more severe anemia being discovered, or whether the hemolysis related to the donkey factor is lower than that generated by the reaction to types Aa and Qa, the main causes of NI in horses.

The clinical manifestations of NI, which begin in the first hours to days of the neonate's life, will be the same as those of a hemolytic anemia: pale and/or icteric mucous membranes, apathy, weakness, dyspnea, with or without hemoglobinuria.

It must be differentiated from anemia due to other causes: umbilical cord rupture; hemophilia; internal or external bleeding; hemoparasites; sepsis. The Coombs test can be confirmatory, but not all laboratories do it and, if the clinical manifestations and other laboratory results are very suggestive, treatment should be started before and regardless of the result of the test (Mendoza and Toribio 2024).

Treatment will vary according to severity and can range from close observation and fluid therapy to prevent kidney damage from hemoglobin to the need for oxygen therapy and blood transfusion (Knottenbelt et al. 2006). In mules, as isoerythrolysis is usually associated with antibodies against the "donkey factor", any horse that has not been previously exposed to this factor (through transfusion, pregnancy or living with donkeys) can be considered a good blood donor (Estell and Magdesian 2015). Before transfusion, the use of corticosteroids can reduce the chance of adverse reactions (Mendoza and Toribio 2024).

In cases where blood transfusion is necessary, red blood cells washed from the mother or blood from another animal can be used, preferably with a compatibility test. In the case of mules, preferably blood from animals that have not been exposed to asinine blood (Radcliffe et al. 2022).

Burton et al. (2014) recommend that all pluriparous mares or mares with an unknown pregnancy history that are pregnant with a mule foal should be tested for the presence of donkey erythrocyte factor antibodies 2–3 weeks before giving birth so that measures can be taken to prevent the disease from occurring, if necessary. However, not all laboratories have this type of test available.

4.4 Hemoparasites

Key Points

- **Mule** neonates, like horses, can have transplacental passage of *Theileria equi.*

Hemoparasite infections are endemic in many parts of the world. Young animals are less susceptible to clinical manifestations than adults and donkeys present with acute disease less frequently than horses (Anzuino 2008). Besides being infected by *Theileria equi* and *Babesia caballi* via ectoparasites (ticks) and showing the same manifestations as adults or becoming persistently infected without clinical disease, neonates can also be infected through the placenta by *Theileria equi*. This has been discovered for both horses and mules (Françoso et al. 2018).

If foals are born infected, they may be born weak or show more apparent signs in the first 2 weeks of life (Knottenbelt et al. 2006).

In cases with obvious clinical signs, treatment with imidocarb may be indicated. The dose for donkeys is higher than that for horses, ranging from 2 to 4 mg/Kg, intramuscularly (IM), with frequency and duration of treatment varying according to the agent and the severity of the manifestations. It is important to note that this treatment is very toxic to donkeys, although to a lesser extent in foals (Knottenbelt et al. 2006). Adverse effects such as diarrhea, excessive salivation and pain at the injection site can occur.

4.5 Immune-Mediated Thrombocytopenia (IMT)

Key Points

- **Mule** neonates have a higher occurrence of alloimmune thrombocytopenia than horses and donkeys
- Thrombocytopenia can be associated with dermatitis, isoerythrolysis or neutropenia

As with NI, IMT occurs due to the female's sensitization to antigenic determinants from the male, but in this case against platelets rather than red blood cells. It is not yet known which antigens on the platelets are responsible for this sensitization. There is no known breed or sex predilection for IMT in horses, but there is an apparently higher occurrence of this condition in mules (Sellon 2006).

Neonates can develop isolated alloimmune thrombocytopenia or it can occur in association with multiple alloimmune disorders: NI, neutropenia and dermatitis (Sellon 2006), being associated with ulcerative dermatitis the most common (Bernard 2018).

As clinical manifestations, there will be petechiae, ecchymoses, prolonged bleeding at vascular puncture points and even hemorrhages due to thrombocytopenia. In cases of dermatitis, this will mainly occur around the eyes, mouth and perineal region, but crusty dermatitis may also occur all over the body and oral ulcers. Some animals may show weakness and apathy when there is associated anemia, and some may also present with neutropenia. A differential must be made for other causes of thrombocytopenia such as sepsis and viral diseases (Bernard 2018). In addition to low platelets and clinical manifestations, other laboratory tests can be used for diagnosis, when available, such as flow cytometry (which shows antibodies coating the platelets) and immunofluorescence (Mendoza and Toribio 2024).

Treatment in isolated cases of thrombocytopenia are based on the severity, ranging from environmental management to avoid trauma to the use of platelet-rich plasma or whole blood transfusions in the absence of available plasma. Corticosteroids can also be used for a few days or weeks, leading to a rapid increase in the platelet count, however, it should be noted that the use of corticosteroids can predispose the neonate to infections by suppressing the immune response, requiring intensive observation (Sellon 2006). The prognosis is generally good, with the platelet count returning to normal within 10 days (Bernard 2018).

4.6 Failure of Passive Transfer of Immunity (FPT)

Key Points

- **Asinine** neonates are born with more serum antibodies than horses, making them less hypogammaglobulinemia
- In **donkey** neonates, the maximum concentration of antibodies is reached within 12 h, allowing this to be used as the evaluation point instead of 24 h, without loss of information.

Just like in horses, donkey and mule neonates also become more susceptible to infections if they do not achieve adequate passive immunity transfer after birth. Bacterial entry site can be the digestive or respiratory tract, or more commonly, through the umbilical area, especially in cases of omphalitis (Fig. 6). After entering the body, bacteria can cause sepsis, diarrhea, and/or polyarthritis (Fig. 7).

The recommendation for donkeys is to start ingesting colostrum in the first 2–4 h of life, with greatest absorption occurring in the first 12 h of life. 250 ml is recommended every hour for the first 6 h of life, with a total of 1.5 L (Svendsen 2008). Cow colostrum is not ideal, although it is better than none. A colostrum bank

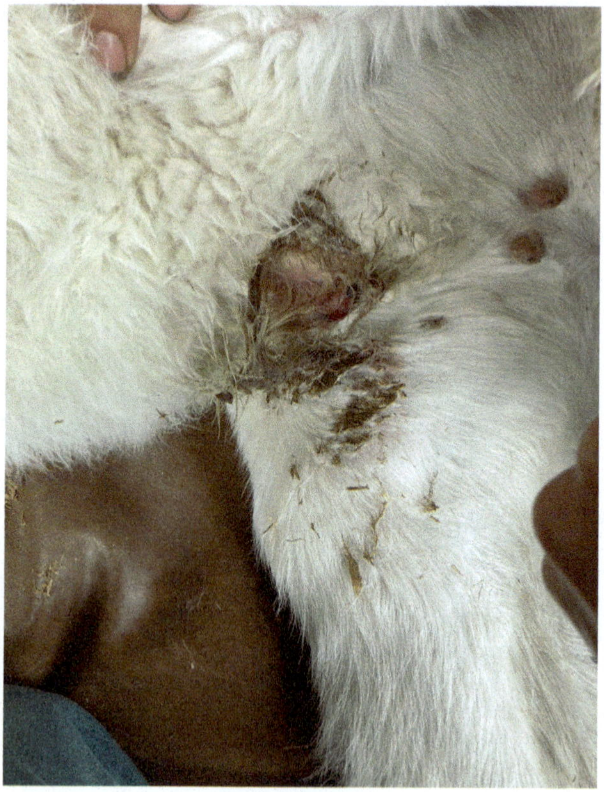

Fig. 6 Asinine neonate with omphalitis. (Source: DVM MsC Fernanda Timbó)

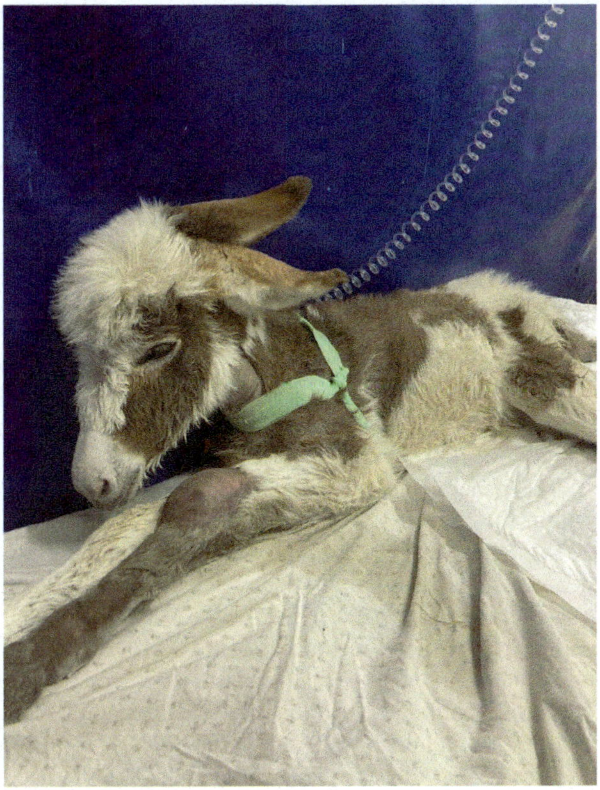

Fig. 7 Asinine neonate with sepsis and polyarthritis. (Source: DVM MsC Fernanda Timbó)

for donkeys can also be made. In donkeys, small volumes in a greater number of colostrum administrations should be used to mimic the species' natural behavior (Mendoza and Toribio 2024).

If the neonate has not received colostrum and it is no longer in the ideal absorption timeframe, plasma should be administered intravenously (Svendsen 2008). The volume will depend on the size of the foal (variable with the breed) and the degree of failure in antibody transfer, ranging from 250 mL (in miniature donkeys) to 1.5 L. In the case of donkeys, extra caution should be taken if it is necessary to use equine plasma to avoid cases of immune-mediated hemolysis or thrombocytopenia (see specific section) in these animals. In some parts of the world, commercial plasma tested for the presence of antibodies against the donkey factor is available for sale (Mendoza and Toribio 2024), but not everywhere. Asinine and mule neonates that receive plasma transfusion have a higher chance of developing thrombocytopenia or hemolytic anemia, so a follow-up blood count should be performed 24 h later (Mendoza and Toribio 2024).

A study that evaluated colostrum absorption by radial immunodiffusion in Amiata donkey foals found that, 24 h postpartum, the average serum IgG level was 14.91 ± 0.50 mg/mL (or 1491 ± 50 mg/dL), with no significant difference from values obtained 12 h after birth. This same study observed that these foals had higher antibody levels at birth (897 ± 50 mg/dL) than what is reported for horse neonates (30 mg/dL), indicating that these animals may be less hypogammaglobulinemic (Turini et al. 2020a). These higher IgG levels at birth in donkeys were also found in other studies involving different breeds, suggesting that this may be a species-specific pattern, such as in Martina Franca donkeys (Veronesi et al. 2014a), with values of 310 mg/dL.

In mules, IgG values at birth were considered close to zero (assessed by radial immunodiffusion), while after 24 h and adequate colostrum intake, the values averaged 1980 ± 620 mg/dL, alike the levels obtained at 12 h (Turini et al. 2024). This indicates that, unlike donkey neonates, mules do not have higher antibody levels at birth, although the reason for this difference is still unknown. Another study on mules (Baptista et al. 2020) also found values within the normal range (910.20 ± 143.10), but with a lower average, possibly indicating local and breed variations.

The general recommendation for asinine foals is also to measure IgG levels within the first 18–24 h of life, with an ideal value >800 mg/dL (Burton et al. 2014). However, based on the newer studies cited, it is observed that, unlike horse neonates, maximum antibody levels are already reached by 12 h postpartum. This means there is no need to wait 18–24 h for the final measurement, allowing for earlier intervention when necessary.

It is also noteworthy that the best way to assess passive immunity transfer is by measuring antibodies. Although there are studies in both horses and mules (Turini et al. 2024) and donkeys (Nervo et al. 2021; Turini et al. 2021) showing a relationship between antibody levels and serum protein concentration, including proposed calculation formulas, these indirect methods are not as precise. They require validation with a larger number of animals and are affected by several interference factors. In addition to radial immunodiffusion, other quantitative and semi-quantitative evaluation methods can be used, such as electrophoresis, zinc sulfate turbidity tests, and others (Baptista et al. 2020; Mendoza and Toribio 2024).

As with horses, if mare's or jennet's milk is not available, it is possible to use goat's milk, homemade preparations with cow's milk (but not pure) or commercial formulas for horse foals, although there are no publications on the latter for asinine and mule neonates.

The average antibody levels in the colostrum of healthy Amiata jennies remained high even 24 h after birth (Turini et al. 2020a). In a study with Martina Franca jennies, IgG levels in colostrum ranged between 1600 and 7500 mg/mL (Veronesi et al. 2014a).

To assess colostrum quality in jennies, the Brix refractometer value can be used, with a cutoff point of 17% for adequate quality (greater than 80 mg/mL) (Turini et al. 2020b), although some literature recommends the same cutoff point (20%) used for horses or a specific density above 1.080 or 1.600 (Evans and Crane 2018; Mendoza and Toribio 2024).

4.7 Orphan or Rejected Foals

As with horses, it is possible to try to get orphaned donkey foals to be adopted by another female. If the foal has not yet suckled colostrum and is still in the ideal timeframe for absorbing antibodies, colostrum should be provided orally, spontaneously or by nasogastric tube (Svendsen 2008).

It is important to remember that these same considerations can be made in cases where the mother does not die but is unable to produce good-quality colostrum and milk due to a pre-existing illness or, especially in donkeys, when hyperlipemia occurs during late gestation or early lactation (Evans and Crane 2018). Maternal rejection can also be observed, for example, in cases of primiparous jennies (Mendoza and Toribio 2024).

For a healthy horse foal, the recommended milk volume is 100 ml/kg per day, or 10% of live weight. Healthy foal's requirement is 30 kcal/kg and, for a sick or premature foal, 36 kcal/kg per day (Svendsen 2008). For asinine and mule neonates, these specifications do not yet exist, but they can be considered equivalent.

A foal naturally suckles its mother around seven times a day. When fed artificially, an interval of 2–3 h can be used between feedings, although in the first week every 1–2 h is preferable (Svendsen 2008).

The nutritional requirement for donkey foals can be considered as 30 Kcal/kg per day, or 100 mL/kg per day, or 10% of body weight in the first 2 days, increasing to 30% of body weight per day in the second week of life (Evans and Crane 2018).

It is also important to pay attention to the general management of the orphaned foal and not just its feeding. Trained and attentive staff are needed to provide the foal with the company of another animal (preferably another equid) if possible, provide an area and time for exercise and rest, protect against the elements and provide good health management (Svendsen 2008).

4.8 Gastrointestinal Tract Disorders

Asinine and mule neonates are also prone to developing colic syndrome. The causes, manifestations, and treatments are similar to those in horses. Donkey neonates are less likely to show signs of pain compared to horses, which may cause initial or milder changes to go unnoticed (Mendoza and Toribio 2024).

As with horses, donkeys can suffer from gastric ulceration, due to stress or anorexia. Treatment can include the use of medications such as omeprazole, with the same dosage used with horses, 2–4 mg/Kg (Duffield 2008). Gastroscopic assessment to diagnose gastric ulcers in donkeys can be more difficult than in horses due to the significant gastric curve, which makes the cardiac and pylorus orifices very close together in the lesser curvature (Wissdorf et al. 2021).

As with horses, donkeys and mules can also develop diarrhea due to various viral, bacterial and parasitic pathogens and non-infectious diarrhea, such as lactose intolerance, foal-heat diarrhea, and sand ingestion (Mendoza and Toribio 2024). There are reports of diarrhea cases caused by Rotavirus, helminths, and bacteria, including *Salmonella* sp., *E. coli*, *Clostridium perfringens*, and others, with mixed infections being common (Haq et al. 2017a, b; Haq et al. 2018). There are also reports of antibiotic-resistant strains (Carneiro et al. 2017b).

Infection by *Eimeria leuckarti* can occur in various equid species. Most of the time, it is only an incidental finding with no clinical relevance, but it has been implicated in some cases of diarrhea and enteritis. Although clinical signs are more common in older animals due to the 30-day prepatent period, foals as young as 28 days have been found shedding oocysts, indicating that infection can occur shortly after birth (Dubey and Bauer 2018).

A survey in Ireland detected equine coronavirus in a donkey foal with enteric alterations, showing a high degree of homology with the strain found in horses in the same study (Nemoto et al. 2019). A recombinant form of equine coronavirus was also found in two donkey foals with diarrhea in China (Qi et al. 2022).

5 Medications

The pharmacokinetics of various medications differ between species, but not all have been studied in donkeys and mules (Burton et al. 2014), and even fewer in their neonates. It is also important to remember that doses should be calculated based on weight, as donkey foals are generally smaller than horse foals. Some dosage tables adapted for neonates, with or without the personal experience of the authors, have been published (Mendoza and Toribio 2024).

Carprofen has a longer half-life in donkeys, unlike other NSAIDs (phenylbutazone and flunixin meglumine) that generally need to be administered more frequently than in horses. The dosage and dose interval should be applied based on the rate of metabolism and the animal's clinical response (Du Toit and Burden 2015). The lower volume of distribution and/or higher clearance or shorter half-life are also demonstrated for other hepatically metabolized medications (such as oxytetracycline and sulfa with trimethoprim), but not for all of them (Horspool 2008). It's important to remember that this information is known for adults, and there is still no data involving the pharmacokinetics for neonates. The same applies to information regarding the use of meloxicam, which is effective against the systemic effects of endotoxemia in adult asinines (Mendoza Garcia et al. 2020).

Donkeys are more sensitive than horses to guaifenesin, a centrally acting muscle relaxant, requiring a 40% lower dose to induce recumbency and the donkey must be monitored carefully (Horspool 2008).

Regarding intestinal antiparasitics, imidazoles are considered safe, while avermectins and milbemycins should be avoided until 6 months of age (Mendoza and Toribio 2024). Moxidectin can produce neurological side effects in equine neonates, due to the lower amount of adipose tissue (Horspool 2008). Ivermectin is also not indicated for foals under 4 months of age, especially at high doses. There is a report of possible ivermectin toxicosis in a 9-week-old miniature mule that received an overdose of a product containing praziquantel and ivermectin and showed acute blindness, ataxia and depression, but recovered after symptomatic and supportive treatment (Plummer et al. 2006).

Lidocaine in continuous infusion in asinines should be started at half the dose used for horses, to avoid ataxia. Some animals may need higher doses, and it is not known what influences this, although the hypothesis of lower liver clearance in animals with lipidosis in the liver has been raised, but not proven (Burton et al. 2014).

Metronidazole should be administered in lower doses and less frequently in horse foals than in adults, but information is lacking for donkeys and mules (Mendoza and Toribio 2024).

6 Others

Donkey neonates are also affected by the influenza virus. The symptoms are similar to those reported in horses. The survival rate is high, but there can be significant mortality in foals born to seronegative females, emphasizing the importance of vaccination in these animals as well (Ahearne et al. 2022). The importance of vaccinating jennies is also noted for equine herpesvirus (EHV-1), as vaccinated females result in greater transfer of specific immunity to their neonates (Francesco et al. 2020).

Although rare, there have been reports of pneumonia caused by *Aspergillus fumigatus* in five albino Asinara donkey neonates on the same property (Stefanetti et al. 2015).

Lungworms rarely cause clinical signs in asinine neonates, with more evident manifestations seen in horse and mule foals, some of which exhibit respiratory issues before the prepatent period (Mendoza and Toribio 2024).

Donkey neonates are highly susceptible to tetanus and can also develop botulism, although the latter is rarer and may occur due to contaminated feed or wounds (Mendoza and Toribio 2024).

It is unclear whether neonatal maladjustment syndrome occurs in donkeys and mules in the same way as in horses. Differences in hormonal dynamics have been observed between horse and mule neonates, possibly due to differences in placentation and the intrauterine environment (Boakari et al. 2022), but whether this also affects the neurosteroid dynamics associated with the syndrome remains unknown.

A case of a 5-day-old donkey neonate diagnosed with *Listeria monocytogenes* encephalitis and hepatic alterations has been reported (Jula et al. 2024).

In an experimental *Trypanosoma evansi* infection study, a pregnant jenny was infected three months before delivery, leading to transplacental transmission of the parasite to the foal. The pregnancy was carried to term, but the foal exhibited clinical signs of apathy, difficulty standing, and inability to nurse after 24 h, dying 36 h after birth. Although this is a single experimental case report, it is important to consider this possible cause of perinatal disease in regions where the parasite is endemic (Kumar et al. 2015).

A case in India reported chronic fluoride poisoning affecting not only adult animals but also two donkey foals under 2 months old. These foals exhibited erupted incisors with hypoplasia and light brown-yellow pigmentation on the enamel surface (Choubisa 2010).

Another crucial factor to consider, especially in regions with poor soils or where adequate supplementation is lacking, is selenium and vitamin E deficiency. These deficiencies have been linked to various equid disorders (white muscle disease, motor neuron disease, etc.). Foals acquire selenium primarily through the placenta and vitamin E through colostrum ingestion.

A study on Miranda jennies and their foals showed that low vitamin E levels in the mothers correlated with foal mortality. Vitamin E and selenium levels were lower in foals (though higher in females) than in adult jennies. Selenium levels in jennies were not linked to foal health issues, but low vitamin E levels were associated with foal weakness, and even slightly higher levels were linked to flexural deformities in the carpus of neonates (Quaresma et al. 2021).

Another study measuring vitamin E and selenium levels in healthy donkeys also found lower values in foals compared to adults and indicated that selenium levels in donkeys were lower than those reported for horses, suggesting a possibly lower requirement for this element in asinines (Bazzano et al. 2019).

7 Conclusion

Neonatal care for donkeys and mules presents unique challenges due to physiological and behavioral differences from horses. Although many neonatal principles apply broadly across equids, specific variations in gestation length, postpartum adaptation, and clinical parameters require tailored approaches in veterinary practice. Studies indicate that mule neonates exhibit higher Apgar scores, faster adaptation to extrauterine life, and distinct metabolic profiles compared to horse foals, while donkey neonates show a predisposition to hypothermia and slower umbilical involution. Despite limited literature on these species—especially hinnies—recent research has provided valuable insights into their peripartum management, neonate examination, and laboratory findings. However, further studies are essential to deepen our understanding and improve evidence-based neonatal care for these often-overlooked equids.

Questions and Answers

1. What role does gestation length play in monitoring equine pregnancies, and how does it differ between donkeys, horses, and mules?
2. How do anecdotal reports about mule pregnancies compare to the findings of Boakari et al. (2019) regarding gestation length?
3. What foaling prediction indicators were studied by Octaviano et al. (2024), and how did mares carrying mule fetuses differ from those carrying horse fetuses?
4. What structural differences in the allantochorion membranes of mule versus horse pregnancies were observed by Tinel et al. (2023)?
5. Why is maintaining routine practices such as vaccination, deworming, colostrum testing, and balanced nutrition particularly important in pregnancies involving donkeys and mules?

Appendix

Supplemental Information: Tables of Blood Values for Mule and Donkey Neonates

Table 1 Blood values for mule foals

Parameters of the neonate examination												
	Ref	0 h	1 h	6 h	12 h	24 h	48 h	7 days	14 days	21 days	30 days	60 days
RBC ($\times 10^6$/mL)	3	9.7 ± 0.86						7.1 ± 1.2	7.8 ± 0.7	8.3 ± 1.2	9 ± 0.9	9.6 ± 0.8
Hb (g/dL)	3	14.8 ± 0.8						10.5 ± 1.4	11 ± 1.4	11.4 ± 1.8	12.3 ± 1.2	12.7 ± 0.8
	1	16 ± 0.1			15.5 ± 1.1	15 ± 1.3	13 ± 2.1					
Hct (%)	3	44.6 ± 3.5						28.9 ± 4.3	30.3 ± 4.4	31.9 ± 5.9	34.1 ± 4.2	34.9 ± 2.7
	1	42.2 ± 6.3			42 ± 2.8	42 ± 3.3	36.3 ± 3.8					
MCV (fL)	3	47.1 ± 2.7						41.4 ± 3.4	38.8 ± 3.1	36.5 ± 5.7	38.1 ± 3.2	36.3 ± 3.2
MCH (pg)	3	15.2 ± 0.8						15.0 ± 1.1	14.2 ± 1	13.8 ± 0.8	13.7 ± 0.7	13.2 ± 0.7
MCHC (%)	3	32.2 ± 0.8						36.3 ± 0.9	37.1 (35–37.6)	36.4 ± 1.6	36 ± 1.4	36.9 (33.1–37.2)
PLT ($\times 10^3$/mL)	3	256.7 ± 62.3						211.3 ± 92.8	286.3 ± 166	276.1 ± 110.9	299.4 ± 96.9	257.3 ± 57.1
WBC (K/μL)	3	6.2 ± 1.9						8.1 ± 1.9	8.9 ± 2.3	10.2 ± 3.2	9.8 (7.8–17)	8.9 ± 2.5
Seg. Ne. (K/μL)	3	4.1 ± 1.7						5.2 ± 1.3	4.3 ± 1.3	4.4 (4.3–12.2)	5 (4.4–12.2)	5 ± 0.9
Limph. (K/μL)	3	1.8 ± 0.4						2.2 ± 0.7	2.9 ± 0.6	3.3 ± 1	3.6 ± 1	4.1(2–5)
Mon. (K/μL)	3	0.2 ± 0.1						0.6 ± 0.2	0.7 ± 0.5	0.5 ± 0.1	0.6 ± 0.2	0.5 ± 0.2
Eo. (K/μL)	3	0a,c (0–0.02)						0.1 ± 0.03	0.1 (0–0.2)	0.1 ± 0	0.1 ± 0	0.3 ± 0.2
Bas. (K/μL)	3	0 (0–0.5)						0 (0–0.2)	0 (0–0.2)	0.1 ± 0	0.1 ± 0.1	0 ± 0
CPK (IU/L)	3	150.9 ± 72.5						167.7 ± 81.1	253.3 ± 154	155.8 ± 55.7	227 ± 86.2	175 (95.8–510.7)

AST (IU/L)	3	48a (35.8–149)						181.9 ± 44.7	188.1 ± 12.9	181.1 ± 20.3	189 ± 40.1	221.3 ± 42.9
GGT (IU/L)	3	10.6 ± 5.1						9.5 (6.4–29.5)	8.5 (6.9–30.8)	12.5 ± 9.7	10.1 ± 4.3	10.3 ± 7.2
ALP (IU/L)	3	3375 ± 1279						829.6 ± 317.4	720.7 ± 261.1	760 (403–3690)	879.6 ± 312.1	1179 ± 594.7
GLDH (U/L)	3	1 (1–12)						3(1–15)	4.9 ± 4.1	3.3 ± 2.1	3.6 ± 2.1	4.5 ± 1.9
T bil (mg/dL)	3	0.2 ± 0.1						0.3 ± 0.1	0.2 ± 0.1	0.2 ± 0.1	0.2 ± 0.1	0.2 ± 0.1
D bil (mg/dL)	3	1.3 (1.1–2.9)						1.7 ± 0.9	1.2 ± 0.4	1.1 ± 0.3	1.2 ± 0.3	1.2 ± 0.3
I bil (mg/dL)	3	1.3 (1.1–1.8)						1.7 ± 0.9	1.2 ± 0.4	1.1 ± 0.3	1.2 ± 0.2	1.2 ± 0.3
BUN (mg/dL)	1	12.2 ± 1.9			13 ± 2.2	12.5 ± 3.5	9.7 ± 3.1					
	3	34 ± 4.6						22.3 ± 8.1	19.6 ± 6.1	20.9 ± 7.5	17 (16–30)	23 ± 7.3
CREA (mg/dL)	1	1.9 ± 0.6			1.4 ± 0.2	1.2 ± 0.2	0.8 ± 0.3					
	3	1.8a (1.1–4.6)						1.1 ± 0.2	1.1 ± 0.3	1.1 ± 0.4	1 (0.9–1.5)	1.1 ± 0.2
TRIG (mg/dL)	3	32.9 ± 13.9						43 (22–220)	62 (29–159	68.4 ± 33.6	50.7 ± 17.3	40 (35–230)
TP (g/L)	3	4.7 ± 0.6						5.9 ± 0.9	5.9 ± 0.7	5.8 ± 0.8	5.8 ± 0.8	5.3 ± 0.4
ALB (g/L)	3	4.2 ± 0.5						3.6 ± 0.4	3.5 ± 0.5	3.5 ± 0.5	3.5 ± 0.6	3.4 ± 0.6
Glucose (mg/dL)	1	81.5 ± 12.1			141.7 ± 10.4	150 ± 32.6	138.5 ± 15.8					
	2	60.61 ± 5.80	101.05 ± 10.87	147.40 ± 7.88	184.90 ± 7.03	196.18 ± 9.53		153.95 ± 4.52			136.68 ± 3.78	
	3	81.5 ± 12.1						147.9 ± 13.3	128 (123–164)	140 (105–153)	142.3 ± 17.8	122.1 ± 12.3
Lactate (mmol/L)	1	3.8 ± 0.6			2.8 ± 1.4	2.6 ± 1.3	1.5 (1.3–1.9)					

(continued)

Table 1 (continued)

Parameters of the neonate examination													
	Ref	0 h	1 h	6 h	12 h	24 h	48 h	7 days	14 days	21 days	30 days	60 days
	3	3.8 ± 0.6						1.5 ± 0.3	1.2 ± 0.3	1.3 ± 0.4	1.3 ± 0.1	1 ± 0.1
K (mmol/L)	1	4.2 ± 0.5			4.1 ± 0.4	4.44 ± 0.2	4.2 ± 0.4					
	3	4.2 (4.2–33.4)						4.2 ± 0.2	4.2 ± 0.3	4.3 ± 0.3	4 ± 0.4	4.5 ± 0.3
Na (mmol/L)	1	132.8 ± 4.4			134.1 ± 3.8	131.6 ± 2.8	130.3 ± 4					
	3	133.1 ± 4.1						130.7 ± 5.5	132.9 ± 1.5	132.8 ± 2.6	131.7 ± 2.2	134.5 ± 1.4
Ca (mmol/L)	1	1.48 ± 0.2			1.51 ± 0.05	1.50 ± 0.03	1.48 ± 0.1					
	3	1.5 ± 0.2						1.6 (1.2–1.6)	1.6 ± 0.1	1.6 ± 0	1.6 ± 0	1.5 ± 0
Cl (mmol/L)	1	108.2 (104.4–121.7)			108.1 ± 1.9	105.5 ± 2.1	105.7 ± 4.1					
	3	107.1 ± 2.4						105.7 ± 2.8	103.5 ± 2.9	103.6 ± 2.6	103.1 ± 1.7	103.4 ± 1.1
Mg (mmol/L)	1	0.5 ± 0.1			0.7 ± 0.1	0.7 ± 0.1	0.5 ± 0.1					
	3	0.5 ± 0.1						0.5 ± 0.1	0.6 (0.4–0.6)	0.6 ± 0.1	0.5 ± 0.1	0.6 (0.4–0.7)
pH	1	7.4 ± 0.6			7.4 ± 0.3	7.4 ± 0.01	7.5 ± 0.03					
pCO_2 (mmHg)	1	50.9 ± 12.2			36.9 (33.2–61.4)	41.4 ± 11.8	39.8 ± 4.3					
pO_2 (mmHg)	1	44.1 ± 7.2			47.6 ± 2.5	48 ± 7.2	44.9 ± 14.9					
SO_2 (%)	1	80.3 ± 6			87.5 ± 5.2	85.2 ± 6.6	82.8 ± 7.3					
BE (mmol/L)	1	3.7 ± 7			3.2 ± 5.8	4.2 ± 7.4	5 ± 4.1					

HCO_3 (mmol/L)	1	29.3 ± 6.7			27.5 ± 6	28.5 ± 7.5	28.9 ± 3.8					
tCO_2 (mmol/L)	1	30.9 ± 7			28.8 ± 6.3	29.8 ± 7.8	30.2 ± 3.9					

References:

1—Bindi et al. (2023)

2—Boakari et al. (2021)

3—Vernaccini et al. (2024)

Abbreviations: *RBC* red blood cell, *Hb* hemoglobin, *Hct* hematocrit, *MCV* mean corpuscular volume, *MCH* mean corpuscular hemoglobin, *MCHC* MCH concentration, *PLT* platelets count, *WBC* white blood cell, *Seg Ne* segmented neutrophils, *Limph* limphocytes, *Mon* Monocytes, *Eo* eosinophils, *Bas* basophils, *CPK* creatine phosphokinase, *AST* aspartate aminotransferase, *GGT* gamma-glutamyl-transpeptidase, *ALP* alkaline phosphatase, *GLDH* glutamate dehydrogenase, *T bil* total bilirubin, *D bil* direct bilirubin, *I bil* indirect bilirubin, *BUN* blood urea, *CREA* creatinine, *TRIG* triglycerides, *TP* total protein, *ALB* albumin, pCO_2 partial pressure of CO_2, pO_2 partial pressure of O_2, SO_2 oxygen saturation, *BE* base excess, HCO_3 carbonate, tCO_2 total CO_2

Table 2 Blood values for asinine foals

	Ref	0 h	12 h	24 h	48 h	7 days	14 days	21 days	30 days	60 days
RBC (×10^6/mL)	1	9.8 ± 1.0	9.42 ± 0.20	8.8 ± 0.8	8.76 ± 0.69	7.5 ± 0.2				
	2	9.9 ± 0.8		8.9 ± 1.9	7.4 ± 1.7	7.4 ± 1.3	6.7 ± 1.4	7.1 ± 1.2	7.7 ± 1.0	7.1 ± 1.0
Hb (g/dL)	1	18.07 ± 0.55	16.45 ± 1.18	15.54 ± 1.39	15.60 ± 1.24	13.2 ± 1.4				
	2	16.8 ± 1.1		15.1 ± 2.3	13.5 ± 0.8	12.6 ± 1.7	11.2 ± 0.8	12.5 ± 1.2	11.4 ± 1.1	10.9 ± 0.1
Hct (%)	1	46.3 ± 5.2	42.55 ± 2.9	41.1 ± 1.4	39.63 ± 2.68	33.4 ± 3.1				
	2	50.2 ± 2.6		42.5 ± 1.3	40.0 ± 2.4	35.3 ± 2.0	31.8 ± 2.4	35.2 ± 4.0	29.3 ± 1.1	30.5 ± 1.4
MCV (fL)	1	47.72 ± 2.44	46.89 ± 2.21	45.87 ± 2.83	45.39 ± 3.08	45.4 ± 2.4				
	2	49.8 ± 2		50.7 ± 2.1	50.3 ± 2.9	49.7 ± 2.3	49.2 ± 2.7	46.3 ± 2.0	42.6 ± 1.5	40.8 ± 1.8
MCH (pg)	1	18.5 ± 1.0	17.96 ± 1.46	17.9 ± 1.4	17.81 ± 1.55	18.0 ± 1.1				
	2	17.1 ± 0.9		17.2 ± 1.1	17.1 ± 1.1	17.0 ± 0.9	17.6 ± 1.7	16.6 ± 0.9	16.2 ± 0.7	15.3 ± 0.7
MCHC (%)	1	38.34 ± 2.52	38.74 ± 1.93	39.01 ± 2.05	39.20 ± 2.05	39.7 ± 1.6				
	2	34.3 ± 2.0		33.9 ± 1.8	34.2 ± 1.3	34.2 ± 1.2	35.1 ± 0.7	35.8 ± 1.9	37.3 ± 1.5	37.4 ± 0.7
PLT (×10^3/mL)	1	235.4 ± 74.3	284.5 ± 94.5	275.1 ± 60.3	316.8 ± 87.9	284 ± 59.1				
	2	342.0 ± 58.2		285.6 ± 43.9	259 ± 27.8	254.7 ± 32.5	263.3 ± 53.9	322.7 ± 59.0	267.7 ± 51.7	180.8 ± 20.3
WBC/μL	1	7213.8 ± 1374.2	6886.8 ± 1582	7286.6 ± 1430.4	8152.4 ± 1362.7	9398.5 ± 1316.2				
	2	5900 ± 1500		7300 ± 900	7900 ± 1700	8400 ± 700	8400 ± 1200	8700 ± 1000	8300 ± 1400	10000 ± 500
Seg. Ne./μL	1	5228.6 ± 1549.1	4817.5 ± 1530.2	4982.9 ± 1160.5	5951.2 ± 1355.1	5989.4 ± 1049.2				
Limph./μL	1	1427.4 ± 441.5	1803 ± 376	2029 ± 631.8	1830 ± 607	2743.6 ± 447.1				
Mon./μL	1	97 ± 48.1	119.8 ± 54.2	127.3 ± 50.7	218.3 ± 71.5	255.5 ± 50.8				
Eo./μL	1	86 ± 57.8	81.5 ± 95.2	73.4 ± 62.9	73.5 ± 52.1	70.5 ± 29.3				
Bas./μL	1	76.3 ± 32.4	55.4 ± 30.4	49.9 ± 28.1	41.6 ± 13.4	40.9 ± 17				
CPK (IU/L)	1	126.1 ± 74	149.4 ± 104.3	108 ± 76.8	123.4 ± 121	76 ± 36.4	104.3 ± 45.1	127.4 ± 82.3		

	2	80.8 ± 41.1		20.2 ± 16.4	17.4 ± 13.5	57.4 ± 24.0	22.9 ± 15.3	40.7 ± 25.5	73.0 ± 36.3	44.6 ± 8.6
AST (IU/L)	1	171.9 ± 114.3	263.7 ± 187.6	336.9 ± 279.1	375.6 ± 326.2	263.7 ± 158.6	222.5 ± 115.9	227.1 ± 104		
	2	60.3 ± 14.6		107.9 ± 26.3	125.1 ± 57.7	109.7 ± 43.4	89.3 ± 21.8	88.6 ± 16.2	114.0 ± 31.9	165.3 ± 38.3
GGT (IU/L)	1	26.9 ± 12.7	32.8 ± 15.8	38 ± 17.1	38.9 ± 17.6	38.7 ± 13.2	42.2 ± 15.9	35.1 ± 12.6		
	2	20.5 ± 7.9		35.5 ± 17.0	29.4 ± 11.3	23.2 ± 11.4	29.6 ± 6.7	25.9 ± 10.2	14.3 ± 10.0	12.5 ± 6.0
ALT (IU/L)	1	16.5 ± 12.1	20.8 ± 14.4	21.9 ± 12	21.8 ± 10.9	12.3 ± 10.2	9.4 ± 7.5	9.9 ± 7.5		
LDH (IU/L)	1	284.5 ± 125.3	450.1 ± 99.2	378.8 ± 137.6	347.6 ± 131.8	325.3 ± 73.6	318 ± 97.2	361.8 ± 126.2		
ALP (IU/L)	1	1065.8 ± 334.5	1273.8 ± 346.8	988.2 ± 211.2	834.6 ± 172.3	560.4 ± 165.9	488.3 ± 174.4	497.6 ±± 174.2		
	2	879.7 ± 658.3		1041.6 ± 408.1	656.0 ± 233.4	378.6 ± 152.9	326.7 ± 111.1	387.8 ± 40.4	372.9 ± 49.0	415.5 ± 57.1
BUN (mg/dL)	1	33.7 ± 8	35.8 ± 8.1	33.8 ± 6.5	24.7 ± 9.5	19.2 ± 8.1	19.3 ± 6.3	20.7 ± 8.5		
	2	38.4 ± 14.1		32.9 ± 9.0	28.3 ± 8.7	23.4 ± 6.3	19.6 ± 5.3	25.0 ± 1.7	24.6 ± 10.6	22.6 ± 4.5
CREA (mg/dL)	1	1.67 ± 0.22	1.51 ± 0.2	1.45 ± 0.24	1.25 ± 0.19	1.02 ± 0.21	0.97 ± 0.17	1.03 ± 0.16		
	2	2.4 ± 0.7		1.3 ± 0.3	1.2 ± 0.3	0.9 ± 0.1	1.0 ± 0.1	1.2 ± 0.2	1.2 ± 0.3	1.2 ± 0.3
TP (g/dL)	1	4.37 ± 1.06	5.03 ± 1.07	4.8 ± 1.02	5 ± 0.88	4.96 ± 0.66	4.82 ± 0.57	5.01 ± 0.3		
ALB. (g/dL)	1	3.03 ± 0.26	2.89 ± 0.34	2.81 ± 0.31	2.94 ± 0.27	2.79 ± 0.34	2.65 ± 0.16	2.75 ± 0.15		
	2	3.2 ± 0.6		3.2 ± 0.6	3.2 ± 0.8	3.1 ± 0.7	3.0 ± 0.9	3.2 ± 0.2	2.8 ± 0.6	3.0 ± 0.8
T bil. (mg/dL)	1	0.23 ± 0.1	0.25 ± 0.11	0.23 ± 0.1	0.19 ± 0.05	0.18 ± 0.03	0.14 ± 0.03	0.16 ± 0.03		
	2	0.3 ± 0.2		0.3 ± 0.1	0.4 ± 0.1	0.5 ± 0.3	0.2 ± 0.1	0.2 ± 0.1	0.1 ± 0.03	0.2 ± 0.07
Glucose (mg/dL)	1	125.8 ± 33.3	150.3 ± 28.8	132.5 ± 20.4	142.1 ± 24.6	127.5 ± 29	123.8 ± 16.6	119.5 ± 17.4		
	2	97.6 ± 37.1		128.6 ± 15.7	126.6 ± 7.4	109.9 ± 22.6	121.2 ± 37.4	86.4 ± 12.4	130.0 ± 19.4	109.6 ± 10.5
Lactate (mmol/L)	1	5.49 ± 1.36	2.54 ± 0.44	1.47 ± 0.39						
Arterial	3	3.78 ± 1.16	2.43 ± 0.58	2.14 ± 0.61	1.58 ± 0.28					
Venous	3	3.86 ± 1.15	2.19 ± 0.46	2.10 ± 0.45	1.43 ± 0.35					

(continued)

Table 2 (continued)

	Ref	0 h	12 h	24 h	48 h	7 days	14 days	21 days	30 days	60 days
TRIG (mg/dL)	2	34.8 ± 7.2		46.0 ± 22.0	58.0 ± 38.6	60.6 ± 30.4	81.6 ± 42.0	59.2 ± 12.7	61.0 ± 14.0	76.6 ± 23.4
Ca (mg/dL)	1	12.1 ± 0.9	11.6 ± 0.6	11.7 ± 0.8	11.9 ± 1	11.7 ± 0.5	11.6 ± 0.8	12 ± 0.8		
	2	10.7 ± 1.6		10.8 ± 0.9	10.9 ± 1.1	11.3 ± 2.3	10.9 ± 1.8	11.6 ± 1.3	12.0 ± 2.1	11.0 ± 1.4
P (mg/dL)	1	4.09 ± 0.78	4.47 ± 0.53	5.05 ± 1.5	6.48 ± 1.04	7.16 ± 1.49	6.60 ± 1.55	7.11 ± 0.56		
	2	4.7 ± 1.4		4.9 ± 1.0	5.6 ± 1.7	7.4 ± 2.4	5.0 ± 2.3	6.5 ± 1.2	6.4 ± 1.7	6.1 ± 1.1
Mg (mg/dL)	1	1.85 ± 0.45	2.12 ± 0.31	1.92 ± 0.24	1.91 ± 0.33	2.13 ± 0.48	2.18 ± 0.41	2.14 ± 0.44		
Na (mEq/L or mMol/L)	1	137.2 ± 3.9	136.4 ± 5.4	135.6 ± 4.2	135.3 ± 3.8	133.6 ± 6	133.6 ± 5.1	135.2 ± 2.2		
	2	132.5 ± 11.6		136.6 ± 3.7	135.0 ± 3.8	133.9 ± 4.1	133.3 ± 2.3	134.0 ± 3.6	132.5 ± 4.7	132.3 ± 1.4
Cl (mEq/L)	1	101.2 ± 4.3	101.9 ± 2.8	100.9 ± 3.1	101.8 ± 4.2	100.7 ± 3.7	100.8 ± 3.7	101.1 ± 2.4		
K (mEq/L or mMol/L)	1	4.34 ± 0.37	4.28 ± 0.45	4.14 ± 0.39	4.53 ± 0.34	4.55 ± 0.19	4.65 ± 0.45	4.71 ± 0.3		
	2	4.5 ± 0.6		4.3 ± 0.8	4.5 ± 0.6	4.2 ± 0.5	4.4 ± 0.3	4.5 ± 0.2	4.5 ± 0.3	4.6 ± 0.1
Se(mg/mL)	4									0.05 ± 0.01[a]
Vit E (mmol/L)	4									5.92 ± 1.24[a]
pH (venous) (	1	7.35 ± 0.05	7.47 ± 0.05	7.44 ± 0.04						
Arterial	3	7.394 ± 0.039	7.453 ± 0.029	7.446 ± 0.021	7.452 ± 0.023					
Venous	3	7.387 ± 0.038	7.450 ± 0.023	7.450 ± 0.020	7.445 ± 0.026					
pCO_2 (mmHg)	1	46.08 ± 3.08	37.11 ± 2.85	38.97 ± 2.77						
Arterial	3	41.4 ± 3.9	39.6 ± 3.5	40.3 ± 2.2	39.6 ± 2.5					
Venous	3	42.2 ± 2.7	40.7 ± 2.2	40.9 ± 3.6	42.9 ± 3.2					
pO_2 (mmHg)	1	32.44 ± 4.88	39.89 ± 4.46	38.78 ± 5.91						
Arterial	3	60 ± 16a	71 ± 9	73 ± 9	70 ± 7					
Venous	3	37 ± 11	40 ± 8	39 ± 9	36 ± 10					
SO_2 (%)	1	59 ± 14.59	78 ± 9.25	76 ± 11.85						
Arterial	3	85 ± 9	95 ± 3	95 ± 3	94 ± 1					
Venous	3	64 ± 6	73 ± 6	72 ± 7	71 ± 6					

BE (mmol/L)	1	−0.44 ± 1.01	2.22 ± 1.64	2.67 ± 2.74						
Arterial	3	−0.1 ± 2	3 ± 1	3 ± 1	3 ± 1					
Venous	3	1 ± 2	3 ± 2	3 ± 1	4 ± 1					
HCO_3 (mmol/L)	1	25.74 ± 0.72	26.58 ± 1.13	27.41 ± 1.34						
Arterial	3	25.3 ± 2.3	27.2 ± 1.7	27.7 ± 1.3	27.7 ± 1.4					
Venous	3	26.4 ± 2.2	27.5 ± 2.1	28.1 ± 1.6	28.9 ± 1.7					
tCO_2 (mmol/L)	1	27 ± 1.12	27.56 ± 1.33	29.67 ± 1.22						
Arterial	3	27 ± 2	28 ± 2	29 ± 1	29 ± 1					
Venous	3	27 ± 3	29 ± 2	29 ± 2	30 ± 2					

References:

1—Veronesi et al. (2014b) (venous blood; Martina Franca)

2—Sgorbini et al. (2013) (Amiata)

3—Carluccio et al. (2017) (Martina Franca)

4—Bazzano et al. (2019)

Abbreviations: *RBC* red blood cell, *Hb* hemoglobin, *Hct* hematocrit, *MCV* mean corpuscular volume, *MCH* mean corpuscular hemoglobin, *MCHC* MCH concentration, *PLT* platelets count, *WBC* white blood cell, *Seg Ne* segmented neutrophils, *Limph* limphocytes, *Mon* Monocytes, *Eo* eosinophils, *Bas* basophils, *CPK* creatine phosphokinase, *AST* aspartate aminotransferase, *GGT* gamma-glutamyl-transpeptidase, *ALT* alanine aminotransferase, *LDH* lactate dehydrogenase, *ALP* alkaline phosphatase, *BUN* blood urea, *CREA* creatinine, *TP* total protein, *ALB* albumin, *T bil* total bilirubin, *TRIG* triglycerides, *Se* selenium, *Vit E* vitamin E, pCO_2 partial pressure of CO_2, pO_2 partial pressure of O_2, SO_2 oxygen saturation, *BE* base excess, HCO_3 carbonate, tCO_2 total CO_2

[a] Animals with 1–4 months of age

References

Abu-Seida AM, Shamaa AA (2022) Treatment of a congenital lateral patellar luxation by recession trochleoplasty in a donkey foal. Equine Vet Educ 34:35–39. https://doi.org/10.1111/eve.13424

Ahearne MM, Pentzke-Lemus LL, Romano AM, Larsen ED, Watson AM, O'Fallon EA, Landolt GA (2022) Disease progression, pathologic, virologic findings of an equine influenza outbreak in rescue donkeys. J Vet Intern Med 36(6):2230–2237. https://doi.org/10.1111/jvim.16563

Alonso MA, Boakari YL, Riccio AV, Belli CB, Fernandes CB (2023) Perinatal parameters of mule and equine foals: similarities and differences. J Vet Behav 63:31–35. https://doi.org/10.1016/j.jveb.2023.04.001

Anzuino J (2008) Exotic infections. In: Svendsen ED (ed) The professional handbook of the donkey. Whittet Books, pp 255–287

Baptista VS, Guttmann PM, Rusca AC, Silva KM, Macieira DB, Alencar NX, Lessa DAB (2020) Evaluation of acquired passive immunity in mule foals up to 60 days of age. J Equine Sci 31(1):1–4. https://doi.org/10.1294/jes.31.1

Barrio E, Rickards KJ, Thiemann AK (2019) Clinical evaluation and preventative care in donkeys. Vet Clin N Am Equine Pract 35:545–560. https://doi.org/10.1016/j.cveq.2019.08.013

Barton AK, Ohnesorge B (2009) Concurrent unilateral choanal atresia and congenital laryngeal deformity in a miniature donkey. Vet Rec 164:93–94

Bazzano M, McLean A, Tesei B, Gallina E, Laus F (2019) Selenium and vitamin E concentrations in a healthy donkey population in Central Italy. J Equine Vet Sci 78:112e116. https://doi.org/10.1016/j.jevs.2019.04.003

Bazzano M, Bonfili L, Eleuteri AM, Serri E, Scollo C, Yaosen Y, Tesei B, Laus F (2022) Assessment of serum amyloid A concentrations and biochemical profiles in lactating jennies and newborn Ragusano donkey foals around parturition and one month after foaling in Sicily. Reprod Domest Anim 57:262–268. https://doi.org/10.1111/rda.14048

Bernard WV (2018) Immunologic and hematologic disorders. In: Bernard WV, Barr BS (eds) equine pediatric medicine, 2nd edn. Taylor & Francis Group, pp 57–72

Bindi F, Vernaccini M, Bonelli F, Nocera I, Fanelli D, Sgorbini M (2023) Apgar score, clinical, hemato-biochemical, and venous blood gas parameters in a cohort of newborn mule foals: preliminary data. J Equine Vet Sci 130:104917. https://doi.org/10.1016/j.jevs.2023.104917

Boakari YL, Alonso MA, Riccio AV, Fernandes CB (2019) Are mule pregnancies really longer than equine pregnancies? Comparison between mule and equine pregnancies. Reprod Domest Anim 54(6):823–827. https://doi.org/10.1111/rda.13423

Boakari YL, Alonso MA, Riccio AV, Affonso FJ, Losano JDA, Nichi M, Belli CB, Fernandes CB (2021) Evaluation of blood glucose and lactate concentrations in mule and equine foals. J Equine Vet Sci 101:103369. https://doi.org/10.1016/j.jevs.2020.103369

Boakari YL, Legacki E, Alonso MA, Santos ACF, Nichi M, Conley AJ, Fernandes CB (2022) Postnatal dynamics of circulating steroid hormones in mule and equine neonates. Vet Sci 9:598. https://doi.org/10.3390/vetsci9110598

Bonelli F, Nocera I, Conte G, Panzani D, Sgorbini M (2020) Relation between Apgar scoring and physical parameters in 44 newborn Amiata donkey foals at birth. Theriogenology 142:310e314. https://doi.org/10.1016/j.theriogenology.2019.10.020

Burton AJ, Mittel LD, Merriam J (2014) Emergency treatment of mules and donkeys. In: Orsini JA, Divers TJ (eds) Equine emergencies: treatment and procedures. Elsevier, pp 646–651

Carluccio A, Contri A, Gloria A, Veronesi MC, Sfirro MP, Parrillo S, Robbe D (2017) Correlation between some arterial and venous blood gas parameters in healthy newborn Martina Franca donkey foals from birth to 96 hours of age. Theriogenology 87:173–178. https://doi.org/10.1016/j.theriogenology.2016.08.021

Carneiro VC, Aquino MHC, Cerqueira AMF, Guttmann PM, Magalhães H, Costa MFM, Cunha LE, Lessa DAB (2017a) Microbiological findings in tracheal wash from mule foals with and without clinical evidence of respiratory disease. J Equine Vet Sci 55:40–43. https://doi.org/10.1016/j.jevs.2017.01.019

Carneiro VC, Lessa DAB, Guttmann PM, Magalhães H, Aquino MHC, Cunha LER, Arais LR, Cerqueira AMF (2017b) Virulence, resistance, and genetic relatedness of *Escherichia coli* and *Klebsiella* sp. isolated from mule foals. Arq Bras Med Vet Zootec 69(5):1073–1082. https://doi.org/10.1590/1678-4162-9115

Chacur MGM, Ruediger FR, Yamasaki LP (2014) Bilateral hydranencephaly in a hybrid foal: obstetric case. Semina: Ciências Agrárias 35(3):1389–1394. https://doi.org/10.5433/1679-0359.2014v35n3p1389

Choubisa SL (2010) Osteo-dental fluorosis in domestic horses and donkeys in Rajasthan, India. Fluoride 43(1):5–12

Crisci A, Rota A, Panzani D, Sgorbini M, Ousey JC, Camillo F (2014) Clinical, ultrasonographic, and endocrinological studies on donkey pregnancy. Theriogenology 81:275–283. https://doi.org/10.1016/j.theriogenology.2013.09.026

Du Toit N, Burden F (2015) Donkey colic. In: Sprayberry KA, Robinson NE (eds) Robinson's current therapy in equine medicine, v. 7. Elsevier, pp 333–335

Dubey JP, Bauer C (2018) A review of Eimeria infections in horses and other equids. Vet Parasitol 256:58–70. https://doi.org/10.1016/j.vetpar.2018.04.010

Duffield H (2008) Colic. In: Svendsen ED (ed) The professional handbook of the donkey. Whittet Books, pp 37–51

Estell KE, Magdesian KG (2015) Use of fresh and frozen blood products in foals. In: Sprayberry KA, Robinson NE (eds) Robinson's current therapy in equine medicine, v. 7. Elsevier, pp 746–749

Evans L, Crane M (eds) (2018) The clinical companion of the donkey, 1st edn. The donkey sanctuary. 265p

Fielding D (1988) Reproductive characteristics of the jenny donkey– Equus asinus: a review. Trop Anim Health Prod 20:161–166. https://doi.org/10.1007/BF02240085

Francesco CE, Smoglica C, De Amicis I, Cafini F, Carluccio A, Contri A (2020) Evaluation of colostral immunity against equine herpesvirus type 1 (EHV-1) in Martina Franca's foals. Front Vet Sci 7:579371. https://doi.org/10.3389/fvets.2020.579371

Françoso R, Riccio AV, Fernandes CB, Alonso MA, Belli CB (2018) Transplacental transmission of Theileria equi in mules: should we worry? Vet Parasitol 264:39–41. https://doi.org/10.1016/j.vetpar.2018.10.017

Galisteo J, Perez-Marin CC (2010) Factors affecting gestation length and estrus cycle characteristics in Spanish donkey breeds reared in southern Spain. Theriogenology 74:443–450. https://doi.org/10.1016/j.theriogenology.2010.02.027

Giger R, Meier HP, Kupfer U (1997) Length of gestation of Freiberger mares with mule and horse foals. Schweizer Archiv Fur Tierheilkunde 139:303–307

Grove V (2008) Conditions of the eye. In: Svendsen ED (ed) The professional handbook of the donkey. Whittet Books, pp 202–221

Haq I, Durrani AZ, Khan MS, Mushtaq MH, Ahmad I (2017a) A study on causes of pathogenic diarrhea in foals in Punjab, Pakistan. J Equine Vet Sci 56:88–92. https://doi.org/10.1016/j.jevs.2017.05.010

Haq I, Durrani AZ, Khan MS, Mushtaq MH, Ahmad I (2017b) Study of antimicrobial resistance and physiological biomarkers with special reference to Salmonellosis in diarrheic foals in Punjab, Pakistan. Acta Trop 176:144–149. https://doi.org/10.1016/j.actatropica.2017.08.003

Haq I, Durrani AZ, Khan MS, Mushtaq MH, Ahmad I, Khan A, Ali M (2018) Identification of bacteria from diarrheic foals in Punjab, Pakistan. Pak J Zool 50(1):381–384. https://doi.org/10.17582/journal.pjz/2018.50.1.sc5

Horspool L (2008) Clinical pharmacology. In: Svendsen ED (ed) The professional handbook of the donkey. Whittet Books, pp 222–238

Johnson J (2006) Foal with isoerythrolysis. In: Paradis MR (ed) Equine neonatal medicine: a case-based approach. Saunders, Philadelphia, pp 40–46

Jula C, Buechner-Maxwell V, Southard T, LeCuyer T (2024) *Listeria monocytogenes* encephalitis in a donkey foal. Equine Vet Educ 36:e79–e84. https://doi.org/10.1111/eve.13915

Knottenbelt D (2008) Skin Disorders. In: Svendsen ED (ed) The professional handbook of the donkey. Whittet Books, pp 124–152

Knottenbelt, D., Holdstock, N., Madigan, J.E. 2006. Equine neonatology: medicine and surgery. Saunders. 508p

Kumar R, Kumar S, Virmani N, Yadav SC (2015) Transplacental transmission of Trypanosoma evansi from experimentally infected donkey mare to neonatal foal. J Equine Vet Sci 35:337–341. https://doi.org/10.1016/j.jevs.2015.02.004

Lassaline M, Latimer C (2018) Ophthalmologic disorders. In: Bernard WV, Barr BS (eds) Equine pediatric medicine, 2nd edn. Taylor & Francis Group, pp 249–267

Lowe JS (1972) Patent ductus arteriosus in a donkey foal. N Z Vet J 20(1–2):15. https://doi.org/10.1080/00480169.1972.33992

McLean A, Varnum A, Ali A, Heleski C, González FJN (2019) Comparing and contrasting knowledge on mules and hinnies as a tool to comprehend their behavior and improve their welfare. Animals 9:488. https://doi.org/10.3390/ani9080488

Mendoza Garcia FJ, Gonzalez-De Cara C, Aguilera-Aguilera R, Buzon-Cuevas A, Perez-Ecija A (2020) Meloxicam ameliorates the systemic inflammatory response syndrome associated with experimentally induced endotoxemia in adult donkeys. J Vet Intern Med. 34, 1631–1641. https://doi.org/10.1111/jvim.15783

Mendoza F, Toribio RE (2024) Special considerations for the neonatal donkey and mule foal. In: Wong DM, Wilkins PA (eds) Equine neonatal medicine. Wilwy Blackwell, pp 1399–1418

Moustafa A, Elmetwally M, EL-Khodery S, Hamed M, Gomaa N, Abdo Rizk M (2022) Abdominal hernia in equine: animal level risk factors and repair using polypropylene mesh. J Equine Vet Sci 111:103889. https://doi.org/10.1016/j.jevs.2022.103889

Nelson BB, Ferris RA, McCue PM, Leise BS (2015) Surgical management of atresia ani and perineal hypospadias in a miniature donkey foal. Equine Vet Educ 27(10):525–529. https://doi.org/10.1111/eve.12292

Nemoto M, Schofield W, Cullinane A (2019) The first detection of equine coronavirus in adult horses and foals in Ireland. Viruses 11:946. https://doi.org/10.3390/v11100946

Nervo T, Bertero A, Donato GG, Suffat EP, Vincentia L (2021) Analysis of factors influencing the transfer of passive immunity in the donkey foal. Ital J Anim Sci 20(1):1947–1956. https://doi.org/10.1080/1828051X.2021.1963863

Nocera I, Sorvillo B, Sgorbini M, Aliboni B, Citi S (2023) Radiographic appearance of the fore digit and carpal joint in the mule foal from birth to 3 months of age. Animals 13:2417. https://doi.org/10.3390/ani13152417

Nocera I, Di Franco C, Sorvillo B, Sgorbini M, Moroni R, Patroncini S, Citi S (2024) Multiple congenital musculoskeletal abnormalities in a mule foal: a case report. J Equine Vet Sci 141:105149. https://doi.org/10.1016/j.jevs.2024.105149

Octaviano JI, Alonso MA, Boakari YL, Gomes V, Mori CS, Fleury PDC, Fernandes CB (2024) A comparative study across mule, equine and equine clone pregnancies regarding the determination of the day of birth. Theriogenology 228:104–109. https://doi.org/10.1016/j.theriogenology.2024.08.006

Paolucci M, Palombi C, Sylla L, Stradaioli G, Monaci M (2012) Ultrasonographic features of the mule embryo, fetus and fetal-placental unit. Theriogenology 77:240–252. https://doi.org/10.1016/j.theriogenology.2011.07.030

Plummer CE, Kallberg ME, Ollivier FJ, Brooks DE, Gelatt KN (2006) Suspected ivermectin toxicosis in a miniature mule foal causing blindness. Vet Ophthalmol 9(1):29–32

Qi P-F, Gao X-Y, Ji J-K, Zhang Y, Yang S-H, Cheng K-H, Cui N, Zhu M-L, Tao H, Dong X, Yan B, Wang C-F, Yang H-J, Shi W-F, Zhang W (2022) Identification of a recombinant equine coronavirus in donkey, China. Emerging Microbes Infect 11(1):1010–1013. https://doi.org/10.1080/22221751.2022.2056522

Quaresma M, Marín C, Bacellar D, Nóvoa M, Navas FJ, McLean A (2021) Selenium and vitamin E concentrations in Miranda jennies and foals (*Equus asinus*) in Northeast Portugal. Animals 11:1772. https://doi.org/10.3390/ani11061772

Radcliffe RM, Bookbinder LC, Liu SY, Tomlinson JE, Cook VL, Hurcombe SDA, Divers TJ (2022) Collection and administration of blood products in horses: transfusion indications, materials, methods, complications, donor selection, and blood testing. J Vet Emerg Crit Care 32(S1):108–122. https://doi.org/10.1111/vec.13119

Riccio AV, Costa BK, Alonso MA, Affonso FJ, França DS, Nichi M, Belli CB, McLean AK, Boakari YL, Fernandes CB (2023) Comparative assessment of oxidative and antioxidant parameters in mule and horse neonates during their first month of extrauterine adaptation. Animals 13:3878. https://doi.org/10.3390/ani13243878

Robbe D, Carluccio A, Gloria A, Contri A, Ventura M, Catacchio CR, Straticò P, Petrizzi L (2012) Digital Agenesia in Martina Franca Donkey Foal: a case report. J Equine Vet Sci 32:844–847. https://doi.org/10.1016/j.jevs.2012.03.009

Sellon DC (2006) Foal with immune-mediated thrombocytopenia. In: Paradis MR (ed) Equine neonatal medicine: a case-based approach. Saunders, Philadelphia, pp 46–50

Sgorbini M, Crisci A, Falconcini A, Rota A, Corazza M (2009) Evaluation of semeiotic data, haematological and biochemical parameters in the Amiata donkey newborn foals. In: Attidel Congresso Nazional e SISVet: proceedings of the LXIII SISVET congress; 2009; Udine, Italy, pp 304–305

Sgorbini M, Bonelli F, Rota A, Baragli P, Marchetti V, Corazza M (2013) Hematology and clinical chemistry in Amiata donkey foals from birth to 2 months of age. J Equine Vet Sci 33:35–39. https://doi.org/10.1016/j.jevs.2012.04.010

Sgorbini M, Bonelli F, Percacini G, Pasquini A, Rota A (2021) Maternal and neonatal evaluation of derived reactive oxygen metabolites and biological antioxidant potential in donkey mares and foals. Animals 11:2885. https://doi.org/10.3390/ani11102885

Shaw SD, Norman TE, Arnold CE, Coleman MC (2015) Clinical characteristics of horses and foals diagnosed with cleft palate in a referral population: 28 cases (1988–2011). Can Vet J 56:756–760

Stefanetti V, Marenzoni ML, Lepri E, Coletti M, Sebastianelli M, Agnetti F, Vuerich M, Passamonti F (2015) Five fatal cases of primary pulmonary aspergillosis in albino Asinara donkey foals. J Equine Vet Sci 35:76–79. https://doi.org/10.1016/j.jevs.2014.10.005

Svendsen ED 2008 The professional handbook of the donkey. Whittet Books. 448p.

Teixeira LG, Spasiani JP, Meirelles AEWB, Garrido E, Nogueira AFS, Alessi AC, Marques JA, Canola JC (2010) Rectal atresia in a newborn donkey. Equine Vet Educ 22(9):434–437. https://doi.org/10.1111/j.2042-3292.2010.00115.x

Tinel JB, Veras MM, Waked D, Alonso MA, Fernandes CB (2023) Comparative stereological evaluation of the term allantochorion membrane in the mare pregnant with mule foals and equine foals. Anim Reprod Sci 250:107201

Turini L, Bonelli F, Nocera I, Battaglia F, Meucci V, Panzani D, Mele M, Sgorbini M (2020a) Evaluation of jennies' colostrum: IgG concentrations and absorption in the donkey foals. A preliminary study. Heliyon 6:e04598. https://doi.org/10.1016/j.heliyon.2020.e04598

Turini L, Nocera I, Bonelli F, Mele M, Sgorbini M (2020b) Evaluation of brix refractometry for the estimation of colostrum quality in jennies. J Equine Vet Sci 92:103172. https://doi.org/10.1016/j.jevs.2020.103172

Turini L, Bonelli F, Nocera I, Meucci V, Conte G, Sgorbini M (2021) Evaluation of different methods to estimate the transfer of immunity in donkey foals fed with colostrum of good IGG quality: a preliminary study. Animals 11:507. https://doi.org/10.3390/ani11020507

Turini L, Bindi F, Vernaccini M, Bonelli F, Meucci V, Sgorbini M (2024) Immunity transfer in mule foals fed with good IgG quality colostrum. Heliyon 10:e25560. https://doi.org/10.1016/j.heliyon.2024.e25560

Vernaccini M, Bindi F, Bonelli F, Nocera I, Sgorbini M (2024) Hematology and clinical chemistry in mule foals from birth to two months of age: a preliminary study. J Equine Vet Sci 132: 104977. https://doi.org/10.1016/j.jevs.2023.104977

Veronesi MC, Dall'Ara P, Alessia G, Servida F, Sala E, Robbe D (2014a) IgG, IgA, and lysozyme in Martina Franca donkey jennies and their foals. Theriogenology 81:825–831. https://doi.org/10.1016/j.theriogenology.2013.12.018

Veronesi MC, Gloria A, Panzani S, Sfirro MP, Carluccio A, Contri A (2014b) Blood analysis in newborn donkeys: hematology, biochemistry, and blood gases analysis. Theriogenology 82: 294–303. https://doi.org/10.1016/j.theriogenology.2014.04.004

Vitale V, Nocera I, Sgorbini M, Aliboni B, Laus F, Mannini A, Bazzano M (2021) Ultrasonography evaluation of umbilical structures in clinically healthy donkey foals during the first week of life. Animals 11:1650. https://doi.org/10.3390/ani11061650

Williamson AJ, Stent AW, Milne M et al (2016) Osteopetrosis in a neonatal donkey. Aust Vet J 94(10):358–361. https://doi.org/10.1111/avj.12492

Wissdorf H, Jerbi H, Meir-Schellersheim M (2021) Anatomical differences of the donkey, mule and horse. Utzverlag. 142p

Zakari FO, Ayo JO, Rekwot PI, Kawu MU, Minka NS (2018) Diurnal rhythms of heart and respiratory rates in donkeys of different age groups during the cold-dry and hot-dry seasons in a tropical savannah. Physiol Rep 6(17):e13855. https://doi.org/10.14814/phy2.13855

Reproductive Management of Donkeys in Commercial Milk Production Programs

Luis Losinno and Melina Soledad Pietrani

Abstract

Domestic donkeys have interacted with human civilizations for thousands of years as pack and transportation animals, and indeed they still are, particularly in the driest and poorest countries of the world. Additionally, the beneficial effects of donkey milk for human health have been known and documented for over 2000 years. This knowledge has laid the foundations for the current growth of commercial production systems, primarily aimed at producing milk for children and elderly adults who are allergic to cow milk proteins (CMA). Additionally, the peptides from donkey milk are used in the human cosmetic industry. As with any milk production system, donkey milk production requires a focus on reproduction. These farms are mostly semi-intensive and based on pasture natural breeding. These conditions align with donkeys' natural behaviors, reducing the need for intensive human handling and thereby decreasing stress, resulting in positive and cost-effective outcomes for both the animals and the system. However, considering future genetic improvement programs and the preservation of lines and breeds, assisted reproduction techniques such as embryo transfer, embryo cryopreservation, semen cryopreservation, and *in vitro* embryo production will be necessary. While these techniques are currently being developed for donkeys, they remain in the experimental stages.

L. Losinno (✉)
Laboratorio de Equidos, Universidad Nacional de Río Cuarto, Río Cuarto, Argentina

Equine Academy, Córdoba, Argentina

M. S. Pietrani
Instituto Académico Pedagógico de Ciencias Básicas y Aplicadas, Universidad Nacional de Villa María, Villa María, Argentina
e-mail: mpietrani@unvm.edu.ar

A. K. McLean et al. (eds.), *Donkey and Mule Medicine: Clinical Medicine and Reproduction*, https://doi.org/10.1007/978-3-032-18540-2_20

Keywords

Dairy donkey · Milk production · Farm management · Reproduction · CMA

Abbreviations

AI	Artificial insemination
ARTs	Assisted reproductive technologies
BP	Before present
CMA	Cow's milk allergy
CMPA	Cow milk protein allergy
CHO	Carbohydrates
COCs	Cumulus-oocyte complexes
EG	Egg yolk
ET	Embryo transfer
FAO	Food and Agriculture Organization
GnRH	gonadotropin-releasing hormone
hCG	Human chorionic gonadotropin
ICSI	Intracytoplasmic Sperm Injection
IM	Intramuscular
IV	Intravenously
IVM	In vitro maturation
OPU	Ovum pick-up
PMIE	Post-mating induced endometritis
PR	Pregnancy rates
SCNT	Somatic cell nuclear transfer

1 Introduction

Over 6000 years ago, donkeys were hunted by humans for meat. They were later domesticated in northern Africa and used mostly as pack and transport animals, playing a crucial role in transportation and agriculture in many historical contexts, thereby greatly facilitating commerce among regions. Over the years, and particularly after the industrial revolution in the nineteenth century, donkeys were slowly replaced by machines in developed countries, which led to a decrease in their use for carrying loads, except in the poorest countries, where they are still involved in daily work and transportation. According to Tibary and Bakkoury (1994), donkeys and mules represent 60–80% of the equid population in arid and semi-arid countries in the world. In economically developed regions like Europe, donkey populations have decreased dramatically and are more often bred as pets, with some exceptions like Italy and other Mediterranean countries where the donkey milk industry is a productive economic niche. In China, for more than 2000 years, donkey production has

Table 1 Estimated world distribution of donkey population

Continent /countries	Donkey population
Asia	15.000.000
África	10.000.000
Middle East	10.000.000
South America	9.000.000
Europe	1.500.000
North America	50.000
Australia	5.000.000
Total	**50.550.000**

Adapted from Polidori and Vincenzetti (2019)

been intended for meat, pharmaceuticals, and milk. In recent years, globally, donkey milk has begun to gain importance due to its qualities for human health and the cosmetic industry. This has led to an interest in donkey production worldwide, necessitating efficiencies in commercial production systems.

The global donkey population is continuously and progressively declining (FAO 2020) (Table 1). Particularly, most of the 189 donkey breeds developed by human selection over thousands of years on almost all continents are categorized as threatened, protected, or directly extinct in the wild (Camillo et al. 2018).

As with other domestic mammals, the diversity of breeds has generated great phenotypic variability in conformational and functional traits. This allows adaptation to different environments and purposes for which these breeds have been developed. An example is the average body weight of adults, ranging between 80 and 480 kg, and height, between 64 cm in miniature donkeys and 170 cm in the Mammoth breed. Donkeys' longevity extends between 30 and 35 years, during which they maintain their capacity for active labor even in advanced age. The majority of donkeys worldwide are not part of a specific breed and lack genealogical registration. In many countries, these donkeys are found in a feral state as they are not viewed as "productive" animals but rather as "invasive". Consequently, there are government-sponsored elimination programs, akin to those for feral horses, dogs, and pigs, aimed at controlling their population in federal lands or national parks. This situation is recurrent, including the authors' home country of Argentina; therefore, our program aims to rehabilitate these animals by integrating them into a formal production system that significantly impacts public health. This initiative provides them with an alternative to being hunted or sent to the slaughterhouse.

Currently, donkey production can be divided into two main categories: (1) leisure, recreational, or educational activities, and (2) provision of services (work, animal assisted therapies) or products (milk, meat) for human use.

2 Donkey Milk and Human Health

There are fossil evidence and written records of the cosmetic and biomedical use of donkey milk dating back 3000 years. More recently, documents in ancient medical treatises such as *Naturalis Historia* by Gaius Plinius Secundus (2000 BP); Valuable Prescriptions for Emergencies by Sun Simiao (Tang Dynasty 1800 BP); Compendium of Materia Medica by Li Shizhen (Ming Dynasty, 1000 BP) reported the use of donkey milk for the treatment of arthritis, cough, surgical wounds, ulcers, and dysentery. In the *Corpus Hippocraticum*, a collection of around 50 books (spanning over a thousand pages) attributed to Hippocrates, the father of contemporary Western medicine, and written in Ionic Greek approximately 2700 years ago, specific reference is made to donkey's milk (raw or boiled, alone or mixed with honey) as a therapeutic and dietary substance. This is alongside numerous elements of natural medicine and nutrition such as honey, herbs, vegetables, and fruits. Textually, the diseases for which donkey's milk is recommended include: pleuropneumonia, empyema, consumption (tuberculosis), melanoma, inflammatory disease, melancholy, lung conditions, spinal dissection, kidney disease, dropsy, liver ailment, jaundice, and dense disease (Tratados Hipocraticos, Libro VI -Enfermedades- Editorial Gredos, Madrid, 1990). Moreover, one of the most renowned and impactful documents in the medical community for hundreds of years, the famous *Al-Qanum*, known as "The Canon of Avicenna", the monumental 14 volumes of medical work written around 1100 BP by the Persian physician Ibn Sina (Avicenna) where he advised the use of donkey milk for the treatment of cough, hemoptysis, ulcers, wounds, ascites, fever, fatigue, and asthma.

There are also graphic records documenting a widespread practice in Europe in the nineteenth century where donkeys were bred adjacent to human hospitals, especially those housing orphaned children. Nurses would feed the babies directly from the jennies, which allows us to assume that donkey milk was considered closely related to human milk (something currently experimentally demonstrated) (Altomonte et al. 2019) and that it had clinically proven beneficial effects (Parrot 1882).

Today, the sale of donkey milk—obtained through manual milking on demand, particularly for children with respiratory problems or elderly individuals—is a common practice in many Mediterranean and Latin American countries. This practice is rooted in ancient cultural traditions and supported by substantial empirical evidence of its beneficial effects (Losinno 2022).

Scientific production relating to donkeys and donkey milk began in the mid-twentieth century growing significantly in recent years (Figs. 1 and 2). Comparative compositional and microbiological studies involving donkey milk and the milk of other domestic mammals emerged at the beginning of this century (D'auria et al. 2005). The first randomized clinical trials in children diagnosed with cow's milk protein allergy (CMPA) were approved and conducted in Europe using hydrolyzed bovine milk formulas as a control. These hydrolyzed bovine formulas are the standard recommendation in most countries that either do not promote or are unaware of the proven benefits of donkey milk. Recently, the results of clinical trials

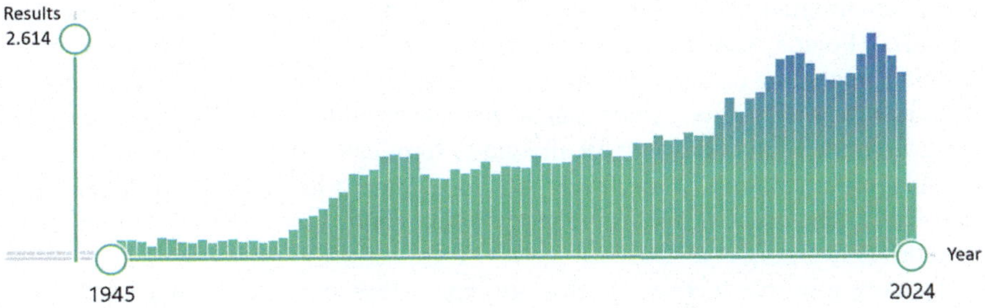

Fig. 1 Scientific publications per year using the keyword DONKEY in the PubMed database between 1945 (45) and 2024 (2138)

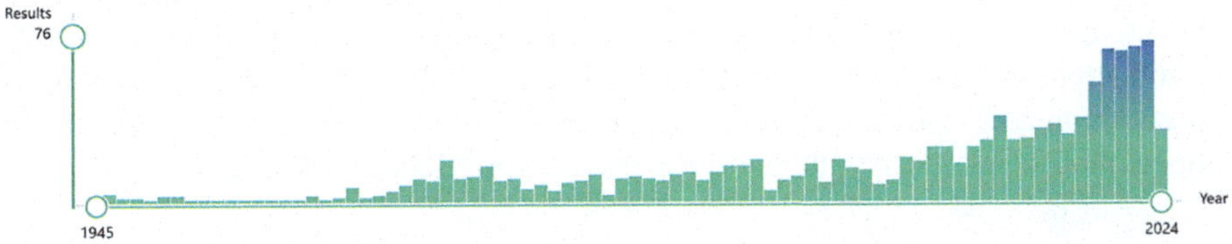

Fig. 2 Scientific publications per year using the keyword DONKEY MILK in the PubMed database between 1945 (0) and 2024 (45)

Table 2 Chemical composition (g/100 g) of donkey, cow, and human milks

Milk	Dry matter	Fat	Total protein	Whey protein (g/L)	Casein (g/L)	Lactose	Energy (KJ/Kg)
Human	12.4	3.3	1.6	7.0–8.3	3.2–4.2	6.7	2855
Jennie	9.61	1.2	1.7	5.0–8.0	6.4	6.3	1939
Cow	12.3	3.4	3.4	4.5	27.2	4.7	2983

Adapted from Polidori et al. (2015), Salimei et al. (2004) and Vincenzetti et al. (2017)

using fortified donkey milk in premature babies (with birth weights of less than 1250 g) with remarkable positive effects have been published (Peila et al. 2020). In the last 5 years, studies have been more focused on metabolomics, proteomics, molecular compositional analysis (protein and fatty acid profiles), exosomes, and *in vitro* effects assays on human cell cultures evaluating gene expression patterns.

The biomedical use of donkey milk is partly due to its compositional similarity to human milk and its marked differences with ruminant milk. Donkey milk has a low protein concentration, especially caseins, considered the main component responsible for food allergies in human infants (CMPA), a high concentration of lactose, which makes it very sweet and palatable for babies, and a low-fat content (Polidori et al. 2015; Salimei et al. 2004; Vincenzetti et al. 2017) (Table 2).

The global donkey milk market was valued at USD 24.79 million in 2022 and is expected to grow to USD 59.07 million at a CAGR (Compound Annual Growth Rate) of 9.1% by 2032, based on the potential benefits of donkey milk with many anti-aging effects, skin-nourishing ingredients, and healing potential including its ability to improve skin hydration and reduce the appearance of fine lines and wrinkles (Polaris Market Research 2023). Donkey milk is rich in vitamins and minerals, including calcium, phosphorus, and vitamin D, and has a lower fat content compared to cow's milk. Additionally, it is easier to digest, making it an excellent alternative for adults who are still allergic to ruminant milk.

This milk contains high levels of lysozyme, α-lactalbumin, β-lactalbumin, lactoferrin, lactoperoxidase, immunoglobulins and more than 1330 bioactive peptides (Piovesana et al. 2015).

Donkey milk production systems are mainly based on "semi-intensive" reproductive management. In order to obtain optimal milk production, and maintain animal health and welfare, donkey farm management is changing from traditional to more organized systems. This involves production programs that consider nutrition for the different categories, amount of milking per day, separation time from the jennies (number of hours and time of the day), specific milking facilities, and also reproductive design. Dairy farm management is highly correlated with welfare. In poorly managed systems, animals will be predisposed to chronic stress which can impact their behavior, health, and milk production (Salari et al. 2022).

3 Donkey Milk Production Programs

According to FAOSTAT (2013), livestock are key for sustainable agriculture, food security and nutrition of billions of people worldwide. Also, working equids provide support and help in a wide range of sectors in developing countries including agriculture, tourism, public transport, construction and mining are often referred to as "invisible workers" (thebrooke.org/invisible-workers).

Donkeys have historically been neglected by science until quite recently, with a notable surge in scientific interest, particularly over the past decade. Moreover, traditionally they have not been viewed as a "productive" animal in the conventional sense, where productivity is typically associated with the generation of tangible products, often *post mortem*, such as meat and skin.

The donkey has been a service animal rather than a product for humans. However, for many years, donkey milk production has been a common productive practice in family farms in Mediterranean countries, the Balkans, Northern Africa, Latin America, and mostly China. Currently, there are specific commercial programs of low, medium, and large scale for milk production destined for direct human consumption, dairy by-products, or the cosmetic industry.

Donkey production programs for commercial purposes have gained special relevance in recent years, mainly due to clinical evidence proving that donkey milk is a natural alternative for children with CMPA and elderly adults. Furthermore, donkey's milk has demonstrated a positive clinical impact on human health due to its bioactive peptide profile, and it is increasingly utilized in the cosmetic industry for products such as soaps and beauty creams. Its composition closely resembles human milk, offering proven tolerability, palatability, and an optimal concentration of nutrients, along with a rich array of bioactive compounds (Sarti et al. 2019; Martini et al. 2021; Di Salvo et al. 2023). This has sparked a notable upsurge in scientific research on a species that has historically been overlooked by the scientific community.

Some specific aspects of this species deserve to be highlighted, to quickly dispel any kind of comparison or free association with horses, much less with a dairy cow.

Donkeys evolved in arid and mountainous environments, with scarce water sources and low-quality forage, so in general, "native" donkey populations, i.e., not belonging to a particular registered breed, tend to be of a medium biotype (110–130 cm tall and 100–200 kg weight). Like many other domestic animals (horses, dogs and cattle), "extreme" phenotypes in both biotype and coat are a product of human domestication and controlled breeding towards "functional" traits (e.g., larger size, sires for mule production) or purely aesthetic (coat color, dwarfs) phenotypes generically called "breeds".

The four classic categories in donkeys related to size, are: (1) large (over 130 cm and in some cases up to 160 cm); (2) Medium (110–130 cm); (3) Small (90–110 cm); (4) Dwarf or Miniature (less than 90 cm). The last two, the smallest ones, have been the most popular for export in recent years, especially to the United States, but they are also in vogue in Europe, especially in France, where the breeding of small donkeys has been developed in a highly selective manner. A poorly studied aspect is the relationship between coat color (black, brown, grey, piebald, skewbald, white) and the aptitudes for adaptation to specific environments or agricultural activities. For example, black donkeys typical from Europe could have a better resistance to solar radiation and high humidity, while grey or skewbald donkeys may have a better adaptation to temperate mountain areas with large temperature variations between day and night.

Ideally, considering animal welfare regulations and characteristics of donkey biology patterns, a pasture-based system (i.e., where animals can graze all day or with certain hourly restrictions) has been proposed as the basis for donkey milk production. They are gregarious migratory herbivores, and their needs for maintaining homeostasis and fitness include choosing and consuming grass, preferably of medium to low quality and lignified, as well as freely roaming and social interaction. As scientists, technicians, and teachers of animal science and production, it is crucial that we strive to enhance the quality of life for these animals, which have served us and provided economic (and often emotional) benefits for thousands of years.

Production scales can be categorized according to investment and potential profitability, considering the number of milking jennies. Low scale -or familiar- (5–30 jennies), medium scale (40–100), and high scale (more than 100). Regarding investment, the objectives of the production system for categorization must be considered; from primary production of raw chilled fluid milk to a complete production cycle with on-site industrialization. It is considered that the majority of the production systems are in the medium scale category, between 30 and 60 milking jennies, under mixed pasture-based systems, with daily milking and an average production that varies between 0.5 and 1.2 L/jennie/day with average lactation lengths of 6 months. Quality standards and procedures are determined by each country, but as a general rule, milking must be carried out mechanically, in approved environments under biosafety standards, pasteurized, applying strict cleaning protocols and standardized animal welfare conditions since it involves the production of food with nutraceutical properties for babies. The product can be marketed as refrigerated, frozen, or lyophilized.

Donkey milk production in European countries has been estimated at 300 tons per year (Valle et al. 2018) and 270 thousand tons/year in China, the world's leading producer, related to a long tradition and its high donkey population (2.7 million) (Seyiti and Kelimu 2021). There are no official reports in Latin America, but there is evidence of commercial production programs on different scales in Argentina and Chile according to unofficial recent data. Donkey milk production is an emerging and innovative productive phenomenon with a strong potential impact on public health, industries, and specific jobs, transforming an excluded animal into a productive one. It is possible to create artificial systems that are sustainable, and environmentally friendly with animals and people working in them in harmony, creating responsible systems as a model for future generations.

4 Breeding in Milk Production Programs

To ensure continuous milk production throughout the year, it is crucial to understand the donkeys' reproductive, social, and nutritional aspects and manage them in an organized and predictable manner. This approach helps ensure that births occur evenly distributed throughout the year. Parameters such as foaling interval, foaling to conception interval, and weaning rate must be regarded as the pillars on which productivity and profitability of the systems are based.

4.1 Reproductive Patterns

Donkeys have a territorial, non-harem type of sociosexual behavior, but patterns similar to horse harems are described in the wild and/or feral donkeys under certain environmental conditions (Moehlman 1998; Linklater 2000). In the classical pattern, territorial jacks spend most of their time away from females. Jennies move freely, usually in groups, and those in estrus may form a sexually active group, similar to cows, displaying estrus behavior, mounting each other to attract the male and approaching him frequently. Females are gregarious and migratory animals, meaning they live in relatively stable social groups with their offspring and prefer to stay in proximity, something very similar to what happens with most zebras. On the contrary, isolation and loneliness cause them anxiety and severe stress that can even lead to depression and anorexia, resulting in illness, so these situations should be avoided. There's an old saying that goes: "If you're going to buy a donkey, it's better to buy two", referring to avoiding isolation (unnatural) and providing a better quality of life in captivity. This situation can even be reversed by placing the donkey with another animal such as a sheep, goat, or even a horse. We have successfully

Fig. 3 Stallion and jack living together in a paddock

Fig. 4 Freely and natural copulative behavior in domestic donkeys in a pasture-based donkey milk production program

experimented by placing two breeding males, a stallion (equine) with a donkey (jack) together in a pen for years, and their behavior and libido greatly improved, decreasing the typical aggressiveness of the isolated animal (Fig. 3).

Unlike females, males are generally territorial, meaning they live relatively solitary in territories they mark as "their own" through defecating and urinating. Females are sexually active and they approach male's territories during estrus, guided by the typical braying of donkeys that can be heard from at least 3 km away due to an anatomical modification in the larynx. Additionally, during the heat or copulatory receptivity period, female donkeys adopt a typical position in front of the male by lowering their heads and opening their mouths wide as if they were chewing (Figs. 4 and 5).

In the donkey, reproductive activity appears to be less affected by season than in horses. Studies on reproductive seasonality indicate that a significant (at least 65%) population of jennies exhibit continuous polyestrous cycles compared to mares under similar environmental conditions, influenced by geographical factors and intrinsic characteristics. Significant influences from factors such as breed, body

Fig. 5 Freely precopulatory sequence in a pasture-based donkey milk production program

condition score, temperature, humidity, latitude, and social stress conditions are proposed. However, the wide variation in the results evidence that this aspect of donkey reproductive biology remains unclear and require further investigation. Therefore, in practical and productive terms, it is recommended to record and analyze reproductive behavior of specific populations in response to the environmental conditions in each production system to achieve more efficient management of annual milk production based on planned pregnancies. Considering this, it is important to have a breeding program design to achieve continuous year-round production, if that is the goal. However, there will be a higher proportion of donkeys foaling in spring and summer, so reproductive management should be strategic to establish a foaling sequence in autumn and winter. As previously mentioned, the nutritional status of donkeys can influence seasonality. Therefore, in a productive system, it is crucial to ensure a consistent supply of appropriate forage throughout the year.

Jennies reach puberty at 18–24 months of age, depending on various factors such as breed, nutrition, health, temperature, and photoperiod. Breeding is not recommended before age 3, when they reach sexual maturity.

Genital organs are similar to those of mares, although their internal genitalia are proportionally larger. The uterus is Y shaped and the tip of the horns are located at the level of the fifth lumbar vertebra. The ovaries are bean shaped, present an ovulation fossa, and are located at the level of the fourth or fifth lumbar vertebra. The cervix points upwards or to the sides, is longer and smaller in diameter than in mares, protrudes 2–3 cm into the vagina, which may preclude intrauterine ejaculation, and also represents a challenge for intrauterine procedures like AI (Artificial insemination) or ET (Embryo Transfer). In the jenny, the pelvis is more sloped downward leading to the vulva being tipped such that the ventral commissure of the vulva is more cranial than the dorsal commissure. This is a more desirable angle because this slightly ventrally tilted slope makes contamination of the reproductive tract less likely. Thus, a vulvoplasty (Caslick's procedure) is less frequently required and performed in jennies than in mares.

Knowledge of the estrus cycle is essential for appropriate breeding management, and although the jenny is similar to the mare in several reproductive aspects, they present some differences, therefore, extrapolating knowledge from the mare is inaccurate. Jennies have a longer estrus cycle, with a range of 20–40 days (average 24 ± 1 days). Inter ovulatory intervals (IOI) tend to be longer in older jennies and shorter in large breeds. Estrus may last 6–9 days but it may be shorter during the spring and summer than in the fall and winter. Diestrus varies from 16 to 20 days. Follicular deviation occurs 8–9 days before ovulation with an approximate follicular diameter of 20 mm. Preovulatory diameter may be influenced by body frame and breed, between 30 and 33 mm in Poitou and North African jennies and 35–40 mm in Pega and Mammoth breeds. Follicular growth averages 2.7 mm/day and ovulation occurs during the last 24 h of heat. Most follicles change shape before ovulation. Double ovulation rates are extremely low (less than 5%) in North African breeds and can be as high as 37–61% in Mammoth and Ragusana dairy breeds. The pattern of estradiol and progesterone plasma concentrations during the estrous cycle is similar to that reported for the mare. Uterine edema may be observed approximately 1 week after the disappearance of the CL and increases towards the day of ovulation. The score of endometrial folds is not a reliable predictor of imminent ovulation as in mares. The best predictors for an imminent ovulation are estrus behavior, follicular size and texture.

Transrectal palpation may be challenging in small framed jennies due to their size, the anatomy of the pelvis and strong rectal contractions. It is important to use appropriate facilities, adapted to their size, introduce two jennies together, and feed them so they remain calm (Fig. 6). The use of lubricant alone or in combination with butyl scopolamine bromide or Vaseline™ may facilitate rectal palpation. Palpation and ultrasound are performed as in mares, with the same applications.

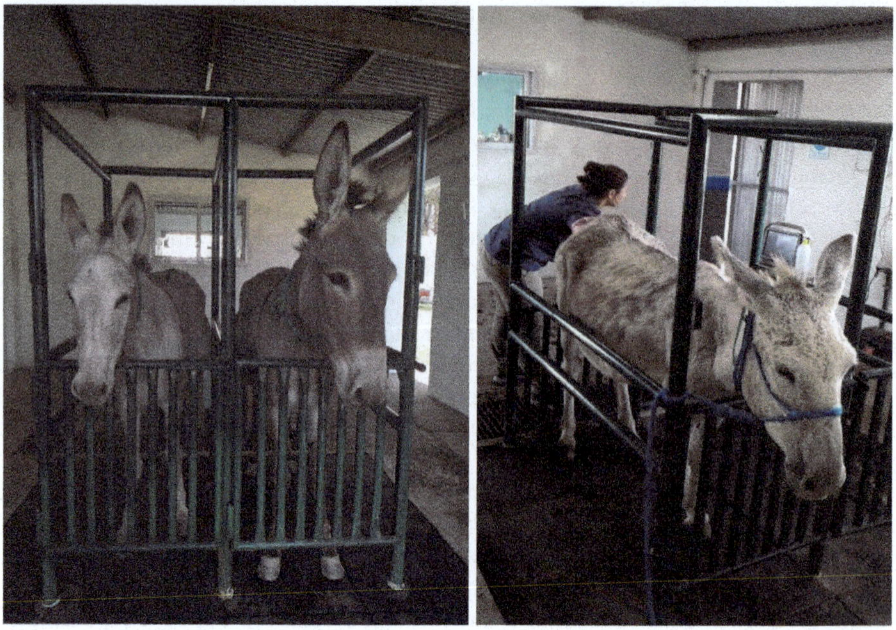

Fig. 6 Milking jennies in double examination stocks with specific sizes for donkeys

Fig. 7 Post-partum jennies in pasture-based programs

The gestation period in donkeys is longer than in mares (on average 371 days) and a very important characteristic, especially in milk production programs, is that jennies only produce milk if they remain with their offspring, which means that if for any reason she loses her foal permanently, she will stop producing milk (Fig. 7). This is why in organized milk production programs; scheduled separations of the jennies and their offspring are carried out for a maximum of 3–6 h to achieve efficient daily milking (Fig. 8).

Fig. 8 (**a**) Jennies immediately after milking process (right) with their offspring's waiting (left); (**b**) Jennies reunited with the foals after milking

Fig. 9 Precopulatory behavior in a hand-bred donkey reproduction program

4.2 Controlled Natural Breeding

4.2.1 Hand Breeding

Jennies in natural breeding programs may be pasture or hand-bred. Hand breeding is less common, due to the unique nature of the jack's sexual behavior (Fig. 9). Jacks, when hand-bred, usually take a long time (20–60 min) to achieve erection and mount, they also tend to perform several mounts without erection and appear to require space and liberty to approach and retreat from jennies before successful mount and ejaculation. Donkeys have long held the reputation of being slow and difficult to hand-breed. For hand-breeding, estrus jennies require minimal to no physical restraint, they tend to be less aggressive than mares, hence, breeding accidents are less common. When a hand-breeding program is employed, jennies should be, if possible, teased and evaluated by manual palpation and ultrasound to enhance success.

This implies more intensive management that may cause stress, but on the other hand, it allows better control of the estrus cycle, reduced number of services per

Fig. 10 Natural breeding in pasture-based program

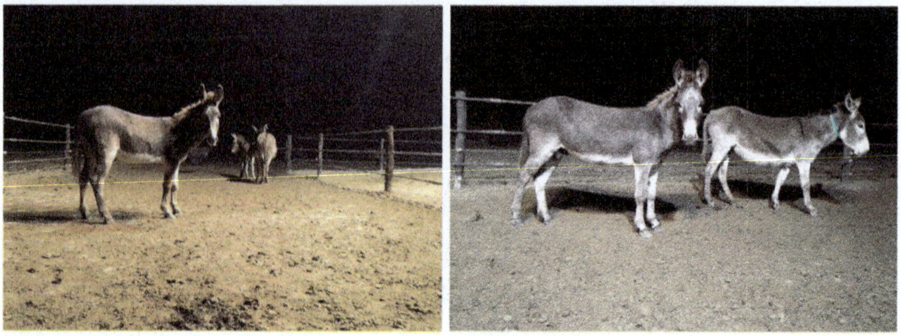

Fig. 11 Natural breeding in a paddock

cycle, and accurate conception date determination, among other things. As a result, most are bred at pasture or in paddocks where they are allowed free interaction with one or more jennies (Figs. 10 and 11).

4.2.2 Teasing

Teasing must be done cautiously, both in the individual or group management of jennies and in the interpretation of the results. Unlike what is generally done with mares, it is advisable to do it in paddocks with groups of females rather than individually. This is the most natural approach where they feel safer. It's important that the males used have a calm and non-aggressive temperament, considering that in this species, females are sexually active and play an important role in seeking out the male when in heat. In China, they often use a system where a male is tied to a stake within a paddock, and a group of jennies is introduced to assess them, observed from outside for at least 30 min for their sexual receptivity, as within the group there is also dominance and subordination that can alter their estrous behavior patterns (Fig. 12) (S. Zeng, personal communication, 2017).

Fig. 12 Teasing with a male tied to a stake within a paddock. (Courtesy Dr. Shenming Zheng)

Fig. 13 Milking jennies and jack in a pasture-based breeding program

4.2.3 Pasture Breeding

In pasture breeding, the jack is turned into pasture with the jennies. Usually, up to 20–35 jennies per jack for a prolonged period of time, approximately 90 days (Fig. 13). Pregnancy exams by ultrasonography can be performed at 15-day intervals and pregnant jennies may be replaced with barren ones but sometimes this is not an ideal strategy because the social groups may be in conflict. It is advisable and very important to place only one male per group of jennies, as having more than one can led to problems such as fighting and aggression while competing for females in heat. This can result in significant harm to their overall condition and health (weight loss, injuries, inactivity of one male, aggression towards females) and reproductive performance of the group. In dairy farms, jennies are grouped according to whether they are being milked or not. It is advisable to group jennies that are not being milked by age, leaving those younger than 30 months out of breeding.

The jack performs estrus detection and may mate with a jenny several times per day. An experienced pasture breeding jack will rarely attempt to breed a jenny that is not in estrus, and in general, is very tolerant with foals although they can become aggressive with male donkeys reaching puberty. The advantages of pasture breeding are decreased personnel and breeding expenses, as well as decreased handling; the latter is of particular importance to reduce stress in donkeys. This minimal handling may become a disadvantage given that we may not be able to detect reproductive problems soon enough in jacks or jennies, leading to decreased pregnancy rates. This can be overcome by examination of jacks and jennies before pasture turnout and

Fig. 14 Electric fence in a pasture breeding program

pregnancy exams every 15 days to determine conception day accurately and obtain a potential foaling date. Also, this allows early detection and reduction of twin pregnancies, which are rare in our population of mixed-breed "native" jennies in Argentina, but can occur. Per-season pregnancy rates in large scale (250 jennies) pasture breeding programs can range between 85% and 90% (personal unpublished data, Equslac™, Argentina). Pregnancy rate (PR) can be affected by several factors such as seasonality and the jennies' age. As mentioned before, reproductive activity appears to be less affected by season in the donkey; although, PR seems to be influenced by season because it appears to be significantly greater during the spring and summer, even in jennies exhibiting a continuous reproductive cycle. PR is also negatively affected by age, showing a decrease from 80% to 60% after the sixth year of age and below 50% in 14–16-old-year jennies (Tosi et al. 2013).

Another advantage of a pasture-based system is that it allows for separating subpopulations of animals by category using electric fencing, which enables visual contact and communication between individuals. This reduces the stress associated with physical separation while allowing for differential management (Fig. 14).

4.2.4 Gestation Control

Pregnancy can be detected by ultrasonography between day 11 and 12 post ovulation. The embryonic vesicle is mobile in the uterine lumen until day 16 post ovulation as in mares. Therefore, maternal recognition of pregnancy in donkeys seems to share similarities with mares. Embryonic vesicle growth rate is 3–3.5 mm per day until day 18 and then it slows from day 19–30 (0.5 mm/day). Heartbeat is detected around day 23. Equine chorionic gonadotropin is secreted after day 36 and secondary *corpora lutea* can be seen ultrasonographically between day 38–45. Chorionic girdles in donkey pregnancies are found to be narrower and generally smaller than horse girdles. Serum progesterone during pregnancy varies from 19–20 ng/ml during the first 3 weeks and decrease to 12 ng/ml on day 30 with another peak (17–18 ng/ml) at days 30–40 declining gradually to days 110–160 to 6 ng/ml (Meira 1998).

Placental and fetal growth monitoring during mid and late pregnancy can be evaluated in the same manner as in mares.

4.2.5 Foaling Control

Pregnancy length in jennies varies significantly, ranging from 333 to 395 days, with an average duration of 371 days. Pregnancy length in mares may be influenced by breed, maternal age, parity, environmental factors (month of conception within the breeding season, month and year of foaling), foal gender and birth weight, but it appears that pregnancy length in jennies is influenced by fewer factors compared to mares. However, foal gender and the timing of foaling within the breeding season have been reported as significant factors. As in mares, jennies appear to have more control on the time of delivery than other species. The majority of jennies foal during the night when disturbances are minimal. This, coupled with the less intensive management of donkeys compared to mares and their tendency to avoid people during birth or postpartum, makes foaling control particularly challenging. However, non-invasive monitoring systems can be applied in order to control foaling. Prepartum jennies should be housed in pens near the dairy staff so that any abnormal behavior or potential issues at the time of birth can be promptly observed and addressed (Fig. 15).

In small-scale production systems, births are usually controlled in stalls where the jennies are placed after the first stage of labor if they are to be intensively monitored. However, in our opinion, this is a tool that may result in more complications than benefits due to the stress of separating the female from her group and to the high level of contamination that is often present in foaling facilities. Based on our experience, we have observed a higher neonatal mortality rate compared to field foaling systems, where animals remain with their group and conditions are more similar to those in nature. Foaling stages' duration and characteristics are similar between donkeys and horses. Although the normal donkey foal may take a little longer to stand, nurse and eliminate meconium than the horse foal.

One of the frequently mentioned problems in pasture births is the attack of wild natural predators, especially pumas (*Puma concolor*) which are widespread in Argentina, during the first week after birth. Although female donkeys have a strong defensive instinct towards their young and will face pumas rather than flee from them (in fact, they are often used as protectors for calves and lambs), when

Fig. 15 Prepartum jennies in a pasture-based donkey milk production program

confronted by multiple predators simultaneously, there are typically significant offspring losses. This not only affects the herd but also results in a substantial negative economic impact in a milk production system. To control this problem, there are many management practices that can be applied such as the use of flashing lights, nighttime confinement of females, etc.

4.2.6 Postpartum Breeding

Efficient postpartum breeding is critical in order to obtain a short foaling interval in dairy jennies. A fertile estrus (foal heat) may occur 8–9 days after foaling, with a wide range of 2–60 days reported. Estrus length, daily growth of the follicle and preovulatory diameter of the follicle are similar to those recorded in normal cycling jennies. Ovulation occurs between day 13 and 17 postpartum in jennies with normal foaling, but it is widely influenced by body score, environmental and social stress. Pregnancy rate can be affected by foal heat timing, season and jennies' age. Pregnancy rate at foal heat may be lower than in the subsequent estrous cycles, although it increases when ovulation occurs with intervals greater than 12 days after foaling. Pregnancy loss in jennies inseminated at foal heat is lower than that reported in mares (2.4% v. 12.4% pregnancy lost between day 15 and 35) (Carluccio et al. 2017). The management and control of foal heat may be challenging in pasture breeding systems; although, postpartum jennies may be placed separately until the reproductive conditions are optimal. A survey report of several donkey farms in China (n = 10,159 animals) reported an overall mean foaling interval of 410.9 days and an overall mean foaling alive rate of 83.3%. Weaning age of donkey foals varies greatly among farms depending on particular farm milking management, breeding systems, facility availability, among others; but it varies between 5 and 9 months. Postpartum reproductive problems in jennies may be similar to those observed in mares. However, since this species is less intensively managed than mares, usually, fewer reproductive problems are seen. Body condition score in jennies must be controlled through all of pregnancy, but especially in the last trimester, since in this stage the capacity of the digestive tract decreases, which may lead to negative energy intake and therefore risk of hyperlipidemia. This also applies to the first lactation trimester.

4.3 Reproductive Technologies and Management

The use of assisted reproductive technologies (ARTs) has become increasingly prevalent in equine species over the past few decades. Most ARTs applied to donkeys are adapted from horses (Oliveira et al. 2006), extrapolating information and knowledge and creating new information specific to donkeys. Within ARTs, the most widely used in donkeys are artificial insemination with fresh or chilled semen; however, there are others much more complex such as semen freezing, embryo transfer and intracytoplasmic sperm injection (ICSI) which are in very early stages of development in donkeys. While AI with fresh and cooled semen has given satisfactory results, the use of frozen semen presents major challenges. Despite advancements, the pregnancy rates achieved with frozen semen are significantly lower compared to fresh or chilled semen.

Although pregnancy rates in semi-intensive pasture breeding programs can be high, and this type of program seems to have more advantages than disadvantages, especially from the animal welfare point of view, in certain situations, ARTs may be necessary to implement.

ARTs can be applied for genetic improvement programs, inbreeding risk avoidance, preserving biodiversity, and preserving certain endangered donkey breeds.

These technologies offer a strategic approach to managing reproductive challenges and advancing the long-term sustainability of donkey populations. More studies and improvements in ARTs are necessary for them to be applied successfully in productive systems.

4.3.1 Ovulation Induction

Ovulation induction agents are routinely used in mares' reproductive cycles to optimize breeding management. Different ovulation induction and estrus synchronization protocols have been studied in jennies to optimize reproductive management, especially in semi-intensive reproduction programs (Table 3).

Estrus duration in jennies varies based on factors such as time of year, body condition score and age. Ovulation induction can aid in narrowing the ovulation window and facilitating breeding management, artificial insemination and/or

Table 3 Ovulation induction agents in jennies

Drug	Jennies (n)	Dose	Follicle diameter at treatment (mm)	Interval to ovulation (h)	% ovulation within 48 h	References
Lecirelin	34	100 μg (IV)	36–40	42.8 ± 14.0	80	Carluccio et al. (2007)
hCG	27	2500 UI (IV)	36–40	42.4 ± 13.0	100	
Buserelin	14	40 μg (SC)	33 ± 1	24–48	71.4	Camillo et al. (2014)
LH	25	400 UI (IM)	35	24–48	92	Chang et al. (2019)
hCG	25	3000 UI (IM)	35	24–48	76	
Histrelin acetate	10	250 μg (IM)	29–32	36–48	100	Oliveira et al. (2020)
hCG	10	2500 UI (IV)	29–32	36–48	90	
Deslorelin acetate	12	0.75 mg (IM)	36.34 ± 0.71	48.79 ± 2.69	78.7	Bottrel et al. (2022)
hCG	13	1500 UI (IM)	37.85 ± 1.27	62.61 ± 7.20	60.9	

embryo transfer. Human chorionic gonadotropin (hCG) and gonadotropin-releasing hormone (GnRH) agonists have been utilized to induce ovulation. As in mares, the recommended follicular size to apply the ovulation induction agent correlates with body size. Small-framed jennies ovulate small follicles (28–32 mm), and larger-framed jennies, such as the Catalan and Mammoth breed, ovulate 40–44 mm follicles. Jennies induced with hCG or lecirelin with periovulatory follicles (36–40 mm) may also ovulate additional smaller follicles (30–35 mm) (Carluccio et al. 2007).

4.3.2 Artificial Insemination with Fresh and Cooled Semen

Artificial insemination is a useful technique for donkeys and can be applied with fresh, cooled, and frozen semen. Jacks, like stallions, can undergo collection procedures; however, it is advisable to allow longer teasing periods for them. A collection session with a jack may last between 30 and 60 min, with younger jacks exhibiting slower collection times compared to mature ones. Various methods, including using an estrous jennie or mare, a dummy, or ground collection, can be utilized for jack semen collection (Fig. 16). Standard equine equipment can be employed for semen collection, evaluation, transportation, and artificial insemination in donkeys (Canisso et al. 2019).

Compared with stallions, donkeys exhibit superior motility parameters, with higher velocities and progressive motility rates. Typically, sperm concentration ranges from 300 to 400 million sperm/mL, with 80–90% total and progressive motility. Younger donkeys tend to produce ejaculates with lower volume but higher concentrations than older jacks.

Artificial insemination with fresh semen in donkeys has been used for more than 50 years with per-season pregnancy rates between 45% and 78% (Svendsen 2008) and different breeding doses and extenders have been tested. (Table 4).

Cooled semen can be stored at 4 °C for up to 48 h with satisfactory fertility rates post-insemination (Rota et al. 2012). High per-cycle pregnancy rates (50%) have

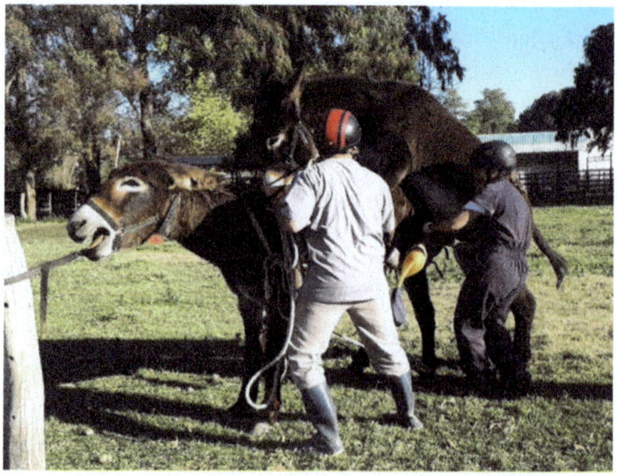

Fig. 16 Jack's semen collection with a Missouri artificial vagina using an estrus jennie. (Photo courtesy of Luisina Chapero and Liliana Rosetto)

Table 4 Pregnancy rates in jennies after artificial insemination with fresh and cooled donkey semen

Type	Breeding dose (million sperm)	Extender	Pregnancy rate (%)	References
Fresh	400	Skimmed milk	86 (6/7)	Vidament et al. (2009)
	500	Botusemen	40 (6/15)	Oliveira et al. (2016)
	1000	Botusemen	73 (11/15)	
	800	INRA 96	81 (25/31)	Carluccio et al. (2017)
Cooled	200	Skimmed milk	45 (9/20)	Vidament et al. (2009)
	460	Skimmed milk	78 (7/9)	
	460	INRA 82 + 2% egg yolk	64% (5/8)	
	500	Equiplus +1% egg yolk	50% (4/8)	Alonso et al. (2023)

Adapted from Canisso et al. (2019)

been achieved in mares inseminated with frozen donkey semen (Oliveira et al. 2006). However, when it is used in jennies, pregnancy rates were 0–36% (Oliveira et al. 2006). Frozen semen induces a more severe inflammatory reaction in the uterus than fresh semen, and it was hypothesized (Canisso et al. 2019), that this may be the cause of the low pregnancy rates obtained with frozen donkey semen in this species.

4.3.3 Artificial Insemination Techniques

Artificial insemination technique is similar in all equine species. The pregnancy rate after AI with fresh semen does not differ significantly between jennies and mares. Jenny's reproductive anatomy is similar to mares, although there are some differences. The cervix in jennies points upwards or to the side, and is longer and smaller in diameter than in mares (Fig. 17). It protrudes 2–3 cm into the vagina and also represents a challenge for intrauterine procedures like AI or embryo transfer.

Due to this, some studies have been carried out to evaluate pregnancy rates in jennies using different AI techniques. In the conventional artificial insemination technique, the jennies' tail is wrapped and tied to clean the perineal area. The pipette is introduced transvaginally and guided through the cervix, one finger is inserted into the cervix and the pipette is passed into the uterine body. The semen-filled syringe is attached to the pipette to deliver semen with a gentle push. If the jennie is too small, the technique becomes more difficult to perform (Figs. 18 and 19).

An alternate technique for insemination could be the use of cervical forceps to hold the cervix in position using the Wilsher technique. A small size Polansky speculum is introduced in the vagina and the cervix is held by the ventral segment with the cervical forceps and pulled backward to straighten it. Finally, an AI catheter is inserted through the cervix into the uterus. Flores Bragulat et al. 2022 reported pregnancy rates per cycle not statistically different using Wilsher and conventional AI techniques (54.5% and 75.0% conventional and Wilsher forceps techniques

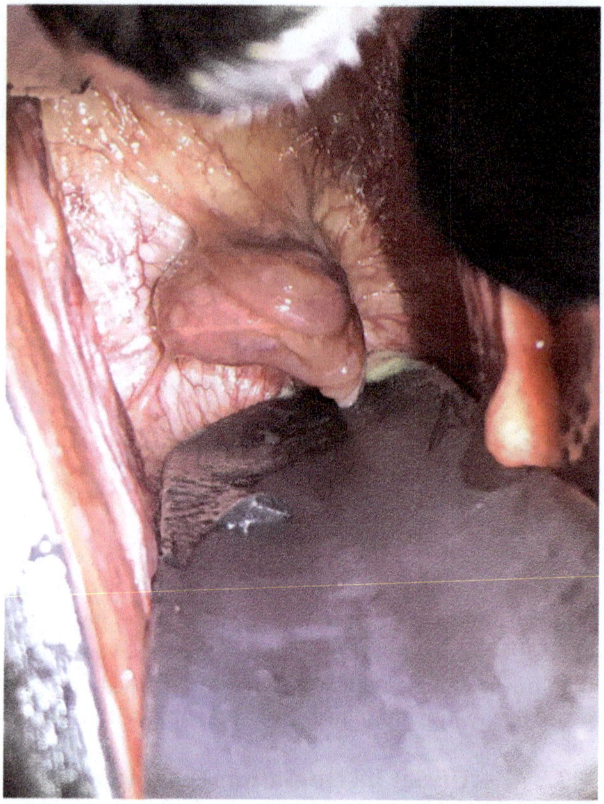

Fig. 17 Jennie cervix observed with the use of a Polansky speculum. Note the cervical os pointing to the side

Fig. 18 Jennies in stocks ready for artificial insemination

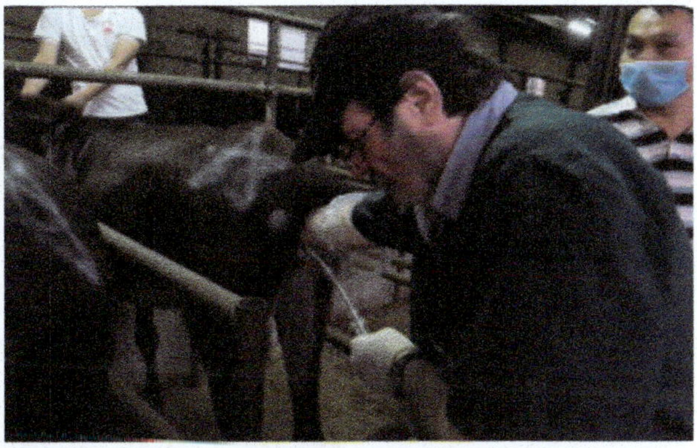

Fig. 19 Artificial insemination in jennies using conventional transvaginal technique

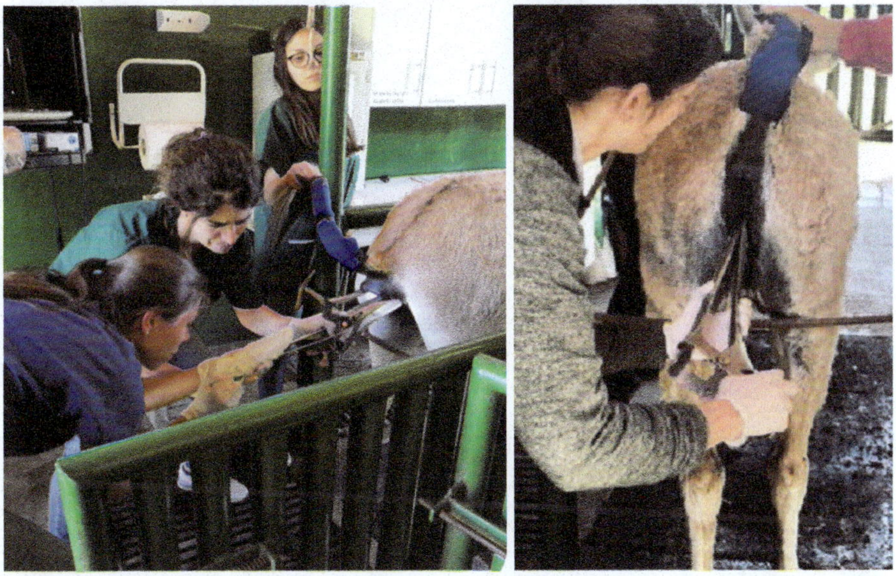

Fig. 20 Artificial insemination in jennies using the Wilsher technique and Polansky speculum. (Photo's courtesy of Carolina Alonso and Ana Flores Bragulat)

respectively). Despite these results, the application of the Wilsher technique can be an especially useful tool in AI of small donkey breeds (Figs. 20 and 21).

Also, the AI technique can vary depending on the place of semen deposition. The breeding dose can be deposited in the uterine body (used for fresh and cooled semen), or deep in the uterine horn. In the last case, semen is released as deep as possible into the uterine horn ipsilateral to the ovary with the dominant follicle and/or ovulation. Oliveira et al. 2016 found differences in pregnancy rates in jennies inseminated with the same breeding dose of frozen semen in the uterine body (0%) and deep horn (28%) (Fig. 22 and Table 5).

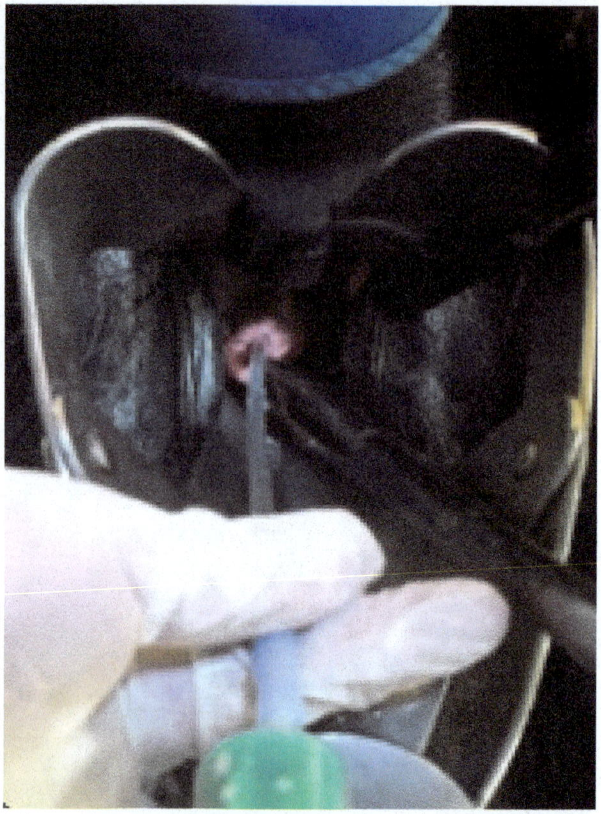

Fig. 21 Artificial insemination in a jennie using the Wilsher cervical forceps technique and standard AI pipette. (Photo's courtesy of Carolina Alonso and Ana Flores Bragulat)

Fig. 22 Deep horn manual transvaginal artificial insemination in a jennie with frozen semen. (Photo's courtesy of Luisina Chapero and Liliana Rosetto)

Table 5 Pregnancy rates in jennies after artificial insemination with frozen and vitrified donkey semen

Type	Breeding dose (million sperm)	Extender	Pregnancy rate	References
Frozen	600	Skimmed milk +4% glycerol +10% EY	0% (0/17)	Trimeche et al. (1998)
		Skimmed milk +4% glycerol +10% EY (glycerol removal after thaw)	38,9% (8/21)	
Frozen	400	INRA 82 + 2,2% glycerol +2% EY	11% (4/38)	Vidament et al. (2009)
Frozen	500	INRA 96 + 2,2% glycerol +2% EY	20% (1/5)	Rota et al. (2012)
		INRA 96 + 2,2% glycerol +2% EY + SP (SP adition after thaw)	61,5% (8/13)	
Frozen	1000 (body)	Botucrio	0% (0/12)	Oliveira et al. (2016)
Frozen	1000 (deep horn)		28,26% (13/46)	
Vitrified	500	Gent	22% (2/9)	Diaz-Jimenez et al. (2020)
Frozen		INRA 96 + 2,2% glycerol +2% EY	10% (1/10)	
Frozen	500	INRA 96 + 2,2% glycerol +2% EY	22% (2/9)	Fanelli et al. (2022)
		INRA 96 + 2,2% glycerol +2% EY + SP (SP adition after thaw)	60% (6/10)	

EG egg yolk, *SP* seminal plasma

5 Practical Reproductive Management

Donkey milk production systems range from intensive to extensive commercial programs and this wide range affects the application of reproductive management directly. Semi-intensive systems will allow greater veterinary intervention and therefore the application of more complex reproductive biotechnologies. Semi-extensive systems commonly use natural pasture breeding and less complex reproductive biotechnologies such as periodic ultrasound examinations and artificial insemination in special situations. To ensure optimal reproductive management, adequate facilities such as stocks specially designed for donkeys are necessary. This allows a safe environment for the operator and donkey allowing them to feel calm and without stressful situations. Donkeys have a great memory, therefore, subtle and progressive handling is recommended to generate a gradual habituation to different situations, such as entering the stocks or the dairy facilities. Donkeys'

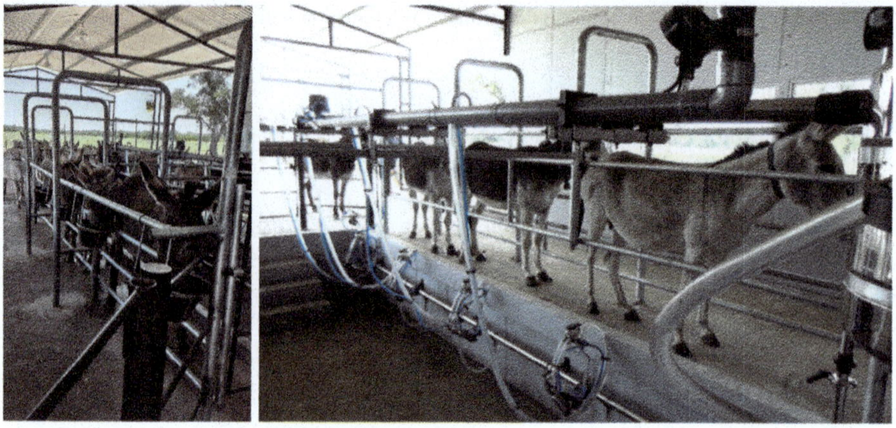

Fig. 23 Jennies during mechanical milking process

habituation to these facilities can be accompanied by stimulation with food, entry of other animals at the same time, correct lighting of spaces, and giving them time to observe. In case of fearful animals or reluctance to handle, sedation or temporary covering of the eyes with a soft mask may be helpful.

Pasture breeding, at the start of a program, allows animals gradual habituation to handling, with check-ups every 15–20 days. Replacement of males every 2 years must be taken into account in this type of program to avoid inbreeding; another option could be to create different groups of jennies to be bred with non-related jacks. Subsequently, more complex reproductive biotechnologies could be applied in special cases with the aim of improving the genetic potential for milk production in future generations using animals with a high genetic potential for milk production as parents.

Feeding of donkeys deserves separate comments, especially when considering production (and reproduction) programs, in which, due to ignorance, improvisation, and analogy, practices applied to other domestic species such as horses or dairy cows are transferred. Donkeys are highly efficient in utilizing dietary fiber for energy production, better than most other herbivores, with low tolerance to excess starch and other carbohydrates (CHO). A low-fiber and/or high-CHO diet is contraindicated as it endangers not only productivity but also the animals' lives through clinical conditions such as colic, laminitis, hyperlipidemia, metabolic syndrome, and dysbiosis.

Even though all aspects of a production system are important, such as genetics, reproduction, and animal health; animal welfare and respect for the biological and behavioral aspects of this species are essential to work properly and obtain optimum results (Fig. 23).

6 Conclusions

Donkeys have been much less genetically manipulated by humans than other species, so intensive management that entails changes in social groups, isolation, aggressiveness, and sudden changes in diets generate significant stress that negatively impacts reproduction and their productive health status, especially in milk production programs. Therefore, it is important to take all these factors into account when designing a reproductive program for donkeys. Changes and management practices, such as serial ultrasounds, insemination, and flushings, should be introduced sequentially and in a gentle manner.

Questions

- Why is a pasture-based system recommended for donkey milk production?
- What are some of the key welfare considerations for donkeys in production systems?
- How can donkey milk production be part of sustainable, environmentally friendly systems?
- What factors influence the reproductive seasonality of donkeys?
- What is the recommended breeding age for jennies
- How does the estrus cycle of a jenny differ from that of a mare?
- Why is it important to keep jennies and their foals together in milk production programs?
- What are the potential effects of isolation on donkeys?
- What is the significance of using animals with high genetic potential for milk production in breeding programs?
- What is the importance of replacing males every 2 years in pasture breeding programs?
- What clinical conditions can arise from a low-fiber or high-carbohydrate diet in donkeys?
- Why is animal welfare and respect for donkeys' biological and behavioral aspects essential in a production system?

References

Alonso CN, Castañeira C, Bragulat APF, Losinno L (2023) Effects of egg yolk extender on fertility of cooled donkey semen. J Equine Vet Sci 125:104575

Altomonte I, Salari F, Licitra R, Martini M (2019) Donkey and human milk: insights into their compositional similarities. Int Dairy J 89:111–118. https://doi.org/10.1016/j.idairyj.2018.09.005

Bottrel M, Ortiz I, Hidalgo M, Díaz-Jiménez M, Pereira B, Consuegra C et al (2022) Hormonal management for the induction of Luteolysis and ovulation in Andalusian jennies: effect on reproductive performance, embryo quality and recovery rate. Animals 12(2):143

Camillo F, Vannozzi I, Tesi M, Sabatini C, Rota A, Paciolla E et al (2014) Induction of ovulation with Buserelin in jennies: in search of the minimum effective dose. Anim Reprod Sci 151(1–2): 56–60

Camillo F, Rota A, Biagini L, Tesi M, Fanelli D, Panzani D (2018) The current situation and trend of donkey industry in Europe. J Equine Vet Sci 65:44–49

Canisso IF, Panzani D, Miró J, Ellerbrock RE (2019) Key aspects of donkey and mule reproduction. Vet Clin North Am Equine Pract 35(3):607–642

Carluccio A, Panzani S, Tosi U, Faustini M, De Amicis I, Veronesi MC (2007) Efficacy of hCG and GnRH for inducing ovulation in the jenny. Theriogenology 68(6):914–919

Carluccio A, Gloria A, Robbe D et al (2017) Reproductive characteristics of foal heat in female donkeys. Animal 11:461–465

Chang ZL, Li BX, Liu B, Yao L, Yu J, Jiang GM et al (2019) Effects of FSH and the weather during induced ovulation and timed artificial insemination to increase jenny conception rates. Sci Rep 9(1):3220

D'auria E, Agostoni C, Giovannini M, Riva E, Zetterstrom R, Fortin R, Greppi GF, Bonizzi L, Roncada P (2005) Proteomic evaluation of milk from different mammalian species as a substitute for breast milk. Acta Paediatr 94:1708–1713. https://doi.org/10.1080/08035250500434793

Diaz-Jimenez A, Rota J, Dorado D, Panzani D, Fanelli M, Tesi F, Camillo D, Monaco B, Pereira C, Consuegra M, Hidalgo (2020) Comparison of uterine inflammatory response of jennies after artificial insemination with vitrified or frozen-thawed donkey sperm Journal of Equine Veterinary Science 89103036. https://doi.org/10.1016/j.jevs.2020.103036

Di Salvo ED, Conte F, Casciaro M, Gangemi S, Cicero N (2023) Bioactive natural products in donkey and camel milk: a perspective review. Nat Prod Res 37(12):2098–2112. https://doi.org/10.1080/14786419.2022.2116706

Fanelli D, Tesi M, Monaco D, Diaz-Jimenez M, Camillo F, Rota A et al (2022) Deep-horn artificial insemination with frozen thawed semen after re-extension in autologous seminal plasma may improve pregnancy rates in jennies. J Equine Vet Sci 112:103932

FAOSTAT (2013) Available: http://faostat3.fao.org/home/E

FAOSTAT database (2020) Food and Agriculture Organization UN Rome Italy. p 1. Available: http://www.fao.org/faostat/en/

Flores Bragulat AP, Alonso C, Castañeira C, Losinno L (2022) Wilsher cervical forceps for artificial insemination technique in jennies. J Equine Vet Sci 113:103993

Flores Bragulat AP, Ortiz I, Catalán J, Dorado J, Hidalgo M, Losinno L et al (2023) Time-lapse imaging and developmental competence of donkey eggs after ICSI: effect of preovulatory follicular fluid during oocyte in vitro maturation. Theriogenology 195:199–208

Linklater WL (2000) Adaptive explanation in socio-ecology: lessons from the Equidae. Biol Rev 75:1–20

Losinno L (2022) Producción de Burros I. Ediciones del Puente, Argentina. ISBN 978-987-48610-92

Martini M, Altomonte I, Tricò D, Lapenta R, Salari F (2021) Current knowledge on functionality and potential therapeutic uses of donkey Milk. Animals 11(5):1382

Meira C, Ferreira JC, Papa FO, Henry M (1998). Ovarian activity and plasma concentrations of progesterone and estradiol during pregnancy in jennies, Theriogenology 49:1465–73

Miragaya MH, Neild DM, Alonso AE (2018) A review of reproductive biology and biotechnologies in donkeys. J Equine Vet Sci 65:55–61

Moehlman PD (1998) Feral asses (Equus africanus): intraspecific variations in social organization in arid and Mesic habitats. Appl Anim Behav Sci 60:171–195

Oliveira JV, Alvarenga MA, Melo CM, Macedo LM, Dell'aqua JAJ, Papa FO (2006) Effect of cryoprotectant on donkey semen freezability and fertility. Anim Reprod Sci 94:82–84

Oliveira JVD, Oliveira PVDLF, Melo E, Oña CM, Guasti PN, Monteiro GA, Sancler Da Silva YFR et al (2016) Strategies to improve the fertility of fresh and frozen donkey semen. Theriogenology 85(7):1267–1273

Oliveira SN, Segabinazzi LGTM, Canuto L, Lisboa FP, Medrado FE, Dell'Aqua JA et al (2020) Comparative efficacy of Histrelin acetate and hCG for inducing ovulation in Brazilian North-eastern Jennies (Equus africanus asinus). J Equine Vet Sci 92:103146

Parrot J (1882) La nourricerie de l'hospice des enfants-assistés. Bull Acad Med 2e série:839–853

Peila C, Spada E, Bertino E, Deantoni S, Percivati F, Moro GE, Giribaldi M, Cavallarin L, Cresi F, Coscia A (2020) The "Fortilat" randomized clinical trial follow-up: auxological outcome at 18 months of age. Nutrients 12(12):3730. https://doi.org/10.3390/nu12123730

Piovesana S, Caprioti L, Cavaliere C, La Barbera G, Samperi R, Zenezini Chiozzi R, Lagana A (2015) Peptidome characterization and bioactivity analysis of donley milk. J Proteome 119:21–29

Polaris Market Research (2023) Donkey Milk market share, size, trends, industry analysis report, by form (Liquid, powder); by application (Food service, food processing, medical applications, others); By Region; Segment Forecast, 2023–2032. https://www.polarismarketresearch.com/industry-analysis/donkey-milk-market

Polidori P, Vincenzetti S. (2019). The therapeutical, nutritional and cosmetic properties of donkey milk; Cambridge Scholars Publishing: Cambridge; pp. 45–68

Polidori P, Ariani A, Vincenzetti S (2015) Use of donkey milk in cases of cow's milk protein allergies. Int J Child Health Nutr 4:174–179

Rota A, Panzani D, Sabatini C, Camillo F (2012) Donkey jack (Equus asinus) semen cryopreservation: studies of seminal parameters, post breeding inflammatory response, and fertility in donkey jennies. Theriogenology 78(8):1846–1854

Salari F, Mariti C, Altomonte I, Gazzano A, Martini M (2022) Impact of variability factors on hair cortisol, blood count and milk production of donkeys. Animals 12:3009. https://doi.org/10.3390/ani12213009

Salimei E, Fantuz F, Coppolac R, Chiofalod B, Polidori P, Variscoe G (2004) Composition and characteristics of ass's milk. Anim Res 53(1):67–78. https://doi.org/10.1051/animres:2003049

Sarti L, Martini M, Brajon G, Barni S, Salari F, Altomonte I et al (2019) Donkey's Milk in the Management of Children with Cow's Milk protein allergy: nutritional and hygienic aspects. Ital J Pediatr 45(1):102

Seyiti S, Kelimu A (2021) Donkey industry in China: current aspects, suggestions and future challenges. J Equine Vet Sci 102:103642. https://doi.org/10.1016/j.jevs.2021.103642

Svendsen ED (2008) The professional handbook of the donkey. Whittet Books Limited. The Donkey Sanctuary, Yatesbury, p 331. https://www.thedonkeysanctuary.org.uk/research/book

Tibary A, Bakkoury M (1994) Particularités de la reproduction chez les autres espèces équines. Reproduction équine, Tome I: La jument. Actes Editions, pp 385–400

Tosi U, Bernabò N, Verni F, Valbonetti L, Muttini A, Mattioli M, Barboni B (2013) Postpartum reproductive activities and gestation length in Martina Franca jennies, an endangered Italian donkey breed. Theriogenology 80:120–124

Trimeche P, Renard D, Tainturier (1998) A procedure for poitou jackass sperm cryopreservation Theriogenology 50(5) 793–806. https://doi.org/10.1016/S0093-691X(98)00184-8

Valle E, Pozzo L, Giribaldi M, Bergero D, Gennero MS, Dezzutto D et al (2018) Effect of farming system on donkey milk composition. J Sci Food Agric 98(7):2801–2808

Vidament M, Vincent P, Martin FX, Magistrini M, Blesbois E (2009) Differences in ability of jennies and mares to conceive with cooled and frozen semen containing glycerol or not. Anim Reprod Sci 112(1–2):22–35

Vincenzetti S, Pucciarelli S, Polzonetti V, Polidori P (2017) Role of proteins and some bioactive peptides on the nutritional quality of donkey milk and their impact on human health. Beverages 3:34. https://doi.org/10.3390/beverages3030034

Advanced *In Vitro* Breeding Technologies for Donkeys and Mules

Joanne Smith, Patricio D. Palacios, Rhiannon Gurkin, and Andres Gambini

Abstract

This chapter provides an overview of advanced *in vitro* breeding technologies for donkeys and mules, focusing on the latest advancements and their potential applications in equid reproduction and conservation. We begin by discussing *ovum* pick-up (OPU) and the techniques for *in vitro* oocyte maturation and vitrification, highlighting their critical roles in enhancing the efficiency of advanced reproductive technologies (ARTs). The development of intracytoplasmic sperm injection (ICSI) is explored in depth, including the first successful production of domestic donkey and mule blastocysts in laboratory settings. These recent breakthroughs mark a significant milestone in long-ear reproductive research and opens new possibilities for improving breeding programs. Challenges and advancements associated with *in vitro* fertilization (IVF) in equids are covered, noting the recent more reliable IVF protocols for horses and their future application in other equids. Cloning is addressed with reference to the first cloned equid, a cloned mule, Idaho Gem, and its implications for research, competition, and conservation of other endangered equids. Finally, the role of newly developed technologies in reproductive medicine associated with stem cells is examined with the possibility of creating gametes or embryos *in vitro* and addressing reproductive challenges in donkeys and mules. In conclusion, this chapter underscores the transformative impact of advanced *in vitro*

J. Smith · A. Gambini (✉)
School of Agriculture and Food Sustainability, The University of Queensland, Gatton, QLD, Australia
e-mail: a.gambini@uq.edu.au

P. D. Palacios · R. Gurkin
School of Agriculture and Food Sustainability, The University of Queensland, Gatton, QLD, Australia

A. K. McLean et al. (eds.), *Donkey and Mule Medicine: Clinical Medicine and Reproduction*, https://doi.org/10.1007/978-3-032-18540-2_21

breeding technologies on donkeys and mules for addressing reproductive challenges and enhancing genetic diversity.

Keywords

Oocyte · Blastocyst · ICSI · OPU · Cloning · Stem cell

1 Introduction

Assisted reproductive technologies (ART) have revolutionized equine breeding practices. Over the past few decades, there has been a surge in the popularity of reproductive biotechnologies among domestic horses, primarily driven by the high economic value of these animals and the growing demand for elite-performance individuals. In equids, ART offers a range of benefits, including improved reproductive efficiency, genetic selection, and genetic conservation efforts (Cabeza and Gambini 2023). Among ART, *in vitro* embryo production techniques such as *in vitro* fertilization (IVF), intracytoplasmic sperm injection (ICSI), and cloning have been successfully applied in various species, including domestic horses, where they play a critical role in genetic improvement and the preservation of elite bloodlines. In recent years, some of these technologies have expanded successfully to other equids, including Przewalski's horses, donkeys, mules, and zebras (Dobbie et al. 2024; Cabeza and Gambini 2023; Arroyo-Salvo et al. 2024). All these species are members of the *Equidae* family, exhibiting unique genetic plasticity among individuals of various taxa within this genus. This capability leads to the production of viable hybrids and successful interspecific gestation. Hinnies and mules serve as exceptional animal models for studying reproduction and epigenetics (Giulotto et al. 2017; Zhang et al. 2023). Furthermore, embryo transfer technology has enabled remarkable outcomes, such as zebra or donkey embryos being gestated by a mare (Summers et al. 1987; Panzani et al. 2018), or a mule carrying a donkey embryo to term and having a successful delivery (González et al. 2015). In 2002, Australia witnessed the first successful birth of a Poitou donkey foal through conventional embryo transfer. The foal, named Quasar, was carried to term by a surrogate Standardbred mare after the embryo, produced via artificial insemination of the biological Poitou mother, was transferred to the recipient mare (The Horse Staff 2002). Additionally, a notable case involved a donkey embryo being successfully carried to term by a mule, which then delivered a healthy donkey foal (González et al. 2015). This has opened the possibility of mules being used as a universal recipient for equid embryo transfer (Camargo and Rechsteiner 2020). In embryo transfer programs in Argentina, naturally cycling mules exhibit higher pregnancy rates than jennies after embryo transfer of *in vivo*-produced donkey embryos (unpublished data). Techniques such as *ovum* pick up (OPU) have developed significantly over the last decade in the mare, becoming a well-established commercial technology utilized in numerous horse breeds. OPU and ICSI are now routine in commercial equine breeding programs, enhancing genetic progress and enabling the conservation of valuable genetic lines

(Stout 2020; Fonte et al. 2024). With recent developments in equine reproductive technologies, significant milestones have been achieved. In recent years, the first domestic donkey (Flores Bragulat et al. 2023) and mule (Arroyo-Salvo et al. 2024) ICSI blastocysts (day 7–10 stage embryos) were reported, including an early embryonic vesicle (present until day 13) of a donkey ICSI embryo transferred in a jenny recipient (Dobbie et al. 2024). Donkeys are gaining importance for their milk production (Bertino et al. 2022) and the application of ART can significantly enhance genetic progress, leading to improved milk yield and quality. Additionally, mules are renowned for their resilience and ability to thrive in harsh environments, making them invaluable for transportation, especially in military operations and mountainous expeditions. The genetic improvement of mules is promoted through advanced reproductive technologies such as artificial insemination and could greatly benefit from *in vitro* breeding technologies. Interestingly, mules hold the distinction of being the first equids to be successfully cloned. The first cloned mule, named Idaho Gem, was born in 2003 (Woods 2003). In some countries, including the USA, cloning is permitted in mule racing, providing a practical application for cloning research in equine species. Cloning, also known as Somatic Cell Nuclear Transfer (SCNT), along with advanced techniques such as oocyte maturation, IVF, and ICSI, holds the potential to preserve genetic material from equids, including donkeys and mules. These technologies can improve genetic diversity within these populations and contribute to the conservation of endangered equids (Cabeza and Gambini 2023).

In this chapter, we will explore the state-of-the-art advancements in *in vitro* breeding technologies, including oocyte collection, maturation, activation, vitrification, ICSI, IVF, and cloning. Figure 1 summarizes the main technologies that will be covered in this manuscript.

However, we will not cover other assisted biotechnologies such as artificial insemination, embryo transfer, semen collection, and freezing. Most of our knowledge has been derived from work with domestic horses, and we will draw comparisons to highlight the advantages and challenges when applying these technologies to donkeys and mules. Moreover, the use of equids as models for advanced ART in human medicine is gaining importance. Horses and donkeys share reproductive similarities with humans, such as gestation periods and ovarian physiology, making them valuable for studying ART outcomes. Their longevity and resilience also allow for extended observation of long-term effects, offering crucial insights for refining ART practices in both veterinary and human medicine.

2 Oocyte Collection in Donkeys and Mules

Cumulus-oocyte complexes (COCs) collection is a cornerstone of ART. There are two methods of COCs collection, including retrieval from ovaries post-mortem and through a technique known as *ovum* pick up (OPU) for harvesting COCs of a live animal. In equids, OPU is performed on the donor female via ultrasound guided transvaginal aspiration (TVA) under standing sedation with or without epidural

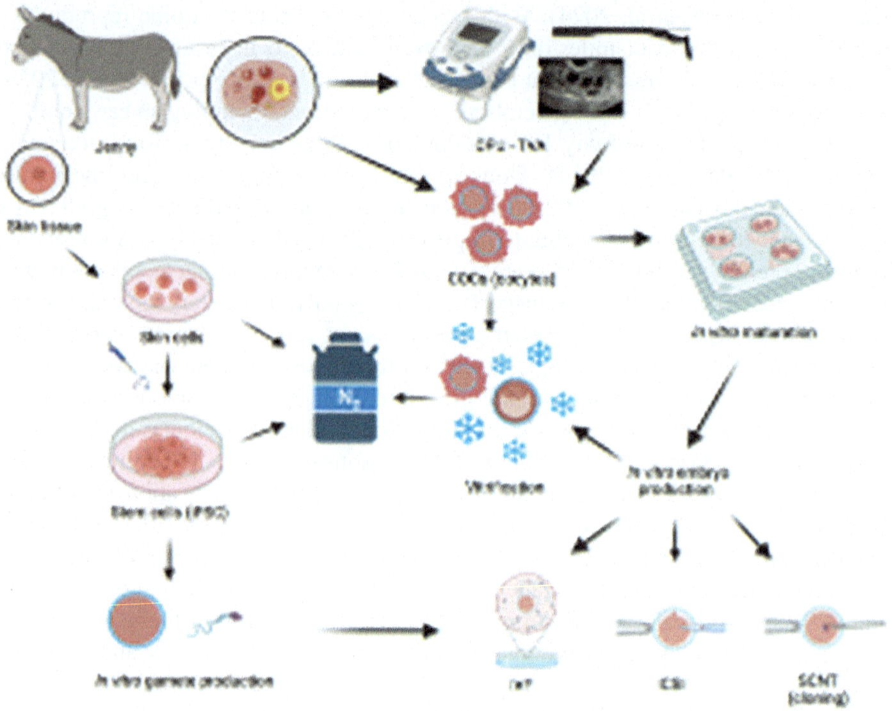

Fig. 1 Overview of advanced *in vitro* breeding technologies in donkeys. This schematic illustrates the integration of various advanced *in vitro* reproductive technologies discussed in the book chapter, specifically focused on their application in donkeys. OPU *ovum* pick up, TVA ultrasound guided transvaginal aspiration, COCs cumulus-oocyte complexes, IVF *in vitro* fertilization, ICSI intracytoplasmic sperm injection, SCNT somatic cell nuclear transfer. (Created with BioRender.com)

anesthesia. COCs are packaged and transported under optimal conditions to an ICSI lab for *in vitro* embryo production. Oocyte collection procedures have been perfected in the last decade in mares, achieving the efficacy to be commercially applied. According to reports on donkeys, the procedure follows a similar protocol, though anatomical and physiological differences between species necessitate slight modifications in technique and will be further discussed in the chapter.

Most studies in the domestic horse suggest that transportation of post-mortem ovaries should be done at room temperature and oocytes retrieved as soon as possible (Love et al. 2003) as *in vitro* production of embryos declines rapidly after 12 h post-mortem (Gambini et al. 2014). Oocyte collection post-mortem has also been described in domestic donkeys (Abdoon et al. 2018; Dai et al. 2023; Flores Bragulat et al. 2023) and in mules (González et al. 2015). Although anatomical differences might exist in follicle-oocyte interaction, the collection procedure was performed similarly among equids. In Fig. 2, we show a jenny ovary and the COCs

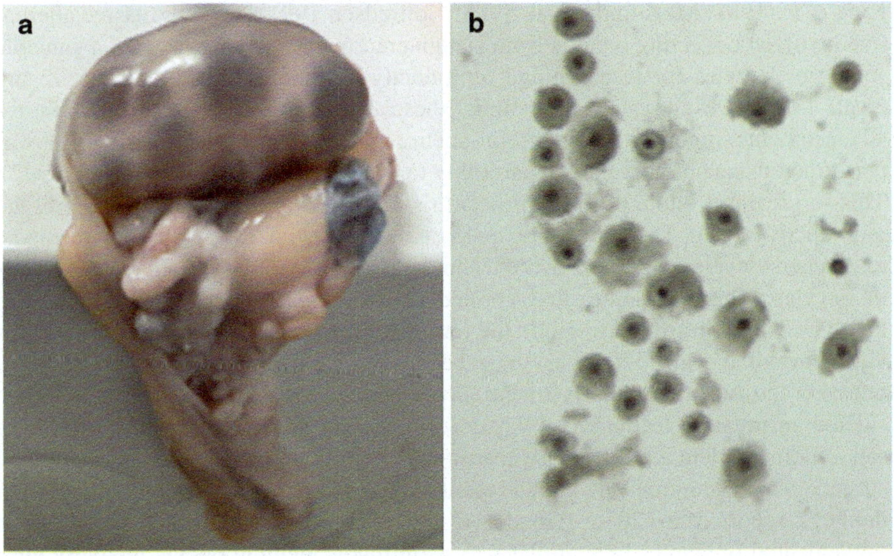

Fig. 2 Oocyte collection from post-mortem ovary (**a**) Jenny ovary collected from an abattoir, showing the presence of multiple antral follicles. (**b**) Cumulus-oocyte complexes (COCs) obtained by aspiration and scraping of the antral follicles of the ovary showed in panel A using a 10 mL syringe and an 18-gauge needle

recovered after follicle aspiration (24 COCs recovered from two ovaries), these oocytes were used for ICSI embryo production (Flores Bragulat et al. 2023).

A study conducted during 2022, compared the number, morphology, and maturation rates of cumulus-oocyte complexes (COCs) collected from abattoir-derived ovaries and live jennies. They concluded that the quality of donkey COCs retrieved through *ovum* pick-up (OPU) was superior to those obtained from abattoir-derived ovaries (Dai et al. 2023).

Interestingly, the presence of COCs in mule follicles was first documented in the 1970s (Taylor and Short 1973). Since then, little research has been published on the use of mule oocytes in ART. A study reported the recovery of 11 COCs from mule follicles after postmortem aspiration, with three of these oocytes classified as good quality (González et al. 2015). This remains an area for further investigation, as mule oocytes could be valuable in advancing our understanding of hybrid fertility in equids and as potential recipients in nuclear transfer embryo production experiments.

Early techniques for obtaining equine oocytes through ovarian follicle aspiration in mares presented significant challenges. One method, colpotomy, involved surgical access through the vaginal wall but could only be performed a limited number of times due to the risk of developing adhesions (Hinrichs and Kenney 1987). Another technique involved aspirating the follicles via the paralumbar fossa, which required puncturing the horse's flank and was explored by several researchers, though it posed difficulties in terms of access and potential complications (Palmer et al. 1986;

King et al. 1987; McKinnon et al. 1988; Gotfredsen 1991). A less invasive alternative involved collecting ovaries from slaughtered mares, which provided a valuable source of oocytes for research and reproductive technology development (Zhang et al. 1990). Despite their limitations, these techniques laid the foundation for advancements in equine assisted reproduction. The first reported equine OPU was performed using surgical flank laparotomy to access the ovaries (Brück et al. 1992). This surgical procedure posed significant risks such as infection and damage to the reproductive tract. To address the risks associated with early reported methods, researchers developed less invasive techniques, leading to the introduction of TVA of oocytes. This technique, first applied to horses in 1992 (Brück et al. 1992) involves using ultrasound to guide the retrieval of oocytes directly from the ovaries via the vaginal wall. TVA has since become a standard procedure in the horse breeding industry, particularly for *in vitro* embryo production. Although still considered an invasive technique, TVA quickly became the gold standard, offering a safer, more efficient alternative to laparotomy by eliminating the need for abdominal surgery. This technique was later adapted for use in donkeys, with the first OPU in donkeys reported in 2016 (Goudet et al. 2016). Since then, OPU has been implemented in jennies in a similar way as in mares yielding oocyte recovery rates between 35% and 75% (Goudet et al. 2016; Dai et al. 2023; Flores Bragulat et al. 2023). In line with these observations, another research group in Spain reported recovering 178 oocytes from 259 follicles (67.93%) out of 15 OPU sessions in 9 donkeys (Mateos et al. 2022).

Preparation for a donkey OPU session is similar to that in the mare. The jenny is placed in stocks and the tail is wrapped to maintain cleanliness in the procedural area. A rectal examination is then performed with the removal of any feces. The perineal area is cleansed using an organic iodine scrub, rinsed with water, and dried thoroughly to ensure sterility. A urinary catheter can be inserted to prevent urine from contaminating the procedural field. Medications are administered to improve safety both for the donkey and veterinarian including sedation and rectal relaxation for the OPU procedure. Table 1 provides a guide to donkey drug dosages and indications during the OPU procedure. Interestingly, a recent study has shown that phenylbutazone can have detrimental effects on oocyte developmental competence in horses (Ramírez-Agámez et al. 2024). It is crucial to consider and investigate the potential effects of any other medications administered during the OPU procedure in equids.

Briefly, the procedure and equipment required for OPU in horses or donkeys involve using a typical transvaginal ultrasound probe housed within a transvaginal guide. A non-embryo-toxic sterile lubricant must be applied. A ~60 cm, 12-gauge, double-lumen needle is inserted into the needle channel of the guide. An arm is placed rectally for ovary manipulation. The ultrasound probe is then manipulated along with the ovary until the ovarian follicles are clearly visible ultrasonographically through the vaginal wall, ensuring that no other tissue is between the probe and the ovary. The needle is then advanced through the vaginal wall into the aligned ovarian follicle, and the follicular fluid is aspirated using a vacuum pump. Once the follicle collapses, the ovary is manipulated around the

Table 1 Recommended drug dosages and their clinical applications for donkeys during OPU

Medication	Dose Rate	Route	Clinical Indication	References
Hyoscine Butyl Bromide	0.35 mg/kg	IV	Smooth muscle relaxation (Rectal relaxation)	Dobbie et al. (2024)
Butylscopolamine	0.12 mg/kg	IV	Smooth muscle relaxation (Rectal relaxation)	Goudet et al. (2016) Deleuze et al. (2018) Flores Bragulat et al. (2023)
Lidocaine hydrochloride 2%	60 mL	Rectal	Pain relief	Dobbie et al. (2024)
Detomidine	0.005–0.5 mg/kg	IV	Sedation	Deleuze et al. (2018) Goudet et al. (2016) Mateos et al. (2022) Dobbie et al. (2024)
Butorphanol	0.025–0.5 mg/kg	IV	Sedation & pain relief	Goudet et al. (2016) Deleuze et al. (2018) Mateos et al. (2022) Dobbie et al. (2024)
Gentamicin	6.6 mg/kg	IV	Antibiotic	Flores Bragulat et al. (2023) Dobbie et al. (2024)
Potassium penicillin	22 IU/kg	IV	Antibiotic	Goudet et al. (2016) Dobbie et al. (2024)
Flunixin meglumine	1–1.1 mg/kg	IV	NSAID	Flores Bragulat et al. (2023) Dobbie et al. (2024)

IV Intravenous, *NSAID* Non-steroidal anti-inflammatory

needle to scrape the follicular wall, facilitating the retrieval of the COCs. The follicle is then refilled using flush media containing heparin to maximize oocyte recovery and prevent blood clot formation. Each follicle is aspirated, scraped, and refilled 8–12 times, similar to the procedure used in mares. This process is repeated for all follicles greater than 5 mm in diameter on both the left and right ovaries. The fluid recovered is filtered through a 75-μm filter and the COCs are identified using a dissecting microscope. The COCs are then washed and transferred into a glass vial containing room-temperature holding medium for transportation to the laboratory. Although overnight incubation of horse oocytes at room temperature is routinely used in many OPU-ICSI laboratories, its impact has not been studied in donkeys. However, ICSI blastocysts can be achieved by holding donkey oocytes overnight at room temperature, as has been reported in horses (Flores Bragulat et al. 2023; Dobbie et al. 2024). Several commercial needles, vacuum pumps, warming devices, and flushing media are now available for TVA in horses and could be utilized for donkeys. Some commercial companies available for these are Minitube, WTA, IVF bioscience, and Stroebech Media.

While the general principles of OPU remain consistent across species, several key differences distinguish the procedure in mares and donkeys. These differences include variations in ovarian size, follicular development patterns, and the ease of transvaginal access. Some specifics of these challenges are described for the Baudet du Poitou jenny (Dobbie et al. 2024). The caudal reproductive tract of this breed of donkey is positioned more acutely (directed from the vulva craniodorsally toward the sacrum) than in the mare (Canisso et al. 2019). The ovaries are also larger and more cranial in the abdomen (Renner-Martin et al. 2009), which may make directing the transvaginal probe and grasping the ovary more difficult than in the mare. It is also more challenging to hold the ovaries pressed adequately (through the vaginal wall) against the face of the intravaginal ultrasound probe (Dobbie et al. 2024). The anatomical differences highlight the need for specialized facilities, training and techniques when performing OPU in donkeys to ensure successful outcomes. This also includes several clinical considerations due to the differences in sedation response and pain behavior compared to mares (Mateos et al. 2022). Veterinarians must be vigilant in observing subtle indicators of discomfort and adjust their medical observations accordingly. In contrast to horses, donkeys exhibit pain and discomfort more subtly, often leading to recognition only at advanced stages. A Donkey Grimace Scale based on facial expressions, body posture, and overall appearance has been developed (Orth et al. 2020). The scale was found to be accurate and effective in identifying pain-related discomfort, particularly following surgical castration. Among the different sedation protocols tested for donkeys undergoing OPU, the combination of 0.5 mg/kg detomidine +0.5 mg/kg Butorphanol was the alternative with the least increased stress (cortisol, C-reactive) hormone during the OPU procedure (Mateos et al. 2022). Animals subjected to OPU should be monitored for at least 24 h, and anti-inflammatory and antibiotic medications are recommended to manage inflammation, infection, and/or pain post-procedure. A physical examination is conducted the day after the TVA to ensure normal recovery before transporting the jennies home. This includes being more aware of pain and measuring it through indirect signs rather than waiting for overt displays of distress. Recognizing subtle pain indicators during and after the OPU procedure in donkeys may be challenging for those with limited experience (van Dierendonck et al. 2020). While facial cues such as ears, eyes, and nostrils can be influenced by external factors, observing the donkey's overall body posture and behavior improves the likelihood of identifying pain. Therefore, it is crucial to assess the donkey's entire body rather than relying solely on facial expressions.

As the popularity of OPU technology grows and becomes more widely adopted, there may be a need to design specialized equipment tailored to the reproductive anatomy of female donkeys. OPU has been performed on the French domestic donkey Grand Noir du Berry (Goudet et al. 2016), Catalonian donkey (Flores Bragulat et al. 2023), mixed-breed Argentinian donkey (Ferrante et al. 2022) and Baudet du Poitou (Dobbie et al. 2024), one of the largest donkey breeds. A shorter probe might be necessary for certain breeds, while larger donkeys may be able to utilize the standard length/size equine TVA probes. Figure 3 shows our research group performing TVA in a small mixed-breed Argentinian donkey and the challenges associated with this. The fluid recovered from donkey follicular aspiration can be filtered through a 75-μm flat-bottom embryo filter. The contents of the filter are then flushed into a gridded sterile 8-cm Petri dish and the Cumulus oocyte complexes (COCs) located using a dissecting microscope at 10–30× magnification. The COCs are transferred into a glass vial containing room-temperature embryo holding medium.

The OPU procedure carries certain risks, which have been documented in mares (Velez et al. 2012). Complications such as ovarian bleeding, abdominal infection, ovarian abscess, abdominal discomfort, pyrexia, rectal bleeding and even death have been reported in mares undergoing OPU (Velez et al. 2012). Although the current data on OPU complications in donkeys is limited, these risks while performing OPU must be considered and monitored closely.

In conclusion, while the principles of donkey OPU are similar to those of mares, the application in donkeys requires specific considerations due to their unique physiological and behavioral traits. Adjustments in sedation protocols, pain management, and equipment are necessary to optimize the procedure for donkeys. Animal welfare must be a top priority and carefully considered before recommending the OPU procedure in donkeys or any other animals. Adjustments in sedation protocols, pain management, and equipment are necessary to optimize the procedure for donkeys. By understanding and addressing these differences, we can improve oocyte retrieval rates, clinical outcomes, and overall success in advanced reproductive technologies for donkeys. Future advancements may also necessitate the development of specialized equipment tailored to the anatomical and breed-specific needs of donkeys, further enhancing the efficiency and effectiveness of OPU in equids.

3 Oocyte Maturation and Cryopreservation

3.1 *In Vitro* Oocyte Maturation

Oocyte maturation is the process by which an immature oocyte undergoes a series of developmental changes to become a "mature" egg that is ready to be fertilized. This process involves the resumption of meiosis where the oocyte progresses from the prophase I stage to metaphase II, evidenced by the extrusion of the first polar body. However, other crucial events, including chemical, molecular, and organelle

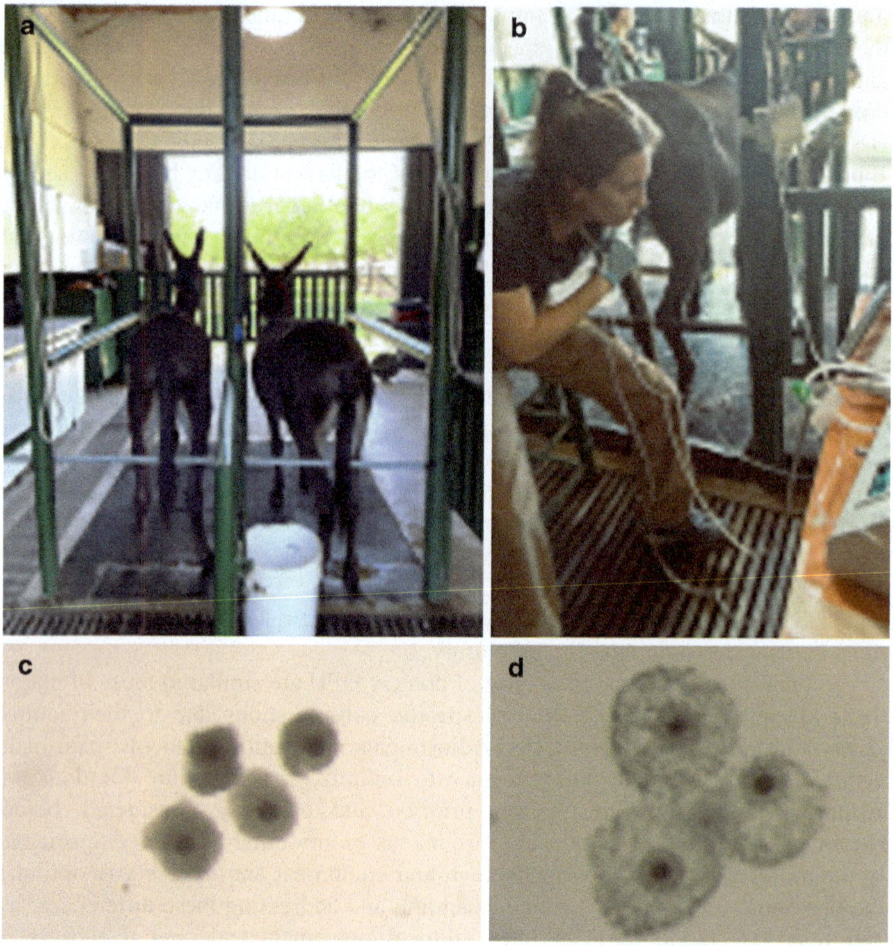

Fig. 3 Oocyte collection by *ovum* pick-up (OPU) (**a**) Two jennies positioned in stocks for transvaginal oocyte retrieval, highlighting the need to develop strategies to perform OPU in species or breeds with different sizes. (**b**) Ultrasound-guided transvaginal aspiration being performed by DVM Ana Flores Bragulat in 2023, from the Universidad Nacional de Rio Cuarto, Argentina, to collect oocytes from an Argentinian mixed jenny. (**c**) Cumulus-oocyte complexes (COCs) retrieved post-aspiration, before *in vitro* maturation (IVM). (**d**) Cumulus-oocyte complexes (COCs) after in vitro maturation (IVM) displaying cumulus cell expansion, a phenomenon typically observed following successful maturation

redistribution, are also taking place in the oocyte. Successful oocyte maturation is essential for fertilization and the subsequent development of a healthy embryo. This process can be facilitated in a controlled laboratory environment by culturing immature COCs in media containing components that support maturation, typically involving growth factors and hormones such as gonadotropins. While *in vitro* maturation (IVM) research has been conducted extensively with horse oocytes,

Table 2 Typical media components used in *In Vitro* maturation media for horse or donkey oocytes

Component	Function	Concentration/ Range
Base Media (TCM-199 or DMEM/F-12)	Provides essential nutrients, amino acids, vitamins, and energy substrates	Base solution
FBS, donkey fetal serum or serum replacement	Provides essential nutrients and growth factors	10% v/v
FSH and LH	Stimulates oocyte maturation	0.05–0.1 IU/mL
ITS	Supports growth, antioxidant	1% v/v
Sodium Pyruvate	Serves as an energy source for metabolism	0.2 mM
Penicillin-Streptomycin or Gentamicin	Antibacterial protection	50 IU/mL (Penicillin), 50 μg/mL (Streptomycin)
Cysteamine	Enhances cytoplasmic maturation by supporting GSH	100 μM
EGF	Promotes oocyte maturation and cumulus expansion	10–50 ng/mL

TCM Tissue Culture Medium, *DMEM/F-12* Dulbecco's Modified Eagle Medium/Nutrient Mixture F-12, *FBS* Fetal Bovine Serum, *FSH* Follicle Stimulating Hormone, *LH* Luteinizing Hormone, *ITS* Insulin-Transferrin-Selenium, *GSH* selenium-dependent glutathione peroxidase, *EGF* Epidermal Growth Factor

IVM of donkey oocytes remains an area requiring further investigation. The first successful IVM report in horses was achieved in 1981 (Fulka and Okolski 1981), paving the way for the application of IVM in other equids, including donkeys. One of the first reports of IVM of donkey oocytes was conducted in 2011 (Zhao et al. 2011) demonstrating the slow progress in the long-eared equids. While *in vitro* maturation of oocytes in horses is now an established procedure, developmental oocyte competence after *in vitro* maturation of donkey oocytes remains to be improved. The culture media used for equine oocytes has been commonly used for donkey oocytes (Zhao et al. 2011; Deleuze et al. 2018; Mateos et al. 2022; Flores Bragulat et al. 2023; Dobbie et al. 2024). In Table 2, we detail the typical components used in *in vitro* maturation media to support maturation of horse and donkey oocytes.

Despite the similarity in maturation rates between donkeys and horses (~50–70%), donkey oocytes require more time to mature *in vitro*. Our own experience and other recent reports (Dai et al. 2023; Dobbie et al. 2024) suggest that donkey oocytes require between 36 and 46 h (about 2 days) for the extrusion of the first polar body (nuclear maturation). This time is longer than the typical maturation (24–32 h) time for horse oocytes (Rodríguez et al. 2019). This extended maturation period may be influenced by the differences in reproductive physiology between species in the genus *Equus*. Further understanding the particularities of the estrous cycle, ovulation timing and oocyte developmental competence of donkeys and how it compares to that of mares could provide insights into optimizing IVM protocols and help to determine if specific adjustments to the IVM media can reduce maturation time and improve outcomes.

Our group contributed on improving donkey maturation rates and demonstrated that supplementing IVM media with donkey preovulatory follicular fluid supports oocyte maturation and early preimplantation embryo development following ICSI in donkeys (Flores Bragulat et al. 2023). In Fig. 3, we present donkey COCs recovered by OPU before and after 38 h of IVM. Interestingly, a recent study revealed the presence of at least 12 metabolites found in jenny follicular fluid but not in the mare (Fernández-Hernández et al. 2020; Catalán et al. 2022).

Moreover, another study determined that fetal bovine serum could be effectively replaced by fetal donkey serum (FDS) in the IVM medium (Dai et al. 2023). Finally, the advent of RNA sequencing (RNA-seq) technologies has enabled the comprehensive analysis of all transcribed RNAs in tissues or cells under specific conditions, a method widely used in both animal and plant research. Recent studies using RNA-seq have begun to uncover the unique features of donkey oocytes identifying specific genes such as IL10, CXCL10, and WGCNA that may play a role in oocyte maturation in donkeys, distinct from other eutherian mammals (Li et al. 2021; Zhang et al. 2022). All these recent findings are expected to contribute to the improvement of *in vitro* maturation conditions for donkey oocytes.

Maturing donkey oocytes *in vitro* presents unique challenges and opportunities. While the technology has advanced significantly, continued research is essential to refine IVM protocols and media specifically for donkeys. Understanding the species-specific requirements for oocyte maturation will enhance the efficiency and success rates of IVM, contributing to the broader application of assisted reproductive technologies in donkeys. These advancements will not only improve reproductive outcomes but also aid in the conservation and genetic improvement of wild donkey populations.

3.2 Oocyte Vitrification

The possibility of preserving oocytes has the potential to revolutionize not only the horse breeding industry but also facilitate the conservation of endangered equid female genetics. This technique has revolutionized human ART. Unfortunately, cryopreserving equine oocytes has proven to be difficult due to several contributing factors (De Coster et al. 2020). Nevertheless, oocyte vitrification has become increasingly promising for the equine breeding industry. The first successful pregnancies resulting from the transfer of vitrified equine oocytes were reported in 2002. In this study, oocytes obtained through OPU were transferred to an artificially inseminated mare, resulting in the birth of two healthy foals (Maclellan et al. 2002). Later, in 2017, the birth of a foal from a vitrified oocyte following ICSI was reported (Ortiz-Escribano et al. 2017). This was followed by the birth of two additional foals in Argentina in 2020 (Clérico et al. 2021).

To date, no *in vitro* embryos have been produced from vitrified donkey or mule oocytes. The first attempts to vitrify immature donkey oocytes were only reported in 2019 (Douet et al. 2019). In the mentioned study, vitrification was attempted on immature donkey oocytes with corona radiata using two different methods, with horse oocytes serving as the control group. Conventional cryoprotectants were employed, and the most successful protocol exposed oocytes to dimethyl sulfoxide, ethylene glycol and sucrose in a solution of TCM199-Hepes supplemented with bovine serum albumin which resulted in approximately 33% of the donkey oocytes maturing *in vitro* after thawing. Although some oocytes resumed meiosis following *in vitro* maturation, their competence for fertilization and subsequent development was not evaluated. Further research is necessary to develop more effective oocyte vitrification protocols and to assess the fertilization and developmental competence of vitrified-warmed donkey oocytes. If successful, oocyte vitrification in donkeys could significantly contribute to genetic preservation technologies in these species.

3.3 *In Vitro* Embryo Production

In vitro embryo production encompasses several advanced reproductive technologies, including ICSI, IVF, and nuclear transfer technologies, including somatic cell nuclear transfer (SCNT) or cloning. These technologies offer significant potential for genetic improvement and are now integrated in programs in different livestock species, particularly cattle but also in the horse breeding industry. *In vitro* embryo production is particularly promising for preserving endangered equids by enabling the production of multiple offspring from a single female, increasing genetic diversity, and reducing the need for extensive captive breeding programs.

Although some progress has been made in recent years preserving the genetics of equids (Cabeza and Gambini 2023), the vitrification of *in vivo* produced embryos in horses and donkeys remains challenging. Interestingly, horse embryos produced in the laboratory are smaller and have less developed embryo capsules, which allows them to tolerate vitrification without significant impact on their development. This presents a promising new strategy for genetic conservation currently expanding in the horse breeding industry. However, the application of *in vitro* embryo technologies in other domestic or wild equids has been limited due to restricted access to biological samples and funding constraints in this specialized field of research (Table 3). This section of our chapter explores recent advancements, and the challenges associated with using *in vitro* embryo technologies for the genetic improvement or preservation of equids, including mules and donkeys.

3.3.1 *In Vitro* Fertilization

In vitro fertilization (IVF) is an ART involving the ex vivo fertilization of a matured oocyte by sperm within a controlled laboratory environment. The IVF process typically includes the retrieval and IVM of oocytes, as described earlier in this chapter, followed by fertilization with *in vitro* capacitated sperm resulting in zygotes that will be cultured *in vitro* for 7–10 days to achieve blastocyst stage, these embryos can then be transferred into a recipient female or vitrified. This technique is widely used in humans ART and in the cattle industry. Historically, conventional IVF via gamete coincubation has demonstrated limited success in the domestic horse, mostly attributed to an inadequate *in vitro* capacitation of the horse sperm (Leemans et al.

Table 3 List of *in vitro* embryo reports for endangered domestic or wild equids and mules

Species	Technique	Oocyte data	Sperm/cell data	Key findings	Year	References
Equus quagga (Plains zebra)	IVF	*Equus zebra hartmannae*	Epididymal semen *(Equus quagga)*	Zebra IVF blastocyst	1997	Meintjes et al. (1997)
	ICSI	Domestic horse	Epididymal semen	Zebroid ICSI blastocyst	2020	Gambini et al. (2020)
	ICSI	Domestic horse	Electro-ejaculated and epididymal semen	Zebroid ICSI blastocysts	2020	Gerits et al. (2020)
	SCNT	Domestic horse	Male skin fibroblast	Zebra cloned blastocyst	2020	Gambini et al. (2020)
Equus africanus asinus (Donkey)	ICSI	Domestic donkey (Catalan, endangered breed)	Ejaculated (Andalusian domestic donkey)	Vitrified blastocysts	2023	Flores Bragulat et al. (2023)
	ICSI	Domestic donkey (Baudet du Poitou, (endangered breed)	Frozen semen (Baudet du domestic Poitou donkey)	Day 12 pregnancy	2024	Dobbie et al. (2024)
Equus ferus przewalskii (Przewalski's horse)	SCNT	Domestic horse	Przewalski's horse fibroblast	Birth of first endangered equid spp.	2020	Revive & Restore[a]
Equus ferus caballus x Equus africanus asinus (Mule)	ICSI	Domestic horse	Ejaculated Chilled or Frozen (Remonta Argentino -domestic donkey)	Vitrified blastocyst	2024	Arroyo-Salvo et al. (2024)

IVF in vitro fertilization, *ICSI* intracytoplasmic sperm injection, *SCNT* somatic cell nuclear transfer
[a]Milestones not from peer-reviewed articles

2016). As a result, efforts have been redirected towards developing *in vitro* embryos using ICSI, which has now become an established commercial procedure in the horse breeding industry. The first successful IVF in horses was reported in 1990 (Palmer et al. 1991), but it wasn't until recently that this technology became more reproducible and applicable, leading to the birth of IVF foals in the United States (Felix et al. 2022) and Europe (Papas et al. 2024).

The main advantage of developing a successful IVF protocol for equids lies in the relative simplicity and the more 'natural' approach of the technology compared to ICSI. However, many of the advantages offered by ICSI (see later section), such as the maximization of limited or poor-quality sperm, cannot be replicated by IVF. The current IVF protocol in horses developed by the team led by Dr. Katrin Hinrichs at the University of Pennsylvania, USA, involves pre-incubating fresh sperm in a specific media for ~22 h followed by co-incubation with oocytes for 3 h. This resulted in the birth of three healthy foals after embryo transfer (Felix et al. 2022). The high fertilization rate and production of viable embryos in this report herald new and exciting possibilities for the horse breeding community. Although current modification of the IVF protocol is underway to improve the efficiency using frozen-thawed semen (Felix and Hinrichs 2023), a significant limitation is the need of high-quality semen or strict semen selection process when starting the procedure.

In donkeys, one study reported an attempt at IVF without achieving blastocyst development (Goudet et al. 2016). For wild equids, the only report of IVF was published in 1997, with one blastocyst obtained by *in vitro* fertilization of Hartmann's zebra oocytes with epididymal fresh sperm cells from Burchell's zebra (Meintjes et al. 1997). The first attempts at IVF in donkeys following protocols like the one described by Felix et al. 2022 will hopefully be published soon. The primary challenge is determining whether the protocols for horse sperm capacitation can be successfully applied to donkey sperm. If successful, an IVF protocol with donkey sperm could be used to produce donkey and mule embryos in a laboratory setting without the need for micromanipulation of the gametes, facilitating its application and popularization. Moreover, it will be a crucial step towards translating IVF procedures to wild or endangered equids, providing a valuable tool for their conservation.

3.3.2 Intracytoplasmic Sperm Injection

Intracytoplasmic sperm injection (ICSI) involves the injection of a single sperm cell directly into the cytoplasm of a matured oocyte using a narrow micropipette. This technique has revolutionized human ART and has had a significant impact on equine reproduction. Initially, the limited number of oocytes obtained via transvaginal oocyte aspiration posed a major barrier to the widespread application of ICSI in equids. However, advancements over the past decade have established OPU and ICSI as commercially viable technologies, rapidly transforming the horse breeding industry (Claes and Stout 2022). These advancements have improved the efficiency of ICSI to levels that surpass those of more conventional ART. In combination with recent progress in the use of sexed semen in horse ICSI (Dini et al. 2023) this is

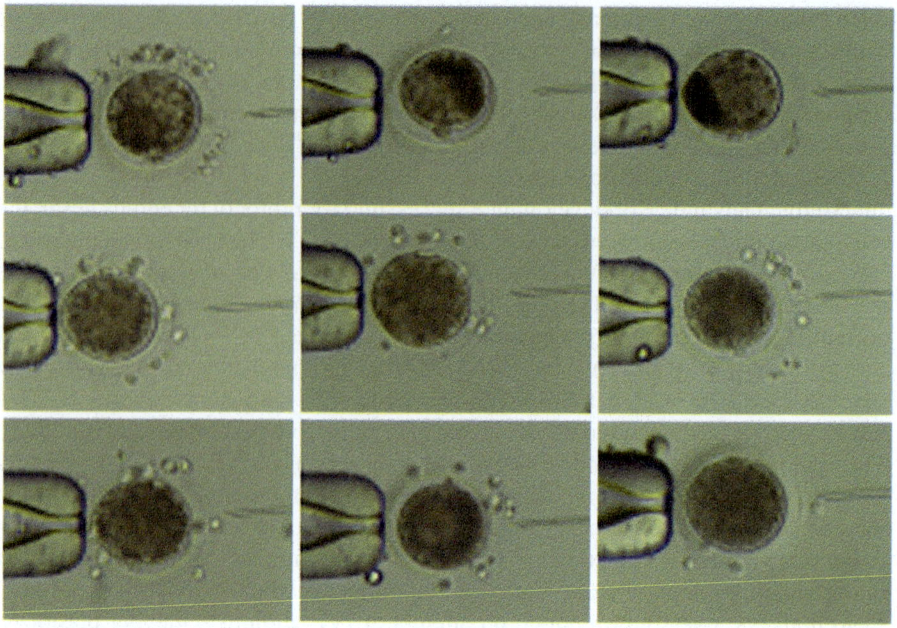

Fig. 4 Sequence of microinjection photographs depicting matured donkey oocytes with visible first polar bodies, captured immediately before injection with a donkey spermatozoon. The matured oocytes are secured by the holding pipette, with the oocyte rotated to position the polar body away from the injection needle, thereby minimizing the risk of damaging the associated metaphase II (MII) spindle. Note that some cumulus cells remain attached after mechanical removal; however, the low number of remaining cumulus cells in some oocytes does not impact the visualization of the first polar body or the sperm injection procedure

pushing the technology to its maximum potential in the current horse breeding industry.

Despite advancements in horses, the application of ICSI to other equids has been slow to develop. The first domestic donkey blastocyst produced using ICSI was reported in 2023 by our research group (Flores Bragulat et al. 2023), though developmental rates were lower compared to those in horses. This disparity underscores the differences in gamete and embryo culture conditions between equid species. Nonetheless, similar morphokinetic dynamics were observed between horse and donkey ICSI embryos using time-lapse imaging by our group. Figure 4 displays a photograph of donkey oocytes right before sperm injection that were part of this study. More recently, successful production of Poitou donkey blastocysts following IVM and ICSI was reported (Dobbie et al. 2024), including the first transfer of ICSI-produced donkey embryos into recipient jennies and the establishment of a very early vesicle that was not detected later.

This group reported a high blastocyst development rate in Poitou donkeys, suggesting that culture conditions established for horses can indeed support donkey embryo development at similar rates. In Argentina, our research group, in collaboration with researchers from the Universidad Nacional de Rio Cuarto, Universidad Católica Argentina, and the Argentinian Army, is currently producing donkey ICSI embryos and setting the conditions for transferring these embryos into recipient mules (unpublished data). The results of this experiment will contribute to understanding the factors limiting pregnancy rates in donkey ICSI embryos.

The first mule ICSI embryos were reported in 2024 (Arroyo-Salvo et al. 2024). In this study, one of the most critical proteins in sperm cells that triggers oocyte activation, phospholipase C zeta (PLCζ), was described in donkey sperm, highlighting differences compared to domestic horses. To assess the efficiency of the mentioned protein, donkey sperm was injected into pig oocytes, resulting in reduced oocyte activation. In contrast, when the sperm was injected into horse oocytes, it demonstrated significant activation capability, highlighting the protein's effectiveness in species within the same family. No differences in cleavage or blastocyst rates were observed between mule and horse ICSI embryos. These findings emphasize the differences in the ability of donkey sperm to induce oocyte activation and demonstrate that mule ICSI embryos can be successfully produced *in vitro* at similar rates to those reported in horses (Arroyo-Salvo et al. 2024).

Continued research is essential to refine ICSI protocols for donkeys and other equids, improve developmental rates, and address factors affecting pregnancy rates after embryo transfer. Researchers of our group, in collaboration with researchers at the Universidad de Buenos Aires, Facultad de Veterinaria (Argentina), led by Dr. Marcelo Miragaya, has also successfully attempted the injection of horse sperm into donkey oocytes to produce hinny blastocysts (unpublished data). This is a fascinating area for further development, as ICSI facilitates the production of hybrids in the laboratory, contributing to a deeper understanding of the genetic plasticity within the genus *Equus*, as demonstrated by the production of zebroid blastocysts, a zebra sperm cell injected into a matured horse oocyte (Gambini et al. 2020).

3.3.3 Cloning

Nearly 30 years after the birth of Dolly, more than 25 mammalian species have been cloned through SCNT technology (Gambini et al. 2022). Interestingly, in equids, the first cloned animal was a mule, followed shortly by the birth of the first cloned horse in 2003 (Woods 2003; Galli et al. 2003). Since then, interest in equine cloning has grown substantially (Gambini and Maserati 2017). Domestically, SCNT cloning technology is widely applied in pets, agriculture, and the sport horse industry, with the majority of cloned animals being produced in Argentina and the United States. During the SCNT procedure, a differentiated somatic cell, typically a skin fibroblast, is transferred into an enucleated oocyte, creating an embryo that is genetically identical to the donor animal. This method bypasses the traditional fertilization process, enabling the propagation of specific genotypes and the preservation of the genetics of elite, castrated, or deceased animals with desirable traits or significant value. Cryopreserving somatic cells, particularly skin fibroblasts, which can be obtained through minimally invasive procedures from donor animals, has greatly facilitated the establishment of genetic banks. These banks are essential for the preservation of valuable genetics, and efforts should be made to be implemented for endangered domestic donkey breeds and wild species like the critically endangered Somalian Wild Ass. The straightforward process of culturing and freezing fibroblasts ensures that these genetic materials can be effectively preserved for future use in reproductive technologies, such as cloning or *in vitro* gamete production.

The first equid clone using SCNT was derived from a cell from a 45-day-old fetus created using *in vivo*-matured oocytes and immediate oviducal embryo transfer into a mare (Woods 2003). Named Idaho Gem, this mule was born on May 5, 2003, following a normal 346-day gestation in a surrogate mare. As mules are typically infertile hybrids, this technology was groundbreaking in cloning a sterile animal. This cloned mule was produced by a large team from Utah State University who worked from 1999 to 2009 to produce the clone. The team, consisting of Gordan Woods and Dirk Vanderwall, focused on equine reproduction, while the USU team, consisting of Ken White and Tom Bunch, focused on the cloning process. Idaho Gem, and siblings named Utah Pioneer, and Idaho Star, all cloned mules have identical DNA and therefore the same capabilities athletically, but with different temperaments. These cloned mules have all raced (Dr. Dirk Vanderwall, personal communication). During 2006, the two mule clones Idaho Gem and Idaho Star raced at Winnemucca, mule racing's triple crown. The mules were the first cloned athletes to participate in any sport (Science Daily 2024) In 2009, a famous champion racing mule named Sarah Nelson was cloned, marking the first time a mule was cloned from a live animal.

SCNT has been specifically utilized for conservation purposes in the critically endangered Przewalski's wild horse (*Equus ferus przewalskii*). A significant milestone was reached with the birth of the first Przewalski's clone, named Kurt, at the Viagen facility in Texas in August 2020, using a cell line cryopreserved since 1980. A second cloned Przewalski's horse, Ollie, was born on February 17, 2023, from the same cell line, in honor of Dr. Oliver Ryder. Both Kurt and Ollie hold the potential to restore lost genetic variation within their species (Revive and Restore 2020). However, it is important to consider the presence of mitochondrial DNA from different species when performing interspecific SCNT, as this can have implications for species preservation. While selecting male individuals, who do not transmit mitochondria to their offspring, can mitigate this issue, the possibility of biparental inheritance of mitochondria remains controversial (Pagnamenta et al. 2021). This breakthrough is critical for conservation, as cloned animals like Kurt and Ollie could significantly increase genetic diversity within the Przewalski's horse population.

Despite successes in cloning domestic horses, Przewalski's horses, and mules, to date, there are no published reports of successful cloning in donkeys. In 2019, trials using domestic horse oocytes as recipients for donkey fibroblasts were conducted by the authors of this chapter in collaboration with researchers from the Universidad Nacional de Rio Cuarto in Argentina, but these attempts were unsuccessful in producing blastocysts (unpublished data). Nevertheless, we recently demonstrated that SCNT embryos produced by using zebra fibroblasts and horse oocytes reached the blastocyst stage at day 8, with no discernible differences in embryo quality compared to cloned horse embryos (Gambini et al. 2020). The successful vitrification of these zebra-cloned embryos, along with the successful cloning of

Przewalski's horses and mules, provides further evidence that horse ooplasm can reprogram somatic cells from other equid species, suggesting that donkey cloning using horse ooplasm should be feasible. Moreover, if OPU advances in domestic donkeys, SCNT of donkey embryos using donkey oocytes could be attempted. This would significantly contribute to the genetic preservation of donkeys, with particular interest in preserving the genetics of the Somalian Wild Ass (Cabeza and Gambini 2023).

The advancements in cloning technology, particularly SCNT, have significant implications for both the equine industry and conservation efforts. The ability to produce genetically identical individuals not only aids in preserving valuable genetics in sport and agriculture but offers a viable method for rescuing endangered species. Continued research and development in cloning technologies will undoubtedly enhance their efficiency and applicability, further contributing to the preservation of equids species.

4 Future *In Vitro* Breeding Technologies

Emerging assisted breeding technologies in equid reproduction and conservation, such as stem cell research, vitro gamete production, and novel methods of sample preservation, offer significant promise. Techniques like sperm freeze-drying and *in vitro* follicle growth are opening new avenues for enhancing fertility and improving genetic conservation. Stem cell technology, such as Spermatogonial Stem Cells (SSC) and particularly the use of induced pluripotent stem cells (iPSCs) (Takahashi and Yamanaka 2006), presents a transformative opportunity for equid reproduction. These new technologies hold significant potential for enhancing breeding programs and conservation efforts by enabling the production and preservation of viable sperm or eggs, thereby improving genetic diversity and preserving valuable genetic material.

Progress with these technologies has been more pronounced in horses, particularly in areas such as preantral follicle growth (Gomes et al. 2015; Haag et al. 2013), the successful isolation and growth of follicles from cryopreserved ovarian tissue (Gastal et al. 2017) and advancements in SSC transfer (Guilherme et al. 2017). However, preantral follicle growth to achieve successful oocyte growth and development is still needed and, although horse SSCs have been successfully collected, identified, and cryopreserved attempts to transfer SSC in equids have not yet been successful, mainly due to the anatomical challenges involved in identifying the rete testis in stallions via ultrasonography (Costa et al. 2017). The mule is an especially promising candidate for SSC transfer technology, as its testes remain functional despite the rare progression of spermatogenesis beyond the spermatocyte stage (Neves et al. 2005). We hope to see some of this research translated toward donkeys in future literature.

The development of methods to convert skin fibroblasts into iPSCs (Takahashi and Yamanaka 2006) has gained a lot of attention. These cells have great potential in research as they are self-renewing and pluripotent with low ethical concerns. Protocols for their production have been developed for many domestic animal species, which have since been used to further our knowledge in the progression and treatment of diseases. Importantly, iPSCs can be reprogrammed to differentiate into gametes in the laboratory, a process known as *in vitro* gamete production (Moradi et al. 2019). In 2011, the first successful derivation of iPSC lines from equine fibroblasts was reported (Nagy et al. 2011), and more recently, the generation of iPSCs from the endangered Somali wild ass (Ben-Nun et al. 2015), Grevy's zebra (Endo et al. 2022) and mules (Zhang et al. 2023). To date, no reports on iPSCs have been found in other equid species. Although the successful generation of germ cells from iPSCs has primarily been demonstrated in mouse models (Imamura et al. 2013) the implications of extending this technology to other species are profound implications and require further research. This approach offers the possibility to harness reproductive potential from non-reproductive bio-banked tissues or cell lines. Consequently, *in vitro* gametogenesis should be considered in discussions about cloning and related technologies, particularly in the contexts of animal breeding and species conservation (Cowl et al. 2024).

Taking this technology a step further, advancements in stem cell research have led to a significant breakthrough with the development of synthetic embryo-like structures, known as iBlastoids. The potential of these structures to revolutionize reproductive biology and research is immense. Derived from iPSCs, iBlastoids can self-organize into structures resembling early embryos. Given the successful creation of these synthetic embryos from iPSCs in other species, this technology will soon be applied to members of the equid family too. Although challenges remain, stem cell technology offers significant potential for equid conservation by facilitating the development of advanced reproductive techniques and potentially increasing both population size and genetic diversity.

Despite the potential of these technologies, successful application to equids remains limited. From improving oocyte and sperm quality to generating gametes and embryos *in vitro*, these advancements can address numerous reproductive challenges in equids, particularly in wild equids. While significant progress has been made in other species, particularly in stem cell research and *in vitro* culture techniques, further research is essential to fully realize their potential in horses and donkeys. The integration of these technologies into equine reproductive programs could revolutionize breeding strategies, enhance genetic diversity, and contribute to the conservation of endangered species (Cabeza and Gambini 2023).

Acknowledgments We extend our deepest gratitude to DVM Ana Flores Bragulat for her invaluable contributions to this chapter. Her insights and expertise were pivotal to the success of this work. We also express our sincere thanks to Dr. Angus McKinnon for his unwavering support and guidance throughout this process. We are very grateful to Dr. Dirk Vanderwall for his significant contributions to the chapter; we deeply appreciate his assistance.

Glossary

OPU	*Ovum* pick up
ART	Advanced reproductive technologies
ICSI	Intracytoplasmic sperm injection

IVF	*In vitro* fertilization
SCNT	Somatic cell nuclear transfer
COC	Cumulus-oocyte complexes
TVA	Ultrasound guided transvaginal aspiration
IVM	*In vitro* maturation
TCM	Tissue culture medium
DMEM/F-12	Dulbecco's Modified Eagle Medium/Nutrient Mixture F-12
FBS	Fetal Bovine Serum
FSH	Follicle stimulating hormone
LH	Luteinizing hormone
ITS	Insulin-Transferrin-Selenium
GSH	Selenium-dependent glutathione peroxidase
EGF	Epidermal Growth Factor
PLCζ	Phospholipase C zeta
SSC	Spermatogonial Stem Cells
iPSC	Induced pluripotent stem cells

References

Abdoon AS, Fathalla SI, Shawky SM, Kandil OM, Kishta AA, Masoud SR (2018) In vitro maturation and fertilization of donkey oocytes. J Equine Vet Sci 65:118–122

Arroyo-Salvo C, Villarreal C, Clérico G et al (2024) The ability of donkey sperm to induce oocyte activation and mule embryo development after ICSI. Theriogenology 218:200–207

Ben-Nun I, Montague SC, Houck ML, Ryder O, Loring JF (2015) Generation of induced pluripotent stem cells from mammalian endangered species. Methods Mol Biol 1330:101–109

Bertino E, Agosti M, Peila C, Corridori M, Pintus R, Fanos V (2022) The donkey Milk in infant nutrition. Nutrients 14:403

Brück I, Raun K, Synnestvedt B, Greve T (1992) Follicle aspiration in the mare using a transvaginal ultrasound-guided technique. Equine Vet J 24:58–59

Cabeza JP, Gambini A (2023) Advancements and challenges in in vitro reproductive technologies for the conservation of equine species. Theriogenol Wild 2:100036

Camargo C, Rechsteiner (2020) The mule (Equus mulus) as a recipient of horse (Equus caballus) embryos: comparative aspects of early pregnancy with mares. Theriogenology 145:217–225

Canisso I, Panzani D, Miró J, Ellerbrock R (2019) Key aspects of donkey and mule reproduction. Vet Clin N Am Equine Pract 35:607–642

Catalán J, Martínez-Rodero I, Yánez-Ortiz I, Mateo-Otero Y, Bragulat AF (2022) Metabolic profiling of preovulatory follicular fluid in jennies. Res Vet Sci 153:127–136

Claes A, Stout TAE (2022) Success rate in a clinical equine in vitro embryo production program. Theriogenology 187:215–218

Clérico G, Taminelli G, Veronesi JC, Polola J, Pagura N, Pinto C, Sansinena M (2021) Mitochondrial function, blastocyst development and live foals born after ICSI of immature vitrified/warmed equine oocytes matured with or without melatonin. Theriogenology 160:40–49

Costa GM, Avelar G, Lacerda SM, Figueiredo AF, Tavares AO, Rezende-Neto JV, Martins FG, França LR (2017) Horse spermatogonial stem cell cryopreservation: feasible protocols and potential biotechnological applications. Cell Tissue Res 370:489–500. https://doi.org/10.1007/s00441-017-2660-3

Cowl VB, Comizzoli P, Appeltant R, Bolton RL, Browne RK, Holt WV, Penfold LM, Swegen A, Walker SL, Williams SA (2024) Cloning for the twenty-first century and its place in endangered species conservation. Annu Rev Anim Biosci 12:91–112

Dai S, Di Z, Li N, Zeng S (2023) Optimization of recovery and maturation methods for cumulus-oocyte complexes in jennies. Reprod Domest Anim 58:168–175

De Coster T, Velez DA, Van Soom A, Woelders H, Smits K (2020) Cryopreservation of equine oocytes: looking into the crystal ball. Reprod Fertil Dev 32:453–467

Deleuze S, Douet C, Couty I, Moros-Nicolás C, Barrière P, Blard T, Reigner F, Magistrini M, Goudet G (2018) Ovum pick up and in vitro maturation of jennies oocytes toward the setting up of efficient in vitro fertilization and in vitro embryos culture procedures in donkey (Equus asinus). J Equine Vet Sci 65:111–117

Dini P, de la Fuente A, Samper JC, Cabrera C, Holyoak G, Megehee S, Meyers SA (2023) Successful use of sexed stallion semen for in vitro embryo production; a preliminary study. J Equine Vet Sci 125:104645

Dobbie T, Felix MR, Gleason K, Hinrichs K (2024) Transvaginal aspiration, oocyte maturation and production of blastocysts and a pregnancy via ICSI in an endangered breed: Baudet du Poitou donkey. Theriogenology. https://doi.org/10.1016/j.theriogenology.2024.04.021

Douet C, Reigner F, Barrière P, Blard T, Deleuze S, Goudet G (2019) First attempts for vitrification of immature oocytes in donkey (Equus asinus): comparison of two vitrification methods. Theriogenology 126:261–265

Endo Y, Kamei KI, Hasegawa K, Okita K, Ito H, Terada S, Inoue-Murayama M (2022) Generation and gene expression profiles of Grevy's zebra induced pluripotent stem cells. Stem Cells Dev 31(9–10):250–257

Felix M, Hinrichs K (2023) Selection of frozen-thawed semen for standard in vitro fertilization. J Equine Vet Sci 125:104648

Felix M, Turner RM, Dobbie T, Hinrichs K (2022) Successful in vitro fertilization in the horse: production of blastocysts and birth of foals after prolonged sperm incubation for capacitation. Biol Reprod 107:1551–1564

Fernández-Hernández P, María Jesús Sánchez-Calabuig, Luis Jesús García-Marín, Bragado MJ, Gutiérrez-Adán A, Millet Ó, Bruzzone C, Lauro González-Fernández, Macías-García B (2020) Study of the metabolomics of equine Preovulatory follicular fluid: a way to improve current in vitro maturation media. Animals 10:883–883

Ferrante AA, Arraztoa CC, Plaza JP, Gambini A, Baca Castex C, Rodriguez MB, Brisky O, Bruno S, Olivieri G, Miragaya MH, Neild DM (2022) The efficiency of transvaginal oocyte aspiration in Remonta Argentino jennies. J Equine Vet Sci 113:103996. https://doi.org/10.1016/j.jevs.2022.103996

Flores Bragulat AP, Ortiz I, Catalán J et al (2023) Time-lapse imaging and developmental competence of donkey eggs after ICSI: effect of preovulatory follicular fluid during oocyte in vitro maturation. Theriogenology 195:199–208

Fonte J, Alonso M, Peter M, Gonçalves M, Pontes J, Bordignon V, Dagher P, Fernandes C (2024) Successful equine in vitro embryo production by ICSI – effect of season, mares' age, breed, and phase of the estrous cycle on embryo production. Theriogenology 223:47–52. https://doi.org/10.1016/j.theriogenology.2024.04.007

Fulka J, Okolski A (1981) Culture of horse oocytes in vitro. J Reprod Fertil 61(1):213–215

Galli C, Lagutina I, Crotti G, Colleoni S, Turini P, Ponderato N, Duchi R, Lazzari G (2003) A cloned horse born to its dam twin. Nature 424:635–635

Gambini A, Maserati M (2017) A journey through horse cloning. Reprod Fertil Dev 30:8–17

Gambini A, Javier J, Florencia K, De Stéfano A, Salamone DF (2014) Effect of collection–maturation interval time and pregnancy status of donor mares on oocyte developmental competence in horse cloning. J Anim Sci 92:561–567

Gambini A, Rodriguez MD, Rodriguez MB, Briski O, Paula A, Demergassi N, Losinno L, Salamone D (2020) Horse ooplasm supports in vitro preimplantation development of zebra

ICSI and SCNT embryos without compromising YAP1 and SOX2 expression pattern. PLoS One 15:e0238948

Gambini A, Briski O, Canel NG (2022) State of the art of nuclear transfer technologies for assisting mammalian reproduction. Mol Reprod Dev 89:230–242

Gastal GD, Aguiar FL, Alves BG, Alves KA, de Tarso SG, Ishak GM, Cavinder CA, Feugang JM, Gastal EL (2017) Equine ovarian tissue viability after cryopreservation and in vitro culture. Theriogenology 97:139–147

Gerits I, Angel-Velez D, Ververs C, Govaere J, Van Soom A, Smits K (2020) First blastocyst production after ICSI with zebra semen. J Equine Vet Sci 89:103053. https://doi.org/10.1016/j.jevs.2020.103053

Giulotto E, Raimondi E, Sullivan K (2017) The unique DNA sequences underlying equine centromeres. Prog Mol Subcell Biol 56:337–354

Gomes RG, Lisboa LA, Silva CB, Max MC, Marino PC, Oliveira RL, González SM, Barreiros TRR, Marinho LSR, Seneda MM (2015) Improvement of development of equine preantral follicles after 6 days of in vitro culture with ascorbic acid supplementation. Theriogenology 84: 750–755

González SM, Gomes RG, Souza AK, Silva CB, Silva-Santos KC, Seneda MM (2015) Evidence of regular estrous cycles in mules and successful use of these animals as recipients for donkey embryos. J Equine Vet Sci 35:869–872

Gotfredsen P (1991) Periovulatory ultrasonography, endocrinology and oocyte morphology in the mares. ProQuest Dissertations & Theses

Goudet G, Douet C, Kaabouba-Escurier A, Couty I, Moros-Nicolás C, Barrière P, Blard T, Reigner F, Deleuze S, Michèle Magistrini (2016) Establishment of conditions for ovum pick up and IVM of jennies oocytes toward the setting up of efficient IVF and in vitro embryos culture procedures in donkey (Equus asinus). Theriogenology 86:528–535

Guilherme FG, Almeida F, Tavares A, Rezende-Neto J, França L (2017) Horse spermatogonial stem cell cryopreservation: feasible protocols and potential biotechnological applications. Cell Tissue Res 370:489–500

Haag KT, Magalhães-Padilha D, Fonseca GR, Wischral A, Gastal M, King SS, Jones KL, Figueiredo JR, Gastal EL (2013) In vitro culture of equine preantral follicles obtained via the biopsy pick-up method. Theriogenology 79:911–917

Hinrichs K, Kenney R (1987) A colpotomy procedure to increase oocyte recovery rates on aspiration of equine preovulatory follicles. Theriogenology 27:237

Imamura M, Hikabe O, Lin ZY-C, Okano H (2013) Generation of germ cells in vitro in the era of induced pluripotent stem cells. Mol Reprod Dev 81:2–19

King W, Bezard J, Bousquet D, Palmer E, Betteridge K (1987) The meiotic stage of preovulatory oocytes in mares. Genome 29:679–682

Leemans B, Gadella BM, Stout TAE, De Schauwer C, Nelis H, Hoogewijs M, Van Soom A (2016) Why doesn't conventional IVF work in the horse? The equine oviduct as a microenvironment for capacitation/fertilization. Reproduction 152:R233–R245

Li Z, Song X, Yin S, Yan J, Peiru Lv, Shan H, Cui K, Liu H, Liu Q (2021) Single-cell RNA-Seq revealed the gene expression pattern during the in vitro maturation of donkey oocytes. Genes 12: 1640

Love LB, Choi YH, Love CC, Varner DD, Hinrichs K (2003) Effect of ovary storage and oocyte transport method on maturation rate of horse oocytes. Theriogenology 59:765–774

Maclellan LJ, Carnevale EM, Coutinho da Silva MA, Scoggin CF, Bruemmer JE, Squires EL (2002) Pregnancies from vitrified equine oocytes collected from super-stimulated and non-stimulated mares. Theriogenology 58:911–919

Mateos SS, de Miguel IS, Rollán NH, Margallo FM (2022) Optimizations of an ovum pick up protocol in Donkey. Anim Reprod 19(2):e22177

McKinnon AO, Squires EL, Carnevale EM, Voss JL, Seidel GE (1988) Heterogenous and xenogenous fertilization of equine oocytes. Theriogenology 29:278

Meintjes M, Bezuidenhout C, Bartels PJ, Visser DS, Meintjes JJ, Loskutoff NM, Fourie F, Barry DM, Godke RA (1997) In vitro maturation and fertilization of oocytes recovered from free-ranging Burchell's zebra (Equus burchelli) and Hartmann's zebra (Equus zebra hartmannae). J Zoo Wildl Med 28(3):251–259

Moradi S, Mahdizadeh H, Šarić T, Kim J, Harati J, Shahsavarani H, Greber B, Moore JB (2019) Research and therapy with induced pluripotent stem cells (iPSCs): social, legal, and ethical considerations. Stem Cell Res Ther 10(1):341. https://doi.org/10.1186/s13287-019-1455-y

Nagy K, Sung H-K, Zhang P et al (2011) Induced pluripotent stem cell lines derived from equine fibroblasts. Stem Cell Rev Rep 7:693–702

Neves ES, Chiarini-Garcia H, França LR (2005) Ultrastructural observation of the mule testis indicates normal function of somatic cells. Anim Reprod 2(4):263–271

Orth EK, Navas González FJ, Iglesias Pastrana C, Berger JM, le Jeune SS, Davis EW, McLean AK (2020) Development of a donkey grimace scale to recognize pain in donkeys (Equus asinus) post castration. Animals 10:1411

Ortiz-Escribano N, Bogado Pascottini O, Woelders H et al (2017) An improved vitrification protocol for equine immature oocytes, resulting in a first live foal. Equine Vet J 50:391–397

Pagnamenta AT, Wei W, Rahman S, Chinnery PF (2021) Biparental inheritance of mitochondrial DNA revisited. Nat Rev Genet 22:477–478

Palmer E, Duchamp G, Bezard J, Magistrini M, King A, Bousquet D, Betteridge K (1986) Recovery of follicular fluid and oocytes of mares by non-surgical puncture of the preovulatory follicle. Theriogenology 25:178

Palmer E, Bézard J, Magistrini M, Duchamp G (1991) In vitro fertilization in the horse. A retrospective study HAL (Le Centre pour la Communication Scientifique Directe). J Reprod Fertil Suppl 44:375–384

Panzani D, Rota A, Tesi M, Fanelli D, Camillo F (2018) Update on donkey embryo transfer and cryopreservation. J Equine Vet Sci 65:50–54

Papas M, Govaere J, Peere S, Gerits I, Frenández-Montoro A, De Coster T, Hedia M, Angel-Velez D, Van Soom A, Smits K (2024) Development of the first equine blastocyst produced by conventional IVF and in vitro culture in Europe resulting in the birth of a foal. Vlaams Diergeneeskd Tijdschr. https://doi.org/10.21825/vdt.91527

Ramírez-Agámez L, Hernández-Avilés C, Whitfield-Cargile C, Whitfield-Cargile C, Coleman M, Coleman M, Love C (2024) Treatment of mares with the non-steroidal anti-inflammatory drug (NSAID) phenylbutazone transiently affects in vitro maturation of equine oocytes and blastocyst development after Intracytoplasmic Sperm Injection (ICSI). Theriogenology 223:53–58. https://doi.org/10.1016/j.theriogenology.2024.04.017

Renner-Martin T, Forstenpointner G, Weissengruber GE, Eberhardt L (2009) Gross anatomy of the female genital organs of the domestic donkey (Equus asinus Linné, 1758). Anat Histol Embryol 38:133–138

Revive & Restore (2020) Przewalski's horse project. Reviverestore.org. https://reviverestore.org/projects/przewalskis-horse/#:~:text=On%20Feb.%2017%2C%202023%2C%20a%20second%20cloned%20Przewalski%E2%80%99s. Accessed 22 Aug 2024

Rodríguez MB, Gambini A, Clérico G, Ynsaurralde-Rivolta AE, Briski O, Largel H, Sansinena M, Salamone DF (2019) Time of first polar body extrusion affects the developmental competence of equine oocytes after intracytoplasmic sperm injection. Reprod Fertil Dev 31:1805

Science Daily (2024) World's first cloned equine athletes training for races. ScienceDaily. https://www.sciencedaily.com/releases/2006/05/060517082254.htm. Accessed 22 Aug 2024

Stout TAE (2020) Clinical application of in vitro embryo production in the horse. J Equine Vet Sci 89:103011

Summers PM, Shephard AM, Hodges JK, Kydd J, Boyle MS, Allen WR (1987) Successful transfer of the embryos of Przewalski's horses (Equus przewalskii) and Grant's zebra (E. burchelli) to domestic mares (E. caballus). Reproduction 80:13–20

Takahashi K, Yamanaka S (2006) Induction of pluripotent stem cells from mouse embryonic and adult fibroblast cultures by defined factors. Cell 126:663–676

Taylor MJ, Short RV (1973) Development of the germ cells in the ovary of the mule and hinny. Reproduction 32:441–445

The Horse Staff (2002) Rare donkey born through embryo transfer in Australia. The Horse. https://thehorse.com/132665/rare-donkey-born-through-embryo-transfer-in-australia/. Accessed 21 Aug 2024

van Dierendonck MC, Burden FA, Rickards K, van Loon JP (2020) Monitoring acute pain in donkeys with the equine utrecht university scale for donkeys composite pain assessment (EQUUS-DONKEY-COMPASS) and the equine Utrecht University scale for donkey facial assessment of pain (EQUUS-DONKEY-FAP). Animals 10(2):354. https://doi.org/10.3390/ani10020354

Velez I, Arnold C, Jacobson CC, Norris JP, Choi Y, Edwards JF, Hayden SS, Hinrichs K (2012) Effects of repeated transvaginal aspiration of immature follicles on mare health and ovarian status. Equine Vet J Suppl 43:78–83

Woods GL (2003) A mule cloned from fetal cells by nuclear transfer. Science 301:1063–1063

Zhang J, Muzs L, Boyle M (1990) In vitro fertilization of horse follicular oocytes matured in vitro. Mol Reprod Dev 26:361–365

Zhang F-L, Zhang S-E, Sun Y-J, Wang J-J, Shen W (2022) Comparative Transcriptomics uncover the uniqueness of oocyte development in the donkey. Front Genet 13:839207. https://doi.org/10.3389/fgene.2022.839207

Zhang J, Zhao L, Fu Y et al (2023) Reprogramming efficiency and pluripotency of mule iPSCs over its parents. Biol Reprod 108:887–901

Zhao G, Wu K, Cui L, Zhao L, Liu Y, Tan X, Zhou H (2011) In vitro maturation and artificial activation of donkey oocytes. Theriogenology 76:700–704

Publisher Correction to: Anesthesia and Pain Management in Donkeys and Mules

Johannes P. A. M. van Loon

Publisher Correction to:
Chapter 14 in: A. K. McLean et al. (eds.), ***Donkey and Mule Medicine: Clinical Medicine and Reproduction***,
https://doi.org/10.1007/978-3-032-18540-2_14

The original version of Chap. 14 was inadvertently published without the inclusion of Annexure written by Dr. Fernanda Castro Dosal. The text has now been included, and the chapter has been approved by the authors. The inclusion of the Annexure has affected the pagination of the book.

The updated version of this chapter can be found at
https://doi.org/10.1007/978-3-032-18540-2_14

A. K. McLean et al. (eds.), *Donkey and Mule Medicine: Clinical Medicine and Reproduction*, https://doi.org/10.1007/978-3-032-18540-2_22

The manufacturer's authorised representative in the EU is Springer Nature Customer Service Centre GmbH, Europaplatz 3, 69115 Heidelberg, Germany. If you have any concerns regarding our products, please contact ProductSafety@springernature.com

Printed and bound by CPI Group (UK) Ltd, Croydon, CR0 4YY
05/10/2026
02228035-0004